Oxford Graduate Texts in Mathematics

Series Editors
R. Cohen   S.K. Donaldson   S. Hildebrandt
T.J. Lyons   M.J. Taylor

OXFORD GRADUATE TEXTS IN MATHEMATICS

Books in the series

1. Keith Hannabuss: *An introduction to quantum theory*
2. Reinhold Meise and Dietmar Vogt: *Introduction to functional analysis*
3. James G. Oxley: *Matroid theory*
4. N.J. Hitchin, G.B. Segal, and R.S. Ward: *Integrable systems: twistors, loop groups, and Rieman surfaces*
5. Wulf Rossmann: *Lie groups: An introduction through linear groups*
6. Qing Liu: *Algebraic geometry and arithmetic curves*
7. Martin R. Bridson and Simon M. Salamon (eds): *Invitations to geometry and topology*
8. Shmuel Kantorovitz: *Introduction to modern analysis*
9. Terry Lawson: *Topology: A geometric approach*
10. Meinolf Geck: *An introduction to algebraic geometry and algebraic groups*
11. Alastair Fletcher and Vladimir Markovic: *Quasiconformal maps and Teichmüller theory*
12. Dominic Joyce: *Riemannian holonomy groups and calibrated geometry*
13. Fernando Villegas: *Experimental Number Theory*
14. Péter Medvegyev: *Stochastic Integration Theory*
15. Martin Guest: *From Quantum Cohomology to Integrable Systems*
16. Alan Rendall: *Partial Differential Equations in General Relativity*
17. Yves Félix, John Oprea and Daniel Tanré: *Algebraic Models in Geometry*

Algebraic Models in Geometry

Yves Félix
Département de Mathématiques
Université Catholique de Louvain-la-Neuve
Louvain-la-Neuve
Belgium
felix@math.ucl.ac.be

John Oprea
Department of Mathematics
Cleveland State University
Cleveland, USA
j.oprea@csuohio.edu

Daniel Tanré
Département de Mathématiques
Université des Sciences et Technologies de Lille
Lille, France
Daniel.Tanre@univ-lille1.fr

OXFORD
UNIVERSITY PRESS

Great Clarendon Street, Oxford, OX2 6DP,
United Kingdom

Oxford University Press is a department of the University of Oxford.
It furthers the University's objective of excellence in research, scholarship,
and education by publishing worldwide. Oxford is a registered trade mark of
Oxford University Press in the UK and in certain other countries

© Y. Felix, J. Oprea, and D. Tanré

The moral rights of the authors have been asserted

First published in 2008

All rights reserved. No part of this publication may be reproduced, stored in
a retrieval system, or transmitted, in any form or by any means, without the
prior permission in writing of Oxford University Press, or as expressly permitted
by law, by licence or under terms agreed with the appropriate reprographics
rights organization. Enquiries concerning reproduction outside the scope of the
above should be sent to the Rights Department, Oxford University Press, at the
address above

You must not circulate this work in any other form
and you must impose this same condition on any acquirer

Published in the United States of America by Oxford University Press
198 Madison Avenue, New York, NY 10016, United States of America

British Library Cataloguing in Publication Data
Data available

Library of Congress Cataloging in Publication Data
Data available

ISBN 978-0-19-920652-0

To Agnès.

To mon Papillon: Kathy.

To the Magnificent Seven: Camille, Elsa, Amélie, Mathieu, Thibault, Lucie, Maude.

Preface

Rational homotopy theory originated with the work of D. Quillen and D. Sullivan in the late 1960s. In particular, Sullivan defined tools and models for rational homotopy inspired by already existing geometrical objects. Moreover, he gave an explicit dictionary between his minimal models and spaces, and this facility of transition between algebra and topology has created many new topological and geometrical theorems in the last 30 years. An introduction to rational homotopy whose main applications were in algebraic topology was written some years ago. Because of recent developments, it is clear that now is the time for a global presentation of some of the more representative geometrical applications of minimal models. That is the theme of this book.

Before giving an overview of its content, we present the basic philosophy behind the theory of minimal models. As Sullivan wrote in the introduction of *Infinitesimal Computations in Topology*:

We have suggested here that one might therefore recall the older methods of differential forms, which are evidently quite powerful.

When de Rham proved that $H^*(A_{DR}(M)) \cong H^*(M;\mathbb{R})$ for the differential algebra of differential forms $A_{DR}(M)$ on a manifold M, it immediately provided a link between the analysis on and the topology of the manifold. Sullivan is suggesting in his remark that even within the world of topology, there is more topological information in $A_{DR}(M)$ (henceforth called the de Rham algebra of M) than simply the real cohomology.

For a compact connected Lie group G, there exists a subdifferential algebra of bi-invariant forms, $\Omega_I(G)$, inside the de Rham algebra $A_{DR}(G)$, such that the canonical inclusion $\Omega_I(G) \hookrightarrow A_{DR}(G)$ induces an isomorphism in cohomology. This is the prototype of the process for models: namely, we look for a simplification $\mathcal{M}_M$ of the de Rham algebra with an explicit differential morphism $\mathcal{M}_M \to A_{DR}(M)$ inducing an isomorphism in cohomology, exactly as bi-invariant forms do in the case of a compact connected Lie group.

In order to implement this strategy, we first have to make precise what a "simplification" means. In the de Rham algebra, we might suspect that some information is contained in two different entities: the product of forms, which tells us how two forms can be combined together to give a third one and the exterior derivative of a form. In a model, we kill the information coming from the product structure by considering free algebras $\wedge V$ (in the commutative graded sense) where V is an $\mathbb{R}$-vector space. This pushes the corresponding information into the differential and into V where it is easier to detect. More precisely, we look for a cdga (for *commutative differential graded algebra*) free as a commutative graded algebra $(\wedge V, d)$ and a morphism $\varphi \colon (\wedge V, d) \to A_{DR}(M)$ inducing an isomorphism in cohomology.

The first question is, can one build such a model for any manifold? The answer is yes for connected manifolds and in fact, there are many ways to do this. So, we have to define a standard way, which we call *minimal*. With this in mind, we again look to Sullivan's introduction:

One proceeds degree by degree to construct a smallest possible sub-differential algebra of forms with the same cohomology. Forms are chosen in each degree to add cohomology not already achieved or to create necessary relations among cohomology classes.

Once we have this minimal model, we may ask what geometrical invariants can be detected in it. In fact, there is a functor from algebra to geometry that, together with forms, creates a dictionary between the algebraic and the geometrical worlds. *But for this we have to work over the rationals and not over the reals.* As a consequence, we have to replace the de Rham algebra by other types of forms. At first glance, this seems to be a disadvantage because we are switching from a well-known object to an unfamiliar one. But this new construction is very similar to the de Rham algebra and will allow the extension of the usual theory from manifolds to topological spaces, which is a great advantage. Denote by $A_{PL}(X)$ this analogue of the de Rham algebra for a topological space X. Since the minimal model construction also works perfectly well over $\mathbb{Q}$, we have the notion of a minimal model $\mathcal{M}_X \to A_{PL}(X)$ of a path connected space X.

Conversely, from a cdga (A, d) we have a topological realization $\langle (A, d) \rangle$ which is the return to Topology we hinted at above. If we apply this realization to a minimal model $\mathcal{M}_X$ of a space X (which is nilpotent with finite Betti numbers), then we get a continuous map $X \to \langle \mathcal{M}_X \rangle$ which induces an isomorphism in rational cohomology. The space $\langle \mathcal{M}_X \rangle$ is what, in homotopy theory, is called a rationalization of X. What must be emphasized in this process is the ability to create topological realizations of any algebraic constructions.

Such a theory begs for applications and examples and we describe models for spheres, homogeneous spaces, biquotients, connected sums, nilmanifolds, mapping spaces, configuration spaces and subspace arrangements. We give geometrical applications in several directions: to complex and symplectic manifolds, the closed geodesic problem, curvature problems, actions of tori, complements of submanifolds, symplectic blow-ups, and the Chas–Sullivan product, for instance.

Roughly, this book is composed of three parts. The first part, consisting of Chapters 1–3, contains the classical theory and the main geometrical examples. These chapters are self-contained except for certain proofs for which we provide references.

Chapters 4–8 are the second part. Each of them is devoted to a particular topic in differential topology or geometry and they are mostly independent.

The third part is the florilège of Chapter 9 where we give short presentations of particular subjects, chosen to illustrate the evolution of applications of minimal models from the theory's inception to the present day. Evidently we have been obliged to make choices in these applications and, therefore, many other interesting applications of algebraic models are not covered.

The following brief description of the material in each of the chapters makes the outline above more precise.

- *Chapter 1.* Throughout this book, Lie groups and homogeneous spaces are used to give foundational examples and to show that some of the basic ideas of Sullivan's rational homotopy theory were already present in this particular case years earlier. As well as describing certain basic structure results about Lie groups, this chapter gives a complete treatment of the computation of the cohomology algebra of a compact connected Lie group and recalls the basic facts about homogeneous spaces. We also look at the Cartan–Weil model and see it as the prototype for models of fibrations.
- *Chapter 2* is concerned with the basic definitions and properties of our algebraic tools: cdga's, models, minimal models, homotopy between morphisms of cdga's and the link between topological spaces and cdga's. When we construct a minimal model for a cdga (A, d), it is possible that we do not have to consider the whole algebra of forms, but rather only the cohomology $H(A, d)$. Although this is not true in general, it *is* true for spheres and Lie groups. This leads us to distinguish special types of spaces, called formal spaces, whose minimal models are determined by cohomology alone. This notion will be of great importance in applications and we delineate its properties.
- *Chapter 3.* Since the main theme of this book is the geometrical aspect of algebraic models, a first question is, how special is a minimal model of a compact simply connected manifold? At the very least, its cohomology must satisfy Poincaré duality. In fact, it was proved recently that there is a

model for X (and not just its cohomology) that satisfies Poincaré duality. What about the converse? How can we detect that the realization of a model contains a manifold in its rational homotopy type? Surprisingly, the conditions that are necessary for this also prove to be sufficient.

Formality in the case of manifolds entails certain properties. We prove here the theorems of Miller and Stasheff giving particular instances when manifolds are formal. Notably, we show, as Stasheff did using another method, that a compact simply connected manifold M is formal if and only if $M\backslash\{*\}$ is formal.

We also extend the construction of models to the case of cdga's equipped with the action of a finite group and apply it to the explicit construction of models of homogeneous spaces and biquotients.

- In *Chapter 4*, we study the link between the Dolbeault and de Rham algebras of a complex manifold M as well as the relationship between the respective models. We carefully consider the topological consequences of the existence of a Kähler metric on M, and, in particular, we prove the *formality of compact Kähler manifolds*. We also consider the Dolbeault model of a complex manifold in detail and compute it in many particular cases, including the case of Calabi-Eckmann manifolds. For that, we use a perturbation theorem which allows the construction of a model of a filtered cdga starting from a model of any stage of the associated spectral sequence. Applications to the Frölicher spectral sequence of a complex manifold are given.

  In the last part of this chapter, we describe some of the implications of models for symplectic topology. For a compact symplectic manifold, we compare the hard Lefschetz property with other properties that appear in the complex situation. In particular, we recall results of Mathieu and Merkulov concerning the relation of the hard Lefschetz property to the existence of symplectically harmonic forms, as defined by Brylinski, in each cohomology class.

- *Chapter 5*. For a smooth Riemannian manifold, the Riemannian structure of the manifold is reflected in its geodesics. The geodesics on a manifold may be viewed as the motion of a physical system, so in some sense, the study of geodesics exemplifies the paradigm expressing the relationship between mathematics and physics. Of course, the motions that are most important in physics are the periodic ones, so we begin by studying the geometric counterpart, closed geodesics.

  The main problem in this area is then: does every compact Riemannian manifold M of dimension at least two admit infinitely many geometrically distinct geodesics? The solution to this problem involves an essential analysis of the rational homotopy type of the free loop space of the manifold.

We give a minimal model of the free loop space and then prove the Vigué–Poirrier–Sullivan theorem, which solves most cases of the closed geodesic problem.

We also present several other connections between the structure of models and properties of a manifold's geodesics. Information about geodesics can often be codified by the dynamical system known as the geodesic flow and we shall see that the flow also holds rational homotopy information within it.

- *Chapter 6.* In the last decade, algebraic models have proven to be useful tools in the study of various differential geometric questions involving curvature. A basic problem is whether curvature and diameter constraints limit to a finite number the possible rational homotopy types of manifolds satisfying those constraints. We describe the use of models in the construction of counterexamples to this question. We also show how models can be used to give a general analysis of the failure of the converse of the Soul theorem of Cheeger and Gromoll.
- *Chapter 7.* The topological qualities of a space are often reflected in its intrinsic symmetries. These symmetries, in turn, may be formalized as the actions of groups on the space. Intuitively, most manifolds are asymmetric, so the existence of a nontrivial group action on a manifold implies that the manifold is special topologically. The properties of a manifold with group action may be gleaned from various topological constructions and their algebraic reflections. Indeed, this chapter focuses on what can be said about group actions from the viewpoint of algebraic models. For instance, there is a longstanding conjecture called the toral rank conjecture which is usually attributed to S. Halperin. The conjecture says, in particular, that if there is a free action of a torus T^r on a space X, then the dimension of the rational cohomology of X must be at least as large as the dimension of the rational cohomology of the torus. We give proofs for homogeneous spaces and Kähler manifolds.

  We also discuss the Borel localization theorem and apply it to the study of the rational homotopy and the rational cohomology of fixed point sets. Finally, we discuss the notion of Hamiltonian action in symplectic geometry and use models to prove a special case of the Lalonde–McDuff question about Hamiltonian bundles.
- *Chapter 8.* The process of taking a blow-up has proven to be extremely useful in complex and symplectic geometry. In order to consider various questions on the interface between geometry and algebraic topology, it is necessary to understand algebraic models of blow-ups. This entails a panoply of related questions which all serve as testaments to the efficacy of rational homotopy theory in geometry. In this chapter we consider two types of questions.

First of all, let $f\colon N \hookrightarrow M$ be a closed submanifold of a compact orientable manifold and denote by C its complement. The natural problem is to know if the rational homotopy type of C is completely determined by the rational homotopy type of the embedding, and in that case to describe a model for the injection $C \hookrightarrow M$ from a model of the initial embedding. We use it to describe nonformal simply connected symplectic manifolds that are blow-ups.

Our second question concerns the geometric intersection theory of cycles in a compact manifold. M. Chas and D. Sullivan have extended the standard intersection theory to an intersection theory of cycles in the free loop space LN for any compact oriented manifold N. More precisely, they define a product on $H_*(LN)$ that combines the intersection product on the chains on N and the Pontryagin composition of loops in ΩN. We present a more homological re-interpretation of the Chas–Sullivan product and, as a corollary, obtain the well-known theorem of Cohen and Jones.

- In *Chapter 9*, we consider various types of geometric situations where algebraic models are useful. Models make their presence felt in the study of configuration spaces, arrangements, smooth algebraic varieties, mapping spaces, Gelfand–Fuchs cohomology, and iterated integrals. Of course, it is impossible to prove everything about such an array of topics, so this chapter is simply a survey of these applications of models. We endeavor to describe and explain the relevant models and then refer to the appropriate literature for details.
- Finally, there are three appendices that recall basic facts about de Rham forms, spectral sequences and homotopy theory.

It is our hope (and *we believe*) that this book will prove enlightening to both geometers and topologists. It should be useful to geometers because of concrete examples showing how algebraic techniques can be used to help solve geometric problems. For topologists, on the other hand, it is important to see what kind of concrete geometrical questions can be studied from a topological point of view.

A project such as this requires a great deal of support and we would like to acknowledge this here. First, this book would never have seen the light of day without Research in Pairs grants from the Mathematisches Forschungsinstitut Oberwolfach in 2003 and 2006. These stays at the MFO were essential to our collaboration and it is a pleasure to acknowledge the generosity of this mathematical haven. Various portions of the book were read by Agusti Roig and he provided many insightful comments and suggestions. We also thank P. Lambrechts and G. Paternain for discussions on several topics. Finally, the support of the University of Louvain-La-Neuve and of the CNRS for the Summer School on Algebraic Models, held at Louvain-La-Neuve in June 2007, was essential to the completion of this work.

Let's now begin.

Contents

Preface   vii

1 Lie groups and homogeneous spaces   1

    1.1   Lie groups   2
    1.2   Lie algebras   3
    1.3   Lie groups and Lie algebras   5
    1.4   Abelian Lie groups   8
    1.5   Classical examples of Lie groups   8
        1.5.1   Subgroups of the real linear group   9
        1.5.2   Subgroups of the complex linear group   10
        1.5.3   Subgroups of the quaternionic linear group   10
    1.6   Invariant forms   11
    1.7   Cohomology of Lie groups   16
    1.8   Simple and semisimple compact connected
           Lie groups   21
    1.9   Homogeneous spaces   26
    1.10 Principal bundles   32
    1.11 Classifying spaces of Lie groups   38
    1.12 Stiefel and Grassmann manifolds   42
    1.13 The Cartan–Weil model   47

2 Minimal models   56

    2.1   Commutative differential graded algebras   57
    2.2   Homotopy between morphisms of cdga's   61
    2.3   Models in algebra   64
        2.3.1   Minimal models of cdga's and morphisms   64
        2.3.2   Relative minimal models   66
    2.4   Models of spaces   67
        2.4.1   Real and rational minimal models   67
        2.4.2   Construction of $A_{PL}(X)$ 69

		2.4.3 Examples of minimal models of spaces	71
		2.4.4 Other models for spaces	74
	2.5	Minimal models and homotopy theory	75
		2.5.1 Minimal models and homotopy groups	75
		2.5.2 Relative minimal model of a fibration	78
		2.5.3 The dichotomy theorem	84
		2.5.4 Minimal models and some homotopy constructions	87
	2.6	Realizing minimal cdga's as spaces	90
		2.6.1 Topological realization of a minimal cdga	90
		2.6.2 The cochains on a graded Lie algebra	91
	2.7	Formality	92
		2.7.1 Bigraded model	95
		2.7.2 Obstructions to formality	96
	2.8	Semifree models	100
3	**Manifolds**		**104**
	3.1	Minimal models and manifolds	105
		3.1.1 Sullivan–Barge classification	105
		3.1.2 The rational homotopy groups of a manifold	106
		3.1.3 Poincaré duality models	109
		3.1.4 Formality of manifolds	110
	3.2	Nilmanifolds	116
		3.2.1 Relations with Lie algebras	117
		3.2.2 Relations with principal bundles	121
	3.3	Finite group actions	123
		3.3.1 An equivariant model for Γ-spaces	123
		3.3.2 Weyl group and cohomology of BG	127
	3.4	Biquotients	133
		3.4.1 Definitions and properties	133
		3.4.2 Models of biquotients	137
	3.5	The canonical model of a Riemannian manifold	139
4	**Complex and symplectic manifolds**		**145**
	4.1	Complex and almost complex manifolds	148
		4.1.1 Complex manifolds	148
		4.1.2 Almost complex manifolds	150
		4.1.3 Differential forms on an almost complex manifold	152
		4.1.4 Integrability of almost complex manifolds	154

4.2		Kähler manifolds	156
	4.2.1	Definitions and properties	156
	4.2.2	Examples: Calabi–Eckmann manifolds	159
	4.2.3	Topology of compact Kähler manifolds	162
4.3		The Dolbeault model of a complex manifold	168
	4.3.1	Definition and existence	169
	4.3.2	The Dolbeault model of a Kähler manifold	172
	4.3.3	The Borel spectral sequence	173
	4.3.4	The Dolbeault model of Calabi–Eckmann manifolds	175
4.4		The Frölicher spectral sequence	178
	4.4.1	Definition and properties	178
	4.4.2	Pittie's examples	179
4.5		Symplectic manifolds	182
	4.5.1	Definition of symplectic manifold	182
	4.5.2	Examples of symplectic manifolds	183
	4.5.3	Symplectic manifolds and the hard Lefschetz property	184
	4.5.4	Symplectic and complex manifolds	187
4.6		Cohomologically symplectic manifolds	187
	4.6.1	C-symplectic manifolds	187
	4.6.2	Symplectic homogeneous spaces and biquotients	188
	4.6.3	Symplectic fibrations	189
	4.6.4	Symplectic nilmanifolds	191
	4.6.5	Homotopy of nilpotent symplectic manifolds	194
4.7		Appendix: Complex and symplectic linear algebra	196
	4.7.1	Complex structure on a real vector space	196
	4.7.2	Complexification of a complex structure	197
	4.7.3	Hermitian products	198
	4.7.4	Symplectic linear algebra	200
	4.7.5	Symplectic and complex linear algebra	201
	4.7.6	Generalized complex structure	202

5	Geodesics		205
	5.1	The closed geodesic problem	207
	5.2	A model for the free loop space	210
	5.3	A solution to the closed geodesic problem	213
	5.4	A-invariant closed geodesics	215
	5.5	Existence of infinitely many A-invariant geodesics	222

5.6	Gromov's estimate and the growth of closed geodesics	223
5.7	The topological entropy	227
5.8	Manifolds whose geodesics are closed	232
5.9	Bar construction, Hochschild homology and cohomology	234

6 Curvature 239

6.1	Introduction: Recollections on curvature	239
6.2	Grove's question	243
	6.2.1 The Fang–Rong approach	243
	6.2.2 Totaro's approach	249
6.3	Vampiric vector bundles	252
	6.3.1 The examples of Özaydin and Walschap	253
	6.3.2 The method of Belegradek and Kapovitch	259
6.4	Final thoughts	265
6.5	Appendix	266

7 G-spaces 271

7.1	Basic definitions and results	273
7.2	The Borel fibration	275
7.3	The toral rank	276
	7.3.1 Toral rank for rationally elliptic spaces	278
	7.3.2 Computation of $\mathrm{rk}_0(M)$ with minimal models	280
	7.3.3 The toral rank conjecture	283
	7.3.4 Toral rank and center of $\pi_*(\Omega M) \otimes \mathbb{Q}$	287
	7.3.5 The TRC for Lie algebras	289
7.4	The localization theorem	291
	7.4.1 Relations between G-manifold and fixed set	292
	7.4.2 Some examples	295
7.5	The rational homotopy of a fixed point set component	298
	7.5.1 The rational homotopy groups of a component	298
	7.5.2 Presentation of the Lie algebra $L_F = \pi_*(\Omega F) \otimes \mathbb{Q}$	303
	7.5.3 $\mathbb{Z}/2\mathbb{Z}$-Sullivan models	305
7.6	Hamiltonian actions and bundles	306
	7.6.1 Basic definitions and properties	306
	7.6.2 Hamiltonian and cohomologically free actions	308
	7.6.3 The symplectic toral rank theorem	312
	7.6.4 Some properties of Hamiltonian actions	312
	7.6.5 Hamiltonian bundles	314

8	Blow-ups and Intersection Products		317
	8.1	The model of the complement of a submanifold	318
		8.1.1 Shriek maps	319
		8.1.2 Algebraic mapping cones	321
		8.1.3 The model for the complement C	324
		8.1.4 Properties of Poincaré duality models	328
		8.1.5 The configuration space of two points in a manifold	329
	8.2	Symplectic blow-ups	330
		8.2.1 Complex blow-ups	331
		8.2.2 Blowing up along a submanifold	332
	8.3	A model for a symplectic blow-up	334
		8.3.1 The basic pullback diagram of PL-forms	334
		8.3.2 An illustrative example	334
		8.3.3 The model for the blow-up	335
		8.3.4 McDuff's example	337
		8.3.5 Effect of the symplectic form on the blow-up	339
		8.3.6 Vanishing of Chern classes for KT	339
	8.4	The Chas-Sullivan loop product on loop space homology	341
		8.4.1 The classical intersection product	341
		8.4.2 The Chas–Sullivan loop product	342
		8.4.3 A rational model for the loop product	344
		8.4.4 Hochschild cohomology and Cohen–Jones theorem	346
		8.4.5 The Chas-Sullivan loop product and closed geodesics	348
9	A Florilège of geometric applications		350
	9.1	Configuration spaces	351
		9.1.1 The Fadell–Neuwirth fibrations	352
		9.1.2 The rational homotopy of configuration spaces	353
		9.1.3 The configuration spaces $F(\mathbb{R}^n, k)$	354
		9.1.4 The configuration spaces of a projective manifold	355
	9.2	Arrangements	358
		9.2.1 Formality of the complement of a geometric lattice	361
		9.2.2 Rational hyperbolicity of the space $M(\mathcal{A})$	362

	9.3	Toric topology	363
	9.4	Complex smooth algebraic varieties	364
	9.5	Spaces of sections and Gelfand–Fuchs cohomology	367
		9.5.1 The Haefliger model for spaces of sections	367
		9.5.2 The Bousfield–Peterson–Smith model	371
		9.5.3 Configuration spaces and spaces of sections	373
		9.5.4 Gelfand–Fuchs cohomology	375
	9.6	Iterated integrals	376
		9.6.1 Definition of iterated integrals	376
		9.6.2 The cdga of iterated integrals	379
		9.6.3 Iterated integrals and the double bar construction	381
		9.6.4 Iterated integrals, the Hochschild complex and the free loop space	384
		9.6.5 Formal homology connection and holonomy	385
		9.6.6 A topological application	387
	9.7	Cohomological conjectures	388
		9.7.1 The toral rank conjecture	388
		9.7.2 The Halperin conjecture	388
		9.7.3 The Bott conjecture	389
		9.7.4 The Gromov conjecture on LM	390
		9.7.5 The Lalonde–McDuff question	390
A	De Rham forms		392
	A.1	Differential forms	392
	A.2	Operators on forms	398
	A.3	The de Rham theorem	402
	A.4	The Hodge decomposition	404
B	Spectral sequences		409
	B.1	What is a spectral sequence?	409
	B.2	Spectral sequences in cohomology	411
	B.3	Spectral sequences and filtrations	412
	B.4	Serre spectral sequence	413
	B.5	Zeeman–Moore theorem	416
	B.6	An algebraic example: The odd spectral sequence	419
	B.7	A particular case: A double complex	420

C	Basic homotopy recollections	423
	C.1 n-equivalences and homotopy sets	423
	C.2 Homotopy pushouts and pullbacks	424
	C.3 Cofibrations and fibrations	428

References 433

Index 451

Lie groups and homogeneous spaces

Lie groups and homogeneous spaces form an important family of examples of manifolds. We will use them systematically in many different parts of this book, to give an illustration for a specific method or, as well, to show that new viewpoints can be obtained by using algebraic models. Therefore, it is important for us to understand basic ideas and results about Lie groups in order to appreciate the development and application of algebraic models to geometry in general.

What we also want to articulate in the next several chapters is that some of the basic ideas of Sullivan's rational homotopy theory were already present in this particular case. Of course, the (rational or real) cohomology algebra is a fundamental example of an algebraic model, but it was understood from the start that only certain information was reflected in it. Nevertheless, computing cohomology for such geometric building blocks as Lie groups became an important goal in the early years of algebraic topology. An algebraic model that proved to be effective in reaching this goal was the Cartan–Weil model. Nowadays, we look at this model and see it as the direct ancestor of Sullivan's minimal models. Indeed, the Cartan–Weil model is the prototype for models of fibrations (see Chapter 2) and, thus holds within it homotopical, rather than just cohomological, information.

A complete treatment of Lie group theory covers several books, so we need to make precise the philosophy of this chapter. Basic properties and definitions on Lie groups and Lie algebras are recalled without proofs in Sections 1.1–1.4. The classical examples of groups of matrices are described in Section 1.5. From that point on, we wish to be complete, with proofs, when we discuss the things that form the main focus of the book: here, that means in particular, the de Rham cohomology of Lie groups and homogeneous spaces. Moreover, and we will discover the importance of this in Chapter 2, we wish to have a "computable" cochain complex Ω linked to the de Rham complex by an algebraic map, $\Omega \to A_{DR}$, which induces an isomorphism in cohomology. For instance, in the case of Lie groups, we prove in Section 1.6 that the inclusion $\Omega_I(G) \hookrightarrow A_{DR}(G)$ of the bi-invariant forms induces an isomorphism in cohomology. We continue in Section 1.7 by providing two structure theorems for the cohomology of Lie groups, including the Hopf theorem which expresses the cohomology algebra as an exterior algebra on generators of odd degrees. We will use this result in Section 1.8 for the computation of the second and third Betti numbers for simple Lie groups.

Since fibrations are the basic ingredients in the theory of minimal models, we develop the notions of principal bundles and classifying spaces. Sections 1.9–1.10 contain definitions of bundles, principal bundles and homogeneous spaces. In Section 1.11, we define and characterize the notion of classifying space for Lie groups. The classical Stiefel and Grassmann manifolds provide us with classifying spaces for the various orthogonal groups and they are studied in Section 1.12. Finally, in Section 1.13, we give the prototype for the theory of minimal models (see Chapter 2), the Cartan–Weil model for equivariant cohomology.

We assume that the reader possesses a good knowledge of the basic concepts about manifolds, as in [28], [104], [226], or (for surface theory) [215] for instance. We also assume the reader has had a classical homology course. Appendices A, B and C supply the basic recollections on some crucial parts of these subjects.

In this chapter, *manifold* means a Hausdorff space with a countable basis of open sets (i.e. a separable space), endowed with a differentiable structure of class C^∞ over the reals.

1.1 Lie groups

In this section, we give the definitions and basic properties of Lie groups. General references are [2], [136] and [199] for instance.

Definition 1.1 *A Lie group is a set G which is both a manifold and a group and for which the multiplication, $(g, g') \mapsto gg'$, and the inverse map, $g \mapsto g^{-1}$, are smooth. The dimension of a Lie group is its dimension as a manifold.*

A homomorphism of Lie groups is a homomorphism of groups which is also a smooth map. An isomorphism of Lie groups is a homomorphism f which admits an inverse f^{-1} as maps and such that f^{-1} is also a homomorphism of Lie groups.

For instance, $\mathbb{R}$ and S^1 are Lie groups for the usual structures of manifolds and groups. One can observe directly from the definition that a product of two Lie groups is a Lie group for the two canonical structures, product of groups and product of manifolds. So $\mathbb{R}^n$ and $(S^1)^n$ are Lie groups. The tori

$T^n = (S^1)^n$ are studied in Section 1.4, and Section 1.5 is devoted to classical examples of Lie groups, such as the different groups of matrices. Observe now that, if V is a real vector space, the set $\mathrm{Gl}(V)$ of linear isomorphisms is a Lie group.

Remark 1.2 Some properties required in Definition 1.1 are in fact automatic:

- if the multiplication is a smooth map, then the inverse is also a smooth map (use the implicit function theorem);
- a bijective homomorphism of Lie groups is an isomorphism of Lie groups (see [214, page 18]);
- a map $f: G \to H$ between Lie groups which is a homomorphism of groups and a continuous map is a homomorphism of Lie groups (see [199, page 44]).

A *Lie subgroup* of a Lie group G is a subgroup which is a submanifold of G. Lie subgroups can be determined easily by the following theorem of Elie Cartan.

Theorem 1.3 ([2, page 17], [214, page 47]) *A subgroup H of a Lie group G is a Lie subgroup of G if and only if H is a closed subgroup of G.*

In this book, we will be especially interested in the homotopy types of Lie groups and, for that, we can reduce the study to compact connected Lie groups as shown by the following result called the *polar decomposition or Iwasawa decomposition*.

Theorem 1.4 ([147]) *Any connected Lie group G admits a maximal compact subgroup H (unique up to conjugacy) such that G is isomorphic to the product $H \times \mathbb{R}^m$. In particular G and H have the same homotopy type.*

Compact connected Lie groups will be classified in Theorem 1.52.

1.2 Lie algebras

We introduce here the notion of Lie algebras and the example of main interest for us, the tangent space $T_e(G)$ of a Lie group G at the identity.

Definition 1.5 *A Lie algebra over $\mathbb{R}$ is a vector space $\mathfrak{l}$ together with a bilinear homomorphism, called the bracket,*

$$[-,-]: \mathfrak{l} \times \mathfrak{l} \to \mathfrak{l}$$

such that, for any $l_1 \in \mathfrak{l}$, $l_2 \in \mathfrak{l}$, $l_3 \in \mathfrak{l}$, one has:

- $[l_1, l_2] = -[l_2, l_1]$ *(skew symmetry)*,
- $[l_1, [l_2, l_3]] + [l_2, [l_3, l_1]] + [l_3, [l_1, l_2]] = 0$ *(Jacobi identity)*.

A *homomorphism of Lie algebras* is a linear map, $\varphi \colon \mathfrak{l} \to \mathfrak{l}'$, preserving the bracket. This means that $\varphi[l_1, l_2] = [\varphi(l_1), \varphi(l_2)]$ for any $(l_1, l_2) \in \mathfrak{l} \times \mathfrak{l}$.

A *Lie subalgebra* of a Lie algebra $\mathfrak{l}$ is a sub-vector space $\mathfrak{n}$ such that $[\mathfrak{n}, \mathfrak{n}] \subseteq \mathfrak{n}$. An *ideal* of $\mathfrak{l}$ is a Lie subalgebra $\mathfrak{n}$ such that $[\mathfrak{n}, \mathfrak{l}] \subseteq \mathfrak{n}$.

Any structure of associative algebra on a vector space A gives a canonical structure of Lie algebra $\mathfrak{l}_A$ on the same vector space by $[a_1, a_2] = a_1 a_2 - a_2 a_1$. Our most interesting example comes from the structure of a Lie group G.

Observe that, because of the group structure, any phenomenon at a particular point of G can be translated everywhere in G by composing with the elements of the group. For instance, a connected Lie group is generated, as a topological space, by any neighborhood of the identity e. We formalize this remark by introducing the notion of left and right translations.

Definition 1.6 *A fixed element $g \in G$ gives the left translation $L_g \colon G \to G$ with $L_g(h) = g \cdot h$ for all $h \in G$. Similarly, we define right translations R_g by $R_g(h) = h \cdot g$.*

Recall first that, if $p \colon T(G) \to G$ is the tangent bundle of the manifold G, a *vector field* X on G is a smooth section of p (see Appendix A).

Definition 1.7 *Denote by $DL_g \colon T(G) \to T(G)$ the map induced by the left translation L_g. A vector field X on G is called left invariant if $DL_g(X) = X$, for any $g \in G$.*

Remark 1.8 Left invariance of objects other than vector fields also turns out to be very important in understanding Lie groups. In particular, if we define a function f on a Lie group G to be left invariant when $f(gh) = f(h)$ for all $g, h \in G$, then clearly we see that f is a constant function. Later (see Definition 1.25), we will phrase this by saying that any left invariant 0-form is a constant.

If G is a Lie group, we denote by $\mathfrak{g}$ the vector space of left invariant vector fields on G. If X and Y are vector fields, then their *bracket* is defined to be the vector field $[X, Y]f = X(Yf) - Y(Xf)$ for all functions f. The bracket is anti-commutative and satisfies a Jacobi identity (see Section A.2). If X and Y are left invariant vector fields, their bracket $[X, Y]$ is also left invariant. Therefore, the vector space $\mathfrak{g}$ has the structure of a Lie algebra, called *the Lie algebra associated to the Lie group G*.

If X is a vector field on a Lie group G, we see directly from the definition that X is left invariant if and only if

$$X_g = (DL_g)(X_e).$$

Therefore *the vector space $\mathfrak{g}$ is isomorphic to the tangent space, $T_e(G)$, at the identity e of G.* We will not make any distinction between these two characterizations of the Lie algebra $\mathfrak{g}$.

From this observation, we deduce that the left invariant vector fields provide n linearly independent sections of the tangent bundle TG for an n-dimensional Lie group G. Therefore, *any Lie group G is parallelizable*; that is, $TG \cong G \times T_e(G) = G \times \mathfrak{g}$. As a consequence, the real Lie algebra of vector fields on G (see Section A.2) is the tensor product $C^\infty(G) \otimes \mathfrak{g}$ where $C^\infty(G)$ is the algebra of smooth real valued functions on G.

Proposition 1.9 *There is a morphism of Lie groups* $\mathrm{Ad}\colon G \to \mathrm{Gl}(\mathfrak{g})$ *given by* $\mathrm{Ad}(g)(X) = ((DR_g)^{-1} \circ (DL_g))(X)$, *where $\mathrm{Gl}(\mathfrak{g})$ is the group of linear isomorphisms of the Lie algebra $\mathfrak{g}$.*

Definition 1.10 *The map* $\mathrm{Ad}\colon G \to \mathrm{Gl}(\mathfrak{g})$ *is called the adjoint representation of the Lie group G.*

Proposition 1.11 *Denote by $\mathfrak{gl}(\mathfrak{g})$ the Lie algebra of the Lie group $\mathrm{Gl}(\mathfrak{g})$. Then, the derivative of* $\mathrm{Ad}\colon G \to \mathrm{Gl}(\mathfrak{g})$ *is the morphism of Lie algebras,* $\mathrm{ad}\colon \mathfrak{g} \to \mathfrak{gl}(\mathfrak{g})$, *defined by* $\mathrm{ad}(X)(Y) = [X, Y]$, *where $[-, -]$ is the bracket of $\mathfrak{g}$.*

1.3 Lie groups and Lie algebras

It is now time to make various *relations between Lie groups and Lie algebras* precise. A basic reference is [214, Section I-&2].

First, by definition of the Lie algebra associated to a Lie group, if $f\colon G \to H$ is a homomorphism of Lie groups, its differential $Df_e\colon \mathfrak{g} \to \mathfrak{h}$ is a homomorphism of Lie algebras. We would like to know if Df_e gives us information about the homomorphism f. For that, we need a better correspondence between Lie algebras and Lie groups which comes from the notion of one-parameter subgroup.

Definition 1.12 *For any Lie group G, a homomorphism of Lie groups, $\theta\colon \mathbb{R} \to G$, is called a* one-parameter subgroup *of G.*

Observe that, by definition, such a θ satisfies $\theta(s+t) = \theta(s) \cdot \theta(t)$, for any s and t in $\mathbb{R}$. It can be shown that any one-parameter subgroup is the integral curve of a left invariant vector field and, reciprocally, that any left invariant vector field admits a one-parameter subgroup as a maximal integral curve. So, we get a new characterization of the Lie algebra associated to a Lie group.

Theorem 1.13 *There is an isomorphism between the Lie algebra $\mathfrak{g}$ associated to a Lie group G and the set of one-parameter subgroups of G.*

With this isomorphism, one can construct a map from $\mathfrak{g}$ to G.

Definition 1.14 *If $X \in \mathfrak{g}$, we denote by θ_X the one-parameter subgroup associated to X as in Theorem 1.13. The exponential from $\mathfrak{g} = T_e(G)$ to G is defined by*

$$\exp(tX) = \theta_X(t).$$

Observe that, by uniqueness of the integral curve, one has $\theta_{\lambda X}(t) = \theta_X(\lambda t)$ and the exponential is well defined. In fact, the exponential is a smooth map which induces the identity on the tangent space at $e \in G$; that is, $D\exp = \mathrm{id}: \mathfrak{g} \to \mathfrak{g}$. It can also be shown that the exponential is an epimorphism if the group G is compact and connected. Moreover, if $f: G \to G'$ is a homomorphism of Lie groups, then one has $f \circ \exp_G = \exp_{G'} \circ Df$.

This process of integration is the key for the two next results. The first one concerns the link between a homomorphism of Lie groups and its induced differential.

Theorem 1.15 *Let G and H be two Lie groups with G connected. Then a homomorphism from G to H is uniquely determined by its differential $Df_e : \mathfrak{g} \to \mathfrak{h}$.*

The second result concerns the realization of morphisms between Lie algebras.

Theorem 1.16 *Let G and H be two Lie groups with G simply connected. Then, for every homomorphism of Lie algebras $\psi: \mathfrak{g} \to \mathfrak{h}$, there exists a homomorphism of Lie groups $f: G \to H$ such that $Df_e = \psi$.*

The theory is very powerful. For instance, the *third Lie theorem* gives a converse to the construction of the Lie algebra of a Lie group (see [229, Leçon 6]).

Theorem 1.17 *Every finite dimensional Lie algebra is the tangent space algebra of some Lie group.*

1.3 Lie groups and Lie algebras

This association can be made more precise:

- every finite dimensional Lie algebra is the tangent space algebra of a unique simply connected Lie group;
- if a homomorphism of Lie groups, $f: G \to G'$, with G simply connected, induces an isomorphism between the associated Lie algebras, then f is a universal cover.

These results imply a correspondence between sub-Lie groups and sub-Lie algebras which can be made explicit for some objects of interest. For the rest of this section, let G be a connected Lie group with associated Lie algebra $\mathfrak{g}$ and let H be a Lie subgroup of G. Then the Lie algebra $\mathfrak{h}$ associated to H is a Lie subalgebra of $\mathfrak{g}$ and the subgroup H is normal if and only if the subalgebra $\mathfrak{h}$ is an ideal. Now let's recall some classical definitions (see [113, page 69]) which will be useful in the rest of this section.

Definition 1.18 *The* centralizer *of a subset A of G is the subgroup*

$$Z(A) = \{x \in G \mid xa = ax \text{ for any } a \in A\}.$$

The centralizer of G is called the center *of G. The* centralizer *of a subset $\mathfrak{m}$ in $\mathfrak{g}$ is the Lie subalgebra*

$$\mathcal{Z}(\mathfrak{m}) = \{l \in \mathfrak{g} \mid [l, m] = 0 \text{ for any } m \in \mathfrak{m}\}.$$

The centralizer of $\mathfrak{g}$ is called the center *of $\mathfrak{g}$.*

It can be shown that *the centralizer of H in G is a Lie group with associated Lie algebra the centralizer of $\mathfrak{h}$ in $\mathfrak{g}$.*

Definition 1.19 *The* normalizer *of a subset A of G is the subgroup of G given by*

$$N(A) = \{x \in G \mid xA = Ax\}.$$

The normalizer *of a subset $\mathfrak{m}$ of $\mathfrak{g}$ is the Lie subalgebra*

$$\mathfrak{n}(\mathfrak{m}) = \{x \in \mathfrak{g} \mid [x, y] \in \mathfrak{m} \text{ for any } y \in \mathfrak{m}\}.$$

It can be shown that the normalizer of H in G is a Lie group with associated Lie algebra the normalizer of $\mathfrak{h}$ in $\mathfrak{g}$. Observe that, for any $x \in N(H)$ and any $h \in H$, we have $xhx^{-1} \in H$. We deduce that there is an *action* (see Definition 1.23) of the Lie group $N(H)$ on the manifold H, called the *conjugation action*.

1.4 Abelian Lie groups

It turns out that it is essential to first study the abelian Lie groups and the abelian Lie subgroups of a compact connected Lie group. A reference for this section is [113, Chapters 1 and 2]. A basic example of an abelian compact Lie group is the circle S^1 endowed with the commutative multiplication of complex numbers. More generally, we have the

Definition 1.20 *An* abelian Lie group *is a Lie group G satisfying $gg' = g'g$ for any $(g,g') \in G \times G$. An* abelian Lie algebra *is a Lie algebra $\mathfrak{l}$ such that $[l,l'] = 0$ for any $(l,l') \in \mathfrak{l} \times \mathfrak{l}$.*

One can prove that *a Lie group G is abelian if and only if its Lie algebra $\mathfrak{g}$ is abelian*. A product of n circles is an abelian Lie group, called an *n-torus* (or simply a torus) and denoted by T^n. Tori are the prototypes of abelian Lie groups.

Theorem 1.21 *Any connected abelian Lie group G is isomorphic to the direct product of Lie groups $T^p \times \mathbb{R}^q$.*

As a consequence, any connected abelian Lie subgroup of a compact connected Lie group G is a torus T. Call a subtorus $T \subset G$ a *maximal torus* in G if it is not properly contained in another torus. One can then prove the following.

Theorem 1.22 *Every element of a compact connected Lie group is contained in a maximal torus. Two maximal tori are conjugate.*

The dimension of a maximal torus is called *the rank* of the Lie group. The normalizer of a maximal torus T of G is a compact Lie group denoted $N(T)$. The quotient $W(G) = N(T)/T$ is called *the Weyl group of G*. Up to isomorphism, this group does not depend on the choice of a maximal torus in G. In fact, $W(G)$ is a finite group. Observe that, since T is abelian, the restriction of the conjugation action of $N(T)$ to T is trivial on T, so it gives an action of $W(G)$ on T. We will come back to the study of maximal tori in Subsection 3.3.2.

1.5 Classical examples of Lie groups

We now describe the classical examples of Lie groups which come from groups of matrices. A reference for this section is [199, Chapter I].

1.5.1 Subgroups of the real linear group

Start with the set of real numbers $\mathbb{R}$ and denote by $\mathrm{Gl}(n,\mathbb{R})$ the group of invertible $n \times n$-matrices with entries in $\mathbb{R}$. Endowed with the canonical structure of a manifold (as an open subset of $\mathbb{R}^{n^2}$), the group $\mathrm{Gl}(n,\mathbb{R})$ is a Lie group, called the *real linear Lie group*. The associated Lie algebra, $\mathfrak{gl}_n(\mathbb{R}) = M(n,\mathbb{R})$, is the vector space of all $n \times n$ matrices with the bracket being the commutator of matrices. The dimension of $\mathrm{Gl}(n,\mathbb{R})$ is n^2.

Any closed subgroup of $\mathrm{Gl}(n,\mathbb{R})$ is a Lie group. In particular, we have the *orthogonal group* $\mathrm{O}(n)$ consisting of the orthogonal linear transformations u of the euclidean space $\mathbb{R}^n$. Recall that, in the canonical basis, this is equivalent to the fact that the matrix A of u satisfies ${}^tAA = I_n$. The associated Lie algebra $\mathfrak{o}(n)$ of $\mathrm{O}(n)$ is the vector space of alternating (or skew-symmetric) matrices, ${}^tA = -A$. The dimension of $\mathrm{O}(n)$ is $\dfrac{n(n-1)}{2} = \binom{n}{2}$ and $\mathrm{O}(n)$ is a maximal compact subgroup of $\mathrm{Gl}(n,\mathbb{R})$.

Since the continuous map $\det: \mathrm{O}(n) \to \{-1,+1\}$ is surjective, one sees that the space $\mathrm{O}(n)$ is not connected. We denote by $\mathrm{SO}(n)$ the subgroup of $\mathrm{O}(n)$, consisting of linear transformations of determinant 1 and call it the *special orthogonal group*. Since it is a connected component of $\mathrm{O}(n)$, the group $\mathrm{SO}(n)$ has the same tangent space at the neutral element e, therefore the same Lie algebra, by definition. As we will see, the group $\mathrm{SO}(n)$ is not simply connected if $n \geq 2$. The universal cover of $\mathrm{SO}(n)$ is called *the nth-spinor group* and denoted by $\mathrm{Spin}(n)$.

The orthogonal group $\mathrm{O}(n)$ is the prototype of Lie groups. Indeed, it can be proved that any compact Lie group G is isomorphic to a closed subgroup of $\mathrm{O}(n)$ (see [199, Theorem 2.14, Chapter V]).

If θ is a real number, we denote by

$$R(\theta) = \begin{pmatrix} \cos\theta & -\sin\theta \\ \sin\theta & \cos\theta \end{pmatrix}$$

the rotation matrix corresponding to the rotation in $\mathbb{R}^2$ by angle θ. Let $(\theta_1,\ldots,\theta_r)$ be r real numbers. Denote by $(R(\theta_1),\ldots,R(\theta_r),1)$ the matrix having the $R(\theta_i)$ and 1 along the diagonal and 0 entries otherwise. The group of matrices of the form $(R(\theta_1),\ldots,R(\theta_r),1)$ is the *maximal torus* of $\mathrm{SO}(2r+1)$ and the rank of $\mathrm{SO}(2r+1)$ is r. The *Weyl group* of $\mathrm{SO}(2r+1)$ has $2^r r!$ elements and acts on the maximal torus by a permutation of the coordinates composed with the substitutions $(\theta_1,\ldots,\theta_r) \mapsto (\pm\theta_1,\ldots,\pm\theta_r)$.

As for the Lie group $\mathrm{SO}(2r)$, its maximal torus consists of matrices $(R(\theta_1),\ldots,R(\theta_r))$ and the rank of $\mathrm{SO}(2r)$ is r. Its *Weyl group* has $2^{r-1}r!$ elements acting on the maximal torus by a permutation of the coordinates composed with the substitutions $(\theta_1,\ldots,\theta_r) \mapsto (\varepsilon_1\theta_1,\ldots,\varepsilon_r\theta_r)$, with $\varepsilon_i = \pm 1$ and $\varepsilon_1\cdots\varepsilon_r = 1$.

1.5.2 Subgroups of the complex linear group

Denote by $Gl(n,\mathbb{C})$ the group of invertible $n \times n$-matrices with entries in the complex numbers $\mathbb{C}$. Endowed with the canonical structure of a manifold (as an open subset of $\mathbb{R}^{2n^2}$), the group $Gl(n,\mathbb{C})$ is a Lie group, called the *complex linear Lie group*. The associated Lie algebra, $\mathfrak{gl}_n(\mathbb{C}) = M(n,\mathbb{C})$, is the vector space of all $n \times n$-matrices with the bracket being the commutator of matrices. The (real) dimension of $Gl(n,\mathbb{C})$ is $2n^2$.

We now introduce the analogue of the orthogonal group. Recall that, if we write a complex number as $z = x + iy$, with $x \in \mathbb{R}$ and $y \in \mathbb{R}$, the conjugate of z is the complex number $\bar{z} = x - iy$. This induces a norm with $\|z\| = \sqrt{z\bar{z}}$. The *unitary group* $U(n)$ consists of the linear transformations u of $\mathbb{R}^{2n}$ that respect this norm; that is, $\|u(z)\| = \|z\|$. In the canonical basis, this is equivalent to the fact that the matrix A of u satisfies ${}^t\bar{A}A = I_n$. The associated Lie algebra $\mathfrak{u}(n)$ of $U(n)$ is the vector space of alternating (or skew) hermitian matrices, ${}^t\bar{A} = -A$. The dimension of $U(n)$ is n^2. It can be proved that $U(n)$ is a maximal compact subgroup of $Gl(n,\mathbb{C})$ and that $U(n) = SO(2n) \cap Gl(n,\mathbb{C})$.

The subgroup of $U(n)$ consisting of linear transformations of determinant 1 is called the *special unitary group* and denoted by $SU(n)$. The associated Lie algebra, $\mathfrak{su}(n)$, consists of matrices of trace 0 such that ${}^t\bar{A} = -A$. The dimension of $SU(n)$ is $n^2 - 1$. The group $SU(n)$ is simply connected. The group $U(n)$ is not, but its universal cover does not constitute something new because, as a space, $U(n)$ is diffeomorphic to the product $S^1 \times SU(n)$.

The maximal torus of $U(n)$ consists of the set of diagonal matrices having $(e^{i\lambda_1},\ldots,e^{i\lambda_n})$ on the diagonal. The Lie group $U(n)$ has rank n. Its Weyl group is the symmetric group Σ_n acting on the maximal torus by a permutation of the coordinates.

The maximal torus of $SU(n)$ consists of the set of diagonal matrices having $(e^{i\lambda_1},\ldots,e^{i\lambda_n})$ on the diagonal such that $\sum_{i=1}^n \lambda_i = 0$. The Lie group $SU(n)$ has rank $n-1$. The Weyl group and its action are the same as for $U(n)$.

1.5.3 Subgroups of the quaternionic linear group

Now consider the field of quaternions $\mathbb{H}$ and denote by $Gl(n,\mathbb{H})$ the group of invertible $n \times n$-matrices with entries in $\mathbb{H}$. Endowed with the canonical structure of a manifold (as an open subset of $\mathbb{R}^{4n^2}$), the group $Gl(n,\mathbb{H})$ is a Lie group, called the *quaternionic linear Lie group*. The associated Lie algebra, $\mathfrak{gl}_n(\mathbb{H}) = M(n,\mathbb{H})$, is the vector space of all $n \times n$ matrices with the bracket being the commutator of matrices. The (real) dimension of $\mathrm{Gl}(n, \mathbb{H})$ is $4n^2$.

For the associated orthogonal group, we have to define a quaternionic conjugation. Let $z = t + ix + jy + kz$ be a quaternion, with $x \in \mathbb{R}$, $y \in \mathbb{R}$, $z \in \mathbb{R}$, $t \in \mathbb{R}$ and i, j, k obeying the usual relations: $i^2 = j^2 = k^2 = -1$, $ij = k$, $ji = -k$, $jk = i$, $kj = -i$, $ki = j$ and $ik = -j$. The conjugate of z is the quaternion $\bar{z} = t - ix - jy - kz$. This induces a norm with $\|z\| = \sqrt{z\bar{z}}$. The *symplectic group* $\mathrm{Sp}(n)$ consists of the linear transformations u of $\mathbb{R}^{4n}$ that respect this norm, $\|u(z)\| = \|z\|$. In the canonical basis, this is equivalent to the fact that the matrix A of u satisfies ${}^t\bar{A}A = I_n$. The associated Lie algebra $\mathfrak{sp}(n)$ of $\mathrm{Sp}(n)$ is the vector space of alternating (or skew) quaternionic matrices, ${}^t\bar{A} = -A$. The dimension of $\mathrm{Sp}(n)$ is $n(2n+1)$. One can show that $\mathrm{Sp}(n)$ is a maximal compact subgroup of $\mathrm{Gl}(n, \mathbb{H})$ and that $\mathrm{Sp}(n) = \mathrm{SO}(4n) \cap \mathrm{Gl}(n, \mathbb{H})$.

Viewed as a subgroup of $\mathrm{U}(2n)$ (see Exercise 1.3), the Lie group $\mathrm{Sp}(n)$ has for a maximal torus the diagonal matrices $(e^{i\lambda_1}, \ldots, e^{i\lambda_{2n}})$ such that $\lambda_i = \bar{\lambda}_{i+n}$ for any $1 \leq i \leq n$. The Lie group $\mathrm{Sp}(n)$ has rank n. Its Weyl group has $2^n n!$ elements acting on the maximal torus as in $\mathrm{SO}(2n+1)$.

In any of these groups of matrices, the *exponential map*, $\exp : \mathfrak{g} \to G$, is the traditional exponential of a matrix:

$$\exp(A) = 1 + A + \cdots + \frac{A^n}{n!} + \cdots.$$

1.6 Invariant forms

In this section, we define the complex of invariant forms on a left G-manifold M, and prove that the cohomology of this complex is isomorphic to the cohomology of M if the manifold M is compact and the Lie group G compact and connected. As we will see in several places, Lie groups are designed as groups of symmetries of manifolds. With this in mind, we define invariant forms in the general setting of G-manifolds.

Definition 1.23 *A Lie group G acts on a manifold M, on the left, if there is a smooth map $G \times M \to M$, $(g, x) \mapsto gx$, such that $(g \cdot g')x = g(g'x)$ and $ex = x$ for any $x \in M$, $g \in G$, $g' \in G$. Such data endows M with the appellation of a left G-manifold. A left action is called*

- *effective if $gx = x$ for all $x \in M$ implies $g = e$;*
- *free if $gx = x$ for any $x \in M$ implies $g = e$.*

For *right actions* and right G-manifolds, we ask for a smooth map $M \times G \to M$, $(x, g) \mapsto xg$, such that $x(g \cdot g') = (xg)g'$ and $xe = x$ for any $x \in M$, $g \in G$, $g' \in G$.

Example 1.24 Let G be a Lie group. The Lie multiplication gives to G the structure of a

- left G-manifold, with $L\colon G \times G \to G$, $L(g,g') = L_g(g') = g \cdot g'$;
- right G-manifold, with $R\colon G \times G \to G$, $R(g',g) = R_g(g') = g' \cdot g$.

Let G be a Lie group. If M is a left G-manifold, we denote by $g^*\colon A_{DR}(M) \to A_{DR}(M)$ the "pullback" map induced on differential forms by the action of $g \in G$. More specifically, for vector fields $X_1, \ldots, X_k$ and a k-form ω, we define at $m \in M$,

$$g^*\omega(X_1, \ldots, X_k)(m) = \omega_{g \cdot m}(Dg_m X_1(m), \ldots, Dg_m X_k(m)).$$

We sometimes write $\omega_x(X_1, \ldots, X_k) = \omega(X_1, \ldots, X_k)(x)$, $L_g^*\omega = g^*\omega$ and $Dg = DL_g$.

Definition 1.25 *An* invariant form *on a left G-manifold M is a differential form $\omega \in A_{DR}(M)$ such that $g^*\omega = \omega$ for any $g \in G$. We denote the set of invariant forms by $\Omega_L(M)$.*

In the case of a Lie group G, we note that the left invariant forms (right invariant forms) correspond to the left (right) translation action. We denote these sets by $\Omega_L(G)$ and $\Omega_R(G)$ respectively. A form on G that is left and right invariant is called bi-invariant (or invariant if there is no confusion). The corresponding set is denoted by $\Omega_I(G)$.

The aim of this section is to prove that these different sets of invariant forms allow the determination of the cohomology of G-manifolds and Lie groups. First, using the operators $i(X)$ and $\mathcal{L}(X)$ on forms discussed in Appendix A (more specifically in Section A.2), we observe the following.

Proposition 1.26 *Let G be a Lie group and M be a left (or a right) G-manifold. Then the set of invariant forms of M is stable under d. Moreover, the sets of left invariant forms and of right invariant forms on G are invariant under $i(X)$ and $\mathcal{L}(X)$, for X a left invariant vector field.*

Proof Suppose ω is a left invariant form on G and X is a left invariant vector field on G. We have, using the left invariance of X and ω,

$$L_g^* i(X)\omega(Y_1, \ldots, Y_k)(x) = L_g^*\omega(X, Y_1, \ldots, Y_k)(x)$$
$$= \omega_{gx}(DL_g(X)_x, DL_g(Y_1)_x, \ldots, DL_g(Y_k)_x)$$
$$= i(DL_g X)\omega(DL_g(Y_1), \ldots, DL_g(Y_k))(gx)$$
$$= i(X)\omega(Y_1, \ldots, Y_k)(x).$$

Hence, $i(X)\omega$ is left invariant. The verification of the other statements is similar. □

The previous result justifies the following definition.

Definition 1.27 *Let G be a Lie group and M be a left G-manifold. The invariant cohomology of M is the homology of the cochain complex $(\Omega_L(M), d)$. We denote it by $H_L^*(M)$.*

The main result is the following theorem.

Theorem 1.28 *Let G be a compact connected Lie group and M be a compact left G-manifold. Then*

$$H_L^*(M) \cong H^*(M; \mathbb{R}).$$

We will prove that the injection map $\Omega_L(M) \to A_{DR}(M)$ induces an isomorphism in cohomology. For that, we need some results concerning integration on a compact connected Lie group.

Proposition 1.29 *On a compact connected Lie group, there exists a bi-invariant volume form.*

Proof Recall from Section 1.2 that the tangent bundle of G trivializes as $T(G) \cong G \times \mathfrak{g}$. If $\mathfrak{g}^*$ is the dual vector space of $\mathfrak{g}$, we therefore have a trivialization of the cotangent bundle $T^*(G) \cong G \times \mathfrak{g}^*$ and of the differential forms bundle. Exactly as for vector fields, we observe that left (right) invariant forms are totally determined by their value at the unit e and that we have isomorphisms

$$\Omega_L(G) \cong \Omega_R(G) \cong \wedge \mathfrak{g}^*,$$

where $\wedge \mathfrak{g}^*$ is the exterior algebra on the vector space $\mathfrak{g}^*$. To make this space precise, recall that the elements of $\mathfrak{g}^*$ are left invariant 1-forms dual to left invariant vector fields. If we choose a basis $\{\omega_1, \ldots, \omega_n\}$ dual to a basis of left invariant vector fields, an element of $\wedge \mathfrak{g}^*$ may be written

$$\alpha = \sum a_{i_1 \cdots i_p} \omega_{i_1} \cdots \omega_{i_p}$$

where the $a_{i_1 \cdots i_p}$'s are constant. Choose such an α of degree n equal to the dimension of G. We associate to α a unique left invariant form α_L such that $(\alpha_L)_e = \alpha$ and a unique right invariant form α_R such that $(\alpha_R)_e = \alpha$. More precisely, we set:

$$(\alpha_L)_g(X_1, \ldots, X_n) = \alpha((DL_g)^{-1} X_1, \ldots, (DL_g)^{-1} X_n),$$
$$(\alpha_R)_g(X_1, \ldots, X_n) = \alpha((DR_g)^{-1} X_1, \ldots, (DR_g)^{-1} X_n).$$

1 : Lie groups and homogeneous spaces

Recall, from Definition 1.10, the homomorphism of Lie groups $\mathrm{Ad}\colon G \to \mathrm{Gl}(\mathfrak{g})$. As direct consequences of the definitions, we have

$$(L_g^* \alpha_R)_h(X_1,\ldots,X_n) = (\alpha_R)_{gh}(DL_g(X_1),\ldots,DL_g(X_n))$$
$$= \alpha((DR_{gh})^{-1} \circ (DL_g)(X_1),\ldots)$$
$$= \alpha((DR_h)^{-1} \circ (DR_g)^{-1} \circ DL_g(X_1),\ldots)$$
$$= (\alpha_R)_h((DR_g)^{-1} \circ DL_g(X_1),\ldots)$$
$$= (\det(\mathrm{Ad}(g))(\alpha_R)_h(X_1,\ldots,X_n).$$

The composition $\det \circ \mathrm{Ad}\colon G \to \mathbb{R}$ has for image a compact subgroup of $\mathbb{R}$; that is, $\{1\}$ or $\{-1,1\}$. Since the group G is connected, we get $\det(\mathrm{Ad}(g)) = 1$, for any $g \in G$, and α_R is a bi-invariant volume form. $\square$

The previous result can be obtained in a more general context. As the proof shows, it is sufficient to have $(\det \circ \mathrm{Ad})(g) = 1$ for any $g \in G$. This is the definition of a unimodular group.

Proof of Theorem 1.28 Denote by $\iota\colon \Omega_L(M) \hookrightarrow A_{DR}(M)$ the canonical injection of the set of left invariant forms. We choose the bi-invariant volume form on G such that the total volume of G is 1, $\int_G dg = 1$. This volume form allows the definition of $\int_G f\, dg \in \mathbb{R}^k$ for any smooth function $f\colon G \to \mathbb{R}^k$.

Let $\omega \in A_{DR}^k(M)$ and $x \in M$ be fixed. As a function f, we take $G \to \wedge T_x(M)^*$, $g \mapsto g^*\omega(x)$. We get a differential form $\rho(\omega)$ on M defined by:

$$\rho(\omega)(X_1,\ldots,X_k)(x) = \int_G g^*\omega(X_1,\ldots,X_k)(x)\, dg$$
$$= \int_G (L_g)^*\omega(X_1,\ldots,X_k)(x)\, dg.$$

We have thus built a map $\rho\colon A_{DR}(M) \to A_{DR}(M)$ and we now analyze its properties.

Fact 1: $\rho(\omega) \in \Omega_L(M)$.

Let $g' \in G$ be fixed. The map $(DL_{g'})\colon T_x(M) \to T_{g'x}(M)$ induces a map $\wedge (DL_{g'})^*\colon \wedge T_{g'x}(M)^* \to \wedge T_x(M)^*$. Therefore, one has (in convenient shorthand):

$$(DL_{g'})^* \rho(\omega)(x) = \wedge (DL_{g'})^* \int_G (L_g)^*\omega(x)\, dg$$
$$= \int_G (L_{g'\cdot g})^* \omega(x)\, dg$$

$$= \int_G (L_g)^* \omega(x)\, dg$$
$$= \rho(\omega)(x).$$

Fact 2: If $\omega \in \Omega_L(M)$ then $\rho(\omega) = \omega$.

If $(L_g)^* \omega(x) = \omega(x)$, then $\rho(\omega)(x) = \int_G (L_g)^* \omega(x)\, dg = \omega(x) \int_G dg = \omega(x)$.

Fact 3: $\rho \circ d = d \circ \rho$.

This is an easy verification from the definitions of d and ρ.

From Facts 1–3, we deduce that $H(\rho) \circ H(\iota) = \mathrm{id}$ and $H(\iota)$ is injective.

Fact 4: The integration can be reduced to a neighborhood of e.

Let U be a neighborhood of e. We choose a smooth function $\varphi \colon G \to \mathbb{R}$, with compact support included in U, such that $\int_G \varphi\, dg = 1$. Now we denote the bi-invariant volume form dg by ω_{vol}. By classical differential calculus on manifolds, the replacement of ω_{vol} by $\varphi \omega_{\mathrm{vol}}$ leaves the integral unchanged. The fact that $\varphi \omega_{\mathrm{vol}}$ has its support in U allows the reduction of the domain of integration to U. Our construction process can now be seen in the following light.

Let $L \colon G \times M \to M$ be the action of G on M. Denote by $\pi_G^*(\varphi \omega_{\mathrm{vol}})$ the pullback of $\varphi \omega_{\mathrm{vol}}$ to $A_{DR}(U \times M)$ by the projection $\pi_G \colon G \times M \to G$ and by $L^* \colon A_{DR}(M) \to A_{DR}(U \times M)$ the map induced by L. If α is a form on $U \times M$, we denote by $I(\alpha)$ the integration of $\alpha \wedge \pi_G^*(\varphi \omega_{\mathrm{vol}})$ over the U-variables, considering the variables in M as parameters. We then have a map $I \colon A_{DR}(U \times M) \to A_{DR}(M)$ which is compatible with the coboundary d and which induces $H(I)$ in cohomology.

To any $\omega \in A_{DR}(M)$ we associate the form $L^*(\omega) \wedge \pi_G^*(\varphi \omega_{\mathrm{vol}})$ on $U \times M$ and check easily (see [113, page 150]):

$$\rho(\omega) = I(L^*(\omega)).$$

In other words, the following diagram is commutative

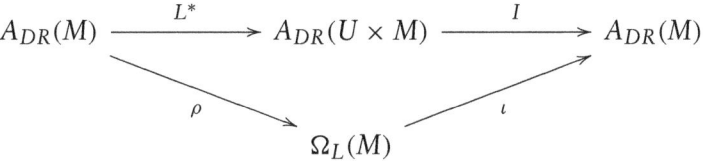

For U, we now choose a contractible neighborhood of e. The identity map on $U \times M$ is therefore homotopic to the composition $U \times M \xrightarrow{\pi} M \xrightarrow{j} U \times M$, where π is the projection and j sends

x to (e,x). By using $I \circ \pi^* = \mathrm{id}$ and the compatibility of de Rham cohomology with homotopic maps, we get:

$$H(I) \circ H(L^*) = H(I) \circ \mathrm{id}_{H(M \times U)} \circ H(L^*)$$
$$= H(I) \circ H(\pi^*) \circ H(j^*) \circ H(L^*)$$
$$= H(j^*) \circ H(L^*) = H((L \circ j)^*) = \mathrm{id}.$$

This implies $\mathrm{id} = H(\iota) \circ H(\rho)$ and $H(\iota)$ is surjective. □

1.7 Cohomology of Lie groups

In this section, we give two structure theorems for the cohomology of a Lie group. The first one comes from the existence of left and right G-manifold structures on G and follows from the results of Section 1.6. The second one, called Hopf's theorem, gives a precise algebra structure for the cohomology.

Recall that $\Omega_L(G)$ ($\Omega_R(G)$, $\Omega_I(G)$) is the set of left invariant (right invariant, bi-invariant) forms on G. Denote by $\Omega_L(G)_{\mathcal{L}=0}$ the set of left invariant forms whose Lie derivative (see Section A.2) by any *left invariant* vector field is zero.

Theorem 1.30 *Let G be a compact connected Lie group with Lie algebra $\mathfrak{g}$. Then we have two series of isomorphisms:*

(1) $\Omega_L(G) \cong \Omega_R(G) \cong \wedge \mathfrak{g}^*$;

(2) $H_L(G) \cong H_R(G) \cong H^*(G; \mathbb{R}) \cong \Omega_I(G) \cong \Omega_L(G)_{\mathcal{L}=0}$.

Remark 1.31 There is one point of view that we do not develop here: the translation of the second line in terms of Lie algebras using the isomorphism $\Omega_L(G) \cong \wedge \mathfrak{g}^*$. For that, one needs to know the image of the coboundary d and the Lie derivative $\mathcal{L}$ through this isomorphism. This theory is well developed in [114]; we give a glimpse of it in Exercise 1.7. Also see Subsection 3.2.1 for the noncompact case of nilpotent Lie groups.

We mention also that the existence of an isomorphism between $H^*(G; \mathbb{R})$ and $\Omega_I(G)$ can be extended to the more general situation of *symmetric spaces* (see Exercise 1.6). In order to prove the theorem, we first need to determine the derivative of the multiplication and the inverse maps.

Lemma 1.32 *Let G be a Lie group. Denote by $\mu \colon G \times G \to G$ the multiplication map and by $\nu \colon G \to G$ the inverse map. Then we have:*

$$D\mu_{(g,g')} = DL_g + DR_{g'} \quad \text{and} \quad D\nu_g = -(DL_g)^{-1} \circ DR_{g^{-1}}.$$

1.7 Cohomology of Lie groups

Proof With the identification $T(G \times G) \cong T(G) \oplus T(G)$, we write a vector field on $G \times G$ as:

$$((g,g'),(X,X')) = (g,(g',X')) + ((g,X),g')$$
$$\in \{g\} \times T_{g'}(G) \oplus T_g(G) \times \{g'\}.$$

Therefore, we have:

$$D\mu_{(g,g')}(X,X') = DL_g(X') + DR_{g'}(X).$$

From this formula and the equality $\mu(g,v(g)) = e$, we deduce:

$$0 = D\mu_{(g,g^{-1})}(X,Dv_g(X)) = DL_g \circ Dv_g(X) + DR_{g^{-1}}(X)$$

and

$$Dv_g = -(DL_g)^{-1} \circ DR_{g^{-1}}.$$

$\square$

Proof of Theorem 1.30 The first series of isomorphisms is clear. It comes from the triviality of the bundle of left (or right) invariant forms on a Lie group G and the identification between $T_e(G)$ and $\mathfrak{g}$.

The first part of (2), $H_L(G) \cong H_R(G) \cong H^*(G;\mathbb{R})$, is a consequence of Theorem 1.28. Observe now that $\Omega_I(G)$ is the set of left invariant forms for the left action of the group $G \times G$ on G defined by $(g_1,g_2)g' = g_1 \cdot g' \cdot g_2^{-1}$. Therefore, Theorem 1.28 implies $H^*(G;\mathbb{R}) = H_I(G)$. The next isomorphism, $H^*(G;\mathbb{R}) \cong \Omega_I(G)$, will follow immediately from the fact that *each bi-invariant form on G is closed*, implying $H_I(G) = \Omega_I(G)$.

To prove this property, let α be a bi-invariant form on G. We compute, using the left and right invariance of α, the inverse image of α by v:

$$v^*\alpha(X_1,\ldots,X_k)(g) = \alpha(-(DL_g)^{-1} \circ DR_{g^{-1}}(X_1),\ldots)(g^{-1}),$$
$$= (-1)^k \alpha(X_1,\ldots,X_k)(g).$$

We then have $v^*\alpha = (-1)^k \alpha$. We now use the fact that $\Omega_I(G) = \Omega_L(G) \cap \Omega_R(G)$ is stable under the coboundary:

$$v^*(d\alpha) = (-1)^{k+1} d\alpha$$
$$d(v^*\alpha) = d((-1)^k\alpha) = (-1)^k d\alpha,$$

which implies $d\alpha = 0$. The last isomorphism is established in the next proposition. $\square$

Now recall from Section A.2 that a form α such that $\mathcal{L}(X)\alpha = 0$ for a vector field X is said to be $\mathcal{L}(X)$-invariant.

Proposition 1.33 *Let G be a compact connected Lie group. The $\mathcal{L}$-invariance of a form with respect to left invariant vector fields corresponds precisely to invariance of the form under right translations. In particular, we have:*

$$\Omega_I(G) = \{\omega \in \Omega_L(G) \mid \mathcal{L}(X)\omega = 0 \text{ for any left invariant vector field } X\}.$$

Proof Let X be a left invariant vector field. We know that X is determined by its value $X_e \in T_e(G)$ and that (see Definition 1.14) the exponential map $\exp\colon T_e(G) \to G$ is defined by $\theta_t(e) = \exp(tX_e)$, where θ is the 1-parameter subgroup associated to X. There is also an exponential map at any $g \in G$ obtained by requiring $\exp(tX_g) = L_g(\exp(tX_e))$, where L_g denotes left translation by g. Using the left invariance of X (i.e. $DL_h(X_e) = X_h$ for all $h \in G$) we have $\theta_t(g) = \exp(tX_g) = L_g(\exp(tX_e)) = g \cdot \exp(tX)$. In conclusion, the flow acts on g by right translation.

On the other hand, if θ is the 1-parameter subgroup associated to X, the Lie derivative satisfies the formula $\theta_t^*\omega - \omega = \int_0^t \theta_s^*\mathcal{L}(X)\omega \, ds$ (see Section A.2 and Exercise A.2). Therefore, the form ω is right invariant for the action of elements in the image of the exponential if and only if $\mathcal{L}(X)\omega = 0$ for any left invariant vector field X, see [113, Proposition VI, page 126]. Since, in a connected compact group, the exponential is an epimorphism, we get the result. □

Theorem 1.34 (Hopf's theorem) *If G is a compact connected Lie group, then there exist elements of odd degree, $x_{2p_i+1} \in H^{2p_i+1}(G; \mathbb{Q})$, such that, as an algebra,*

$$H^*(G; \mathbb{Q}) = \wedge(x_{2p_1+1}, \ldots, x_{2p_r+1}).$$

In fact the number of generators is the rank of the group (see Theorem 3.33).

Remark 1.35 One can determine the elements x_{2p_i+1} from the structure law of G as follows. Denote by $\mu^*\colon H^*(G; \mathbb{Q}) \to H^*(G; \mathbb{Q}) \otimes H^*(G; \mathbb{Q})$ the map induced by the multiplication $\mu\colon G \times G \to G$. An element $h \in H^*(G; \mathbb{Q})$ is called *primitive* if $\mu^*(h) = 1 \otimes h + h \otimes 1$. The set P_G of primitive elements is a vector space and we have $H^*(G; \mathbb{Q}) \cong \wedge P_G$. We will not use this in the sequel and we do not give a proof, instead referring to [113, Theorem IV, page 167], for instance, for an argument with coefficients in $\mathbb{R}$. This approach relies on the general result that a commutative Hopf algebra over a field of characteristic zero is generated, as algebra, by the primitive elements (see [197]).

Proof of Theorem 1.34 Let us denote $H^*(G; \mathbb{Q})$ by H and consider $Q(H) = H^+/(H^+ \cdot H^+)$, the space of indecomposables of the augmentation ideal $H^+ = \sum_{j>0} H^j$ of the algebra H. We choose a homogeneous basis $(x_j)_{1 \leq j \leq r}$

of the $\mathbb{Q}$-vector space $Q(H)$. By fixing a section to the canonical surjection $H^+ \to Q(H)$, we consider the elements x_j as elements of H.

Fact 1: The elements x_j are of odd degree. Suppose that some x_k is of even degree and let $\overline{H}_k$ be the quotient algebra of H by the ideal generated by the elements x_j for $j \neq k$. The canonical map $q_k \colon H \to \overline{H}_k$ is a morphism of algebras.

Denote by $\mu^* \colon H \to H \otimes H$ the map induced by the multiplication of G and by $\overline{\mu} \colon H \to \overline{H}_k \otimes \overline{H}_k$ the composition of μ with $q_k \otimes q_k \colon H \otimes H \to \overline{H}_k \otimes \overline{H}_k$. Since the algebra H is finite, we know that an integer l exists such that $x_k^l \neq 0$ and $x_k^{l+1} = 0$. Let $\rho \colon G \to G \times G$ denote either of the inclusions $g \mapsto (g, e)$ or $g \mapsto (e, g)$, where e is the identity element of G. Then clearly the composition obeys $\mu \rho = \mathrm{id}_G$. Therefore, we have, for any $y \in H$,

$$\mu^*(y) = 1 \otimes y + y \otimes 1 + y'$$

with $y' \in H^+ \otimes H^+$. This implies $\overline{\mu}(x_k) = x_k \otimes 1 + 1 \otimes x_k$. The map $\overline{\mu}$, being a morphism of algebras, gives

$$\overline{\mu}(x_k)^{l+1} = (x_k \otimes 1 + 1 \otimes x_k)^{l+1} = \sum_{i=1}^{l} \binom{l+1}{i} x_k^i \otimes x_k^{l+1-i}.$$

On one side, we have $\overline{\mu}(x_k)^{l+1} = \overline{\mu}((x_k)^{l+1}) = 0$ and, on the other side, thanks to the lack of cross-relations in $\overline{H}_k \otimes \overline{H}_k$, we get $\sum_{i=1}^{l} \binom{l+1}{i} x_k^i \otimes x_k^{l+1-i} \neq 0$. This contradiction implies Fact 1.

By sending the x_j to the chosen elements of the basis and extending this correspondence multiplicatively, we define a morphism of algebras $\phi \colon \wedge (x_1, \ldots, x_r) \to H$.

Fact 2: The morphism ϕ is an isomorphism. By construction, ϕ is surjective so we are reduced to establishing its injectivity. Observe that the restriction of ϕ to $\wedge (x_1)$ is injective. We argue by induction and suppose that its restriction to $\wedge (x_1, \ldots, x_{k-1})$ is injective. Let $a \in \wedge (x_1, \ldots, x_k)$ be such that $\phi(a) = 0$. We decompose a into

$$a = a_1 + x_k a_2,$$

with a_1 and a_2 in $\wedge (x_1, \ldots, x_{k-1})$. We denote by $\overline{\phi}_k$ the following composition

$$\wedge (x_1, \ldots, x_k) \xrightarrow{\phi} H \xrightarrow{\mu^*} H \otimes H \xrightarrow{q_k \otimes \mathrm{id}} \overline{H}_k \otimes H.$$

1 : Lie groups and homogeneous spaces

From $\overline{\phi}_k(x_k) = 1 \otimes x_k + x_k \otimes 1$ and $\overline{\phi}_k(a_i) = 1 \otimes a_i$ for $i = 1, 2$, we deduce

$$0 = \overline{\phi}_k(a) = 1 \otimes a_1 + (x_k \otimes 1 + 1 \otimes x_k)(1 \otimes a_2).$$

This implies $x_k \otimes a_2 = 0$ and $a_2 = 0$. From the induction hypothesis, we now get $a = a_1 = 0$ and the restriction of ϕ to $\wedge(x_1, \ldots, x_k)$ is injective as expected. $\square$

The method used in the previous proof consists essentially in killing elements by taking quotients of algebras by ideals. This is an important technical argument in the theory of minimal models as we will see in the next chapters.

As an illustration, we use Theorem 1.30 to compute the first cohomology group of a compact Lie group.

Proposition 1.36 *Let G be a compact Lie group with associated Lie algebra $\mathfrak{g}$. Then*

$$H^1(G; \mathbb{R}) \cong \mathrm{Hom}(\mathcal{Z}(\mathfrak{g}), \mathbb{R}),$$

where $\mathcal{Z}(\mathfrak{g})$ is the center of the Lie algebra $\mathfrak{g}$.

Proof Let $\omega \in \mathfrak{g}^*$ be a left invariant 1-form. By definition (see Section A.2) we have:

$$d\omega(X, Y) = X\omega(Y) - Y\omega(X) - \omega([X, Y]).$$

Since the form ω is left invariant, this is also true for the functions $\omega(Y)$ and $\omega(X)$. Since left invariant functions are constant, the previous formula reduces to $d\omega(X, Y) = -\omega([X, Y])$. Therefore, the form ω is closed if and only if ω takes the value 0 on $[\mathfrak{g}, \mathfrak{g}]$ (i.e. $\omega \in [\mathfrak{g}, \mathfrak{g}]^\perp$).

Now recall the definition of the center of $\mathfrak{g}$:

$$\mathcal{Z}(\mathfrak{g}) = \{X \in \mathfrak{g} \mid [X, Y] = 0 \text{ for any } Y \in \mathfrak{g}\}.$$

Let F be a positive definite invariant symmetric bilinear form on $\mathfrak{g}$ (which exists by Lemma 1.38). From $F(X, [Y, Z]) = F([X, Y], Z)$, we see that X is in the F-orthogonal complement $[\mathfrak{g}, \mathfrak{g}]^\perp$ of $[\mathfrak{g}, \mathfrak{g}]$ if and only if $X \in \mathcal{Z}(\mathfrak{g})$. Thus, closed left invariant 1-forms are dual to elements of $\mathcal{Z}(\mathfrak{g})$. Since $H_L(G) \cong H^*(G; \mathbb{R})$ and left invariant functions are constant, we have the result. $\square$

Remark 1.37 Observe that, in the previous proof, we have established that $[\mathfrak{g}, \mathfrak{g}] = \mathfrak{g}$ if the center $\mathcal{Z}(\mathfrak{g})$ of the Lie algebra $\mathfrak{g}$ is zero.

Lemma 1.38 *On any Lie algebra $\mathfrak{g}$ of a compact Lie group G, there is a positive definite symmetric bilinear form $F \colon \mathfrak{g} \times \mathfrak{g} \to \mathbb{R}$*

such that:

1. $F(\mathrm{Ad}(g)(X), \mathrm{Ad}(g)(Z)) = F(X, Z)$;
2. $F([X, Y], Z) = F(X, [Y, Z])$ *for any triple* (X, Y, Z) *of elements of* $\mathfrak{g}$.

Such an F is said to be invariant.

Proof Let $\overline{F}$ be any positive definite bilinear form on $\mathfrak{g}$ and set

$$F(X, Z) = \int_G \overline{F}(\mathrm{Ad}(g)(X), \mathrm{Ad}(g)(Z)) dg,$$

where $\mathrm{Ad}(g)(X) = ((DR_g)^{-1} \circ (DL_g))(X)$. This bilinear form satisfies

$$F(\mathrm{Ad}(g^{-1})(X), Z) = F(X, \mathrm{Ad}(g)(Z))$$

for any g in G. The result follows now from the fact that the derivative of $\mathrm{Ad}(g)$ is the bracket in $\mathfrak{g}$ (see Proposition 1.11). □

1.8 Simple and semisimple compact connected Lie groups

We now come to our first concrete application: the vanishing of the second Betti number of a compact semisimple Lie group and the determination of the third Betti number of a simple Lie group. As we note in Remark 1.50, one can, in fact, do better and prove that the second homotopy group is zero and the third homotopy group of a simple Lie group is $\mathbb{Z}$. We will come back to this point in Section 1.11.

Definition 1.39 *A connected compact Lie group is* simple *if it does not contain any nontrivial connected normal subgroups. A Lie algebra is* simple *if it has no proper ideal.*

For instance, $SU(n)$ is simple (see Theorem 1.53) while $U(n)$ is not because it contains $SU(n)$ as a normal subgroup.

Remark 1.40 Let G be a Lie group with a simple Lie algebra $\mathfrak{g}$. From the correspondence between normal Lie subgroups of G and Lie ideals of $\mathfrak{g}$, we see that a normal Lie subgroup $H \neq G$ of G has dimension zero, so it is discrete (remembering that it is closed). With a similar argument for the converse, we have proved that a Lie group is simple if and only if its Lie algebra is simple.

There exist several equivalent definitions of semisimple Lie group. Since we are concerned with real cohomology, we use the following.

Definition 1.41 *A compact connected Lie group is* semisimple *if its first Betti number is zero. That is,* $H^1(G; \mathbb{R}) = 0$.

Proposition 1.42 *For a connected compact Lie group G, the following conditions are equivalent:*

1. *G is semisimple;*
2. *the fundamental group $\pi_1(G) \cong H_1(G; \mathbb{Z})$ is finite;*
3. *the center of G is finite;*
4. *the universal covering group $\widetilde{G}$ of G is compact.*

Proof The equivalence of (1) and (2) is a direct consequence of the universal coefficient theorem which implies $H^1(G; \mathbb{R}) = \text{Hom}(H_1(G; \mathbb{Z}), \mathbb{R})$. As for the equivalence between (2) and (4), observe that a covering space is compact if and only if the fiber is finite. The equivalence of (1) and (3) is a consequence of Proposition 1.36. □

Remark 1.43 Because the Lie algebra associated to the center of G is the center $\mathcal{Z}(\mathfrak{g})$ of the Lie algebra $\mathfrak{g}$, we observe that the center $\mathcal{Z}(\mathfrak{g})$ is zero if $\mathfrak{g}$ is the Lie algebra of a connected compact semisimple Lie group.

Proposition 1.44 *Each simple Lie group is semisimple.*

Proof Since the group is simple, its center must be finite and the result follows from Proposition 1.42. □

A product of two simple Lie groups is an example of a semisimple Lie group that is not simple.

Theorem 1.45 *If G is a compact semisimple Lie group, then the second Betti number $b_2(G)$ is zero. That is, $H^2(G; \mathbb{R}) = 0$.*

Corollary 1.46 *If G is a compact semisimple Lie group, then $\pi_2(G) \otimes \mathbb{Q} = 0$.*

Theorem 1.45 is a *direct consequence* of the definition of semisimple Lie group and of Hopf's theorem (Theorem 1.34). To emphasize the interrelationship between geometry and homotopy theory, we will give two other proofs of the theorem, one using the material we have just discussed on invariant forms and a second one in Section 1.11, using the existence of a universal bundle for a Lie group (see page 40).

Proof 2 of Theorem 1.45 Suppose $\alpha \in \Omega^2_I(G)$. By Theorem 1.30, if we can show that $\alpha = 0$, then this will imply $H^2(G; \mathbb{R}) = 0$. Expressing the $\mathcal{L}$-invariance of α for a left invariant vector field X gives:

$$0 = \mathcal{L}(X)\alpha = (i(X)d + di(X))\alpha = i(X)d\alpha + di(X)\alpha = di(X)\alpha,$$

since α is closed. Hence, $i(X)\alpha$ is a closed 1-form. Since G is semisimple, we have $H^1(G; \mathbb{R}) = 0$, so $i(X)\alpha$ is an exact 1-form. That is, there exists a smooth function $f \colon G \to \mathbb{R}$ such that $i(X)\alpha = df$. By Proposition 1.26, we know that $i(X)\alpha$ is left invariant. By Theorem 1.28, left invariant cohomology is isomorphic to ordinary cohomology, so $i(X)\alpha$ must be exact by a left invariant function as well. Therefore (see Remark 1.8), f is constant and $i(X)\alpha = df = 0$ for any left invariant vector field X. We show now that this implies $\alpha = 0$. For this, let $g \in G$ and choose any two vectors $v_g, w_g \in T_g(G)$. We can find left invariant vector fields V and W with $V_g = v_g$ and $W_g = w_g$ simply by left translating v_g, w_g around G. But because V is left invariant, we then have

$$\alpha_g(v_g, w_g) = \alpha_g(V_g, W_g) = \alpha(V, W)(g) = i(V)\alpha(W)(g) = 0$$

since $i(X)\alpha = 0$ for *all* left invariant X. Since g, v_g and w_g were arbitrary, we have $\alpha = 0$. Therefore, no nonzero invariant 2-forms exist on G and $H^2(G; \mathbb{R}) = 0$. □

Proof of Corollary 1.46 Since a compact semisimple Lie group is finitely covered by a compact simply connected semisimple Lie group, it suffices to prove the result in the simply connected case. The Hurewicz theorem then implies that $\pi_2(G) \cong H_2(G)$, so from $H^2(G; \mathbb{R}) = \text{Hom}(H_2(G), \mathbb{R}) = 0$, we deduce that $\pi_2(G)$ is finite. □

Theorem 1.47 *If G is a compact semisimple Lie group, then the third Betti number $b_3(G)$ is greater than or equal to one. In the simple case, we have $H^3(G; \mathbb{R}) = \mathbb{R}$.*

Observe that, as a direct consequence, we have the following.

Corollary 1.48 *The only spheres which have a Lie group structure are S^0, S^1, S^3.*

With the same argument as in the proof of Corollary 1.46, we also have the following.

Corollary 1.49 *If G is a compact semisimple Lie group, then $\pi_3(G) \otimes \mathbb{Q} \neq 0$. In the simple case, we have $\pi_3(G) \otimes \mathbb{Q} = \mathbb{Q}$.*

Remark 1.50 Since we are interested only in the rational or real world, we are satisfied with the results of Corollaries 1.46 and 1.49. In fact, however, it is possible to prove that $\pi_2(G) = 0$, in the semisimple compact case, and $\pi_3(G) = \mathbb{Z}$ in the simple case. The proofs need material that we do not introduce here:

- for $\pi_2(G) = 0$, see [32] or [50];
- for $\pi_3(G) = \mathbb{Z}$ in the simple compact case, see [36] or [199, Theorem 4.17, page 335].

We will come back to the case of $\pi_2(G) = 0$ in Remark 1.82 with a more homotopical argument.

Proof of Theorem 1.47 Recall first, from Theorem 1.30, that $H^3(G;\mathbb{R})$ is isomorphic to the vector space of invariant 3-forms on G. We follow the proofs of [107, Problem IV-B] and [41, Section V-12]. The idea is to transform the problem of finding an invariant 3-form on G into the problem of finding an invariant symmetric bilinear form on $\mathfrak{g}$. Because such a form exists by Lemma 1.38, we will get the first part of the statement.

Let $\mathcal{B}(\mathfrak{g})$ be the set of symmetric bilinear forms on $\mathfrak{g}$ such that $F([Z,X],Y) = F(X,[Z,Y])$ for any triple (X,Y,Z) of elements of $\mathfrak{g}$. For such an F, we define ω_F by $\omega_F(X,Y,Z) = F([X,Y],Z)$. We check easily that $\omega_F \in \wedge^3 \mathfrak{g}^* \cong \Omega_L^3(G)$. We prove now that $\omega_F \in \Omega_I^3(G)$.

Let $g \in G$. We have, by invariance of F and the definition of ω_F:

$$\omega_F(\mathrm{Ad}(g)(X), \mathrm{Ad}(g)(Y), \mathrm{Ad}(g)(Z)) = F([\mathrm{Ad}(g)(X), \mathrm{Ad}(g)(Y)], \mathrm{Ad}(g)(Z))$$
$$= F(\mathrm{Ad}(g)([X,Y]), \mathrm{Ad}(g)(Z))$$
$$= \omega_F(X,Y,Z).$$

Therefore, we have constructed a linear map χ from $\mathcal{B}(\mathfrak{g})$ to $\Omega_I^3(G) \cong H^3(G;\mathbb{R})$. Since the group G is semisimple, we know (see Remark 1.37 and Remark 1.43) that $[\mathfrak{g},\mathfrak{g}] = \mathfrak{g}$, which gives the injectivity of χ.

We now have to prove that χ is onto. For that, let $\omega \in \Omega_I^3(G)$ and let $X \in \mathfrak{g}$. From $\mathcal{L}(X) = di(X) + i(X)d$, $\mathcal{L}(X)\omega = 0$ and $d\omega = 0$, we deduce $di(X)\omega = 0$ with $i(X)\omega \in A_{DR}^2(G)$. By Theorem 1.45, this implies the existence of a 1-form α_X such that $i(X)\omega = d\alpha_X$. We define $F \colon \mathfrak{g} \times \mathfrak{g} \to \mathbb{R}$ by $F(X,Y) = \alpha_X(Y)$. From the definition of the coboundary d and the construction of F, we deduce:

$$F(X,[Y,Z]) = \alpha_X([Y,Z]) = d\alpha_X(Y,Z) = i(X)\omega(Y,Z) = \omega(X,Y,Z).$$

The invariance of ω and the fact that $[\mathfrak{g},\mathfrak{g}] = \mathfrak{g}$ implies the invariance of F; that is,

$$F(\mathrm{Ad}(g)(X), \mathrm{Ad}(g)(Y)) = F(X,Y).$$

We are reduced to proving the symmetry of F. For that, observe that F can be uniquely decomposed in $F_1 + F_2$ with F_1 symmetric and F_2 skew. The F_2 gives an invariant 2-form on G which must be zero by Theorem 1.45 and therefore F is symmetric.

From this first part, we deduce an isomorphism $\mathcal{B}(\mathfrak{g}) \cong \Omega_I^3(G)$. Since $\mathcal{B}(\mathfrak{g})$ contains a nontrivial element, we have $b_3(G) \geq 1$.

We suppose now that G is simple and we prove $b_3(G) = 1$. Let F be a positive definite invariant bilinear form on $\mathfrak{g}$ and let $F' \in \mathcal{B}(\mathfrak{g})$. We consider the least value λ of the $F'(X,X)$ when X is such that $F(X,X) = 1$. Set $F''(X,X) = \lambda F(X,X) - F'(X,X)$. The kernel of F'' is a subspace of $\mathfrak{g}$, invariant under the bracket (since F'' is invariant). This kernel is not equal to 0 because λ is reached by F', so it must be equal to all of $\mathfrak{g}$ since $\mathfrak{g}$ has no proper ideal. We get $F' = \lambda F$, which means that the dimension of $\mathcal{B}(\mathfrak{g})$ is 1. □

Remark 1.51 Finally, observe that, in the case of a semisimple Lie group G, the negative of the Killing form

$$(X, Y) \mapsto -\text{trace}\,(\text{ad}(X) \circ \text{ad}(Y))$$

is a nondegenerate bilinear symmetric form on the Lie algebra $\mathfrak{g}$. If G is compact, then it can be shown that this symmetric bilinear form is positive definite as well. Since G is parallelizable, this then defines a metric on the tangent bundle. With respect to this metric we can define a Hodge star operator and obtain a Hodge decomposition of forms on G (see Section A.4). From [112] and [107], we can see that *if G is a compact semisimple Lie group, then the harmonic forms are the invariant forms*.

So, together with Proposition 1.33, the harmonic forms are the left invariant forms which are also $\mathcal{L}$-invariant. Because, $\mathcal{L}(X)$ is a derivation, this implies that, on a compact semisimple Lie group, the wedge product of harmonic forms is harmonic. This is an unusual property that can be extended to symmetric spaces (and, in the language of Chapter 2, imbues them with the property of *formality*) (see Exercise 3.8). We do not go further in this direction, instead referring the reader to [107, Propositions 4.4.2 and 4.4.3] for more details in the semisimple Lie group case.

We end this section with Cartan's theorem on the classification of simple Lie groups. Since this lies outside the main subject of this book, we do not give proofs, and, rather, refer the reader to [2], [112], [199], [214], [229] and [262] for presentations of this theory. Cartan's theorem gives a decomposition of compact connected Lie groups into a product of particular Lie groups and gives a complete classification of these factors. For this, semisimple Lie algebras are the key notion because semisimple Lie algebras allow a global Jordan decomposition for any representation (see [103, page 129]). Since the Lie algebra $\mathfrak{g}$ of a Lie group G can be decomposed into

$$\mathfrak{g} = \mathcal{Z} \oplus \sum_{i=1}^{k} \mathfrak{g}_i,$$

where $\mathcal{Z}$ is the center of $\mathfrak{g}$ and $\mathfrak{g}_i$ are simple ideals, the following can be shown (see [199, page 282]).

Theorem 1.52 *For an arbitrary compact connected Lie group G, there exists a torus T, compact simple simply connected Lie groups $G_1, \ldots, G_k$ and a finite group K contained in the center of $T \times G_1 \times \cdots \times G_k$ such that the quotient $(T \times G_1 \times \cdots \times G_k)/K$ is isomorphic to G.*

In other words, any compact connected Lie group admits a finite sheeted covering group which is the product of a torus and of simple Lie groups. For instance, any connected compact Lie group with a trivial center is isomorphic to a product of simple compact Lie groups. This is the beginning of the root system construction. The second part is the classification of compact connected simple Lie groups.

Theorem 1.53 *The following groups are simple:*

1. *the special unitary groups $SU(n_1)$;*
2. *the special orthogonal groups $SO(n_2)$;*
3. *the symplectic groups $Sp(n_3)$;*
4. *the exceptional Lie groups G_2, E_6, E_7, E_8, F_4;*

and any compact connected simple Lie group is isomorphic to one group of this list. If we insist that $n_1 \geq 2$, $n_2 \geq 5$, $n_3 \geq 2$, there are no isomorphisms between two elements of this list.

1.9 Homogeneous spaces

We present the definition of locally trivial fiber bundles here and an important tool for their construction, the pullback along a map with values in the base. Homogeneous spaces are then introduced and studied as examples of bundles. General references are [112], [145] and [242].

Definition 1.54 *A locally trivial fiber bundle of fiber F is a continuous map $p: E \to B$ together with a space F such that B admits a numerable open cover (U_i) with homeomorphisms φ_i making commutative the following diagram*

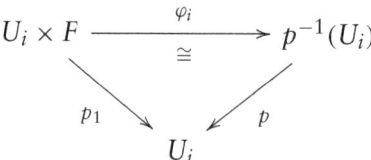

Here p_1 is the canonical projection. The space B is called the base *and E is called the* total space. *The collection (U_i) is called a* trivializing open cover *of the base and the homeomorphisms φ_i the* local trivializations. *If all spaces*

are manifolds and all maps are smooth maps, we say that we have a smooth locally trivial fiber bundle.

Let $p\colon E \to B$ and $p'\colon E' \to B'$ be two locally trivial fiber bundles. A morphism *between p and p' is a pair of maps, (ψ, Ψ), such that the following diagram commutes*

$$\begin{array}{ccc} E' & \xrightarrow{\Psi} & E \\ p' \downarrow & & \downarrow p \\ B' & \xrightarrow{\psi} & B \end{array}$$

If ψ is the identity on B and Ψ a homeomorphism, we say that p and p' are in the same isomorphism class of bundles.

Example 1.55

- For any pair of topological spaces, the canonical projection $B \times F \to B$ is a locally trivial fiber bundle. We call it *the trivial bundle*.
- If the base B of the bundle $E \to B$ is paracompact Hausdorff, then any open cover of B is numerable, so the restriction in the definition is really not important for us throughout the book.

Example 1.56 (Pullback of a locally trivial fiber bundle) If $p'\colon E' \to B'$ is a locally trivial fiber bundle and $\psi\colon B \to B'$ is a continuous map, then we define a topological space

$$\psi^*E' = \{(b, x') \in B \times E' \mid \psi(b) = p'(x')\} = \bigcup_{b \in B} \left(\{b\} \times p'^{-1}(\psi(b)) \right)$$

as a subspace of the product $B \times E'$. The two projections $\psi^*E' \xrightarrow{j} E'$, $(b, x') \mapsto x'$, and $\psi^*E' \xrightarrow{p} B$, $(b, x') \mapsto b$, are obviously continuous. By construction, for any commutative diagram

$$\begin{array}{ccc} \widehat{E} & \xrightarrow{\Psi} & E' \\ \widehat{p} \downarrow & & \downarrow p' \\ B & \xrightarrow{\psi} & B' \end{array}$$

there exists a unique continuous map $\overline{\Psi}\colon \widehat{E} \to \psi^*E$ such that $j \circ \overline{\Psi} = \Psi$ and $p \circ \overline{\Psi} = \widehat{p}$. This map $\overline{\Psi}$ is defined by $\overline{\Psi}(y) = (\widehat{p}(y), \Psi(y))$. Therefore, our construction ψ^*E' satisfies a universal property. Moreover, if the map $p': E' \to B'$ is the canonical projection $B' \times F \to B'$, then we have $\psi^*E = B \times F$ and the map $p: \psi^*E' \to B$ is the trivial bundle on B. We show now that $p: \psi^*E' \to B$ is a locally trivial fiber bundle with fiber F.

Since p' is locally trivial, we may choose a trivializing open cover (U'_i) of B' together with homeomorphisms φ'_i such that the following triangle commutes

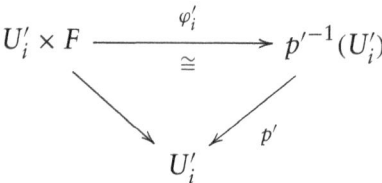

Set $U_i = \psi^{-1}(U'_i)$ and consider the diagram,

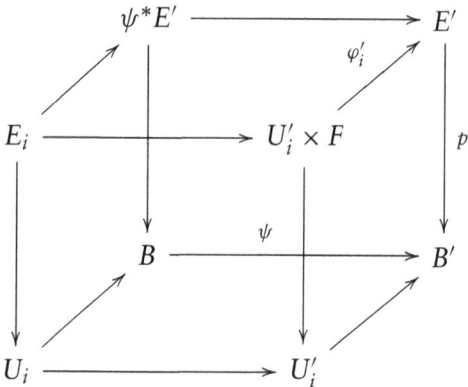

where E_i is the pullback construction applied to the trivial fibration $U'_i \times F \to U'_i$ and the map $U_i \to U'_i$ induced by ψ. We know, from the remark above, that the map $E_i \to U_i$ can be identified with the canonical projection $U_i \times F \to U_i$. We deduce that the (U_i) define a trivializing open cover of B and that $p: \psi^*E' \to B$ is a locally trivial fiber bundle, called *the pullback of $p': E' \to B'$ by the map ψ*.

In the case of a *smooth* locally trivial fiber bundle, one has to define a manifold structure on ψ^*E' such that the various maps are smooth. This is a standard procedure (see [112, Proposition VII, page 29]).

Below, we look at certain geometrically important examples of locally trivial fiber bundles. We shall freely use various properties of bundles such as the homotopy lifting property and the existence of the long exact homotopy sequence. See Exercises 1.9 and 1.10.

In the context of Lie groups, the main examples of locally trivial fiber bundles come from the notion of *homogeneous spaces* that we introduce now. Let H be a closed subgroup of a Lie group G. We denote by G/H the set of left cosets of H; that is, G/H is the quotient of G by the equivalence relation

$$x \sim y \text{ if and only if } x^{-1}y \in H.$$

The elements of G/H are denoted by xH for $x \in G$. In particular, H is the class of the neutral element e of G. We denote by $q\colon G \to G/H$ the canonical projection.

Definition 1.57 *The space G/H constructed above is called a* homogeneous space.

Definition 1.58 *Let H be a closed subgroup of G. A local section of H in G is a continuous map $\sigma\colon U \to G$, defined on an open neighborhood U of $H \in G/H$ such that $q \circ \sigma = \mathrm{id}_U$.*

Proposition 1.59 *If H is a closed subgroup of G admitting a local section, then, for any closed subgroup K of H, the canonical projection $p\colon G/K \to G/H$, $gK \mapsto gH$ is a locally trivial bundle of fiber H/K.*

Remark 1.60 The canonical map $\mathrm{O}(n) \to \dfrac{\mathrm{O}(n)}{\mathrm{O}(n-k)}$ has a local section. Consider $(e_1, \ldots, e_n) \in \mathrm{O}(n)$; we define an open set U of the quotient as

$$U = \left\{ (v_1, \ldots, v_k) \in \frac{\mathrm{O}(n)}{\mathrm{O}(n-k)} \,\bigg|\, (e_1, \ldots, e_{n-k}, v_1, \ldots, v_k) \text{ is a basis of } \mathbb{R}^n \right\}.$$

From the Gram–Schmidt orthonormalization procedure, if $(u_1, \ldots, u_n)$ is a basis, an orthonormal basis is constructed by $GS(u_1, \ldots, u_n) = (u'_1, \ldots, u'_n)$ with

$$u'_n = \frac{u_n}{\|u_n\|}, \quad u'_{n-1} = \frac{u_{n-1} - \langle u_{n-1}, u'_n \rangle u'_n}{\|u_{n-1} - \langle u_{n-1}, u'_n \rangle u'_n\|}, \quad \ldots \;.$$

Therefore, we obtain a section $\sigma\colon U \to \mathrm{O}(n)$ of the canonical projection defined by $\sigma(v_1, \ldots, v_k) = GS(e_1, \ldots, e_{n-k}, v_1, \ldots, v_k) = (e'_1, \ldots, e'_{n-k}, v_1, \ldots, v_k)$.

More generally, in the case of a closed subgroup of a Lie group, there always exists a local section (see [60, 12, Proposition 1]). We do not give

1: Lie groups and homogeneous spaces

the proof here. In every concrete example, a local section can be easily constructed as above.

Proof of Proposition 1.59 Observe first that any local section $\sigma: U \to G$, defined on a neighborhood U of H in G/H, gives a local section $\sigma_g: gU \to G$, $\sigma(g'H) = g.\sigma(g^{-1}g'H)$, with domain the neighborhood gU of gH.

Let $x \in G/H$ with a local section (U, σ). To satisfy the requirements of the definition of a locally trivial bundle, one needs maps φ and ψ such that $\varphi \circ \psi = \mathrm{id}_{p^{-1}(U)}$, $\psi \circ \varphi = \mathrm{id}_{U \times G/K}$ and the following diagram commutes

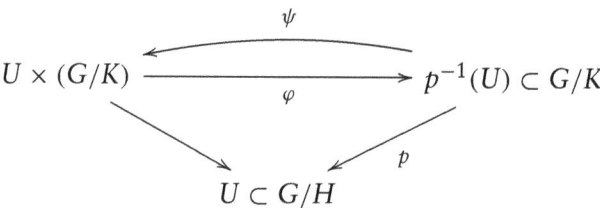

We define $\varphi(y, hK) = \sigma(y)hK$, $\psi(gK) = (gH, (\sigma(gH))^{-1}gK)$ and check easily that

$$\varphi(\psi(gK)) = \varphi(gH, (\sigma(gH))^{-1}gK) = \sigma(gH).(\sigma(gH))^{-1}gK = gK,$$
$$\psi(\varphi(y, hK)) = \psi(\sigma(y)hK) = (\sigma(y)hH, (\sigma(\sigma(y)H))^{-1}\sigma(y)hK),$$
$$= (p\sigma(y), (\sigma(p(\sigma(y))))^{-1}\sigma(y)hK) = (y, hK),$$
$$p\varphi(y, hK) = p(\sigma(y)hK) = \sigma(y)hH = \sigma(y)H = p\sigma(y) = y.$$

□

Our goal later will be to create algebraic models for manifolds and Lie groups and homogeneous spaces in particular. But the algebraic models we will consider work best in the simply connected world – or at least in the nilpotent world (see Definition 2.32 and the discussion that follows). For Lie groups and homogeneous spaces, this will not present a problem, for we prove at the end of this section that Lie groups and homogeneous spaces are simple, and therefore nilpotent, spaces.

Definition 1.61 *A space X is said to be* simple *if its fundamental group is abelian and acts trivially on the higher homotopy groups of X.*

For completeness, we recall the definition of the action of the fundamental group on the other homotopy groups. Let $\alpha \in \pi_1(X)$, $\xi \in \pi_n(X)$. The inclusion of the basepoint $s_0 \hookrightarrow S^n$ is a cofibration (see [265] or [240]), so there exists a commutative diagram as follows

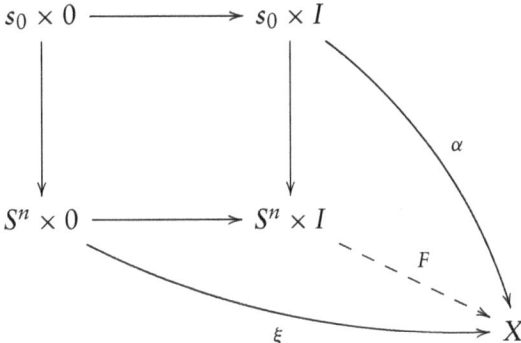

Now, the action of α on ξ, denoted $\alpha \cdot \xi$ is defined to be $\alpha \cdot \xi = F(-, 1)$, and its homotopy class does not depend on the choice of F.

Proposition 1.62 *Suppose H is a connected closed subgroup of a compact connected Lie group G. Then the homogeneous space G/H is a simple space. In particular, G itself is a simple space.*

In fact, this result holds for any topological group or, more generally, for any so-called H-space.

Proof Denote the quotient map by $q\colon G \to G/H$. We use the trivial class H as base point for G/H. Let $\alpha \in \pi_1(G/H)$, $\xi \in \pi_n(G/H)$. Since H is connected, there is a surjection $q_\#\colon \pi_1(G) \to \pi_1(G/H)$, so choose some $\tilde{\alpha}\colon I \to G$ with $\tilde{\alpha}(0) = \tilde{\alpha}(1) = e$ and $q(\tilde{\alpha}(t)) = \alpha(t)$. (The equality can be achieved since q satisfies the homotopy lifting property, see Exercise 1.9.) Now, by using the left action of G on G/H, we *define* a map F making the diagram commute: $F(x, t) = \tilde{\alpha}(t)\xi(x)$. Observe $F(x, 0) = \xi(x)$, $F(x, 1) = \xi(x)$ and

$$F(s_0, t) = \tilde{\alpha}(t)\xi(s_0) = \tilde{\alpha}(t)H = \alpha(t).$$

The last equality shows that F makes the diagram commute. Moreover, we see that $F(-, 1) = \xi$. Hence, $\alpha \cdot \xi = \xi$ and the action is trivial. □

Example 1.63 Consider the following examples:

1. Tori T^n have $\pi_1(T^n) = \mathbb{Z}^n$, but are simple spaces.
2. The special orthogonal groups $SO(n)$ have $\pi_1(SO(n)) = \mathbb{Z}/2$, but are simple spaces.
3. The projective space $\mathbb{R}P(2n)$ is known not to be simple (i.e. the antipodal map on the universal covering S^{2n} has degree -1), but we have $\mathbb{R}P(2n) = O(2n+1)/(O(2n) \times O(2))$. Note that the subgroup H is not connected.

Remark 1.64 The special case $H = T$, where T is a maximal torus of G, is easier to handle because there is a Bruhat decomposition of G/T showing that a CW-structure for G/T has cells only in even dimensions. Hence, *the homogeneous space G/T is simply connected.* A proof using regular elements can be found in [199, page 277]. Also, compare Exercise 1.5.

1.10 Principal bundles

In this section, we define principal bundles with structure group a Lie group G (also called principal G-bundles). Since the pullback of a principal G-bundle is a principal G-bundle, we look for a principal G-bundle $p: E \to B$ which is universal in the following sense: any principal G-bundle over a space B' can be obtained as a pullback of p along a map $B' \to B$. A characterization of such bundles is proved by using the notion of CW-complexes. Other examples will be developed in Section 1.12.

Definition 1.65 *Let G be a Lie group. A principal bundle with structure group G (or principal G-bundle) is a locally trivial bundle $p: E \to B$ with fiber the Lie group G, together with a right action $E \times G \to E$, $(x, g) \mapsto xg$, of G on E and a trivializing open cover of B, (U_i, φ_i), such that $\varphi_i(x, g \cdot g') = \varphi_i(x, g)g'$, for any $x \in E$, $g, g' \in G$.*

If all spaces are manifolds and all maps are smooth maps, we call the principal bundle a smooth principal bundle with structure group G *(or smooth principal G-bundle).*

Observe that the previous definition makes sense for a topological group instead of a Lie group G.

Remark 1.66 As the reader can easily check, if $p: E \to B$ is a principal G-bundle then the right action of G on E is free. Reciprocally, if G is a Lie group that acts freely and properly on a manifold M, the canonical projection $M \to M/G$ is a principal G-bundle, see [113, pages 193 and 229].

Example 1.67 (1) For any space B, the canonical projection $B \times G \to B$ is a principal G-bundle, called *the trivial principal G-bundle*.

(2) If H is a closed subgroup of a Lie group G, the canonical map $G \to G/H$ is a principal H-bundle.

Definition 1.68 *A morphism between two principal G-bundles is a pair of maps, (ψ, Ψ), such that Ψ is compatible with the G-action (i.e. $\Psi(xg) =$*

$\Psi(x)g)$ and the following diagram commutes

$$\begin{array}{ccc} E' & \xrightarrow{\Psi} & E \\ {\scriptstyle p'}\downarrow & & \downarrow{\scriptstyle p} \\ B' & \xrightarrow{\psi} & B \end{array}$$

Definition 1.69 *A principal G-bundle $p\colon E \to B$ is* trivial *if there exists a morphism of principal bundles*

$$\begin{array}{ccc} B \times G & \xrightarrow{f} & E \\ {\scriptstyle p'}\downarrow & & \downarrow{\scriptstyle p} \\ B & = & B \end{array}$$

where $B \times G \to B$ is the canonical projection.

Observe that, together with Exercise 1.11, if a principal G-bundle is trivial, the map f of Definition 1.69 is a homeomorphism.

Proposition 1.70 *Let G be a Lie group and let $p\colon E \to B$ be a principal G-bundle. If p admits a section, then p is trivial.*

Proof Denote by σ the section of p and let $B \times G \to B$ be the canonical projection. We construct a morphism of principal G-bundles

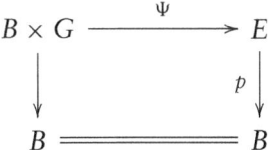

by $\Psi(x,g) = \sigma(x) \cdot g$. The triviality follows by definition. $\square$

Example 1.71 (Pullback of a principal G-bundle) Let G be a Lie group and $p\colon E \to B$ be a principal G-bundle. Let $\psi\colon B' \to B$ be a continuous map. Recall from Example 1.56 the construction of the pullback of p, denoted $p'\colon \psi^*E \to B'$. This is a locally trivial fiber bundle with fiber G. We define a free right action of G on ψ^*E by $(x,g)h = (x, g \cdot h)$. This gives ψ^*E the structure of principal G-bundle and we call $p'\colon \psi^*E \to B'$ the *pullback principal G-bundle of $p\colon E \to B$ along the map ψ*. The same construction gives a smooth principal G-bundle if $p\colon E \to B$ and $\psi\colon B' \to B$ are smooth.

Observe, from Exercise 1.11, that for any morphism of principal G-bundles,

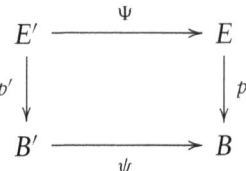

the domain E' is in the isomorphism class of ψ^*E.

Let $p\colon E \to B$ be a fixed principal G-bundle. Using Example 1.71, to any map $f\colon X \to B$ we can associate a principal G-bundle over X. In fact, if we consider the isomorphism classes of principal G-bundles, this association depends only on the homotopy classes of maps (see Exercise 1.8). Denote by $k_G(X)$ the set of isomorphism classes of principal G-bundles over X. What we have described above is a correspondence

$$[X, B] \to k_G(X), \quad [f] \mapsto f^*E.$$

We can ask whether there exist principal G-bundles $p\colon E \to B$ for which this association is an isomorphism (compare with [145], [74, Classification Theorem], [242]). This will elicit the notion of classifying space BG of G developed in Section 1.11. Because classifying spaces are not naturally manifolds (although, by [77], they *can* be viewed as manifolds of infinite dimension), we must leave the framework of manifolds and enter that of CW-complexes. This is still convenient for our study because any compact manifold *is* a CW-complex, as classical Morse theory shows (see [195]), and also because two simply connected compact Lie groups are isomorphic if they are homotopy equivalent (see [231]).

Definition 1.72 *Let G be a Lie group. A principal G-bundle $p\colon E \to B$ is an n-universal G-bundle if the association given above, $[X, B] \to k_G(X)$, is an isomorphism for any CW-complex X of dimension $\leq n$. A universal principal G-bundle is an n-universal principal G-bundle for every n.*

Universal principal G-bundles can be characterized by the following result.

Theorem 1.73 *A principal G-bundle $p_0\colon E_0 \to B_0$ is n-universal if and only if the space E_0 is $(n-1)$-connected.*

This criterion will give us the existence of universal G-bundles for the classical examples of Lie groups (see Section 1.12). For instance, in the case of $G = S^1$ or $G = S^3$ we can explicitly find universal bundles as follows.

Example 1.74 Consider the classical action of the circle S^1 on $S^{2n+1} \subset \mathbb{C}^{n+1}$ given by the complex multiplication $(z_1, \ldots, z_{n+1})z = (z_1 \cdot z, \ldots, z_{n+1} \cdot z)$. The quotient is the complex projective space and we have a principal S^1-bundle, $S^{2n+1} \to \mathbb{C}P(n)$, which is $(2n+1)$-universal by Theorem 1.73. The canonical inclusions $\mathbb{C}P(1) \subset \mathbb{C}P(2) \subset \cdots \subset \mathbb{C}P(n) \subset \cdots$ and $S^1 \subset S^2 \subset \cdots \subset S^n \subset \cdots$ define spaces $\mathbb{C}P(\infty) = \cup_n \mathbb{C}P(n)$ and $S^\infty = \cup_n S^n$. The principal S^1-bundle $S^\infty \to \mathbb{C}P(\infty)$ is a universal S^1-bundle since S^∞ is contractible.

By using the quaternionic multiplication on S^3, we get a $(4n+3)$-universal principal S^3-bundle, $S^{4n+3} \to \mathbb{H}P(n)$, and a universal principal S^3-bundle, $S^\infty \to \mathbb{H}P(\infty)$.

Proof of Theorem 1.73 *Part 1.* Suppose that $p_0 \colon E_0 \to B_0$ is n-universal and let $f \colon S^k \to E_0$ be a representative of a homotopy class in $\pi_k(E_0)$, $k \leq n-1$. We consider the following morphism of principal G-bundles (which is therefore a pullback),

$$\begin{array}{ccc} S^k \times G & \xrightarrow{\Psi} & E_0 \\ \downarrow & & \downarrow p_0 \\ S^k & \xrightarrow{\psi} & B_0 \end{array}$$

defined by $\psi = p_0 \circ f$ and $\Psi(x, g) = f(x)g$. The pullback of p_0 along ψ being trivial, the map ψ must be homotopically trivial by the injectivity part of the hypothesis and there is an extension $\overline{\psi} \colon D^{k+1} \to B_0$ of ψ. From the surjectivity part of the hypothesis, we now get the following morphism of principal G-bundles

$$\begin{array}{ccc} D^{k+1} \times G & \xrightarrow{\overline{\Psi}} & E_0 \\ \downarrow & & \downarrow p_0 \\ D^{k+1} & \xrightarrow{\overline{\psi}} & B_0 \end{array}$$

Using the injectivity part again gives the following sequence of morphisms of principal G-bundles where the maps $S^k \to D^{k+1}$ and $S^k \times G \to D^{k+1} \times G$

are the canonical injections:

$$
\begin{array}{ccccc}
S^k \times G & \longrightarrow & D^{k+1} \times G & \xrightarrow{\overline{\Psi}} & E_0 \\
\downarrow & & \downarrow & & \downarrow p_0 \\
S^k & \longrightarrow & D^{k+1} & \xrightarrow{\overline{\psi}} & B_0
\end{array}
$$

The restriction of $\overline{\Psi}$ to $D^{k+1} \times \{e\}$ is an extension of $f: S^k \to E$ which implies that f is nullhomotopic.

Part 2. Suppose now that $p_0: E_0 \to B_0$ is a principal G-bundle such that $\pi_k(E_0) = 0$ for $k \leq n-1$. We proceed by induction on the dimension of the CW-complex X, the statement obviously being true if X is of dimension 0. We suppose the result to be true for any CW-complex L of dimension less than or equal to $k-1$ with $k \leq n$. We attach a k-cell to L by a map χ:

$$S^{k-1} \xrightarrow{\chi} L \xrightarrow{j} X = L \cup D^k.$$

For proving the surjectivity part of the statement, we consider a principal G-bundle $p: E \to X$. The pullback $(j \circ \chi)^* E$ being trivial, we have a morphism $(\chi, \overline{\chi})$ from the trivial bundle to $j^* E$. The induction hypothesis (surjective part) applied to $j^* E \to L$ gives a morphism (ψ_L, Ψ_L) of principal G-bundles as in the following diagram

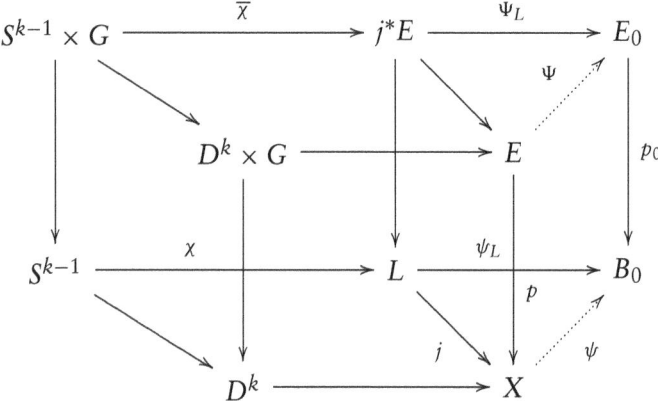

Since the map $\Psi_L \circ \overline{\chi}(-, e): S^{k-1} \to E_0$ is trivial by assumption, there exists an extension $\overline{\Psi}: D^k \to E_0$ of it. We now extend $\overline{\Psi}$ to a map $\widetilde{\Psi}: D^k \times G \to E_0$ using the action of G (i.e. $\widetilde{\Psi}(x, g) = \overline{\Psi}(x)g$). Since the total space E is the pushout of j^*E and $D^k \times G$ over $S^{k-1} \times G$, we construct a morphism

(Ψ, ψ) from p to p_0 as follows.

- if $x \in p^{-1}(y)$, $y \in L$, then $\Psi(x) = \Psi_L(x)$, $\psi(y) = \psi_L(y)$;
- if $x \in p^{-1}(y)$, $y \in D^k$, then $\Psi(x) = \widetilde{\Psi}(\chi^{-1}(x))$, $\psi(y) = p_0(\overline{\Psi}(y))$.

Now, Exercise 1.11 implies that $\psi^* E_0$ and E are in the same isomorphism class.

For proving the injectivity part of the statement we consider two maps $\psi_1, \psi_2 \colon X \to B_0$ giving two isomorphic bundles $\psi_1^* E_0, \psi_2^* E_0$. By induction, the maps $\psi_1 \circ j$ and $\psi_2 \circ j$ are homotopic. The homotopy lifting property gives a map $F_L \colon L \times [0,1] \to E_0$ such that $p_0 \circ F_L$ is the previous homotopy between $\psi_1 \circ j$ and $\psi_2 \circ j$.

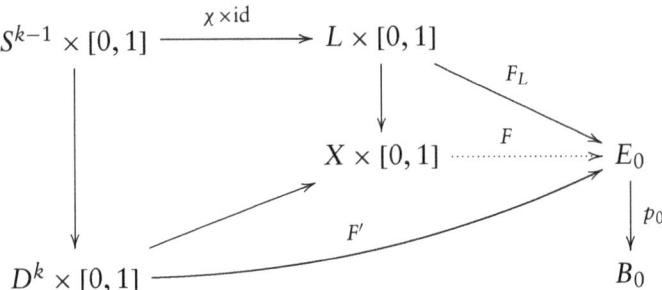

From $\pi_{k-1}(E_0) = 0$, we deduce the homotopy triviality of the composition $F_L \circ (\chi \times \mathrm{id})$ and, therefore, the existence of an extension $F' \colon D^k \times [0,1] \to E_0$. By the construction of X as a pushout, we get a map $F \colon X \times [0,1] \to E_0$ making the previous diagram commutative. The composition $p_0 \circ F$ is a homotopy between ψ_1 and ψ_2. □

Remark 1.75 John Milnor constructed a universal bundle for any topological group. In short, the construction goes like this:

- the total space EG is the infinite join, $EG = G * G * G * \cdots$;
- G acts on EG diagonally by $(g_1, g_2, \ldots)g = (g_1 \cdot g, g_2 \cdot g, \ldots)$. By definition, the space BG is the quotient EG/G.

Milnor proves that $EG \to BG$ is a universal G-bundle. For more details, see [145, Chapter 4, Section 11] or [193]. Here, we do not need this generality. In Section 1.12, we explicitly construct the classifying spaces for $G = O(n)$, $G = U(n)$ and $G = Sp(n)$. Since any compact Lie group is isomorphic to a subgroup of the orthogonal group, the existence of a classifying space for $O(n)$ implies the existence of a classifying space for any compact Lie group.

Remark 1.76 In terms of G-spaces and in the case $n = \infty$, Theorem 1.73 can also be proved by establishing the following property (see [74, 7.7]): If

E is a free G-space and E' a contractible G-space, then there exists a G-map $E \to E'$ and any two such maps are G-homotopic.

1.11 Classifying spaces of Lie groups

Here we give the definition of the classifying space of a compact Lie group and determine its cohomology algebra with rational coefficients. We use this classifying space to give a new, more homotopical, proof of the fact that the second Betti number of a semisimple Lie group is always equal to 0 (see Theorem 1.45).

Definition 1.77 *If $G \to EG \to BG$ is a universal principal G-bundle, the space BG is called the classifying space of the Lie group G.*

Note that, because EG is contractible, the long exact homotopy sequence of the bundle gives $\pi_i(G) \cong \pi_{i+1}(BG)$. In fact, $\Omega(BG) \simeq G$, where $\Omega(BG)$ here denotes the loop space of BG.

Remark 1.78 From Remark 1.75, one knows that such spaces exist. The uniqueness of their homotopy type is guaranteed by the following result. Two compact Lie groups G and H are isomorphic if and only if the classifying spaces BG and BH are homotopy equivalent (see [204], [210, Section 6], [211], [218]).

For concrete constructions of classifying spaces, the following observation will be of use.

Example 1.79 Let G be a compact connected Lie group and H be a closed subgroup of G. Denote by $G \to EG \to BG$ a universal principal G-bundle. The subgroup H acts freely on EG and gives a principal H-bundle $H \to EG \to EG/H$. Since the homotopy groups of EG are trivial, this fibration is a universal principal H-bundle. That means we can take EG/H to be the classifying space BH.

Proposition 1.80 $G/H \to BH \to BG$ *is a fibration.*

Proof As in Proposition 1.59, we have a fibration $G/H \to EG/H \to EG/G$. The result follows now with $BH = EG/H$ and $BG = EG/G$. $\square$

Theorem 1.81 *Let G be a compact connected Lie group with cohomology algebra an exterior algebra, $H^*(G; \mathbb{Q}) = \wedge(u_1, \ldots, u_r)$, where $u_i \in H^{2n_i-1}(G; \mathbb{Q})$. Then the classifying space for G has for cohomology algebra a polynomial algebra, $H^*(BG; \mathbb{Q}) = \mathbb{Q}[v_1, \ldots, v_r]$ with $v_i \in H^{2n_i}(BG; \mathbb{Q})$.*

Recall from Hopf's theorem (Theorem 1.34) that the cohomology algebra $H^*(G; \mathbb{Q})$ is always an exterior algebra on odd degree generators for any compact connected Lie group G. We present two different proofs for

Theorem 1.81. The first one, due to Borel [33], consists of a careful analysis of a spectral sequence. In the second one, we keep track of spaces more carefully. As the reader will see, this process simplifies the technical part of the argument. A third approach, using the construction of the minimal model of a loop space ΩX from the minimal model of X, will be presented in Example 2.67.

Proof 1 of Theorem 1.81 Consider the Serre spectral sequence, with coefficients in $\mathbb{Q}$, of the principal bundle $G \to EG \to BG$. Note that the hypotheses of the theorem guarantee that BG is simply connected. Therefore, the Serre spectral sequence has simple coefficients and the second page has the form $E_2^{p,q} = E_2^{p,0} \otimes E_2^{0,q} = (\wedge(u_1,\ldots,u_r))^p \otimes H^q(BG;\mathbb{Q})$ which satisfies the hypothesis of the Zeeman–Moore theorem (see Theorem B.15). We know from Theorem 1.73 that the page E_∞ is reduced to $\mathbb{Q}$ in degree 0.

Suppose, for a moment, that the u_i satisfy the following property:

$(\mathcal{T})$ $\qquad d_j u_i = 0$ for $j \leq 2n_i - 1$ and $d_{2n_i}(u_i) = v_i \neq 0$,

where the d_j are the differentials in the spectral sequence.

Now define a cochain algebra $\wedge(\bar{u}_1,\ldots,\bar{u}_r) \otimes \mathbb{Q}[\bar{v}_1,\ldots,\bar{v}_r]$ with $d\bar{u}_i = \bar{v}_i$, $|\bar{u}_i| = 2n_i - 1$, $|\bar{v}_i| = 2n_i$. Observe that the sub-differential algebras $(\wedge(\bar{u}_i) \otimes \mathbb{Q}[\bar{v}_i], d)$ are acyclic. Therefore the cohomology of $\wedge(\bar{u}_1,\ldots,\bar{u}_r) \otimes \mathbb{Q}[\bar{v}_1,\ldots,\bar{v}_r]$ is reduced to $\mathbb{Q}$ in degree 0. We filter by the degree in the $\bar{v}_i$'s and get a spectral sequence $\bar{E}_r^{p,q}$ whose second page is $\bar{E}_2^{p,q} = (\wedge(\bar{u}_1,\ldots,\bar{u}_r))^p \otimes (\mathbb{Q}[\bar{v}_1,\ldots,\bar{v}_r])^q$.

The canonical map sending $\bar{u}_i$ to u_i and $\bar{v}_i$ to v_i gives a morphism of spectral sequences $(\bar{E}_n^{p,q}, \bar{d}_n) \to (E_n^{p,q}, d_n)$. This morphism satisfies conditions (1) and (3) of Theorem B.15. Therefore, condition (2) is satisfied also and $H^*(BG;\mathbb{Q}) \cong \mathbb{Q}[v_1,\ldots,v_r]$.

Elements which satisfy property $\mathcal{T}$ are called *transgressive* and the proof of the fact that the u_i's are transgressive is one important point in the proof of Borel [33]. We do not go further in this direction. $\qquad\square$

In the second proof we deal with spaces and do not work only at the level of spectral sequences. Recall that each space X has an associated path fibration $\Omega X \to PX \xrightarrow{p} X$, where PX consists of all continuous paths $\gamma: I \to X$ having $\gamma(0) = x_0$, for a fixed basepoint $x_0 \in X$, and $p(\gamma) = \gamma(1)$. Then the fiber $p^{-1}(x_0)$ consists of all paths with $\gamma(0) = x_0 = \gamma(1)$. This is the loop space of X, ΩX. It is easy to see that PX is contractible, so the long exact homotopy sequence of the fibrations (see Exercise 1.10) gives $\pi_i(\Omega X) \cong \pi_{i+1}(X)$. The following proof applies the Serre spectral sequence to the path fibration.

Proof 2 of Theorem 1.81 We prove the next property by induction on r: Let X be a connected simple space whose loop space ΩX has rational cohomology algebra an exterior algebra $H^*(\Omega X; \mathbb{Q}) = \wedge(u_1, \ldots, u_r)$ with u_i of odd degree $2n_i - 1$. Then $H^*(X; \mathbb{Q}) = \mathbb{Q}[v_1, \ldots, v_r]$ is a polynomial algebra with $|v_i| = |u_i| + 1$.

Since the space X is simple (see Proposition 1.62), it admits a rationalization (see Subsection 2.6.1) and, since we are interested only in rational cohomology, we may replace X by its rationalization. However, for convenience, we still denote the rationalization by X in this proof.

If $r = 1$, we take the argument of the first proof above. Obviously the element u_1 must satisfy the property T and we get the result. Suppose now that the result is true for $r - 1$ generators and let X be as in the statement. If u_1 is an element of lowest positive degree among the u_i's, it must be a transgressive element, and we denote by $v_1 \in H^{2n_1}(X; \mathbb{Q})$ the image of u_1 by d_{2n_1}. This class v_1 corresponds to a map $\varphi(v_1): X \to K(\mathbb{Q}, 2n_1)$. Denote by Y the homotopy fiber of $\varphi(v_1)$ and take the following Puppe fibration sequence (see [137] or [265], for instance)

$$\Omega Y \longrightarrow \Omega X \xrightarrow{\Omega \varphi(v_1)} K(\mathbb{Q}, 2n_1 - 1) \longrightarrow Y \longrightarrow X \xrightarrow{\varphi(v_1)} K(\mathbb{Q}, 2n_1).$$

Now, the fibration $\Omega Y \longrightarrow \Omega X \longrightarrow K(\mathbb{Q}, 2n_1 - 1)$ is trivial because the map $\Omega \varphi(v_1)$ admits a section (see Exercise 1.12).

We thus get $H^*(\Omega X; \mathbb{Q}) = \wedge(u_1) \otimes H^*(\Omega Y; \mathbb{Q})$. By quotienting out the ideal generated by $\wedge(u_1)$, we deduce that $H^*(\Omega Y; \mathbb{Q}) = \wedge(u_2, \ldots, u_r)$.

We apply the induction hypothesis to Y and obtain $H^*(Y; \mathbb{Q}) = \mathbb{Q}[v_2, \ldots, v_r]$. Consider the Serre spectral sequence of the fibration $Y \longrightarrow X \longrightarrow K(\mathbb{Q}, 2n_1)$. Its second page is $E_2 = \mathbb{Q}[v_1] \otimes \mathbb{Q}[v_2, \ldots, v_r]$. Being totally concentrated in even degrees, all the differentials must be zero and we see that $H^*(X; \mathbb{Q}) = \mathbb{Q}[v_1] \otimes \mathbb{Q}[v_2, \ldots, v_r] = \mathbb{Q}[v_1, v_2, \ldots, v_r]$. □

We now use the existence of the classifying space of a Lie group for a third proof of Theorem 1.45. Let G be a compact semisimple Lie group. Recall that, in Corollary 1.46, we proved $\pi_2(G) \otimes \mathbb{Q} = 0$. As we will see in Remark 1.82, this third proof, which is more homotopical in spirit, in fact points the way toward the more general result that $\pi_2(G) = 0$.

Proof 3 of Theorem 1.45 Since a compact semisimple Lie group is finitely covered by a compact simply connected Lie group, it suffices to prove the result in the simply connected case. The Hurewicz theorem then implies that $\pi_2(G) \cong H_2(G)$, so in order to show $H^2(G; \mathbb{R}) = \text{Hom}(H_2(G), \mathbb{R}) = 0$, we need only to show that $\pi_2(G)$ is finite.

Now, $\pi_2(G) \cong \pi_3(BG)$, where BG is the classifying space of G. Because $\pi_2(G) = H_2(G)$ is finitely generated, we have a splitting $\pi_2(G) = F \oplus T$, where F is free abelian and T is the torsion subgroup. Let $f \in F$ be a generator of the free abelian part of $\pi_2(G)$ considered as an element in

$\pi_3(BG)$ and use $f \colon S^3 \to BG$ to induce a principal G-bundle $P \to S^3$ by pulling back the universal one. There is an associated Puppe sequence

$$\cdots \to \Omega S^3 \xrightarrow{\partial} G \to P \to S^3 \xrightarrow{f} BG$$

which is exact on homotopy groups. Furthermore, the connecting map ∂ is simply the loop of the classifying map f up to homotopy: $\Omega f \simeq \partial$. We have the following commutative diagram:

$$\begin{array}{ccc} \mathbb{Z} = \pi_3(S^3) & \xrightarrow{f_\#} & \pi_3(BG) \\ {\scriptstyle \cong} \Big\downarrow & & \Big\downarrow {\scriptstyle \cong} \\ \mathbb{Z} = \pi_2(\Omega S^3) & \xrightarrow{\partial_\#} & \pi_2(G) \end{array}$$

from which it follows that $\partial_\#(x) = f$, where x generates $\mathbb{Z} = \pi_2(\Omega S^3) = H_2(\Omega S^3)$. Because f represents a generator of F in $\pi_2(G) = H_2(G)$, we can define a homomorphism $\phi \colon H_2(G) \to \mathbb{Z}$ as follows: Write the generators of F as $\{f = f_1, f_2, \ldots, f_k\}$ and take

$$\phi(f) = 1, \qquad \phi(f_j) = 0 \quad \text{for } j \neq 1, \qquad \phi(T) = 0.$$

Now, $H^2(G; \mathbb{Z}) \cong \mathrm{Hom}(H_2(G), \mathbb{Z})$, so ϕ is an element of degree 2 cohomology. Moreover, the induced homomorphism $\partial^* \colon H^2(G; \mathbb{Z}) \to H^2(\Omega S^3; \mathbb{Z})$ has the following effect on the generator $x \in H_2(\Omega S^3; \mathbb{Z}) \cong \pi_2(\Omega S^3) = \mathbb{Z}$:

$$\partial^*(\phi)(x) = \phi(\partial_*(x)) = \phi(f) = 1,$$

which implies that $\partial^*(\phi) = \bar{x}$, where $\bar{x} \in H^2(\Omega S^3; \mathbb{Z}) = \mathrm{Hom}(H_2(\Omega S^3), \mathbb{Z})$ is defined by $\bar{x}(x) = 1$. But, from Example B.11, we see that the cup product powers $\bar{x}^m$ are nonzero for all $m \geq 1$. Thus

$$0 \neq \bar{x}^m = (\partial^*(\phi))^m = \partial^*(\phi^m)$$

which implies that $\phi^m \neq 0$ for all m in the finite dimensional manifold G. This contradiction then shows that $f = 0$. Hence $H_2(G) = T$, the torsion part, and $H^2(G; \mathbb{R}) = 0$. $\square$

Remark 1.82 In fact, a theorem due to S. Weingram allows us to show the stronger result $\pi_2(G) = 0$. Weingram's theorem (see [263]) says that a map $f \colon \Omega S^{2n+1} \to K(A, 2n)$ which has $0 \neq f_* \colon H_{2n}(\Omega S^{2n+1}; \mathbb{Z}) \to H_{2n}(K(A, 2n); \mathbb{Z})$ is incompressible; that is, f does not factor through a finite complex. In Proof 3 of Theorem 1.45, if the classifying map $\tau \colon S^3 \to BG$ represents a torsion element of $\pi_2(G) = H_2(G)$, then for some p^r, where p is a prime, there is a dual cohomology class $\hat{\tau} \in \mathrm{Hom}(H_2(G), \mathbb{Z}/p^r) \subseteq H^2(G; \mathbb{Z}/p^r)$. Because, in general, $H^k(X; A) \cong [X, K(A, k)]$ (where $[X, K(A, k)]$ denotes the set of homotopy classes of maps), we obtain the following homotopy commutative diagram by focusing on the connecting map in the Puppe sequence

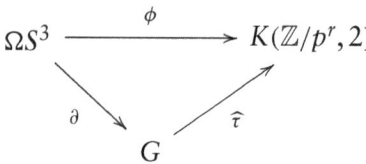

Here, with $\iota \in H^2(K(\mathbb{Z}/p^r, 2); \mathbb{Z}/p^r)$ the fundamental class, we have $\phi^*(\iota) = \partial^* \hat{\tau}^*(\iota) = \partial^* \hat{\tau}$, so ∂^* pulls back $\hat{\tau} \in H^2(G; \mathbb{Z}/p^r)$ into $H^2(\Omega S^3; \mathbb{Z}/p^r)$. By the same argument as before, this class in $H^2(\Omega S^3; \mathbb{Z}/p^r)$ is nontrivial. By Weingram's theorem, this is impossible because G is a compact manifold. Therefore, no such class τ can exist and $\pi_2(G) = 0$.

1.12 Stiefel and Grassmann manifolds

We will now define Stiefel and Grassmann manifolds and compute their cohomology. We deduce from this determination the universal fibrations associated to $O(n)$, $U(n)$, $Sp(n)$. In Proposition 1.87, a particular case of a "model of a fibration" is also given. This is a prototype of the relative models developed in Chapter 2.

Example 1.83 (Stiefel manifolds) We denote the Stiefel manifolds by, for $1 \leq k < n$,

$$V_{n,k}(\mathbb{R}) = \frac{O(n)}{O(n-k)} = \frac{SO(n)}{SO(n-k)},$$

$$V_{n,k}(\mathbb{C}) = \frac{U(n)}{U(n-k)} = \frac{SU(n)}{SU(n-k)},$$

$$V_{n,k}(\mathbb{H}) = \frac{Sp(n)}{Sp(n-k)},$$

and $V_{n,n}(\mathbb{R}) = O(n)$, $V_{n,n}(\mathbb{C}) = U(n)$, $V_{n,n}(\mathbb{H}) = Sp(n)$. As particular cases, we get $V_{n,1}(\mathbb{R}) = S^{n-1}$, $V_{n,1}(\mathbb{C}) = S^{2n-1}$, $V_{n,1}(\mathbb{H}) = S^{4n-1}$. Stiefel manifolds $V_{n,k}(-)$ are isomorphic to the spaces of orthonormal k-frames in the respective $\mathbb{R}^n$, $\mathbb{C}^n$, $\mathbb{H}^n$.

Example 1.84 (Grassmann manifolds) We denote the Grassmann manifolds by, for $1 \leq k < n$:

$$G_{n,k}(\mathbb{R}) = \frac{O(n)}{O(k) \times O(n-k)},$$

$$G_{n,k}(\mathbb{C}) = \frac{U(n)}{U(k) \times U(n-k)},$$

$$G_{n,k}(\mathbb{H}) = \frac{Sp(n)}{Sp(k) \times Sp(n-k)}.$$

As particular cases, we get $G_{n,1}(\mathbb{R}) = \mathbb{R}P(n-1)$, $G_{n,1}(\mathbb{C}) = \mathbb{C}P(2n-1)$, $G_{n,1}(\mathbb{H}) = \mathbb{H}P(4n-1)$. Grassmann manifolds $G_{n,k}(-)$ are isomorphic to the spaces of k-dimensional subspaces in the respective $\mathbb{R}^n$, $\mathbb{C}^n$, $\mathbb{H}^n$.

Before the statement, we recall that $\wedge(V)$ ($\mathbb{Q}[V]$) denotes the exterior (polynomial) algebra generated by the $\mathbb{Q}$-vector space V. Such a vector space is sometimes represented by a basis, $V = (x_1, \ldots, x_k)$.

Proposition 1.85 *The cohomology algebras of complex and quaternionic Stiefel manifolds are described by:*

$$H^*(V_{n,k}(\mathbb{C}); \mathbb{Q}) = \wedge (x_{2(n-k)+1}, \ldots, x_{2n-1}) \text{ with } x_{2i-1} \in H^{2i-1}(V_{n,k}(\mathbb{C}); \mathbb{Q});$$

$$H^*(V_{n,k}(\mathbb{H}); \mathbb{Q}) = \wedge (y_{4(n-k)+3}, \ldots, y_{4n-1}) \text{ with } y_{4i-1} \in H^{4i-1}(V_{n,k}(\mathbb{H}); \mathbb{Q}).$$

By using the particular cases $V_{n,n}(\mathbb{C}) = U(n)$, $V_{n,n}(\mathbb{H}) = Sp(n)$ and Theorem 1.81, one gets the following result immediately.

Corollary 1.86 *The unitary and symplectic groups and their classifying spaces have for cohomology algebras:*

$$H^*(U(n); \mathbb{Q}) = \wedge (x_1, x_3, \ldots, x_{2n-1}) \text{ with } x_{2i-1} \in H^{2i-1}(U(n); \mathbb{Q});$$

$$H^*(SU(n); \mathbb{Q}) = \wedge (x_3, \ldots, x_{2n-1}) \text{ with } x_{2i-1} \in H^{2i-1}(SU(n); \mathbb{Q});$$

$$H^*(Sp(n); \mathbb{Q}) = \wedge (y_3, y_7, \ldots, y_{4n-1}) \text{ with } y_{4i-1} \in H^{4i-1}(Sp(n); \mathbb{Q});$$

$$H^*(BU(n); \mathbb{Q}) = \mathbb{Q}[c_1, c_2, \ldots, c_n] \text{ with } c_i \in H^{2i}(BU(n); \mathbb{Q});$$

$$H^*(BSU(n); \mathbb{Q}) = \mathbb{Q}[c_2, \ldots, c_n] \text{ with } c_i \in H^{2i}(BSU(n); \mathbb{Q});$$

$$H^*(BSp(n); \mathbb{Q}) = \mathbb{Q}[q_1, q_2, \ldots, q_n] \text{ with } q_i \in H^{4i}(BSp(n); \mathbb{Q}).$$

A locally trivial fiber bundle, with a numerable trivializing open cover of the base (see [74]), is a fibration, so the bundle satisfies the homotopy lifting property (see Section 2.5.2 and Exercise 1.9). For the total space of a fibration, Sullivan theory supplies a nice algebraic model which contains models

1 : Lie groups and homogeneous spaces

of the base and of the fiber. As an aperitif, we now present a particular case, sufficient for our study of Stiefel and Grassmann manifolds.

Recall first, from Corollary B.13, that a bundle $S^k \longrightarrow E \xrightarrow{p} B$ induces an exact sequence, called the Gysin sequence,

$$H^j(B;\mathbb{Q}) \xrightarrow{\delta} H^{j+k+1}(B;\mathbb{Q}) \xrightarrow{p^*} H^{j+k+1}(E;\mathbb{Q}) \longrightarrow H^{j+1}(B;\mathbb{Q}) \ldots$$

where $\delta(x) = x \cup e$ for some $e \in H^{k+1}(B;\mathbb{Q})$.

Proposition 1.87 Let $S^{2n+1} \longrightarrow E \xrightarrow{p} B$ be a fibration such that the class $e \in H^{2n+2}(B;\mathbb{Q})$ appearing in the Gysin sequence is zero. Then we have an isomorphism of algebras

$$H^*(B;\mathbb{Q}) \otimes \wedge y \xrightarrow{\cong} H^*(E;\mathbb{Q}),$$

where y is of degree $2n+1$.

Proof The Gysin exact sequence splits into short exact sequences:

$$0 \longrightarrow H^j(B;\mathbb{Q}) \xrightarrow{p^*} H^j(E;\mathbb{Q}) \xrightarrow{\sigma^*} H^{j-2n-1}(B;\mathbb{Q}) \longrightarrow 0.$$

We choose a class $\bar{y} \in H^{2n+1}(E;\mathbb{Q})$ such that $\sigma^*(\bar{y}) = 1$ and define a cochain map

$$\Phi \colon H^*(B) \otimes \wedge y \to H^*(E)$$

by $\Phi(\omega) = p^*(\omega)$ if $\omega \in H^*(B)$ and $\Phi(y) = \bar{y}$. Because y is of odd degree, the map Φ is an isomorphism of algebras. □

Remark 1.88 In [112, page 320], the class e is defined as the *Euler class* of the bundle. In the case of a sphere bundle, this class coincides with the pullback of the *Thom class* along the zero section (see [41, Theorem 13.2, page 390]). A third equivalent definition of Euler class will be given in Example 1.96.

Proof of Proposition 1.85 Let $n \geq 1$ be an integer. For $k = 1$, we have $V_{n,1}(\mathbb{C}) = S^{2n-1}$ and the statement is true. We now use induction on k by supposing that the formula is true for $k-1$. We consider the fibration

$$S^{2(n-k)+1} \longrightarrow V_{n,k}(\mathbb{C}) \longrightarrow V_{n,k-1}(\mathbb{C}).$$

By the induction hypothesis, the Euler class in $H^{2(n-k)+2}(V_{n,k-1}(\mathbb{C});\mathbb{Q})$ is zero. The result follows directly from Proposition 1.87. The argument is similar for the symplectic Stiefel manifolds. □

In the real case, the proof works also by induction on k, using the fibration

$$S^n \to SO(n+k)/SO(n) \to SO(n+k)/SO(n+1),$$

but one needs to determine the Euler class of this fibration. We do not give the proof here, referring the reader to [199, page 121].

Proposition 1.89 *The real Stiefel manifolds have for cohomology algebras:*

$$\begin{aligned}
H^*(V_{n+k,k}(\mathbb{R}); \mathbb{Q}) &= \wedge \left(z_{2n+1}, z_{2n+5}, \ldots, z_{2n+2k-3}\right), \; n \text{ odd}, k \text{ even}, \\
&= \wedge \left(z_{2n+1}, z_{2n+5}, \ldots, z_{2n+2k-5}, z_{n+k-1}\right), \; n \text{ odd}, k \text{ odd}, \\
&= \wedge \left(z_n, z_{2n+3}, z_{2n+7}, \ldots, z_{2n+2k-3}\right), \; n \text{ even}, k \text{ odd}, \\
&= \wedge \left(z_n, z_{2n+3}, z_{2n+7}, \ldots, z_{2n+2k-5}, z_{n+k-1}\right), \\
&\quad n \text{ odd}, k \text{ even},
\end{aligned}$$

with $z_i \in H^i(V_{n+k,k}(\mathbb{R}); \mathbb{Q})$.

Noting the particular cases $V_{n,n}(\mathbb{R}) = O(n)$ and using Theorem 1.81, we deduce the following.

Corollary 1.90 *The orthogonal groups and their classifying spaces have for cohomology algebras:*

$$\begin{aligned}
H^*(SO(2m); \mathbb{Q}) &= \wedge (z_3, z_7, \ldots, z_{4m-5}, z_{2m-1}), \\
H^*(SO(2m+1); \mathbb{Q}) &= \wedge (z_3, z_7, \ldots, z_{4m-1}), \\
H^*(BSO(2m); \mathbb{Q}) &= \mathbb{Q}[p_1, p_2, \ldots, p_{m-1}, \chi], \\
H^*(BSO(2m+1); \mathbb{Q}) &= \mathbb{Q}[p_1, p_2, \ldots, p_m],
\end{aligned}$$

with $z_i \in H^i(SO(-); \mathbb{Q})$, $p_i \in H^{4i}(BSO(-); \mathbb{Q})$, $\chi \in H^{2m}(BSO(2m); \mathbb{Q})$.

Remark 1.91 The results of Proposition 1.85 and Corollary 1.86 are true for cohomology with coefficients in $\mathbb{Z}$, since these cohomology groups have no torsion (see [199, page 119]). This is not the case for the real Stiefel manifolds; their cohomology has 2-torsion. Being concerned here only with coefficients in a field of characteristic zero, we refer the reader to [199, page 121] for an explicit description of the cohomology with coefficients in $\mathbb{Z}_2$.

As a direct consequence of the previous computations and Theorem 1.73, we obtain the following.

Proposition 1.92

1. *The principal $O(k)$-bundle, $V_{n+k,k}(\mathbb{R}) \to G_{n+k,k}(\mathbb{R})$, is n-universal.*
2. *The principal $U(k)$-bundle, $V_{n+k,k}(\mathbb{C}) \to G_{n+k,k}(\mathbb{C})$, is $(2n+1)$-universal.*

3. *The principal* $Sp(k)$-*bundle,* $V_{n+k,k}(\mathbb{H}) \to G_{n+k,k}(\mathbb{H})$, *is* $(4n+3)$-*universal.*

Corollary 1.93 Set $V_{\infty,k} = \cup_n V_{n+k,k}$ and $G_{\infty,k} = \cup_n G_{n+k,k}$ for the fields $\mathbb{R}$, $\mathbb{C}$ and $\mathbb{H}$. Then the three following principal fibrations are universal:

$$V_{\infty,k}(\mathbb{R}) \to G_{\infty,k}(\mathbb{R}), \text{ for } G = O(k);$$
$$V_{\infty,k}(\mathbb{C}) \to G_{\infty,k}(\mathbb{C}), \text{ for } G = U(k);$$
$$V_{\infty,k}(\mathbb{H}) \to G_{\infty,k}(\mathbb{H}), \text{ for } G = Sp(k).$$

Remark 1.94 For the universal fibrations associated to $SO(n)$ and $SU(n)$, it is necessary to introduce the spaces of oriented vector subspaces:

$$SG_{n,k}(\mathbb{R}) = O(n)/SO(k) \times O(n-k) \text{ and } SG_{n,k}(\mathbb{C}) = U(n)/SU(k) \times U(n-k).$$

See [145, Theorem 6.1 of Chapter 4].

Definition 1.95 *Let* $p: E \to B$ *be a principal G-bundle with classifying map* $\psi: B \to BG$. *A characteristic class of p is an element of* $\psi^* H^*(BG; R) \subseteq H^*(B; R)$, *for a commutative ring R. Since we work with coefficients in* $\mathbb{Q}$, *we consider only rational characteristic classes.*

Example 1.96 We list here the characteristic classes corresponding to the different Lie groups of matrices.

- For a $U(n)$-bundle, the characteristic classes are generated by the image of the classes $c_i \in H^*(BU(n); \mathbb{Q}) = \mathbb{Q}[c_1, \ldots, c_n]$. These are called *Chern classes*.
- For a $SO(2m+1)$-bundle, the characteristic classes are generated by the image of the classes $p_i \in H^*(BSO(2m+1); \mathbb{Q}) = \mathbb{Q}[p_1, \ldots, p_m]$. These are called *Pontryagin classes*.
- For a $SO(2m)$-bundle, the characteristic classes are generated by the image of the Pontryagin classes just defined, and the *universal Euler class* $\chi \in H^{2m}(BSO(2m); \mathbb{Q})$.
- For $Sp(n)$, the characteristic classes are generated by the image of the classes $q_i \in H^{4i}(BSp(n); \mathbb{Q})$. These are called *symplectic Pontryagin classes*.

When the bundle $S^r \to E \to B$ has $SO(r+1)$ as structural group, the notion of Euler class introduced in Remark 1.88 from the Gysin sequence may be identified with the definition of this section (see [145, Proof of Theorem 6.5, page 103]). In the case of $S^1 = SO(2)$, one has $BS^1 = \mathbb{C}P^\infty = K(\mathbb{Z}, 2)$ and the Euler class with coefficients in $\mathbb{Z}$ is the image of the generator of $H^2(K(\mathbb{Z},2);\mathbb{Z})$ (see Section 6.5 for a description of the relation between the Euler class and the flatness of the associated vector bundle).

One can also observe some relations between characteristic classes coming from different inclusions between Lie groups (see [198, page 176] for a description of them). For instance, the canonical inclusion $U(m) \subset SO(2m+1)$ gives a map $\psi: BU(m) \to BSO(2m+1)$ which induces in cohomology $\psi^*(p_j) = \sum_{r+s=j}(-1)^r c_r c_s \in H^{4j}(BU(m))$ [199, page 144].

1.13 The Cartan–Weil model

Once the rational cohomology of Lie groups was determined, calculating the cohomology of general homogeneous spaces became an important goal. The combined work of Cartan, Chevalley and Weil after World War II provided an algebraic model whose cohomology was the cohomology of the homogeneous space under consideration. This is the genesis of the theory of minimal models which is the subject of Chapter 2. Indeed, we shall give the minimal model version of the Cartan–Weil model in Theorem 2.71. Here we want to both generalize and particularize the Cartan–Weil model for homogeneous spaces. We generalize the model by describing an algebraic model for an action of a Lie group on a space. We particularize by taking the simplest example: the action of a circle on a manifold. The advantage of the original Cartan–Weil model over newer minimal models is that we can often see more geometry in the Cartan–Weil model because it is constructed from forms. (In particular, there are models based on harmonic forms and the Hodge decomposition which often are more computable; see, for instance, [11].) Our treatment is essentially that of [15].

In Theorem 1.28, we saw that, if a compact connected Lie group G acts on a closed manifold M, then $H^*(M;\mathbb{R}) = H^*_L(M)$, where $H^*_L(M)$ is the cohomology of the complex of left invariant forms. While this is an interesting result, there is a more important cohomology associated to a group action. This is the *equivariant cohomology* defined as follows. As we saw above, there is a universal principal bundle $G \to EG \to BG$ with a free right action of G on EG and $BG = EG/G$. If G also acts on M, then there is a free right action of G on $EG \times M$ given by $g(e,m) = (eg^{-1}, gm)$. The quotient is then

$$M_G \stackrel{\text{def}}{=} EG \times_G M = (EG \times M)/G,$$

and there is a fibration (called the *Borel fibration*),

$$M \to M_G \stackrel{p}{\to} BG$$

where $p([e,m]) = [e]$.

1 : Lie groups and homogeneous spaces

Definition 1.97 *The* equivariant cohomology of M *with respect to an action of G on M is defined to be*

$$H_G^*(M) = H^*(M_G).$$

Any coefficients may be used in the definition. For instance, we will be interested in either $H_G^*(M; \mathbb{Q})$ or $H_G^*(M; \mathbb{R})$. The equivariant cohomology is important because it gives information not only about M, but about the G-action also. While the Borel fibration will be studied extensively from topological and algebraic perspectives in Section 7.2, here we want to hint at how Lie group actions lead to topology and, especially, algebraic models.

Since we are only interested here in motivating the models of Chapter 2, let's take a special case of a smooth action of S^1 on a closed manifold M. For a fixed $m \in M$, the orbit map $S^1 \to M$ given by $g \mapsto gm$ induces a linear map $T_e S^1 \to T_m M$ with $X_\theta \mapsto X_m$, where X_θ is a basis for $T_e S^1 = \mathbb{R}$, the Lie algebra of S^1. Doing this for every $m \in M$ produces a vector field X on M called the *fundamental vector field* of the action. As such, we can form the operators $i(X)$ and $\mathcal{L}(X)$ (see Section A.2).

The *Weil algebra* W of the circle is

$$W = \wedge(\theta, u) = \text{Exterior}(\theta) \otimes \text{Polynomial}(u),$$

where θ is in degree 1 and u is in degree 2. A differential D is defined on W by declaring $D\theta = u$ and $Du = 0$. Notice that the cohomology of this differential graded algebra is zero (except for $H^0(W) = \mathbb{R}$). We mention here that the Weil algebra can be defined for other Lie groups also using the structure constants of the associated Lie algebra to describe the differential. We can extend the action of the operators $i(X)$ and $\mathcal{L}(X)$ from the de Rham algebra $A_{DR}(M)$ to the complex $W \otimes A_{DR}(M)$ as follows. On W, define

$$i(X)\theta = 1, \quad i(X)u = 0, \quad \mathcal{L}(X)\theta = 0, \quad \mathcal{L}(X)u = 0.$$

In fact, the general definition for $\mathcal{L}(X)$ is $\mathcal{L}(X) = i(X)D + Di(X)$ and the equations above hold in the S^1 case.

Definition 1.98 *The basic subcomplex* $\Omega_B(M) \subset W \otimes A_{DR}(M)$ *consists of all elements α with $i(X)\alpha = 0$ and $\mathcal{L}(X)\alpha = 0$.*

Of course, a similar definition holds for general compact connected G. The fundamental result that we shall not prove in the present framework is (in its general form, see [15]),

1.13 The Cartan–Weil model

Theorem 1.99 *If there is a smooth action of a connected compact Lie group G on a closed manifold M, then*

$$H^*(\Omega_B(M)) \cong H_G^*(M; \mathbb{R}).$$

So $\Omega_B(M)$ is, in some sense, a model for the total space of the Borel fibration, M_G. Clearly, however, it is not a particularly tractable model. Below, we shall find a better and more intuitive model.

Let's specialize to an S^1-action for simplicity. Any element of $W \otimes A_{DR}(M)$ has the form $\omega = \sum_k u^k a_k + \sum_j \theta u^j b_j$ where k and j are non-negative integers and $a_k, b_j \in A_{DR}(M)$. The next result identifies the basic elements.

Proposition 1.100 *Using the notation above, an element $\omega \in W \otimes A_{DR}(M)$ is basic if and only if $\mathcal{L}(X)a_k = 0$ and $i(X)a_k = -b_k$ for each k.*

Proof First, recall that we extended $i(X)$ and $\mathcal{L}(X)$ to $W \otimes A_{DR}(M)$ above. Now, since $i(X)$ and $\mathcal{L}(X)$ are graded derivations, we have

$$i(X)\omega = \sum_k u^k(i(X)a_k) + \sum_j (u^j b_j - \theta u^j i(X)b_j) = 0,$$

$$\mathcal{L}(X)\omega = \sum_k u^k(\mathcal{L}(X)a_k) + \sum_j \theta u^j(\mathcal{L}(X)b_j) = 0.$$

Now note that the u^k and $u^k\theta$ are algebraically independent over $A_{DR}(M)$. Hence (using the standard facts that $i(X)^2 = 0$ and $i(X)\mathcal{L}(X) = \mathcal{L}(X)i(X)$), the equations above are seen to be equivalent to those of the statement of the proposition. □

Now we can begin to simplify our model for equivariant cohomology. While the basic forms tell us what we need to know, they are unwieldy because of the condition $i(X)\alpha = 0$. We might reasonably ask what this has to do with the action. The forms that should be important are the forms that are invariant under the flow along orbits. These are, of course, identified with the forms that are annihilated by $\mathcal{L}(X)$. Therefore, let $\Omega_X = \{\alpha \in A_{DR}(M) \mid \mathcal{L}(X)\alpha = 0\}$. As we have said, these are the forms that are invariant under the circle action. Adjoin a degree 2 generator u to Ω_X to obtain an algebra $\Omega_X[u] = \Omega_X \otimes \mathbb{R}[u]$ and an algebra homomorphism $\phi \colon \Omega_X[u] \to W \otimes A_{DR}(M)$ defined by $\phi(\alpha) = \alpha - \theta i(X)\alpha$ and $\phi(u) = u$.

Lemma 1.101 *The image of ϕ lies in the basic subalgebra $\Omega_B(M)$.*

Proof We must show that $\mathcal{L}(X)\phi(\alpha) = 0$ and $i(X)\phi(\alpha) = 0$, for $\alpha \in \Omega_X$. We compute (again using the extensions of $i(X)$ and $\mathcal{L}(X)$ to $W \otimes A_{DR}(M)$

and the relations $i(X)^2 = 0$, $\mathcal{L}(X)i(X) = i(X)\mathcal{L}(X))$.

$$\begin{aligned}\mathcal{L}(X)\phi(\alpha) &= \mathcal{L}(X)(\alpha - \theta i(X)\alpha) \\ &= \mathcal{L}(X)\alpha - (\mathcal{L}(X)\theta)i(X)\alpha - \theta\mathcal{L}(X)i(X)\alpha \\ &= 0 - 0 - \theta i(X)\mathcal{L}(X)\alpha = 0,\end{aligned}$$

where we have used the fact that $\mathcal{L}(X)\alpha = 0$ since Ω_X consists of invariant forms. Similarly, since $i(X)\theta = 1$, we have

$$\begin{aligned}i(X)\phi(\alpha) &= i(X)(\alpha - \theta i(X)\alpha) \\ &= i(X)\alpha - (i(X)\theta)i(X)\alpha + \theta i(X)i(X)\alpha \\ &= i(X)\alpha - i(X)\alpha + 0 = 0.\end{aligned}$$

□

It is not difficult to see that, in fact, the equations of Proposition 1.100 imply that ϕ is an algebra isomorphism onto its image,

$$\phi \colon \Omega_X[u] \xrightarrow{\cong} \Omega_B(M).$$

Therefore, we obtain a differential d_X on $\Omega_X[u]$ by transporting the differential on $W \otimes A_{DR}(M)$ back via ϕ^{-1}. By definition, we have $\phi d_X = D\phi$, where D is a differential defined as follows. On the Weil algebra, D is the differential we defined above: $D\theta = u$ and $Du = 0$. On $A_{DR}(M)$, D is simply the exterior derivative d.

Now let's compute to see what d_X has to be. Suppose $\alpha \in \Omega_X$. Using $\mathcal{L}(X)\alpha = di(X)\alpha + i(X)d\alpha = 0$, we obtain the following.

$$\begin{aligned}D\phi(\alpha) &= D(\alpha - \theta(i(X)\alpha)) \\ &= d\alpha - u(i(X)\alpha) + \theta d(i(X)\alpha) \\ &= d\alpha - ui(X)\alpha - \theta i(X)d\alpha \\ &= \phi(d\alpha - ui(X)\alpha),\end{aligned}$$

since $\phi(d\alpha) = d\alpha - \theta i(X)d\alpha$ and

$$\phi(ui(X)\alpha) = \phi(u)\phi(i(X)\alpha) = u(i(X)\alpha - \theta i(X)i(X)\alpha) = ui(X)\alpha.$$

Therefore, in order for $D\phi = \phi d_X$ to hold, we should define $d_X(u) = 0$ and, for $\alpha \in \Omega_X$,

$$d_X\alpha = d\alpha - ui(X)\alpha.$$

Thus we see that $\phi \colon (\Omega_X[u], d_X) \to (\Omega_B(M), D)$ is an isomorphism of differential graded algebras. Hence, we have the following result.

1.13 The Cartan–Weil model

Theorem 1.102 *Suppose $G = S^1$ acts smoothly on a closed manifold M. The Cartan–Weil model is the complex $(\Omega_X[u], d_X)$, where Ω_X is the subcomplex of $A_{DR}(M)$ consisting of G-invariant forms, u is a degree 2 generator and the differential is defined by $d_X u = 0$ and $d_X \alpha = d\alpha - u i(X)\alpha$, for $\alpha \in \Omega_X$. The Cartan–Weil model is isomorphic to the basic complex $(\Omega_B(M), D)$ of $W \otimes A_{DR}(M)$ and this isomorphism induces an isomorphism*

$$H^*(\Omega_X[u], d_X) \cong H^*(\Omega_B(M), D) \cong H^*(M_{S^1}).$$

Thus, the Cartan–Weil model calculates the equivariant cohomology associated to the action of G on M.

Remark 1.103 The equation $d_X \alpha = d\alpha - u i(X)\alpha$ says something interesting. Although α is an invariant form, even if it is closed, this does *not* mean that α is equivariantly closed (i.e. d_X-closed). For that, we also require $i(X)\alpha = 0$ – and this is exactly a basic form in the model $\Omega_B(M)$.

Remark 1.104 The models can be directly related to the geometry of the Borel fibration. The homomorphism $\Omega_X[u] \to A_{DR}(M)$ obtained by evaluating $u = 0$ is a model for the fiber inclusion $M \to M_{S^1}$. Integration over the fiber is expressed on $\Omega_X[u]$ by $a_k u^k \mapsto (\int a_k) u^k$.

Remark 1.105 The whole discussion above extends to the action of a torus T^n on a closed manifold M. In this case the model for M_{T^n} is $\Omega^*_{X_1, \ldots, X_n}[u_1, \ldots, u_n]$, where the X_i are fundamental vector fields corresponding to a basis for the Lie algebra $\mathbb{R}^n$ of T^n and the u_i are all in degree 2. The differential d_X is given by $d_X \alpha = d\alpha - \sum_k u_k i(X_k) \alpha$.

The differential graded algebra $(\Omega_X[u], d_X)$ is a "model" for the minimal models we shall describe in Chapter 2. Indeed, we shall see a distinct resemblance to the relative minimal models that describe the rational homotopy structure of fibrations. Namely, the algebra (Ω_X, d) comes from the manifold M (while also including information about the action), the degree 2 generator u comes from BS^1 since $H^*(BS^1; \mathbb{R}) = \mathbb{R}[u]$ and the differential d_X on $\Omega_X[u]$ is a perturbation of the differential d by a term involving the action. This is a way to construct a model of the total space of the Borel fibration $M \to M_{S^1} \to BS^1$ starting from the base and fiber; indeed, these are the types of ingredients found in the theory of minimal models that we will present in the next chapter and that will occupy and intrigue us throughout this book.

Exercises for Chapter 1

Exercise 1.1 (1) Prove the existence of the following isomorphisms of Lie groups:

$$\text{Spin}(3) \cong S^3, \quad \text{Spin}(4) \cong S^3 \times S^3, \quad \text{Spin}(5) \cong \text{Sp}(2), \quad \text{Spin}(6) \cong \text{SU}(4),$$

$$\text{SO}(2) \cong S^1, \quad \text{SU}(2) \cong \text{Sp}(1).$$

Hint: [199, page 82–84].

(2) Prove the existence of homeomorphisms of topological spaces:

$$\text{SO}(3) \cong \mathbb{R}P(3), \quad \text{SO}(4) \cong S^3 \times \text{SO}(3).$$

Exercise 1.2 The *special linear group*, $\text{SL}(n, \mathbb{R})$, is the subgroup of $\text{Gl}(n, \mathbb{R})$ formed by the isomorphisms of determinant 1.

(1) Show that $\text{SL}(n, \mathbb{R})$ is a Lie group of dimension $n^2 - 1$ and has for Lie algebra the set, $sl(n, \mathbb{R})$, of $n \times n$ matrices of trace 0.

(2) Prove that $\text{SO}(n, \mathbb{R})$ is the maximal compact subgroup of $\text{SL}(n, \mathbb{R})$. In fact, show

$$\text{SO}(n, \mathbb{R}) \times \mathbb{R}^{\frac{n(n-1)}{2} + n - 1} \cong \text{SL}(n, \mathbb{R}).$$

(3) Define the group $\text{SL}(n, \mathbb{C})$ and prove:

- $\text{SL}(n, \mathbb{C})$ is a Lie group of dimension $2n^2 - 2$;
- $\text{SU}(n) \times \mathbb{R}^{n^2 - 1} \cong \text{SL}(n, \mathbb{C})$.

Hint: see [113, Chapter 2] and [199, page 34].

Exercise 1.3 The *real symplectic group* is defined as

$$\text{Sp}(n; \mathbb{R}) = \{A \in \text{Gl}(2n, \mathbb{R}) \mid {}^tA J_n A = J_n\},$$

where $J_n = \begin{pmatrix} 0 & -I_n \\ I_n & 0 \end{pmatrix}$.

(1) Show that $\text{Sp}(n, \mathbb{R})$ is a Lie group of dimension $n(2n + 1)$ and has for Lie algebra the set, $sp(n, \mathbb{R})$, of $2n \times 2n$ real matrices A such that $J_n A J_n^{-1} = -{}^tA$.

(2) Prove that $U(n)$ is the maximal compact subgroup of $\text{Sp}(n, \mathbb{R})$;

$$U(n) \times \mathbb{R}^{n(n+1)} \cong \text{Sp}(n, \mathbb{R}).$$

(3) Define the group $\text{Sp}(n, \mathbb{C})$ and prove:

- $\text{Sp}(n, \mathbb{C})$ is a Lie group of dimension $2n(n + 1)$;
- $\text{Sp}(n) \times \mathbb{R}^{n(2n+1)} \cong \text{Sp}(n, \mathbb{C})$;
- $\text{Sp}(n) \cong \text{Sp}(n, \mathbb{C}) \cap U(2n)$.

Hint: see [113, Chapter 2] and [199, page 34].

COMMENT: What we call *the symplectic group*, denoted by $\text{Sp}(n)$, is the orthogonal group of the quaternions. The *real and complex symplectic groups*, $\text{Sp}(n, \mathbb{R})$ and

Sp(n, $\mathbb{C}$), are the groups of isomorphisms which respect a symplectic form. Since we are mainly interested in compact Lie groups, these last two will not appear in the rest of the book and there will be no confusion in terminology.

Exercise 1.4 Determine the maximal torus and the Weyl group of the spinor groups, Spin(n). Hint: see [145, 14-8].

Exercise 1.5 Let G be a compact connected Lie group. Denote by $Z(g)_e$ the connected component of the centralizer of $g \in G$ containing the neutral element.

(1) Show that $Z(g)_e$ is the union of the maximal tori containing g.

(2) An element $g \in G$ is called *regular* if g belongs to just one maximal torus. Otherwise, it is called *singular*. Show that g is regular if and only if the dimension of $Z(g)$ is equal to the rank of G. Hint: see [199, page 267].

COMMENT: From the existence of a smooth map $f: M \to G$ such that dim M = dim $G - 3$ and such that $f(M)$ is the set of singular elements of G (this set is not a manifold!), one can deduce that $\pi_2(G) = 0$; see [135, Theorem 4.7, page 260] or [113, Problem 34-35, page 108].

Exercise 1.6 Let G be a compact connected Lie group and let K be a closed connected subgroup of G. Let σ be a smooth automorphism of G such that $\sigma^2 = $ id. Denote by $(G^\sigma)_e$ the connected component of the fixed point set G^σ of σ containing the unit e. The pair (G, K) is called a *symmetric pair* if $(G^\sigma)_e \subset K \subset G^\sigma$. In this case, the quotient G/K is called a *symmetric space*.

(1) Show that $(U(p+q), U(p) \times U(q))$ is a symmetric pair. Give other examples. (See [199, page 147].)

(2) Show that a symmetric space is parallelizable.

(3) Let (G, K) be a symmetric pair. The group G acts on the left on G/K. Denote by $\mathfrak{g}$ and $\mathfrak{k}$ the Lie algebras of G and K respectively and by $\mathfrak{k}^\perp$ the orthogonal complement of $\mathfrak{k}$ in $\mathfrak{g}^*$.

Show that the cohomology $H^*(G/K; \mathbb{R})$ is isomorphic, as an algebra, to the invariant set of $\wedge \mathfrak{k}^\perp$. Hint: in the proof for a semisimple Lie group, replace the inverse map by the automorphism σ.

Exercise 1.7 Let G be a compact connected Lie group with Lie algebra $\mathfrak{g}$. Recall the existence of an isomorphism $\chi: \Omega_L(G) \cong \wedge \mathfrak{g}^*$ defined by $\chi(\omega) = \omega_e$. If X is a left invariant vector field, we also know that $\Omega_L(G)$ is stable by the coboundary d, the interior multiplication $i(X)$ and the Lie derivative $\mathcal{L}(X)$. Let $\Phi \in \wedge^p \mathfrak{g}^*$ and let $(X, X_0, X_1, \ldots, X_p)$ be elements of $\mathfrak{g}$. We set:

$$(i(X)\Phi)(X_1, \ldots, X_{p-1}) = \Phi(X, X_1, \ldots, X_{p-1}).$$

$$(\theta(X)\Phi)(X_1, \ldots, X_p) = -\sum_{i=1}^{p} \Phi(X_1, \ldots, [X, X_i], \ldots, X_p).$$

$$(\delta\Phi)(X_0, \ldots, X_p) = \sum_{i<j}(-1)^{i+j}\Phi([X_i, X_j], X_0, \ldots, \hat{X}_i, \ldots, \hat{X}_j, \ldots, X_p).$$

Prove that these operators are the respective images of d, $i(X)$ and $\mathcal{L}(X)$ by the isomorphism χ. Hint: [113, page 156].

As a consequence, we have $H^*(G;\mathbb{R}) \cong H^*(\wedge\mathfrak{g}^*,\delta) \cong \wedge(\mathfrak{g}^*)_{\theta=0}$. The cohomology $H^*(\wedge\mathfrak{g}^*,\delta)$ is called the *cohomology algebra of the Lie algebra* $\mathfrak{g}$.

Exercise 1.8 Suppose B is paracompact.
(1) If $p': E' \to X \times [0,1]$ is a locally trivial fiber bundle and $i_t: X \to X \times [0,1]$, $x \mapsto (x,t)$ the canonical injections, show that $i_0^* E' \cong i_1^* E'$. Hint: [145, Chapter 4].
(2) Let $p: E \to B$ be a locally trivial fiber bundle and $f_0, f_1: X \to B$ two continuous maps. If f_0 is homotopic to f_1 show that $f_0^* E \cong f_1^* E$.

Exercise 1.9 Let $p: E \to B$ be a locally trivial fiber bundle over a paracompact base B. Suppose we have the following commutative diagram of continuous maps:

$$\begin{array}{ccc} Y \times \{0\} & \xrightarrow{f} & E \\ j \downarrow & {}^{K}\nearrow & \downarrow p \\ Y \times [0,1] & \xrightarrow{H} & B. \end{array}$$

Then prove that there exists a continuous map $K: Y \times [0,1] \to E$ such that $p \circ K = H$ and $K_{|Y \times \{0\}} = f$. This property is called the *homotopy lifting property*. Hint: Use Exercise 1.8.

Exercise 1.10 Let $p: E \to B$ be a locally trivial fiber bundle of fiber F.
(1) Show that the induced map $\pi_i(E,F) \to \pi_i(B)$ is an isomorphism.
(2) Prove the existence of a long exact sequence

$$\cdots \longrightarrow \pi_j(F) \longrightarrow \pi_j(E) \xrightarrow{p_*} \pi_j(B) \xrightarrow{\delta} \pi_{j-1}(F) \longrightarrow \cdots$$

Hint: Use Exercise 1.9 and the long homotopy exact sequence of a pair.

Exercise 1.11 Let G be a Lie group. Consider a morphism between principal G-bundles:

$$\begin{array}{ccc} E' & \xrightarrow{f} & E \\ p' \downarrow & & \downarrow p \\ B & = & B \end{array}$$

Then show that the map f is a homeomorphism. Hint: [145, Theorem 3.2, page 43].

Exercise 1.12 Let $\Omega F \longrightarrow \Omega E \xrightarrow{\Omega p} \Omega B$ be the loop space fibration on a fibration $F \to E \to B$. If Ωp admits a section, show that ΩE has the homotopy type of $\Omega B \times \Omega F$ (simply as spaces). Hint: Use the section to construct a morphism of bundles $\Omega B \times \Omega F \to \Omega E$.

Exercise 1.13 Show that H-spaces are simple (see Example 2.46 for the definition). Hint: Proposition 1.62.

Exercise 1.14 If H is a maximal rank connected subgroup of a compact connected Lie group G, show that the homogeneous space G/H is simply connected. Hint: Consider the commutative diagram of fibrations,

$$\begin{array}{ccccc} T & \longrightarrow & H & \longrightarrow & H/T \\ \Big\| & & \Big\downarrow & & \Big\downarrow \\ T & \longrightarrow & G & \longrightarrow & G/T, \end{array}$$

and use the fact that H/T and G/T are simply connected.

2 Minimal models

A *minimal* model is a particularly tractable kind of commutative differential graded algebra (cdga) that can be associated to any nice cdga or to any nice space. The word "minimal" emphasizes that, at least in many cases of interest, the model is calculable. The amazing feature of minimal models of spaces is their ability to algebraically encode all rational homotopy information about a space. This is, of course, why minimal models are important.

This chapter includes definitions and the main properties of algebraic notions related to minimal models. In particular, the link between Algebra and Topology is explained as it is presented in the Preface. As an illustration, we list some correspondences between a space X (simply connected, for the sake of simplicity, and with finite Betti numbers) and its minimal model $\mathcal{M}_X = (\wedge V, d)$:

- the rationalized homotopy groups, $\pi_i(X) \otimes \mathbb{Q}$, are dual to the graded vector spaces V^i;
- the algebraic construction of $\mathcal{M}_X$ mirrors a well-known topological construction called the Postnikov decomposition of a space;
- the Ganea fibrations which appear in Lusternik–Schnirelmann category correspond to the filtration of $\wedge V$ by the wedge length $\wedge^{\geq n} V$.

Another crucial definition is given in this chapter: the notion of a formal space. Formal spaces are those spaces whose minimal models can be constructed directly from their cohomology algebras. In Chapter 4, we will prove that Kähler manifolds are always formal. In this way, we get an obstruction to the existence of a Kähler metric on a complex manifold M at the level of the rational homotopy type of M. As another geometric example, we will see in Chapter 3 that, among nilmanifolds, only the tori are formal.

Sections 2.1 and 2.2 are concerned with the basic definitions and properties of our algebraic tools: cdga's, models, minimal models and homotopy between morphisms of cdga's. This technical notion of homotopy cannot be avoided. First, we need it to prove the uniqueness up to isomorphism of a minimal model. Secondly, while a rational homotopy type is detected by an isomorphism class of a minimal model, it is not the same for maps. The rational homotopy type of a map corresponds to a homotopy class of algebraic morphisms (and not a class of algebraic morphisms up to isomorphism).

Section 2.3 includes the construction of a minimal model of a cdga and the important notion of relative minimal model. In Section 2.4, we provide the link between topological spaces and cdga's beginning with the $\mathbb{R}$-minimal model of a manifold and giving the analogue of the de Rham algebra for topological spaces, the $A_{PL}(-)$ construction. Subsection 2.4.3 contains a list of concrete examples of minimal models of spaces.

Section 2.5 is devoted to some particular relations between homotopy invariants and minimal models. This includes the tensoring with $\mathbb{Q}$ of the homotopy groups, the Postnikov tower, and the statement of the Dichotomy theorem. Further, we relate the material of Chapter 1 to that of the present chapter by showing how to calculate models of locally trivial bundles, principal bundles and homogeneous spaces in Subsection 2.5.2.

The notion of formality is defined and studied in Section 2.7. We give there the construction of the bigraded model and of the obstructions to formality. These results and techniques will be used several times in the following chapters, particularly in Chapters 3 and 4.

In this chapter, we do not give the full proofs of all results, but, rather, concentrate on the main ideas, concrete examples and applications. For details of these proofs, the reader is referred to the original paper of Sullivan [246], or to expository books on the subject such as [38], [87], [116], [170] and [248].

2.1 Commutative differential graded algebras

We begin with a series of definitions that form the foundation for everything that we will do concerning algebraic models.

Definition 2.1 *Let $\Bbbk$ be a field of characteristic zero. A graded $\Bbbk$-vector space is a family of vector spaces $A = \{A^n\}_{n \geq 0}$ indexed by the non-negative integers. The elements belonging to A^n are called homogeneous elements of degree n and we write $|x| = n$ if $x \in A^n$. We say that A is of finite type if each A^n is finite dimensional.*

Definition 2.2 *The suspension sV of the graded vector space V is the graded vector space defined by $(sV)^n = V^{n+1}$ for all n. More generally, for any $r \in \mathbb{Z}$, $s^r V = s(s^{r-1}V)$ satisfies $(s^r V)^n = V^{n+r}$.*

2 : Minimal models

Definition 2.3 *A* differential graded $\Bbbk$-algebra *(a $\Bbbk$-dga or a dga* for short*), (A,d), is a graded $\Bbbk$-vector space $A = \{A^n\}_{n\geq 0}$ together with a multiplication $A^p \otimes A^q \to A^{p+q}$ that is associative with a unit 1 in A^0, and a linear differential $d\colon A^n \to A^{n+1}$ that is a derivation:*

$$d(a \cdot b) = d(a) \cdot b + (-1)^{|a|} a \cdot d(b).$$

A morphism of dga's $f\colon (A,d) \to (B,d)$ consists of a family of linear maps $f\colon A^n \to B^n$ that satisfy $df = fd$ and $f(a \cdot b) = f(a) \cdot f(b)$.

Definition 2.4 *A* commutative graded algebra *(a cga for short) is a graded algebra A whose multiplication is commutative in the graded sense; that is, for homogeneous elements a and b,*

$$a \cdot b = (-1)^{|a| \cdot |b|}\, b \cdot a.$$

Definition 2.5 *A* commutative differential graded algebra *(a cdga for short) is a differential graded algebra (A,d) whose multiplication is commutative.*

For instance the de Rham algebra of differential forms on a manifold M, $A_{DR}(M)$, is an $\mathbb{R}$-cdga. Any smooth map between manifolds $f\colon M \to N$ induces at the level of differential forms a morphism of cdga's: $A_{DR}(f)\colon A_{DR}(N) \to A_{DR}(M)$. An example of a $\mathbb{Q}$-cdga is given by the rational cohomology of a space X equipped with the zero differential, $(H^*(X;\mathbb{Q}),0)$. A third example is given by the cochain algebra $\wedge \mathfrak{g}^*$ on a Lie algebra $\mathfrak{g}$ (Definition 1.5, Exercise 1.7).

Definition 2.6 *The commutative graded algebra A is called* free commutative *if A is the quotient of TV, the tensor algebra on the graded vector space V, by the bilateral ideal generated by the elements $a \otimes b - (-1)^{|a|\cdot|b|} b \otimes a$, where a and b are homogeneous elements of A.*

As an algebra, A is the tensor product of the symmetric algebra on V^{even} with the exterior algebra on V^{odd}:

$$A = \text{Symmetric}(V^{\text{even}}) \otimes \text{Exterior}(V^{\text{odd}}).$$

We denote the free commutative graded algebra on the graded vector space V by $\wedge V$. Note that this notation refers to a free commutative graded algebra and not necessarily to an exterior algebra alone. For instance, when G is a compact connected Lie group, its rational cohomology is a free commutative graded algebra, $H^*(G;\mathbb{Q}) \cong \wedge(x_1,\ldots,x_n)$ (Theorem 1.34) and the cohomology of its classifying space is also a free commutative graded algebra $H^*(BG;\mathbb{Q}) \cong \wedge(v_1,\ldots,v_n) = \mathbb{Q}[v_1,\ldots,v_n]$ (Theorem 1.81).

Note that, in the case of cohomology, or when we want to emphasize that the cga is polynomial, we may write $\mathbb{Q}[V]$ as well as $\wedge V$.

We usually write $\wedge V = \wedge(x_i)$, where x_i is a homogeneous basis of V. We denote by $\wedge^r V$ the vector space generated by the products $x_1 \cdots x_r$ with the x_i in V. We also write $\wedge^+ V = \oplus_{n \geq 1} \wedge^n V$ and $\wedge^{\geq q} V = \oplus_{n \geq q} \wedge^n V$. The elements of $\wedge^{\geq 2} V$ are referred to as *decomposable elements*.

Clearly the cohomology of a dga is a graded algebra, and the cohomology of a cdga is a commutative graded algebra. A morphism of dga's inducing an isomorphism in cohomology will be called a *quasi-isomorphism*. For instance if G is a compact connected Lie group and M is a compact left G-manifold, then by Theorem 1.28, the injection of left invariant forms, $\Omega_L(M) \hookrightarrow A_{DR}(M)$, is a quasi-isomorphism.

Among the cdga's, some have more interesting properties than others. This is the case for the so-called *Sullivan* cdga's and *minimal* cdga's.

Definition 2.7 A Sullivan cdga *is a cdga* $(\wedge V, d)$ *whose underlying algebra is free commutative, with* $V = \{V^n\}$, $n \geq 1$, *and such that V admits a basis* x_α *indexed by a well-ordered set such that* $d(x_\alpha) \in \wedge(x_\beta)_{\beta < \alpha}$.

Definition 2.8 A (Sullivan) minimal cdga *is a Sullivan cdga* $(\wedge V, d)$ *satisfying the additional property that* $d(V) \subset \wedge^{\geq 2} V$.

Note that, if $(\wedge V, d)$ is a cdga such that $V = V^{\geq 2}$, then it is a Sullivan cdga.

It is quite easy to construct minimal cdga's. If $(\wedge V, d)$ is a minimal cdga and $a \in (\wedge V)^n$ is a cocycle and a decomposable element, then we construct a new minimal cdga by introducing a new generator x in degree $n-1$ and putting $dx = a$. This gives the minimal cdga

$$(\wedge(V \oplus \Bbbk x), d).$$

By iterating this process, we can easily construct a lot of minimal cdga's. For instance, $(\wedge(x, y, z), d)$ with $|x| = |y| = 2$, $|z| = 3$, $dx = dy = 0$ and $dz = x^2 - y^2$ is automatically a Sullivan minimal model.

Remark 2.9 Let $(\wedge V, d)$ be a Sullivan cdga. Denote by d_0 the linear part of the differential d, $d_0 \colon V \to V$, and note that d_0 is a differential. If x is a generator with $d_0 x \neq 0$, then the ideal I generated by x and dx is acyclic and the quotient $(\wedge V/I, \bar{d})$ is also a Sullivan cdga of the form $(\wedge W, d)$. Therefore, in this way we can construct, by induction, a sequence of quasi-isomorphisms between a Sullivan cdga and a minimal one. Taking this into account, non-minimal Sullivan cdga's often are sufficient to answer whatever homotopy questions we have in mind.

The prototype of a Sullivan cdga is the tensor product of a minimal cdga with a tensor product of cdga's of the form $(\wedge(x_i, y_i), d)$ where $dx_i = 0$ and $dy_i = x_i$. In fact, each Sullivan cdga has that form.

Proposition 2.10 ([87, Theorem 14.9]) *Each Sullivan cdga is isomorphic to the tensor product of a minimal cdga and a tensor product of acyclic Sullivan cdga's of the form $(\wedge(x_i, y_i), d)$ with $dx_i = 0$ and $dy_i = x_i$.*

Definition 2.11 An *augmented cdga* is a cdga (A, d_A) together with a morphism $\varepsilon \colon (A, d_A) \to (\Bbbk, 0)$ that induces an isomorphism $H^0(\varepsilon; \Bbbk) \colon H^0(A, d_A) \to \Bbbk$. The morphism ε is called the augmentation.

For instance, the choice of a point x in a connected manifold M defines, by evaluation at x, an augmentation

$$\varepsilon_x \colon A_{DR}(M) \to (\mathbb{R}, 0).$$

When (A, d_A) is a cdga with an augmentation ε, then the kernel of ε, denoted $\bar{A}$, is a differential ideal. The *indecomposables* of an augmented cdga (A, d_A) is the quotient complex $Q(A) = \bar{A}/(\bar{A} \cdot \bar{A})$ with the induced differential $Q(d_A)$.

When $A^0 = \Bbbk$, then (A, d_A) admits a unique and canonical augmentation, and $Q(A) = A^+/(A^+ \cdot A^+)$. When $(A, d_A) = (\wedge V, d)$ is a Sullivan cdga, we always consider the evaluation ε defined by $\varepsilon(V) = 0$. We then have an isomorphism $(Q(\wedge V), Q(d)) \cong (V, d_0)$, where d_0 is the linear part of the differential (see Remark 2.9). Each morphism $f \colon (\wedge V, d) \to (\wedge W, d)$ between Sullivan cdga's induces a morphism of complexes

$$Q(f) \colon (Q(\wedge V), Q(d)) \to (Q(\wedge W), Q(d)).$$

We then have the very useful

Proposition 2.12 ([87, Proposition 14.13]) *Let $f \colon (\wedge V, d) \to (\wedge W, d)$ be a morphism between Sullivan cdga's. Then f is a quasi-isomorphism if and only if $Q(f)$ is a quasi-isomorphism.*

Corollary 2.13 *If $f \colon (\wedge V, d) \to (\wedge Z, d)$ is a quasi-isomorphism between minimal cdga's, then f is an isomorphism.*

Proof Since f is a quasi-isomorphism, $Q(f)$ is a quasi-isomorphism. Since the cdga's are minimal, $Q(d) = 0$. This means that the restriction of f to the indecomposable elements is an isomorphism. This implies directly that f is an isomorphism. □

2.2 Homotopy between morphisms of cdga's

We denote by $(\wedge(t,dt), d)$ the cdga generated by two elements t and dt in respective degrees 0 and 1 with differential $d(t) = dt$ and $d(dt) = 0$. This cdga is acyclic; that is, $H^0(\wedge(t,dt), d) = \Bbbk$ and $H^p(\wedge(t,dt), d) = 0$ for $p > 0$. Let $p_i\colon (\wedge(t,dt), d) \to (\Bbbk, 0)$, $i = 0, 1$, denote the quasi-isomorphisms defined by $p_i(t) = i$ and $p_i(dt) = 0$. The cdga $(\wedge(t,dt), d)$ has to be understood as an algebraic analogue of the algebra of de Rham forms on the interval $[0, 1]$. This analogy is the basis for the following definition.

Definition 2.14 *Two morphisms of cdga's, $f, g\colon (A, d) \to (B, d)$, are homotopic (i.e. $f \simeq g$), if there is a map of cdga's*

$$H\colon (A, d) \to (B, d) \otimes (\wedge(t, dt), d)$$

such that $p_0 \circ H = f$ and $p_1 \circ H = g$.

When $(A, d) = (\wedge V, d)$ is a Sullivan cdga, the homotopy relation $\simeq$ is an equivalence relation on the space of maps from $(\wedge V, d)$ to (B, d) ([87, Proposition 12.7]). The set of homotopy classes of maps between the cdga's $(\wedge V, d)$ and (B, d) is denoted by $[(\wedge V, d), (B, d)]$.

Lemma 2.15 (Lifting lemma) *Let $(\wedge V, d)$ be a Sullivan cdga, $f\colon (A, d) \to (B, d)$ be a quasi-isomorphism of cdga's, and $\varphi\colon (\wedge V, d) \to (B, d)$ be a morphism of cdga's. Then there is a morphism of cdga's $\psi\colon (\wedge V, d) \to (A, d)$ such that $f \circ \psi$ is homotopic to φ.*

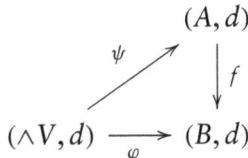

Lemma 2.16 *Let $(\wedge V, d)$ be a Sullivan cdga, $f, g\colon (\wedge V, d) \to (A, d)$ two morphisms of cdga's, and $h\colon (A, d) \to (B, d)$ be a quasi-isomorphism. If $hf \simeq hg$, then $f \simeq g$.*

Lemma 2.15 and Lemma 2.16 can be rephrased as the following global statement.

Theorem 2.17 (see [87, Proposition 12.9]) *Let $(\wedge V, d)$ be a Sullivan cdga and suppose $f\colon (A, d) \to (B, d)$ is a quasi-isomorphism. Then the composition with f induces a bijection on the set of homotopy classes*

of maps
$$[\wedge V, f]\colon [(\wedge V, d), (A, d)] \to [(\wedge V, d), (B, d)].$$

There is another way to define an homotopy relation between morphisms of cdga's. Let $(\wedge V, d)$ be a Sullivan cdga. Define the vector spaces $\bar{V}$ and $\hat{V}$ by $(\bar{V})^n = V^{n+1}$ and $(\hat{V})^n = V^n$. We then form the Sullivan cdga $(\wedge(V \oplus \bar{V} \oplus \hat{V}), D)$ with $D(v) = dv$, $D(\hat{v}) = 0$ and $D(\bar{v}) = \hat{v}$. The injection $i\colon (\wedge V, d) \to (\wedge(V \oplus \bar{V} \oplus \hat{V}), D)$ is then a quasi-isomorphism.

We define a derivation s of degree -1 of the algebra $\wedge(V \oplus \bar{V} \oplus \hat{V})$ by putting $s(v) = \bar{v}$, $s(\bar{v}) = s(\hat{v}) = 0$. The derivation $\theta = sD + Ds$ is a derivation of degree 0 such that for each element u there is some p with $\theta^p(u) = 0$. Therefore we can consider the automorphism e^θ of $\wedge(V \oplus \bar{V} \oplus \hat{V})$,

$$e^\theta = \mathrm{Id} + \theta + \theta^2/2 + \theta^3/6 + \cdots = \sum_n \frac{\theta^n}{n!}.$$

Observe that $e^\theta(z)$ is always a finite sum for any z. Since $\theta(\hat{v}) = 0$, the formula simplifies to

$$e^\theta(v) = v + \hat{v} + \sum_{n \geq 1} \frac{(sd)^n(v)}{n!}.$$

Definition 2.18 *Two morphisms of cdga's, $f, g\colon (\wedge V, d) \to (B, d)$, are* left homotopic, *if there is a map of cdga's $H\colon (\wedge(V \oplus \bar{V} \oplus \hat{V}), D) \to (B, d)$ such that $f = H \circ i$ and $g = H \circ e^\theta$.*

The main advantage of left homotopy is the facility of construction of homotopies. If $f\colon (\wedge V, d) \to (B, d)$ is a morphism of cdga's, and if $g\colon \bar{V} \to B$ is a linear map, then we obtain a morphism of cdga's $F\colon (\wedge(V \oplus \bar{V} \oplus \hat{V}), D) \to (B, d)$ defined by $F(v) = f(v)$, $F(\bar{v}) = g(\bar{v})$ and $F(\hat{v}) = d(g(\bar{v}))$. The map F is a left homotopy between f and $F \circ e^\theta$. Moreover the two definitions of homotopy are the same when the source is a Sullivan cdga.

Proposition 2.19 *Two morphisms of cdga's, $f, g\colon (\wedge V, d) \to (B, d)$, are left homotopic if and only if they are homotopic.*

Proof We first consider the projection $\pi\colon (\wedge(V \oplus \bar{V} \oplus \hat{V}), D) \to (\wedge V, d)$ defined by $\pi(v) = v$, $\pi(\bar{v}) = 0$, and $\pi(\hat{v}) = 0$. Then by Lemma 2.16, the morphisms e^θ and $i\colon (\wedge V) \to (\wedge(V \oplus \bar{V} \oplus \hat{V}), D)$ are homotopic because $\pi \circ e^\theta = \pi \circ i$ and π is a quasi-isomorphism. It follows that two left homotopic maps are homotopic.

Conversely suppose that $f, g\colon (\wedge V, d) \to (B, d)$ are homotopic maps with homotopy given by

$$H\colon (\wedge V, d) \to (B, d) \otimes (\wedge(t, dt), d).$$

2.2 Homotopy between morphisms of cdga's

We denote by F the composition of f with the canonical injection $(B, d) \to (B, d) \otimes (\wedge(t, dt), d)$,

$$F \colon (\wedge V, d) \to (B, d) \otimes (\wedge(t, dt), d).$$

We note that the ideal $I = B \otimes \wedge^+(t, dt)$ is an acyclic ideal and that $\text{Im}(H - F) \subset I$. Therefore, by Lemma 2.20, F and H are left homotopic. The same is true for $f = p_1 \circ F$ and $g = p_1 \circ H$. □

Lemma 2.20 *Let f and g be two morphisms from a Sullivan cdga $(\wedge V, d)$ into a cdga (B, d). Suppose I is an acyclic ideal in B such that for each $v \in V$, $f(v) - g(v) \in I$. Then f and g are left homotopic.*

Proof Suppose $V^1 = 0$ for the sake of simplicity. We construct a left homotopy between f and g by induction on the degree of a homogeneous basis of V. We suppose we have defined $H(v)$ and $H(\bar{v})$ for $v \in V^{\leq n}$ with $H(v) = f(v)$, $H(\bar{v}) \in I$ and $g(v) = H(e^\theta(v))$. Let $x \in V^{n+1}$. The element $g(x) - f(x) - H\left(\sum_{n \geq 1} \frac{(sd)^n(x)}{n!}\right)$ is a well defined cocycle in I. There is therefore an element $u \in I$ such that $d(u) = g(x) - f(x) - H(\sum_{n \geq 1} \frac{(sd)^n(x)}{n!})$. We then extend H linearly to $V^{n+1} \oplus (\bar{V})^n$ by putting $H(x) = f(x)$ and $H(\bar{x}) = u$. This extension has the required properties. □

We end this section with some particular properties of homotopy that will be used later in the book.

Proposition 2.21 (**Lifting of homotopies**) *Let $f \colon (\wedge V, d) \to (A, d)$, $h \colon (A, d) \to (B, d)$ and $g \colon (\wedge V) \to (B, d)$ be morphisms of cdga's with $h \circ f \simeq g$.*

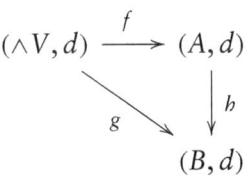

If h is surjective, then there exists a morphism $f' \colon (\wedge V, d) \to (A, d)$ such that $f \simeq f'$ and $h \circ f' = g$.

Proof Let $H \colon (\wedge V \otimes \wedge \bar{V} \otimes \wedge \hat{V}, D) \to (B, d)$ be the homotopy between hf and g. For each element $\bar{v}$ of a basis of $\bar{V}$ we choose an element $u_v \in A$ such that $h(u_v) = H(\bar{v})$. We then define the homotopy $K \colon (\wedge V \otimes \wedge \bar{V} \otimes \wedge \hat{V}, D) \to (A, d)$ by $K(v) = f(v)$ and $K(\bar{v}) = u_v$. The morphism K is a homotopy between f and another morphism f' such that $h \circ f' = g$. More precisely, f' is defined by $f'(v) = K \circ e^\theta(v)$. □

2: Minimal models

Proposition 2.22 (Extension of homotopies) *Let $(\wedge V \otimes \wedge W, D)$ be a Sullivan cdga with $D(V) \subset \wedge V$, let $f: (\wedge V \otimes \wedge W, D) \to (A, d)$ be a morphism of cdga's, and let $g: (\wedge V, D) \to (A, d)$ be a morphism of cdga's homotopic to the restriction of f to $(\wedge V, D)$. Then g extends to a morphism $\hat{g}: (\wedge V \otimes \wedge W, D) \to (A, d)$ homotopic to f.*

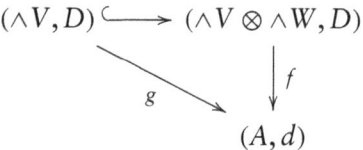

Proof Denote by H the homotopy between the restriction of f and g, $H(v) = f(v)$ and $H \circ e^{\theta} = g$. We extend H by putting $H(\bar{w}) = 0$ for $w \in W$. This gives a new morphism $\hat{g} = H \circ e^{\theta}: (\wedge V \otimes \wedge W, D) \to (A, d)$ that is homotopic to f and extends g. □

Example 2.23 Suppose that the injection $i: (\wedge V, d) \to (\wedge V \otimes \wedge W, D)$ has a homotopy section σ. Then the identity on $(\wedge V, d)$ is homotopic to $\sigma \circ i$ and can be lifted into a map of cdga's $\sigma': (\wedge V \otimes \wedge W, D) \to (\wedge V, d)$ homotopic to σ. We can therefore suppose that $\sigma(v) = v$ for $v \in V$.

2.3 Models in algebra

2.3.1 Minimal models of cdga's and morphisms

For the following, see [87, Section 12].

Theorem 2.24 (Existence and uniqueness of the minimal model) *Let (A, d) be a $\Bbbk$-cdga satisfying $H^0(A, d) = \Bbbk$. Then,*

1. *There is a minimal cdga $(\wedge V, d)$ and a quasi-isomorphism $\varphi: (\wedge V, d) \to (A, d)$.*
2. *The minimal cdga $(\wedge V, d)$ is unique in the following sense: If $(\wedge W, d)$ is a minimal cdga and $\psi: (\wedge W, d) \to (A, d)$ is also a quasi-isomorphism, then there is an isomorphism $f: (\wedge V, d) \to (\wedge W, d)$ such that $\psi \circ f$ is homotopic to φ.*
   The cdga $(\wedge V, d)$ is then called the minimal model *of (A, d). Furthermore,*
3. *If $H^1(A, d) = 0$ and $H^*(A, d)$ is of finite type, then V is also of finite type.*

More generally, a *Sullivan model* of a cdga (A, d) is a Sullivan cdga $(\wedge V, d)$ that is quasi-isomorphic to (A, d).

2.3 Models in algebra

In order to construct the minimal model of a cdga, we define inductively V^n and $\varphi_n = \varphi_{|V^{\leq n}}$ such that $\varphi_n \colon (\wedge V^{\leq n}, d) \to (A, d)$ induces isomorphisms in cohomology in degrees $\leq n$ and induces an injection on H^{n+1}. Suppose $H^1(A, d) = 0$ for simplicity, and suppose that φ_n has been defined. Write $V_1^{n+1} = \operatorname{Coker} H^{n+1}(\varphi_n)$ and $W = \operatorname{Ker} H^{n+2}(\varphi_n)$. Choose cocycles $a_i \in A^{n+1}$ such that their classes $x_i = [a_i]$ form a basis of V_1^{n+1}. Also choose cocycles b_j, for $j \in J$ in $(\wedge V^{\leq n})^{n+2}$ whose classes constitute a basis of W. There are then elements $c_j \in A^{n+1}$ such that $\varphi_n(b_j) = dc_j$. Consider now the vector space V_2^{n+1} generated by the variables y_j, $j \in J$ of degree $n+1$.

If we set $V^{n+1} = V_1^{n+1} \oplus V_2^{n+1}$, then we can extend φ_n to a morphism of cdga's, $\varphi_{n+1} \colon (\wedge V^{\leq n+1}, d) \to (A, d)$, by $dx_i = 0$, $dy_j = b_j$, $\varphi_{n+1}(x_i) = a_i$ and $\varphi_{n+1}(y_j) = c_j$. Then $H^{\leq n+1}(\varphi_{n+1})$ is an isomorphism and $H^{n+2}(\varphi_{n+1})$ is injective. The construction will be done again in the case of an equivariant model, see Subsection 3.3.1.

Example 2.25 Let (A, d) be the cdga $(\mathbb{Q}[x]/x^n, 0)$, with $|x| = 2r$. Following the general process described above, we first introduce a generator y in degree $2r$, with $dy = 0$ and $\varphi(y) = x$ and we have a map $\varphi_{2r} \colon (\wedge y, 0) \to (A, d)$. In cohomology, φ_{2r} induces the surjection

$$\mathbb{Q}[y] \to \mathbb{Q}[x]/x^n, \qquad y \mapsto x.$$

In order to make this map injective, we introduce a new generator z in degree $2rn - 1$, with differential $dz = y^n$, and we define $\varphi(z) = 0$. This gives a new map

$$\varphi_{2rn} \colon (\wedge(y, z), d) \to (A, d).$$

As a vector space, $\wedge(y, z) = \mathbb{Q}[y] \oplus \mathbb{Q}[y] \cdot z$. Since $d(y^k z) = y^{k+n}$, the ideal generated by y^n and z is acyclic, and $H^*(\wedge(y, z), d) = \mathbb{Q}[y]/y^n$. In particular, the morphism φ_{2rn} is a quasi-isomorphism, and the minimal model of (A, d) is the cdga $(\wedge(y, z), dy = 0, dz = y^n)$.

We deduce the existence of minimal models for maps as a corollary of the lifting lemma (Lemma 2.15).

Proposition 2.26 Let $f \colon (A, d) \to (B, d)$ be a morphism of cdga's, and let $\varphi \colon (\wedge V, d) \to (A, d)$ and $\psi \colon (\wedge W, d) \to (B, d)$ be Sullivan models. Then there exists a morphism of cdga's that is unique up to homotopy,

$g: (\wedge V, d) \to (\wedge W, d)$ *such that* $\psi \circ g \simeq f \circ \varphi$.

$$\begin{array}{ccc} (A,d) & \xrightarrow{f} & (B,d) \\ \varphi \uparrow & & \uparrow \psi \\ (\wedge V, d) & \xrightarrow{g} & (\wedge W, d) \end{array}$$

Proof The existence of g follows from Lemma 2.15 because ψ is a quasi-isomorphism. The uniqueness up to homotopy follows directly from Lemma 2.16. □

If $(\wedge V, d)$ and $(\wedge W, d)$ are minimal algebras, then the map $g: (\wedge V, d) \to (\wedge W, d)$ is called *a minimal model* of f. Note that this model is only defined up to homotopy. We will often abuse language and call it *the* minimal model of f.

2.3.2 Relative minimal models

Definition 2.27 *A relative minimal cdga is a morphism of cdga's of the form*

$$i: (A, d_A) \to (A \otimes \wedge V, d),$$

where $i(a) = a$, $d_{|A} = d_A$, $d(V) \subset (A^+ \otimes \wedge V) \oplus \wedge^{\geq 2} V$, *and such that* V *admits a basis* (x_α) *indexed by a well-ordered set such that* $d(x_\alpha) \in A \otimes (\wedge(x_\beta))_{\beta < \alpha}$.

When (A, d_A) is a Sullivan cdga, we have $(A, d_A) = (\wedge Z, d)$. Clearly, a relative minimal cdga $(A \otimes \wedge V, d) = (\wedge(Z \oplus V), d)$ is also a Sullivan cdga, but the cdga $(\wedge(Z \oplus V), d)$ is not necessarily a minimal cdga, even if $(\wedge Z, d)$ is a minimal cdga.

Relative Sullivan cdga's are in some sense the generic models for morphisms of cdga's. We make the role of relative minimal models precise in the following theorem (see [87, Section 14]).

Theorem 2.28 (Relative version of Theorem 2.24) *Let* $f: (A, d) \to (B, d)$ *be a morphism of cdga's. We then have a commutative diagram*

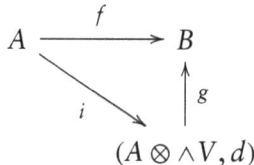

where i is a relative minimal cdga and g is a quasi-isomorphism. This property characterizes $(A \otimes \wedge V, d)$ up to isomorphism.

Under the conditions of Theorem 2.28, the map i is called *the relative minimal model* of f.

2.4 Models of spaces

2.4.1 Real and rational minimal models

To apply minimal models to spaces or manifolds, we need a link between topology and algebra that puts us in the framework of commutative differential graded algebras. Let's begin with manifolds and the field $\mathbb{R}$.

Definition 2.29 *Let M be a connected manifold. The $\mathbb{R}$-minimal model, $(\wedge V, d)$, of the de Rham algebra of forms $A_{DR}(M)$ is called the $\mathbb{R}$-minimal model (or real minimal model) of M.*

If $f: M \to N$ is a smooth map between connected manifolds, the minimal model of $A_{DR}(f)$ is called the $\mathbb{R}$-minimal model of f.

There is, however, a more general construction due to Sullivan that works over the rational numbers. To each space X, Sullivan associated a cdga of forms with rational coefficients, $A_{PL}(X)$ (which we shall discuss in Section 2.4.2), whose cohomology is isomorphic to the cohomology of X with rational coefficients:

$$H^*(A_{PL}(X)) \cong H^*(X; \mathbb{Q}).$$

Definition 2.30 *Let X be a path connected space. The $\mathbb{Q}$-minimal model, $(\wedge V, d)$, of the Sullivan cdga of polynomial forms $A_{PL}(X)$ is called the $\mathbb{Q}$-minimal model (or the rational minimal model) of X.*

If $f: X \to Y$ is a map between path connected spaces, the minimal model of $A_{PL}(f)$ is called the $\mathbb{Q}$-minimal model of f.

In the second part of the definition, we have used the fact that the construction of A_{PL} is natural. Each continuous map $f: X \to Y$ between path connected spaces induces a morphism of cdga's $A_{PL}(f): A_{PL}(Y) \to A_{PL}(X)$ and therefore has a unique (up to homotopy) minimal model $\mathcal{M}_f: (\wedge V, d) \to (\wedge W, d)$. The construction preserves homotopies as follows.

Proposition 2.31 *Two homotopic maps $f, g: X \to Y$ between path connected spaces induce two homotopic morphisms: $\mathcal{M}_f \simeq \mathcal{M}_g$. In other words, if $\mathcal{M}_X$ and $\mathcal{M}_Y$ denote the rational minimal models of X and Y, we* *have a well defined map*

$$[X, Y] \to [\mathcal{M}_Y, \mathcal{M}_X].$$

When M is a manifold, the $\mathbb{R}$-minimal model may be obtained directly from the de Rham algebra or from the $\mathbb{Q}$-minimal model by extension of coefficients. That is, if $(\wedge V, d)$ is the $\mathbb{Q}$-minimal model, then $(\wedge V \otimes_{\mathbb{Q}} \mathbb{R}, d)$ is the $\mathbb{R}$-minimal model.

In the future, except when explicitly indicated otherwise, "minimal model" will always refer to "rational minimal model." Also, each time we speak about the minimal model of a space, the space will be supposed path connected.

Definition 2.32 *A path connected space X is* nilpotent *if its fundamental group, $\pi_1(X)$, is a nilpotent group that acts nilpotently on the higher homotopy groups $\pi_n(X)$, $n \geq 2$ (where the action has been described after Definition 1.61).*

Simply connected spaces and connected Lie groups are prime examples of nilpotent spaces. For instance a simple space (see Definition 1.61) is nilpotent. Therefore homogeneous spaces are nilpotent (see Proposition 1.62). Other examples are given by Eilenberg–Mac Lane spaces $K(G, 1)$, where G is a nilpotent group.

Proposition 2.33 *Let X be a nilpotent space with finite Betti numbers and let $(\wedge V, d)$ be its minimal model. Then V is a graded vector space of finite type.*

Definition 2.34 *The spaces X and Y have* the same rational homotopy type *if there is a finite chain of maps $X \to Y_1 \leftarrow Y_2 \to \cdots \to Y$ such that the induced maps in rational cohomology are isomorphisms.*

In fact, for a nilpotent space X, the rational homotopy type is manifested as a space X_0, called the rationalization of X (see Subsection 2.6.1). The minimal model of X then characterizes the homotopy type X_0. The importance of Definition 2.32 is apparent from the following result.

Proposition 2.35 *Two nilpotent spaces with finite Betti numbers have the same rational homotopy type if and only if they admit isomorphic rational minimal models.*

Remark 2.36 To be clear, let us emphasize two points. First, while minimal models exist for all path connected spaces X of finite type, Proposition 2.35 does not hold in general for non-nilpotent spaces. Second, for a path connected X admitting a minimal model $(\wedge V, d)$, the Sullivan condition (see Definition 2.7) on the existence of a particular basis of V can look strange, but it is necessary. Consider, for instance, the case where X is a wedge of infinitely many copies of the sphere S^2. Since the dimension of $H^2(X;\mathbb{Q}) = \operatorname{Hom}_{\mathbb{Q}}(H_2(X;\mathbb{Q}),\mathbb{Q})$ is not countable, V^2 does not admit a basis indexed by the natural numbers. We *can* however find a basis indexed by a well-ordered set satisfying Sullivan's condition. This also shows the importance of the finite type hypothesis in Proposition 2.33 and in future applications.

For manifolds, we often consider minimal models coming directly from the de Rham algebra. With this in mind, we make the

Definition 2.37 *Two cdga's (or spaces) have the same* real homotopy type *if they have isomorphic $\mathbb{R}$-minimal models.*

Example 2.38 Consider the family of Sullivan minimal cdga's

$$A_a = (\wedge(e_2, x_4, y_7, z_9), D_a), \qquad a \in \mathbb{Q}^+,$$

where subscripts denote degrees, with the differential D_a given by

$$D_a(e) = 0, \ D_a(x) = 0, \ D_a(y) = x^2 + ae^4, \ D_a(z) = e^5.$$

We prove that A_a and $A_{a'}$ have the same rational homotopy type if and only if a/a' is a square in $\mathbb{Q}$.

Suppose first that A_a and $A_{a'}$ are quasi-isomorphic. There then exists by Corollary 2.13 an isomorphism $\varphi\colon A_a \to A_{a'}$. For degree reasons we have

$$\varphi(e) = \lambda e, \varphi(x) = \mu x + \alpha e^2, \varphi(y) = \beta y, \varphi(z) = \gamma z + \delta y e,$$

with $\alpha, \beta, \gamma, \delta, \mu, \lambda \in \mathbb{Q}$, $\lambda \neq 0$, $\mu \neq 0$, $\beta \neq 0$, $\gamma \neq 0$. From the equations $D_{a'}\varphi = \varphi D_a$, we deduce

$$\beta x^2 + \beta a' e^4 = \mu^2 x^2 + \alpha^2 e^4 + 2\mu\alpha x e^2 + a\lambda^4 e^4.$$

Therefore $\beta = \mu^2, \alpha = 0$ and $a/a' = (\lambda^2/\mu)^2$.

Conversely if a/a' is the square of τ, then we define an isomorphism $\varphi\colon A_a \to A_{a'}$ by putting $\varphi(e) = e$, $\varphi(x) = \tau x$, $\varphi(y) = \tau^2 y$, $\varphi(z) = z$.

However, since square roots exist in $\mathbb{R}$, we see that there is always an isomorphism *over* $\mathbb{R}$ between the cdga's A_a and $A_{a'}$. Hence, A_a and $A_{a'}$ always have the same real homotopy type. We will return to this example in Chapter 6 (Example 6.15).

2.4.2 Construction of $A_{PL}(X)$

Let's briefly outline the construction of $A_{PL}(X)$. In fact, the details of the construction are essential for the derivation of certain basic properties of

2: Minimal models

models, but in what follows (i.e. the construction of minimal models and relative minimal models), we will only use the existence of the cdga $A_{PL}(X)$ and properties of minimal models without making explicit the construction of $A_{PL}(X)$.

Recall first that the standard n-simplex Δ^n is the convex hull of the standard basis $e_0, e_1, \ldots, e_n$ in $\mathbb{R}^{n+1}$:

$$\Delta^n = \left\{ (t_0, t_1, \ldots, t_n) \in \mathbb{R}^{n+1} \,\bigg|\, \sum_{i=0}^n t_i = 1,\ t_j \geq 0,\ j = 0, \ldots, n \right\}.$$

Denote the set of singular n-simplices on X (i.e. continuous maps from $\Delta^n \to X$) by $S_n(X)$. The sets $S_n(X)$ constitute a simplicial set whose boundary operators ∂_i and degeneracy operators s_j are defined by:

$$\partial_i : S_n(X) \to S_{n-1}(X)$$
$$\partial_i(\sigma)(t_0, \ldots, t_{n-1}) = \sigma(t_0, \ldots, t_{i-1}, 0, t_i, \ldots, t_{n-1})$$
$$s_j : S_n(X) \to S_{n+1}(X)$$
$$s_j(\sigma)(t_0, \ldots, t_{n+1}) = \sigma(t_0, \ldots, t_j + t_{j+1}, \ldots, t_{n+1}).$$

The simplicial cdga A_{PL} is defined by:

$$(A_{PL})_n = \frac{\wedge(t_0, \ldots, t_n, dt_0, \ldots, dt_n)}{(\sum t_i - 1, \sum dt_i)},$$

where the elements t_i are in degree 0, the dt_i are in degree 1, and the differential d is defined by $d(t_i) = dt_i$. This is an acyclic cdga that can be viewed as an algebra of polynomial $\mathbb{Q}$-forms on Δ^n. The face and degeneracy operators of the simplicial cdga A_{PL} are the morphisms of cdga's defined by

$$\partial_i : (A_{PL})_n \to (A_{PL})_{n-1}, \partial_i(t_k) = \begin{cases} t_k, & k < i \\ 0, & k = i \\ t_{k-1}, & k > i \end{cases}$$

$$s_j : (A_{PL})_n \to (A_{PL})_{n+1}, s_j(t_k) = \begin{cases} t_k, & k < j \\ t_j + t_{j+1}, & k = j \\ t_{k+1}, & k > j. \end{cases}$$

The cdga $A_{PL}(X)$ is then defined as a set of simplicial maps

$$A_{PL}(X) = \mathrm{Hom}_{\mathrm{Simplicial}}(S_*(X), A_*).$$

More precisely, a q-form ω is a correspondence that assigns to each singular n-simplex σ an element $\omega_\sigma \in (A_{PL})_n^q$, such that $\omega_{\partial_i \sigma} = \partial_i \omega_\sigma$ and $\omega_{s_j \sigma} = s_j \omega_\sigma$.

2.4.3 Examples of minimal models of spaces

Example 2.39 (Lie groups) Let G be a compact connected Lie group. By Hopf's theorem (Theorem 1.34), $H^*(G; \mathbb{R})$ is an exterior algebra on a finite set of variables $x_1, \ldots, x_n$ in odd degrees. Therefore by choosing closed forms $\omega_i \in A_{DR}(G)$ representing the x_i, we define a quasi-isomorphism of cdga's

$$\varphi: (\wedge(x_i), 0) \to A_{DR}(G), \qquad \varphi(x_i) = \omega_i.$$

This shows that $(\wedge(x_i), 0)$ is the $\mathbb{R}$-minimal model of G. Notice that bi-invariant forms are a natural choice for the forms ω_i (see Theorem 1.30).

The computation of the rational minimal model is very similar. In fact, we have not used the fact that $A_{DR}(G)$ is the algebra of de Rham forms on G, but rather, only that $H(A_{DR}(G), d) \cong H^*(G; \mathbb{R})$.

We proceed in the same way for the rational minimal model and we keep the same notation for the generators x_i. We choose cocycles ω_i in $A_{PL}(G)$ representing the x_i, and we obtain a quasi-isomorphism of cdga's

$$\varphi: (\wedge(x_i), 0) \to A_{PL}(G), \qquad \varphi(x_i) = \omega_i.$$

This shows that $(\wedge(x_i), 0)$ is the minimal model of G. In particular, by Corollary 1.86, the minimal models of $U(n)$ and $Sp(n)$ are respectively

$$(\wedge(x_1, x_3, \ldots, x_{2n-1}), 0) \quad \text{and} \quad (\wedge(y_3, y_7, \ldots, y_{4n-1}), 0).$$

In the same way, by Corollary 1.90, the minimal model of $SO(n)$ is

$$\begin{cases} (\wedge(z_3, \ldots z_{2n-3}), 0), & \text{if } n \text{ is odd,} \\ (\wedge(z_3, \ldots, z_{2n-5}, z_{n-1}), 0), & \text{when } n \text{ is even}. \end{cases}$$

Here the subscripts indicate the degrees.

Example 2.40 (Stiefel manifolds) If X is a simply connected space whose cohomology $H^*(X; \mathbb{Q})$ is free commutative (i.e. $H^*(X; \mathbb{Q}) = \wedge(x_i)$), then the same procedure as above shows that $(\wedge(x_i), 0)$ is the minimal model of X. In particular, by Proposition 1.85, the minimal model of the complex Stiefel manifold $V_{n,k}(\mathbb{C})$ is

$$(\wedge(x_{2(n-k)+1}, \ldots, x_{2n-1}), 0),$$

and the minimal model of the quaternionic Stiefel manifold $V_{n,k}(\mathbb{H})$ is

$$(\wedge(y_{4(n-k)+3}, \ldots, y_{4n-1}), 0).$$

Here again the degrees of the elements are given by the subscripts.

Example 2.41 (The torus T^n) The minimal model of the torus T^n is the cdga $(\wedge(x_1, \cdots, x_n), 0)$ where all the x_i have degree 1.

Example 2.42 (Classifying spaces) Let G be a compact simply connected Lie group. Then by Theorem 1.81, the rational cohomology of the classifying space BG is a polynomial algebra on a finite number of generators v_i in even degree. Therefore, the choice of cocycles defines a quasi-isomorphism

$$(\wedge(v_i), 0) \to A_{PL}(BG).$$

This shows that $(\wedge(v_i), 0)$ is the minimal model for BG. We will prove this directly in Example 2.67, thus giving a third proof of Theorem 1.81.

Example 2.43 (The spheres S^n) The rational cohomology of the sphere S^n is an exterior algebra on one generator in degree n. Denote by ω a cocycle in degree n in $A_{PL}(S^n)$ representing the fundamental class. Then we get a morphism of cdga's

$$\varphi \colon (\wedge(x), 0) \to A_{PL}(S^n)$$

defined by $\varphi(x) = \omega$. When n is odd, $\wedge(x)$ is an exterior algebra on one generator and φ is clearly a quasi-isomorphism.

When n is even, $\wedge(x)$ is the polynomial algebra $\mathbb{Q}[x]$ and $H^*(\varphi) \colon \mathbb{Q}[x] \to \mathbb{Q}[x]/x^2$ is not an isomorphism. For degree reasons ω^2 is then a coboundary, $\omega^2 = d\alpha$. We then add a new generator y to $\wedge x$ of degree $2n - 1$ with $dy = x^2$, and define

$$\varphi \colon (\wedge(x, y), d) \to A_{PL}(S^n)$$

by putting $\varphi(x) = \omega$ and $\varphi(y) = \alpha$. Since $H^*(\wedge(x,y), d) \cong \mathbb{Q}[x]/x^2$, the new map φ is a quasi-isomorphism. In conclusion, the minimal model of S^n is

$$\begin{cases} (\wedge x, 0) \text{ with } |x| = n, \text{ when } n \text{ is odd} \\ (\wedge(x, y), d), dx = 0, dy = x^2, |x| = n, |y| = 2n - 1, \text{ when } n \text{ is even}. \end{cases}$$

Example 2.44 (The complex projective space $\mathbb{C}P(n)$) Since the rational cohomology, $H^*(\mathbb{C}P(n); \mathbb{Q})$, of the complex projective space of dimension n is isomorphic to $\mathbb{Q}[x]/x^{n+1}$, with $|x| = 2$, we can choose, in $A_{PL}(\mathbb{C}P(n))$, elements α and β of respective degrees 2 and $2n+1$ such that the class of α is x and $d\beta = \alpha^{n+1}$. We then construct a morphism of cdga's

$$\varphi \colon (\wedge(x, y), d) \to A_{PL}(\mathbb{C}P(n))$$

defined by $|x|=2$, $|y|=2n+1$, $dx=0$, $dy=x^{n+1}$, $\varphi(x)=\alpha$ and $\varphi(y)=\beta$. This morphism is clearly a quasi-isomorphism. Therefore $(\wedge(x,y),d)$ is the minimal model of $\mathbb{C}P(n)$.

Example 2.45 (Product of manifolds) If M and N are connected manifolds, then the multiplication map $A_{DR}(M)\otimes A_{DR}(N)\to A_{DR}(M\times N)$ is a quasi-isomorphism. Therefore the $\mathbb{R}$-minimal model of $M\times N$ is the tensor product of the $\mathbb{R}$-minimal models of M and N.

In the same way, if X and Y are path connected spaces, then there is a quasi-isomorphism between $A_{PL}(X\times Y)$ and $A_{PL}(X)\otimes A_{PL}(Y)$. Therefore the minimal model of $X\times Y$ is the tensor product of the minimal models of X and Y.

Example 2.46 (H-spaces) An H-space X is a space with a multiplication $\mu\colon X\times X\to X$ that is associative up to homotopy and admits a unit up to homotopy. Examples are given by Lie groups and loop spaces. The Hopf theorem (Theorem 1.34) can be generalized to H-spaces: The minimal model of an H-space X has the form $(\wedge V,0)$ and its rational cohomology is a free graded algebra.

Example 2.47 (Wedge of two spaces) Let X and Y be spaces with minimal models $(\wedge V,d)$ and $(\wedge W,d)$. We denote their wedge by $X\vee Y$ and the minimal model of the wedge by $(\wedge Z,d)$. Then the injections $i_1\colon X\hookrightarrow X\vee Y$ and $i_2\colon Y\hookrightarrow X\vee Y$ induce a map

$$\varphi\colon (\wedge Z,d)\to (\wedge V,d)\oplus_{\mathbb{Q}}(\wedge W,d),$$

where the cdga $(\wedge V,d)\oplus_{\mathbb{Q}}(\wedge W,d)$ is obtained from the direct sum $(\wedge V,d)\oplus(\wedge W,d)$ by identifying the units, and requiring the multiplication to obey $a\cdot a'=0$ if $a\in\wedge V$ and $a'\in\wedge W$. Since the injections i_j admit retractions, $H^*(\varphi)$ is a surjection. Note now that $H^*((\wedge V,d)\oplus_{\mathbb{Q}}(\wedge W,d))=H^*(X;\mathbb{Q})\oplus_{\mathbb{Q}}H^*(Y;\mathbb{Q})\cong H^*(X\vee Y;\mathbb{Q})$. Therefore φ is a quasi-isomorphism, and a minimal model for $X\vee Y$ is obtained by taking a minimal model of $(\wedge V,d)\oplus_{\mathbb{Q}}(\wedge W,d)$.

Example 2.48 (The diagonal map $\Delta\colon X\to X\times X$) Let X be a nilpotent space with finite Betti numbers and $\mathcal{M}_X$ its minimal model. We note that the composition of the diagonal map $\Delta\colon X\to X\times X$ with the projection onto one component is the identity. It follows that a minimal model for Δ is the multiplication

$$\mu\colon \mathcal{M}_X\otimes\mathcal{M}_X\to\mathcal{M}_X,\qquad \mu(a\otimes b)=a\cdot b.$$

2: Minimal models

A relative model for the diagonal Δ is thus provided by a relative model for the multiplication μ. We prove that this model has the form

$$(\wedge V, d) \otimes (\wedge V, d) \xrightarrow{\mu} (\wedge V, d)$$

with a map φ from $(\wedge V \otimes \wedge V \otimes \wedge(sV), D)$ to $(\wedge V, d)$, and the diagonal composition from $(\wedge V, d) \otimes (\wedge V, d)$ to $(\wedge V \otimes \wedge V \otimes \wedge(sV), D)$.

where $(sV)^n = V^{n+1}$, $\varphi(sx) = 0$ and $D(sx) = (x \otimes 1 \otimes 1) - (1 \otimes x \otimes 1) + \alpha_x$ where α_x is a decomposable element, with $x \in V$. We construct the differential D and the morphism φ by induction on the degree of the generators. Suppose that D and φ have been defined on $(sV)^{<n}$. Then the restriction of φ,

$$\varphi_n \colon (\wedge V^{\leq n} \otimes \wedge V^{\leq n} \otimes \wedge(sV)^{<n}, D) \to (\wedge V^{\leq n}, d)$$

is a quasi-isomorphism because $Q(\varphi_n)$ is a quasi-isomorphism. We now take $x \in V^{n+1}$. The element $y = (dx \otimes 1 \otimes 1) - (1 \otimes dx \otimes 1)$ is a cocycle that is sent to zero by φ_n. There is therefore a decomposable element $z \in (\wedge V^{\leq n} \otimes \wedge V^{\leq n} \otimes \wedge(sV)^{<n})^{n+1}$ such that $D(z) = y$. The element $\varphi_n(z)$ is then a cocycle, and since φ_n is a quasi-isomorphism, there is a cocycle $u \in \wedge V^{\leq n} \otimes \wedge V^{\leq n} \otimes \wedge(sV)^{<n}$ and an element $a \in \wedge V$ such that $\varphi_n(z) = \varphi_n(u) + da$. Since φ_n is surjective, we now choose an element a' with $\varphi_n(a') = a$ and we put $z' = z - u - D(a')$. We also set $D(sx) = (x \otimes 1 \otimes 1) - (1 \otimes x \otimes 1) - z'$. Since $\varphi_n(z') = 0$, the inductive step has been realized.

This computation illustrates the fact that minimal models for spaces and maps are generally built by induction, generator by generator.

2.4.4 Other models for spaces

The minimal model of a space is unique up to isomorphism. This is part of its power. In fact, however, certain other models are also very useful as we will see later. We begin by giving a

Definition 2.49 *A model for a space X is a cdga (A, d) quasi-isomorphic to the minimal model $(\wedge V, d)$ of X,*

$$(A, d) \xleftarrow{\simeq} (\wedge V, d) \xrightarrow{\simeq} A_{PL}(X).$$

A model for a continuous map $f \colon X \to Y$ is a morphism of cdga's $\varphi \colon (B, d) \to (A, d)$ such that there is a diagram, commutative up to *homotopy, where the vertical arrows are quasi-isomorphisms, of the form*

$$\begin{array}{ccc} A_{PL}(Y) & \xrightarrow{A_{PL}(f)} & A_{PL}(X) \\ \simeq \uparrow & & \uparrow \simeq \\ (\wedge W, d) & \longrightarrow & (\wedge V, d) \\ \simeq \downarrow & & \downarrow \simeq \\ (B, d) & \xrightarrow{\varphi} & (A, d) \end{array}$$

To give an example, let $(\wedge V, d)$ be the minimal model of an m-dimensional compact connected manifold M. Take a decomposition of $(\wedge V)^m$ as a direct sum $(\wedge V)^m = (\text{Ker } d)^m \oplus S$. Then the ideal $I = S \oplus (\wedge V)^{>m}$ is acyclic, the quotient map $(\wedge V, d) \to (\wedge V/I, \bar{d})$ is a quasi-isomorphism and $(\wedge V/I, \bar{d})$ is a model of M.

2.5 Minimal models and homotopy theory

2.5.1 Minimal models and homotopy groups

Let X be a simply connected (or more generally, a nilpotent) space and let $(\wedge V, d)$ be its minimal model. Then, in a natural way, from $(\wedge V, d)$ we can obtain the rational cohomology of X, the vector space $\pi_*(X) \otimes \mathbb{Q}$ and the homotopy Lie algebra $\pi_*(\Omega X) \otimes \mathbb{Q}$. We now explain this in detail.

First, by construction, $H^*(\wedge V, d) \cong H^*(X; \mathbb{Q})$. Recall as well that the homotopy groups of a nilpotent space with finite Betti numbers are all finitely generated. For $n \geq 2$, we have $\pi_n(X) = \mathbb{Z}^{\alpha_n} \oplus T_n$ where T_n is a finite group and α_n a finite integer. The integer α_n is called the rank of $\pi_n(X)$.

Theorem 2.50 (see [87, Theorem 15.11]) *Let X be a nilpotent space with finite Betti numbers, and let $(\wedge V, d)$ be its minimal model. Then, for $n \geq 2$, we have a natural isomorphism*

$$V^n \xrightarrow{\cong} \text{Hom}(\pi_n(X) \otimes \mathbb{Q}, \mathbb{Q}) = \text{Hom}(\pi_n(X), \mathbb{Q}).$$

In particular, for $n \geq 2$,

$$\dim V^n = \text{rank } \pi_n(X).$$

Here, "naturality" means that if we have a continuous map $f: X \to Y$ with minimal model $\varphi: (\wedge V, d) \to (\wedge W, d)$ then we have a commutative diagram

$$\begin{array}{ccc} \mathrm{Hom}(\pi_n(Y),\mathbb{Q}) & \xrightarrow{f^*} & \mathrm{Hom}(\pi_n(X),\mathbb{Q}) \\ \cong \uparrow & & \uparrow \cong \\ V^n & \xrightarrow{Q(\varphi)} & W^n \end{array}$$

For recall, $Q(\varphi)$ denotes the map induced by φ on the indecomposable elements $Q(\wedge V) \to Q(\wedge W)$. Modulo the isomorphisms $Q(\wedge V) \cong V$ and $Q(\wedge W) \cong W$, the map $Q(\varphi)$ can be described as the composition $V \xrightarrow{\varphi} \wedge^+ W \to \wedge^+ W / \wedge^{\geq 2} W \cong W$.

Example 2.51 (Lie groups) Let G be a compact connected Lie group. Then, by Example 2.39, the generators of the minimal model are only in odd degrees. Therefore, $\dim \pi_{\mathrm{even}}(G) \otimes \mathbb{Q} = 0$.

Example 2.52 (Nilpotent groups) Let G be a nilpotent group and let $G_{(r)}$ be the lower central series

$$G_{(1)} = G, \ G_{(2)} = [G,G], \ \text{and} \ G_{(r)} = [G, G_{(r-1)}], \text{ for } r \geq 2.$$

The first rational invariant associated to G is the rank of G defined by

$$\mathrm{rank}\, G = \sum_p G_{(p)}/G_{(p+1)}.$$

The rank of the fundamental group of a nilpotent space can be deduced from its minimal model as follows.

Proposition 2.53 *If X is a nilpotent space whose minimal model is $(\wedge V, d)$, then*

$$\mathrm{rank}\, \pi_1(X) = \dim V^1.$$

We will now describe the rational homotopy Lie algebra of a space X. Denote by ΩX the space of based loops on X. As we will see in Subsection 2.5.2, the homotopy groups of ΩX are the homotopy groups of X shifted in degree by one; that is, there is an isomorphism s,

$$s \colon \pi_n(\Omega X) \xrightarrow{\cong} \pi_{n+1}(X).$$

When X is simply connected, the graded vector space $\pi_*(\Omega X) \otimes \mathbb{Q}$ inherits a natural graded Lie algebra structure, called the rational homotopy Lie algebra of X. The Lie algebra structure arises as follows. Let $f \colon S^p \to \Omega X$ and $g \colon S^q \to \Omega X$ be continuous maps. We then consider the map

$$h \colon S^p \times S^q \to \Omega X$$

defined by
$$h(x,y) = f(x) \cdot g(y) \cdot f(x)^{-1} \cdot g(y)^{-1}.$$

It is clear that the restriction of h to the wedge $S^p \vee S^q$ is homotopically trivial, so h induces a quotient map
$$\bar{h}: S^{p+q} \cong \frac{S^p \times S^q}{S^p \vee S^q} \to \Omega X.$$

The class of $\bar{h}$ in $\pi_{p+q}(\Omega X)$ is denoted by $[f,g]$ and is called the Lie bracket of f and g.

Proposition 2.54 *This bracket defines a graded Lie algebra structure, denoted by $(L_X)_* = \pi_*(\Omega X) \otimes \mathbb{Q}$.*

A graded Lie algebra is a graded generalization of a Lie algebra (Definition 1.5) since a Lie algebra can always be viewed as a graded Lie algebra concentrated in degree 0.

Definition 2.55 *A graded Lie algebra L over $\mathbb{Q}$ is a graded vector space together with a linear map*
$$[-,-]: L_p \otimes L_q \to L_{p+q}$$
such that
$$\begin{cases} [a,b] = -(-1)^{|a| \cdot |b|}[b,a] \\ [a,[b,c]] = [[a,b],c] + (-1)^{|a| \cdot |b|}[b,[a,c]]. \end{cases}$$

Denote by $(\wedge V, d)$ the minimal model of the simply connected space X. The differential d decomposes into the sum
$$d = d_1 + d_2 + \cdots$$
where d_k is the component of the differential that increases the length by k: $d_k: V \to \wedge^{k+1} V$. It is easy to see that d_1 is a derivation with $d_1^2 = 0$, and we have the following fundamental result.

Theorem 2.56 ([87, Proposition 13.16]) *Let X be a simply connected space with finite Betti numbers which has minimal model $(\wedge V, d)$ and rational homotopy Lie algebra $L_* = (L_X)_* = \pi_*(\Omega X) \otimes \mathbb{Q}$. Then $V^{p+1} \cong \mathrm{Hom}(L_p, \mathbb{Q})$ and the bracket can be read off from d_1 by the formula*
$$\langle x, s[f,g] \rangle = (-1)^{|g|} \langle d_1 x, sf \wedge sg \rangle, \qquad x \in V, f, g \in L,$$
where s is the isomorphism $\pi_q(\Omega X) \otimes \mathbb{Q} \to \pi_{q+1}(X) \otimes \mathbb{Q}$.

Example 2.57 Let $X = S^{2n}$ be a $2n$-dimensional sphere. Since the minimal model of X is $(\wedge(x,y),d)$, $dx = 0$, $dy = x^2$, $|x| = 2n$ (Example 2.43), the Lie algebra L_X has two generators, a, and $[a,a]$, with a in degree $2n-1$. This graded Lie algebra is the free graded Lie algebra on one generator a of odd degree:

$$L_{S^{2n}} \cong \mathbb{L}(a).$$

Example 2.58 The rational homotopy Lie algebra of a wedge of spaces is the sum in the category of graded Lie algebras. This sum is called the *free product of Lie algebras*. We then write $L_{X \vee Y} = L_X \coprod L_Y$. An important property is that if $L_X \to L_Z$ and $L_Y \to L_W$ are surjective, then $L_{X \vee Y} \to L_{Z \vee W}$ is surjective. We shall use this in Corollary 3.4.

Example 2.59 The minimal model of the projective space $\mathbb{C}P(n)$ is given by $(\wedge(x,y),d)$ with $dx = 0$ and $dy = x^{n+1}$, $|x| = 2$, $|y| = 2n+1$ (Example 2.44). When $n \geq 2$, the quadratic part of the differential, d_1, is zero, and therefore the Lie algebra is abelian, which means that all the brackets are zero.

Denote by $\mathcal{Z}(L_X)$ the center of the Lie algebra L_X. We deduce from Theorem 2.56 the following proposition

Proposition 2.60 *Let $(\wedge V, d)$ be the minimal model of a simply connected space X with finite Betti numbers.*

1. *Suppose $V = W \oplus S$ with $d_1 V \subset \wedge^2 W$. Define the subspace E of $\pi_*(X) \otimes \mathbb{Q}$ by*

$$E = \{ x \in \pi_*(X) \otimes \mathbb{Q} \,|\, [w,x] = 0, \, \forall w \in W \}.$$

Then $s^{-1}E$ is contained in $\mathcal{Z}(L_X)$ and $\dim \mathcal{Z}(L_X) \geq \dim S$.
2. *If $W \subset V = \mathrm{Hom}\,(\pi_*(X), \mathbb{Q})$ denotes the sub-vector space generated by the linear forms that vanish on $s\mathcal{Z}(L_X)$, then $d_1 V \subset \wedge^2 W$.*

2.5.2 Relative minimal model of a fibration

In Chapter 1, we recalled the definition of locally trivial fiber bundles and principal G-bundles. We now introduce a more general concept. A (Hurewicz) *fibration* is a map $p \colon E \to B$ which has the property that, for each space X, and for each commutative diagram

$$\begin{array}{ccc} X \times \{0\} & \xrightarrow{g} & E \\ {\scriptstyle i_0}\downarrow & & \downarrow {\scriptstyle p} \\ X \times [0,1] & \xrightarrow{f} & B, \end{array}$$

there is a continuous map $h\colon X \times [0,1] \to E$ satisfying $p \circ h = f$ and $h \circ i_0 = g$.

What is important for geometers is that a locally trivial fiber bundle with a paracompact base (for instance a manifold, a CW complex or a compact space) is a Hurewicz fibration [240, Chapter 2, Section 7].

When B is path connected, all the fibers $p^{-1}(x)$, $x \in B$, have the same homotopy type. The *fiber*, F, of p is by definition any particular one of these $p^{-1}(x)$. We usually write the fibration in the form $F \to E \xrightarrow{p} B$. Perhaps the most important feature of a fibration is that the homotopy groups of the spaces in the fibration fit into a long exact sequence (see also Exercise 1.10),

$$\ldots \to \pi_n(F) \to \pi_n(E) \to \pi_n(B) \xrightarrow{\partial} \pi_{n-1}(F) \to \pi_{n-1}(E) \to \ldots .$$

An important example of a fibration is the path space fibration on a pointed space X, written

$$\Omega X \to PX \xrightarrow{p} X,$$

where PX is the set of continuous maps $c\colon [0,1] \to X$ with $c(0) = x_0$, the base point of X, and $p(c) = c(1)$. It is easy to see that PX is contractible, so the long exact homotopy sequence gives $\pi_n(X) \cong \pi_{n-1}(\Omega X)$ for all n. Notice that we have already used this isomorphism in the preceding section.

Just as nilpotent spaces have algebraic models that reflect their homotopical qualities and are (in principle) calculable, certain types of fibrations have analogous algebraic models.

Definition 2.61 A fibration $F \to E \to B$ is called quasi-nilpotent if B and F are path connected and the natural action of $\pi_1(B)$ on the homology groups of F is nilpotent. (This is always the case when the base B is simply connected.)

Remark 2.62 The reason we do not call the fibrations of Definition 2.61 *nilpotent* is because there is already a notion of *nilpotent fibration* that we will need only peripherally. A fibration $F \to E \to B$ is *nilpotent* if, in the standard fashion, $\pi_1(E)$ acts nilpotently on $\pi_k(F)$ for all $k \geq 1$. This implies that F is a nilpotent space. Furthermore, it can be shown that the nilpotency

2 : Minimal models

of the spaces E and B implies the nilpotency of the fibration. For more about nilpotent fibrations, see [138, page 67].

A quasi-nilpotent fibration not only provides the right hypothesis for Theorem 2.64 below, but also jibes with the notion of nilpotent space. To see this, for a space X, consider the fibration $\widetilde{X} \to X \to K(\pi_1(X), 1)$, where $\widetilde{X}$ is the universal cover of X. It is a standard result that X is a nilpotent space if and only if $\pi_1(X)$ is nilpotent and $\pi_1(X)$ acts nilpotently on $H_*(\widetilde{X})$, where $\widetilde{X}$ is the universal cover. Therefore, it is clear that X is nilpotent if and only if $\pi_1(X)$ is nilpotent and $\widetilde{X} \to X \to K(\pi_1(X), 1)$ is quasi-nilpotent.

Now consider the lifting homotopy property of the fibration $p \colon E \to B$ in the case $X = \Omega B \times F$ where ΩB is the space of based loops at the point b_0 and $F = p^{-1}(b_0)$. The commutativity of the diagram

$$\begin{array}{ccc} \Omega B \times F \times \{0\} & \xrightarrow{\iota'} & E \\ {\scriptstyle i_0}\downarrow & & \downarrow {\scriptstyle p} \\ \Omega B \times F \times [0,1] & \xrightarrow{f} & B \end{array}$$

where $\iota'(\omega, x, 0) = x$ and $f(\omega, x, t) = \omega(t)$, gives a map $h \colon \Omega B \times F \times [0,1] \to E$ with $ph = f$ and $h i_0 = \iota'$.

Definition 2.63 *The* holonomy representation *of the fibration is the restriction, v, of h to $\Omega B \times F \times \{1\}$. The image of v is contained in F:*

$$v \colon \Omega B \times F \to F.$$

The connecting map *of the fibration is the restriction of v,*

$$\delta = v|\colon \Omega B \times \{x_0\} \to F$$

where x_0 is a point in F and, in the fibration's long exact homotopy sequence

$$\cdots \to \pi_n(E) \to \pi_n(B) \xrightarrow{\partial} \pi_{n-1}(F) \to \cdots,$$

the morphism ∂ is the composition $\pi_n(B) \xrightarrow{\cong} \pi_{n-1}(\Omega B) \xrightarrow{\pi_*(\delta)} \pi_{n-1}(F)$.

Now let's come back to models. Let $F \to E \xrightarrow{p} B$ be a quasi-nilpotent fibration. We form the following commutative diagram

$$\begin{array}{ccc} A_{PL}(B) & \xrightarrow{p} & A_{PL}(E) & \longrightarrow & A_{PL}(F) \\ {\scriptstyle \varphi}\uparrow & & {\scriptstyle \psi}\uparrow & & \uparrow {\scriptstyle \bar{\psi}} \\ (\wedge V, d) & \xrightarrow{i} & (\wedge V \otimes \wedge W, d) & \xrightarrow{\rho} & (\wedge W, \bar{d}) \end{array}$$

Here the morphism $\varphi\colon (\wedge V, d) \to A_{PL}(B)$ is the minimal model of B, ψ is a quasi-isomorphism and $(\wedge V, d) \to (\wedge V \otimes \wedge W, d)$ is a relative minimal cdga. The cdga $(\wedge W, \bar d)$ is the quotient cdga $(\wedge V \otimes \wedge W, d)/(\wedge^+(V) \otimes \wedge W)$ and the map ρ is the quotient map. The map $\bar\psi$ is induced by the commutativity of the left-hand square of the diagram.

Theorem 2.64 ([87, Theorem 15.3]) *Suppose $F \to E \xrightarrow{p} B$ is a quasi-nilpotent fibration. If B and F have finite Betti numbers and $H^1(p)$ is injective, then the map $\bar\psi$ is a quasi-isomorphism, and the cdga $(\wedge W, \bar d)$ is the minimal model of the fiber F.*

Proposition 2.65 (Long exact homotopy sequence of a fibration) *With the above notation, we have a commutative diagram of long exact sequences*

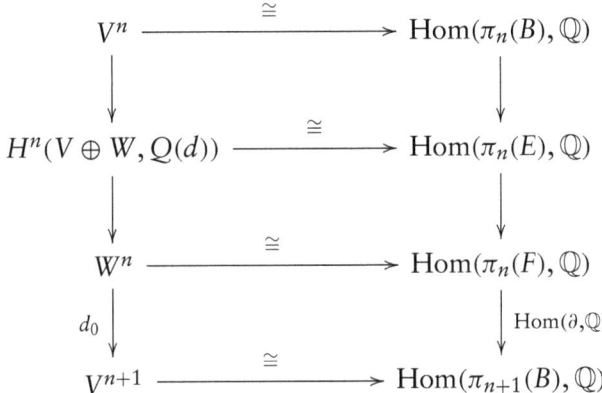

where $\partial\colon \pi_{n+1}(B) \to \pi_n(F)$ denotes the connecting map of the fibration and d_0 is the linear part of the differential d given by the composition

$$W \xrightarrow{d} \wedge^+(V \oplus W) \longrightarrow \wedge^+(V \oplus W)/(\wedge^+ W, \wedge^{\geq 2} V) \cong V.$$

Example 2.66 (The path space fibration) Let $(\wedge V, d)$ be the minimal model of a simply connected space X. Then a model for the path space fibration $\Omega X \to PX \to X$ is given by the relative model

$$(\wedge V, d) \to (\wedge V \otimes \wedge sV, d) \to (\wedge sV, \bar d),$$

where $|sv| = |v| - 1$. Since PX is a contractible space, the linear part d_0 of the differential d in $(\wedge V \otimes \wedge sV, d)$ gives an isomorphism $d_0\colon sV \to V$.

By Theorem 2.64 the cdga $(\wedge sV, \bar d)$ is the minimal model of the loop space ΩX. Since ΩX is an H-space, its cohomology is a free graded algebra. Therefore $\bar d = 0$ and $(\wedge sV, 0)$ is the minimal model of ΩX.

2 : Minimal models

Example 2.67 (The universal bundle) Let $G \to EG \to BG$ be the universal principal bundle associated to the Lie group G. The total space EG is contractible, so as we saw above, $\pi_{n-1}(G) \cong \pi_n(BG) \cong \pi_{n-1}(\Omega BG)$. Indeed, there is a mapping $\Omega BG \to G$ inducing these isomorphisms, so G has the homotopy type of ΩBG. Now, we have seen that the minimal model for G is an exterior algebra on odd degree generators with trivial differential. We shall denote this by $(\wedge sV, 0)$ using the notation of Example 2.66. By what we saw in Example 2.66, the minimal model for BG must have only even degree generators. But then the differential of this model must be trivial also since there are no elements in odd degrees. Thus, the minimal model for BG is $(\wedge V, 0)$ with an isomorphism $sV \cong V$. Note that we have obtained another proof of Theorem 1.81, for $H^*(\wedge V, 0)$ is a polynomial algebra on V. A relative minimal model for the universal principal bundle is given by

$$(\wedge V, 0) \to (\wedge V \otimes \wedge sV, d) \to (\wedge sV, 0),$$

with $d(sv) = v$.

Example 2.68 (The Hopf fibration) Let $S^3 \to S^7 \xrightarrow{H} S^4$ be the Hopf fibration coming from the action of $\mathbb{H}$ on $\mathbb{H}^2$ by multiplication in each component. A relative model for H has the form

$$(\wedge(x, y), d) \to (\wedge(x, y) \otimes \wedge V, d) \to (\wedge V, \bar{d}),$$

with $|x| = 4$, $|y| = 7$, $dx = 0$ and $dy = x^2$. Since $(\wedge V, \bar{d})$ is the minimal model for the fiber S^3, $\wedge V = \wedge u$, with $|u| = 3$, $\bar{d}u = 0$. Now since $H^4(\wedge(x, y, u), d) = 0$, we must have $du = x$, and we have a complete description of the relative minimal model.

Example 2.69 (Model of a sphere bundle) Let $S^{2n+1} \to E \xrightarrow{p} B$ be a sphere bundle. Denote by $(\wedge V, d)$ a model of B. Since the model of S^{2n+1} is $(\wedge u, 0)$, $|u| = 2n + 1$, a relative minimal model for the sphere bundle is given by

$$(\wedge V, d) \to (\wedge V \otimes \wedge u, d) \to (\wedge u, 0).$$

Here du is a cocycle $a \in (\wedge V)^{2n+2}$, and $\text{Ker } H^{2n+2}(p) = \mathbb{Q} \cdot [a]$. The class $[a]$ is the Euler class of the sphere bundle, and the model of the sphere bundle is completely determined by its Euler class.

Let $F \to E \xrightarrow{p} B$ be a quasi-nilpotent fibration, $f \colon X \to B$ a continuous map, and

$$F \to E' \xrightarrow{p'} X$$

the pullback fibration of p along f, defined as in Example 1.56 for locally trivial bundles. We denote by $(\wedge V, d) \to (\wedge V \otimes \wedge W, d) \to (\wedge W, \bar{d})$ a relative minimal model for p and by $\varphi \colon (\wedge V, d) \to (\wedge Z, d)$ a minimal model for f. Note that the naturality of the action of the fundamental group of the base on the homology of the fiber implies that the pullback of a quasi-nilpotent fibration is quasi-nilpotent. Then we have the following.

Theorem 2.70 *With the above notation, the relative minimal cdga*

$$(\wedge Z, d) \to (\wedge Z \otimes \wedge W, D) \overset{\text{def}}{=} (\wedge Z, d) \otimes_{\wedge V} (\wedge V \otimes \wedge W, D) \to (\wedge W, \bar{D})$$

is the relative minimal model of the fibration p'. The differential D is defined by $D(w) = (\varphi \otimes 1)d(w)$, where $\varphi \otimes 1$ is the natural multiplicative map $\varphi \otimes 1 \colon \wedge V \otimes \wedge W \to \wedge Z \otimes \wedge W$.

We deduce the following result.

Theorem 2.71 *Let H be a closed connected subgroup of a compact connected Lie group G. We denote by $\iota \colon H \to G$ the canonical inclusion and by $B\iota \colon BH \to BG$ the induced map. Let $H^*(BG; \mathbb{Q}) = \wedge V$ and $H^*(BH; \mathbb{Q}) = \wedge W$ the respective cohomology algebras of BG and BH. We denote by sV a copy of the vector space V shifted by one degree, $|sv| = |v| - 1$ if $v \in V$ and define a differential d on $\wedge W \otimes \wedge (sV)$ by $dw = 0$ if $w \in W$ and $d(sv) = H^*(B\iota)(v)$ if $sv \in sV$. Then the cdga $(\wedge W \otimes \wedge (sV), d)$ is a Sullivan model for the homogeneous space G/H. In particular $H^*(G/H; \mathbb{Q}) = H(\wedge W \otimes \wedge (sV), d)$.*

These are the relative minimal model versions of the Cartan–Weil models of the homogeneous spaces G/H.

Proof From Example 2.67, the relative minimal model of the fibration $G \to EG \xrightarrow{p} BG$ is

$$(\wedge V, 0) \to (\wedge V \otimes \wedge sV, d) \to (\wedge sV, 0), \qquad d(sv) = v,$$

and the model of $B\iota$ is $H^*(B\iota; \mathbb{Q}) \colon (\wedge V, 0) \to (\wedge W, 0)$. The pullback fibration of p along $B\iota$ has the form $G \to E \xrightarrow{q} BH$. Since EG is contractible, q has the homotopy type of the inclusion of the fiber of $B\iota$. Now by Proposition 1.80, this pullback fibration is the fibration $G \to G/H \to BH$. Therefore the relative minimal model of the fibration $G \to G/H \to BH$ is given by

$$(\wedge W, 0) \to (\wedge W, 0) \otimes_{\wedge V} (\wedge V \otimes \wedge sV, d) \to (\wedge sV, 0).$$

Now $(\wedge W, 0) \otimes_{\wedge V} (\wedge V \otimes \wedge sV, d)$ is a Sullivan model for G/H. We can describe this model: $(\wedge W, 0) \otimes_{\wedge V} (\wedge V \otimes \wedge sV, d) = (\wedge W \otimes \wedge sV, d)$ with $dw = 0$ and $d(sv) = H^*(B\iota)(v)$. □

Example 2.72 (Model of a principal G-bundle) A principal G-bundle $X \to B$ is the pullback of the universal bundle $EG \to BG$ along a continuous map $f : B \to BG$, so its model has the form

$$(\wedge W, d) \to (\wedge W \otimes \wedge sV, d) \to (\wedge sV, 0),$$

where $(\wedge W, d)$ is the minimal model of B and $(\wedge sV, 0)$ is the minimal model of G. Moreover, by Theorem 2.70, we can understand the differential $d(sv)$ in terms of the classifying map f. Namely, $d(sv)$ is a cocycle in $(\wedge W, d)$ representing $f^*(v)$, where $v \in \wedge V \cong H^*(BG; \mathbb{Q})$.

There are thus two ways for computing a Sullivan model for the fiber of a locally trivial bundle, or more generally the fiber of a fibration, $p : E \to B$. The first way constructs a relative minimal model of p, $(\wedge V, d) \to (\wedge V \otimes \wedge W, d) \to (\wedge W, \bar{d})$, and a minimal model for the fiber F is then $(\wedge W, \bar{d})$.

The second way constructs the relative minimal model of the path fibration on B, $(\wedge V, d) \to (\wedge V \otimes \wedge sV, d) \to (\wedge sV, 0)$, a minimal model of p, $(\wedge V, d) \to (\wedge Z, d)$, and forms the tensor product

$$(\wedge Z \otimes \wedge sV, D) \stackrel{\text{def}}{=} (\wedge Z, d) \otimes_{\wedge V} (\wedge V \otimes \wedge sV, d).$$

2.5.3 The dichotomy theorem

Rational homotopy theory provides a surprising way to "classify" simply connected manifolds which has often proved useful in geometry (e.g. see Chapter 6 and Section 5.7). The criterion for classification applies equally to all nilpotent spaces, so we will formulate it in this more general framework. We first divide the family of nilpotent spaces with finite dimensional rational cohomology (i.e. $\sum_q \dim H^q(X; \mathbb{Q}) < \infty$) into two classes as follows.

Definition 2.73 Let X be a nilpotent space with finite dimensional rational cohomology. Then X is called a rationally elliptic space if $\sum_{p \geq 2} \dim \pi_p(X) \otimes \mathbb{Q} < \infty$. Otherwise X is called a rationally hyperbolic space.

The properties of the two classes are very different, so let's begin to catalogue these differences. Let n be the maximal integer such that $H^n(X; \mathbb{Q}) \neq 0$ and say that n is the *dimension* of X. For a closed manifold, this is, of course, the usual dimension of the manifold.

2.5 Minimal models and homotopy theory

Theorem 2.74 (The dichotomy theorem: Hyperbolic case [87]) *Let X be a rationally hyperbolic space of dimension n. Then:*

1. *The sequence $\sum_{p \leq m} \dim \pi_p(X) \otimes \mathbb{Q}$ has exponential growth: There are constants $A > 1$ and $C > 0$ such that, for m large enough,*

$$\sum_{p \leq m} \dim \pi_p(X) \otimes \mathbb{Q} \geq C \cdot A^m.$$

   In particular, for each integer q, there is an odd integer m such that $\dim \pi_m(X) \otimes \mathbb{Q} \geq q$.
2. *The sequence of Betti numbers of the loop space ΩX on X has exponential growth. This means that there is an integer $A > 1$ such that $\sum_{i=0}^{k} b_i(\Omega X) \geq A^k$ for k large enough.*
3. *There is no large gap in the sequence of homotopy groups. More specifically, for each integer q there is some integer p in the interval $(q, q+n)$ such that $\dim \pi_p(X) \otimes \mathbb{Q} \neq 0$.*
4. *There are infinitely many nonzero brackets in the rational homotopy Lie algebra L_X. More specifically, there is an integer N such that, for each α in L_X with $|\alpha| > N$, there is another element $\beta \in L_X$ such that $[\alpha, [\alpha, \ldots, [\alpha, \beta] \ldots] \neq 0$.*

For the elliptic case, denote the *homotopy* Euler characteristic of the space X by

$$\chi_\pi(X) = \sum_{q \geq 0} (\text{rank } \pi_{2q}(X) \otimes \mathbb{Q} - \text{rank } \pi_{2q+1}(X) \otimes \mathbb{Q}).$$

Theorem 2.75 (The dichotomy theorem: Elliptic case [129],[87]) *Let X be a rationally elliptic space of dimension n. Then:*

1. *The homotopy groups $\pi_q(X)$ are finite groups for $q \geq 2n$.*
2. *$\chi_\pi(X) \leq 0$ (so $\dim \pi_{even}(X) \otimes \mathbb{Q} \leq \dim \pi_{odd}(X) \otimes \mathbb{Q}$), and $\chi(X) \geq 0$.*
3. *The rational cohomology of X satisfies Poincaré duality.*
4. *$\dim H^*(X; \mathbb{Q}) \leq 2^n$.*
5. *The three following properties are equivalent:*
   - *$\chi(X) > 0$.*
   - *The rational cohomology is concentrated in even degrees; $H^q(X; \mathbb{Q}) = 0$ if q is odd.*
   - *$\chi_\pi(X) = 0$.*

2 : Minimal models

6. *The dimension of X depends only on the ranks of the rational homotopy groups.*

$$n = \left(\sum_q (-1)^{q+1} q \cdot \dim \pi_q(X) \otimes \mathbb{Q}\right) + \dim \pi_{\text{even}}(X) \otimes \mathbb{Q}$$
$$= \sum_q (2q-1) \dim \pi_{2q-1}(X) \otimes \mathbb{Q} - \sum_q (2q-1) \dim \pi_{2q}(X) \otimes \mathbb{Q}.$$

7. $\sum_q 2q \dim \pi_{2q}(X) \otimes \mathbb{Q} \le n$. *In particular, there is no even homotopy in degrees greater than the dimension.*
8. $\sum_q (2q-1) \dim \pi_{2q-1}(X) \otimes \mathbb{Q} \le 2n-1$; *in particular, there can only be at most one nontrivial* $\pi_{2q-1}(X) \otimes \mathbb{Q}$ *with* $n \le 2q-1 \le 2n-1$ *and, necessarily,* $\dim \pi_{2q-1}(X) \otimes \mathbb{Q} = 1$.
9. $\dim \pi_*(X) \otimes \mathbb{Q} \le n$.
10. *The sequence of Betti numbers of ΩX has polynomial growth. This means that there is an integer m such that* $\sum_{i=0}^{k} b_i(\Omega X) \le k^m$ *for all* k.

A useful criterion for rational ellipticity is given in Exercise 2.4. We can also characterize a rationally elliptic space in terms of its minimal model. A finite dimensional space X, with minimal model $(\wedge V, d)$ is rationally elliptic if and only if V is finite dimensional. For instance, tori T^n, spheres S^n, complex projective spaces $\mathbb{C}P(n)$, Lie groups G and homogeneous spaces G/H are rationally elliptic spaces. On the other hand, a connected sum X (see Example 3.6) of q copies of $S^3 \times S^3$ is rationally hyperbolic for $q \ge 2$, because $\chi(X) < 0$ and this violates Theorem 2.75 (2).

Of course, one distinct advantage of having an algebraic model at our disposal is that we can put algebraic constraints on the model and see the geometric reflection of the algebra. One important instance of this is the following.

Definition 2.76 *A Sullivan cdga is said to be a* pure model *if it has the form* $(\wedge Q \otimes \wedge P, d)$ *with Q concentrated in even degrees, P in odd degrees and*

$$d(Q) = 0, \qquad d(P) \subset \wedge Q.$$

By Proposition 2.10, the minimal model of a pure Sullivan cdga is a pure model. Hence, the minimal model of a homogeneous space is a pure model by Theorem 2.71. The elliptic spaces having pure models constitute a special family of elliptic spaces. Part (5) of Theorem 2.75 is proved for that family of spaces in Theorem B.18. There is also a general structure theorem for pure models.

Theorem 2.77 Let $(\wedge Q \otimes \wedge P, d)$ be a pure cdga and suppose that $P_0 \subset P$ has $d(P_0) = 0$. Then

$$(\wedge Q \otimes \wedge P, d) \cong (\wedge Q \otimes \wedge P_1, d) \otimes (\wedge P_0, 0),$$

where $P_0 \oplus P_1 = P$.

The idea of the proof is simple. Since $d(P) \subset \wedge Q$ and $d(Q) = 0$, it is clear that the elements of P_0 never appear in any differentials. Therefore, $(\wedge P_0, 0)$ splits off from $(\wedge Q \otimes \wedge P, d)$. A geometric consequence is the following.

Corollary 2.78 Let X be a space whose minimal model is pure and suppose that $\alpha \in H_{2k+1}(X; \mathbb{Q})$ is in the image of the rational Hurewicz map $h \colon \pi_{2k+1}(X) \otimes \mathbb{Q} \to H_{2k+1}(X; \mathbb{Q})$. Then $X \simeq_{\mathbb{Q}} Y \times S^{2k+1}$, where Y also has a pure model.

Proof The hypotheses imply that α corresponds to an odd degree generator of X's minimal model which is also a cocycle. By Theorem 2.77, the model splits off $(\wedge \alpha, 0)$ and this corresponds to X splitting off a sphere factor S^{2k+1}. □

Pure models provide a very interesting family of elliptic spaces:

Theorem 2.79 ([87, Proposition 32.16]) *A rationally elliptic space X with $\chi(X) > 0$ admits a pure model. Moreover we have the following formula connecting the Betti numbers to the rank of the homotopy groups, $r_q = \operatorname{rank} \pi_q(X)$:*

$$\sum_q \dim H_q(X; \mathbb{Q}) t^q = \prod_i (1 - t^{2i})^{r_{2i-1} - r_{2i}}.$$

2.5.4 Minimal models and some homotopy constructions

Let X be a nilpotent space with minimal model $\mathcal{M}_X = (\wedge V, d)$. We already know how to derive the rational cohomology and the rational homotopy Lie algebra of X from $(\wedge V, d)$. In this section we explain the relations between the minimal model of X and its Postnikov tower. We then explain how to derive from the minimal model a good approximation for the Lusternik–Schnirelmann category of X. We also show how to obtain the minimal model of the homotopy cofiber of a map f from the minimal model of f.

First of all, let G be an abelian group, and let $n \geq 2$. The Eilenberg–Mac Lane space $K(G, n)$ is a CW complex such that $\pi_n(K(G, n)) = G$, and such that the other homotopy groups are zero. Denote by $(\wedge W, d)$ the minimal model of $K(G, n)$. Since $W^* \cong \operatorname{Hom}(\pi_*(K(G, n)), \mathbb{Q})$, $d = 0$, $W = W^n$ and $\dim W = \operatorname{rank} G$.

2 : Minimal models

To a simply connected space X, we can associate a sequence of fibrations

$$K(\pi_n(X),n) \to X_n \xrightarrow{p_n} X_{n-1},$$

and maps $f_n \colon X \to X_n$ with $p_n \circ f_n = f_{n-1}$,

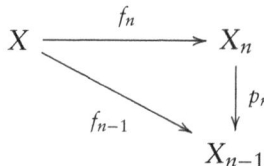

satisfying the following properties:

- $\pi_q(X_n) = 0$, $q > n$;
- $\pi_q(f_n)$ is an isomorphism for $q \leq n$;
- $p_n \colon X_n \to X_{n-1}$ is a principal fibration obtained as a pullback of the path fibration $K(\pi_n(X),n) = \Omega(K(\pi_n(X),n+1)) \to PK(\pi_n(X),n+1) \to K(\pi_n(X),n+1)$ along a map $k_n \colon X_{n-1} \to K(\pi_n(X),n+1)$.

The sequence of fibrations is called *the Postnikov tower* of X and the maps k_n are called *the associated k-invariants* ([240]).

Let $(\wedge V, d)$ be the minimal model of X. We then have the following properties.

- The minimal model of X_n is given by the sub-cdga $(\wedge V^{\leq n}, d)$.
- The minimal models of f_n and p_n are, respectively, given by the injections

$$(\wedge V^{\leq n}, d) \hookrightarrow (\wedge V, d), \qquad (\wedge V^{\leq n-1}, d) \hookrightarrow (\wedge V^{\leq n}, d).$$

- The minimal model of the nth k-invariant $k_n \colon X_{n-1} \to K(\pi_n(X), n+1)$ is given by the map $\tilde{k}_n \colon (\wedge (s^{-1}V^n), 0) \to (\wedge V^{\leq n-1}, d)$, where $(s^{-1}V^n)^{n+1} = V^n$, $(s^{-1}V^n)^q = 0$, $q \neq n+1$, $\tilde{k}_n(s^{-1}v) = dv$.

The *Lusternik–Schnirelmann category* of a space X, cat X, is the least integer n such that X can be covered by $n+1$ open sets, each contractible *in* X. For instance the category of a point is zero, and the category of a sphere is one. The properties of the Lusternik–Schnirelmann category, its description in terms of minimal models and its role in algebraic and differential topology are described in [66] and [87]. Here we only recall the main points.

A lower bound for cat X is given by the rational cup length

$$\mathrm{cup}_0(X) = \max\{n \mid \exists \alpha_1, \ldots, \alpha_n \in H^+(X; \mathbb{Q}) \text{ such that } \alpha_1 \cdots \alpha_n \neq 0\},$$

and an upper bound is given by the dimension of X.

$$\mathrm{cup}_0(X) \leq \mathrm{cat}\, X \leq \dim X.$$

2.5 Minimal models and homotopy theory

Suppose now that X is simply connected. By a result of Toomer ([253]), the category of the rationalization X_0 (see Subsection 2.6.1) is less than or equal to the category of X,

$$\operatorname{cat} X_0 \leq \operatorname{cat} X.$$

The integer $\operatorname{cat} X_0$ is called *the rational category* of X, and is denoted $\operatorname{cat}_0(X)$. Because the rational homotopy type of X is encoded in its minimal model, it is no surprise that this invariant can be calculated from the minimal model $(\wedge V, d)$ of X.

Theorem 2.80 ([86]) *The integer* $\operatorname{cat}_0(X)$ *is the least integer m such that the minimal model of the projection* $q_m \colon (\wedge V, d) \to (\wedge V/(\wedge^{>m}V), \bar{d})$ *admits a homotopy retraction.*

Denote by $\rho_m \colon (\wedge V, d) \to (\wedge W_m, d)$ the minimal model of q_m.

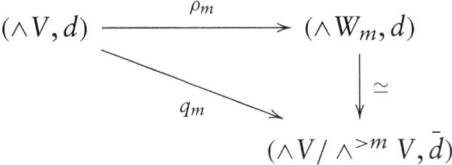

The rational category of X is the least integer m such that there is a morphism $r \colon (\wedge W_m, d) \to (\wedge V, d)$ with $r \circ \rho_m = id_{\wedge V}$. One of the main results concerning rational category is the so-called *mapping theorem*.

Theorem 2.81 (The mapping theorem [87, Theorem 28.6]) *Let $f \colon X \to Y$ be a continuous map between nilpotent spaces. If $\pi_*(f) \otimes \mathbb{Q}$ is injective, then* $\operatorname{cat}_0(X) \leq \operatorname{cat}_0(Y)$.

The rational homotopy Lie algebra of a space of finite rational category satisfies properties that are similar to the properties of finite dimensional spaces.

Theorem 2.82 ([87, Theorem 33.6 and 36.8]) *Let X be a simply connected space with finite Betti numbers and finite rational category. When $\dim \pi_*(X) \otimes \mathbb{Q} < \infty$, then $\dim H^*(X; \mathbb{Q}) < \infty$ and X is a rationally elliptic space. When $\dim \pi_*(X) \otimes \mathbb{Q} = \infty$, then*

1. *There is $B > 1$ such that $\sum_{i=1}^{k} \operatorname{rank} \pi_i(X) > B^k$ for k enough large.*
2. $\dim \pi_{\mathrm{odd}}(X) \otimes \mathbb{Q} = \infty.$
3. *There is an integer N such that for all $\alpha \in \pi_q(\Omega X) \otimes \mathbb{Q}$ with $q > N$, there is $\beta \in \pi_*(X) \otimes \mathbb{Q}$ such that $\operatorname{ad}(\alpha)^n(\beta) \neq 0$ for all n.*

Minimal models are also a very good tool for describing the rational homotopy type of the homotopy cofiber of a map. If $f\colon X \to Y$ is a continuous map, its homotopy cofiber C_f is the space $Y \cup CX/\sim$, where $CX = X \times [0,1]/X \times \{1\}$ and where $\sim$ identifies $(x,0) \in CX$ to $f(x) \in Y$. Recall that if f denotes the inclusion of a sub-CW complex or the inclusion of a submanifold, then C_f is homotopy equivalent to the usual quotient Y/X.

Denote by $\mathcal{M}_f$ the minimal model of f and suppose the following diagram is commutative up to homotopy, where the vertical arrows are quasi-isomorphisms.

$$\begin{array}{ccc} \mathcal{M}_Y & \xrightarrow{\mathcal{M}_f} & \mathcal{M}_X \\ {\scriptstyle\simeq}\downarrow & & \downarrow{\scriptstyle\simeq} \\ (A,d) & \xrightarrow{g} & (B,d). \end{array}$$

Then we have,

Theorem 2.83 ([87, Proposition 13.6]) *If g is surjective, then the minimal model of $\mathbb{Q} \oplus \operatorname{Ker} g$ is the minimal model of C_f.*

In particular the geometric realization of the minimal model of $\mathbb{Q} \oplus \operatorname{Ker} g$ is the rationalization of C_f.

2.6 Realizing minimal cdga's as spaces

In order to apply the algebra of minimal cdga's to geometry, we have to understand how algebraic data can be realized geometrically. We begin by considering how a minimal cdga may be realized topologically.

2.6.1 Topological realization of a minimal cdga

Consider on one side the category of nilpotent spaces with finite Betti numbers Top_N and, on the other side, the category $\mathcal{A}_N$ composed of the cdga's (A,d) with $H^0(A,d) = \mathbb{Q}$, and which admit a finite type minimal model. There is a realization functor $\langle\ \rangle$ going from $\mathcal{A}_N$ to Top_N (see [38] and [246]) that has the following properties:

- The realization of a minimal cdga $(\wedge V, d)$, $\langle(\wedge V, d)\rangle$, is a *rational space*; that is, its homotopy groups are rational vector spaces.
- The correspondences $X \mapsto \mathcal{M}_X$ and $(\wedge V, d) \mapsto \langle(\wedge V, d)\rangle$ are inverse to each other up to rational homotopy equivalence. The minimal model of

$\langle(\wedge V, d)\rangle$ is $(\wedge V, d)$; moreover, there is a map $X \to \langle \mathcal{M}_X \rangle$ that induces an isomorphism in rational cohomology.

The space $\langle \mathcal{M}_X \rangle$ is called the rationalization of X, and is denoted by X_0. This space is characterized by the following properties:

1. X_0 is a rational space.
2. There is a map $f \colon X \to X_0$ inducing an isomorphism in rational cohomology.
3. If Y is a rational space and $g \colon X \to Y$ is a continuous map, then there is a map, unique up to homotopy, $h \colon X_0 \to Y$ such that $g \simeq h \circ f$.

The correspondence between A_N and Top_N extends to maps.

- If $X, Y \in Top_N$, then rationalization induces a bijection

$$[X, Y] \to [X_0, Y_0] \cong [\mathcal{M}_Y, \mathcal{M}_X].$$

- The realization of a relative minimal cdga $(\wedge V, d) \to (\wedge V \otimes \wedge W, D)$ is a fibration $p \colon E \to B$.

The rationalization of a nilpotent finite type CW complex is not a finite type CW complex because the homotopy groups of the rationalization are $\mathbb{Q}$-vector spaces, while the homotopy groups of a finite type CW complex are finitely generated. However, the realization described above can, in general, be improved, in a nonfunctorial way, to give a finite type CW complex whose minimal model is the original one.

- Each finite type minimal cdga $(\wedge V, d)$ is the minimal model of a nilpotent space X with finite Betti numbers. If $H^*(\wedge V, d)$ is finite dimensional, then we can choose X to be a finite CW complex.
- Each map between finite type minimal cdga's is the minimal model of a continuous map between finite type CW complexes.
- Each finite type relative minimal cdga $(\wedge V \otimes \wedge W, D)$ is the relative minimal model of a fibration $p \colon E \to B$ where E and B are finite type CW complexes.

2.6.2 The cochains on a graded Lie algebra

Let L be a graded Lie algebra. We can associate to L its cochain algebra $C^*(L) = (\wedge V, d)$ with $V^{p+1} \cong \mathrm{Hom}(L_p, \mathbb{Q})$ and differential $d = d_1$ defined by

$$\langle x, s[f, g] \rangle = (-1)^{|g|} \langle d_1 x, sf \wedge sg \rangle, \qquad x \in V, f, g \in L.$$

The next proposition follows directly.

Proposition 2.84 *Every finite type graded Lie algebra L such that $L_0 = L_1 = 0$ can be realized as the rational homotopy Lie algebra of a simply connected space.*

The cochain algebra $(\wedge V, d)$ admits a bigradation defined by $(\wedge V)^{p,q} = (\wedge^p V)^{q-p}$. Since $d(\wedge V)^{p,q} \subset (\wedge V)^{p+1,q}$, this bigradation induces a bigradation on the cohomology. The qth rational cohomology vector space of a graded Lie algebra L, $H^q(L; \mathbb{Q})$, is by definition the graded vector space $H^{q,*}(C^*(L))$.

Every short exact sequence of Lie algebras $0 \to L_1 \to L_2 \to L_3 \to 0$ induces a relative minimal cdga

$$C^*(L_3) \to C^*(L_2) \to C^*(L_1),$$

and is therefore realized by a fibration.

A minimal presentation of a graded Lie algebra is a short exact sequence

$$0 \to \mathbb{L}(W) \xrightarrow{f} \mathbb{L}(S) \to L \to 0$$

where $\mathbb{L}(\ )$ denotes the free graded Lie algebra functor and where $f(W) \subset \mathbb{L}^{\geq 2}(S)$. By [171] and [248], we then have isomorphisms of graded vector spaces

$$S \cong H^1(C^*(L)), \qquad W \cong H^2(C^*(L)).$$

When L is a finite dimensional nilpotent Lie algebra concentrated in degree 0, the cochain algebra $C^*(L)$ is finite dimensional. Its realization is a nilmanifold. The theory of nilmanifolds will be described in Section 3.2.

2.7 Formality

Definition 2.85 *A nilpotent space X, with minimal model $(\wedge V, d)$, is called* formal *if there is a quasi-isomorphism*

$$\varphi : (\wedge V, d) \to (H^*(X; \mathbb{Q}), 0).$$

Remark 2.86

- We can also define a cdga (A, d) to be *formal* if there is a chain of quasi-isomorphisms

$$(A, d) \leftarrow (B_1, d_1) \to \cdots (B_k, d_k) \to (H^*(A), 0).$$

We can take the minimal models of (A,d), the minimal models of the (B_i, d_i) and the minimal models of the morphisms by Theorem 2.26. By applying Corollary 2.13, we then see that the condition above is equivalent to Definition 2.85.
- By a result of Sullivan (see Proposition 2.101, [206] and [246]), the definition of formality given above is equivalent to the existence of a quasi-isomorphism over the reals,

$$\varphi: (\wedge V, d) \otimes_{\mathbb{Q}} \mathbb{R} \to (H^*(X; \mathbb{R}), 0).$$

For a manifold M, this is also equivalent to the existence of a sequence of quasi-isomorphisms of $\mathbb{R}$-cdga's connecting $A_{DR}(M)$ and $(H^*(M; \mathbb{R}), 0)$. We will come back to this point in Proposition 2.101. For a classical criterion guaranteeing formality, see Exercise 2.3.

It follows directly from the definition that Lie groups, complex and quaternionic Stiefel manifolds and loop spaces are formal spaces because their respective cohomologies are isomorphic to their minimal models.

Example 2.87 (Product of formal spaces) Let X and Y be formal spaces with respective minimal models $(\wedge V, d)$ and $(\wedge W, d)$. By the formality of X and Y, we have quasi-isomorphisms

$$\varphi: (\wedge V, d) \to (H^*(X; \mathbb{Q}), 0) \quad \text{and} \quad \psi: (\wedge W, d) \to (H^*(Y; \mathbb{Q}), 0).$$

The minimal model of $X \times Y$ is the tensor product $(\wedge V, d) \otimes (\wedge W, d)$ and we have a quasi-isomorphism

$$\varphi \otimes \psi: (\wedge V, d) \otimes (\wedge W, d) \to (H^*(X; \mathbb{Q}) \otimes H^*(Y; \mathbb{Q}), 0)$$
$$\cong (H^*(X \times Y; \mathbb{Q}), 0).$$

This shows that $X \times Y$ is formal.

Example 2.88 (Retract of a formal space) If X is a formal space and Y is a retract of X, then Y is a formal space. Denote i the injection and r the retraction

$$Y \xrightarrow{i} X \xrightarrow{r} Y, \qquad r \circ i \simeq \mathrm{id}_Y.$$

By the formality of X, we have a quasi-isomorphism $\varphi: \mathcal{M}_X \to (H^*(X; \mathbb{Q}), 0)$. Then the following composition of morphisms is a quasi-isomorphism

$$\mathcal{M}_Y \xrightarrow{\mathcal{M}_r} \mathcal{M}_X \xrightarrow{\varphi} (H^*(X; \mathbb{Q}), 0) \xrightarrow{H^*(i; \mathbb{Q})} (H^*(Y; \mathbb{Q}), 0).$$

This shows that Y is a formal space. In particular, if a product $X \times Y$ is formal, the spaces X and Y are formal.

An important property of formal spaces is the vanishing of all their Massey products. Let's recall the definition.

Definition 2.89 *Let (A, d) be a cdga with cohomology H^*. Let a, b, c be cohomology classes whose products ab and bc are zero. Choose cocycles (or closed forms if we work with the de Rham complex) x, y and z representing a, b and c. By definition, there are elements v and w such that $dv = xy$ and $dw = yz$. The element*

$$vz - (-1)^{|x|} xw$$

is then a cocycle whose cohomology class depends on the choice of v and w. The set $\langle a, b, c \rangle$ formed by the cohomology classes constructed using all the possible choices of v and w is called the triple Massey product *of a, b and c. The triple Massey product is said to be trivial if the element 0 belongs to the set $\langle a, b, c \rangle$.*

Denote by I the ideal of H^* generated by the classes of a and c. The set $\langle a, b, c \rangle$ projects to a single element in H^*/I. Moreover, this element is 0 if and only if the triple Massey product is trivial.

Proposition 2.90 *Let X be a formal space. Then all triple Massey products vanish.*

Proof Denote by $(\wedge V, d)$ the minimal model of X. By definition we have a quasi-isomorphism

$$\varphi \colon (\wedge V, d) \to (H^*(X; \mathbb{Q}), 0).$$

Now take three cohomology classes a, b and c represented by the cocycles x, y and z in $(\wedge V, d)$, and suppose that there are elements v and w such that $dv = xy$ and $dw = yz$. The cocycle $vz - (-1)^{|x|} xw$ is mapped by φ to a cocycle in the ideal generated by the elements $\varphi(x) = a$ and $\varphi(z) = c$. Since φ is a quasi-isomorphism, this implies that the triple Massey product set $\langle a, b, c \rangle$ belongs to the ideal generated by a and c. This Massey product is thus zero. □

Example 2.91 (A nonformal manifold) Denote by $q \colon S^2 \times S^2 \to S^4$ the map obtained by collapsing the wedge $S^2 \vee S^2$ to a point, and take the pullback of the Hopf fibration $S^3 \to S^7 \to S^4$ along q (see Example 2.68). We obtain in this way a principal bundle $S^3 \to M \to S^2 \times S^2$, whose relative minimal model is given by

$$(\wedge(a, b, u, v), d) \to (\wedge(a, b, u, v, t), d)$$

with $|a| = 2$, $|b| = 2$, $|u| = 3$, $|v| = 3$, $|t| = 3$, $da = 0$, $du = a^2$, $db = 0$, $dv = b^2$, $dt = ab$. We can compute the cohomology in low degrees. A basis of $H^2(M; \mathbb{Q})$ is given by the classes a and b; $H^3(M; \mathbb{Q}) = H^4(M; \mathbb{Q}) = 0$. The cohomology in degree 5 has dimension two, and a basis is given by the two nontrivial Massey products

$$\langle a, a, b \rangle = [ub - at], \quad \text{and} \quad \langle a, b, b \rangle = [tb - av].$$

Because these Massey products do not vanish, the manifold M is not formal.

Definition 2.92 (Higher order Massey products) *Let $u_1, \ldots, u_p$ be cohomology classes. A defining system for the Massey product $\langle u_1, \ldots, u_p \rangle$ is a collection of cochains m_{ij}, $1 \leq i \leq j \leq p$, $(i,j) \neq (1,p)$, with m_{ii} a cocycle representative for u_i, $i = 1, \ldots, p$ and*

$$d(m_{ij}) = \sum_{k=i}^{j-1} (-1)^{|m_{ik}|} m_{ik} \cdot m_{k+1,j}.$$

We then form the cocycle

$$\alpha_{\{m_{ij}\}} = \sum_{k=1}^{p-1} (-1)^{|m_{1k}|} m_{1k} \cdot m_{k+1,p}.$$

The Massey product is trivial if there is a system m_{ij} for which the cohomology class of $\alpha_{\{m_{ij}\}}$ is zero.

2.7.1 Bigraded model

The main tools for the study of formality of spaces and manifolds are bigraded and filtered models (see [132]).

Theorem 2.93 (Bigraded model of a graded commutative algebra) *Let A be a finitely generated graded commutative algebra. We suppose that $A^0 = \Bbbk$. Then the cdga $(A, 0)$ admits a minimal model $\varphi \colon (\wedge V, d) \to (A, 0)$ where V is equipped with a lower gradation $V = \oplus_{p \geq 0} V_p$ extended in a multiplicative way to $\wedge V$ and where the following properties hold.*

1. *$d(V_p) \subset (\wedge V)_{p-1}$. In particular, $d(V_0) = 0$. Therefore the cohomology is a bigraded algebra $H^*(\wedge V, d) = \oplus_{p \geq 0} H_p(\wedge V, d)$.*
2. *$H_q(\wedge V, d) = 0$ for $q > 0$, and $H_0(\wedge V, d) \cong A$.*
3. *$\varphi(V_p) = 0$ for $p > 0$.*

The cdga $(\wedge V, d)$ is called the bigraded model *of the graded algebra A.*

The construction of the bigraded model begins with the generators (or indecomposables) of A, $V_0 = A^+/(A^+ \cdot A^+)$. We define $d|_{V_0} = 0$ and

$\varphi \colon \wedge V_0 \to A$ by extending a vector space splitting $V_0 \to A$. Thus, at the very first stage, we obtain a surjection from $\wedge V_0$ to A. We form $\wedge(V_0 \oplus V_1)$ and define the differential on V_1 to kill the kernel of $\wedge V_0 \to A$. Note that $d(V_1) \subset \wedge V_0$. The addition of V_1 may introduce yet another kernel in cohomology and this must be killed by adding in V_2 with $d(V_2) \subset (\wedge(V_0 \oplus V_1))_1$. The process goes on, eventually resulting in the bigraded model.

When X is a formal space, the minimal model of X is the minimal model of the cdga $(H^*(X; \mathbb{Q}), 0)$, so we can therefore choose the bigraded model of the algebra $H^*(X; \mathbb{Q})$ as a minimal model for X.

Proposition 2.94 *Let M be a formal nilpotent manifold. Then all the Massey products $\langle u_1, \ldots, u_p \rangle, p \geq 3$, are trivial in the de Rham complex $A_{DR}(M)$.*

Proof Consider the Massey product $\langle u_1, \cdots, u_p \rangle$ with $p \geq 3$. In the bigraded model $(\wedge V, d)$ we can choose a defining system m_{ij} with $m_{ij} \in (\wedge V)_{j-i}$. This shows that the element $\alpha_{\{m_{ij}\}}$ is a cocycle in $(\wedge V)_{p-2}$ and therefore is a coboundary because of Theorem 2.93 (3). Now, since $(\wedge V, d)$ is the minimal model of M, we have a quasi-isomorphism $\psi \colon (\wedge V, d) \to A_{DR}(M)$ and the family $\psi(m_{ij})$ is a trivial defining system for the Massey product in $A_{DR}(M)$. This shows that all Massey products are zero. $\square$

2.7.2 Obstructions to formality

Let (A, d_A) be a cdga. We suppose that $H^0(A, d_A) = \mathbb{Q}$ and that each $H^p(A, d_A)$ is finite dimensional. We denote by

$$\mu \colon (\wedge V, d) \to (H^*(A, d_A), 0)$$

a bigraded model of the cohomology. There then exists a perturbation of $(\wedge V, d)$ that is a (not necessarily minimal) Sullivan model of (A, d_A). More precisely, we have the following.

Theorem 2.95 ([132]) *With the previous notation, there is a differential D on $\wedge V$ such that*

1. $(D - d)(V_p) \subset (\wedge V)_{\leq p-2}$ *(In other words the differential can be written $D = d + d_2 + d_3 + \cdots$ where $d_q(V_p) \subset (\wedge V)_{p-q}$).*
2. $(\wedge V, D)$ *is a Sullivan model for (A, d_A).*
3. (A, d) *is formal if and only if there is an isomorphism $\varphi \colon (\wedge V, D) \to (\wedge V, d)$ of the form $\varphi = \mathrm{id} + \varphi_1 + \varphi_2 + \cdots$ with $\varphi_q(V_p) \subset (\wedge V)_{p-q}$.*

Definition 2.96 The Sullivan model $(\wedge V, D)$ is called *the filtered model* of the cdga (A, d_A).

We want to use Theorem 2.95 (3) to develop an obstruction theory for formality, so we note that it follows from a more general uniqueness result

2.7 Formality

for filtered models (see [132, Theorem 4.4]) and the fact that, if (A, d_A) is formal, then the bigraded model is a minimal model and a filtered model for (A, d_A).

Now let's see what necessary condition Theorem 2.95 (3) leads to. Suppose $\varphi \colon (\wedge V, D) \to (\wedge V, d)$ is an isomorphism with φ of the form $\varphi = \mathrm{id} + \varphi_1 + \varphi_2 + \cdots$. From the relations $D^2 = 0$, $\varphi(ab) = \varphi(a)\varphi(b)$ and $d\varphi = \varphi D$ we obtain

$$\begin{cases} d_2^2 = 0, \\ \varphi_1(ab) = \varphi_1(a)b + a\varphi_1(b), \\ d\varphi_1 = d_2 + \varphi_1 d. \end{cases}$$

In particular, d_2 and φ_1 are derivations that satisfy $\varphi_1(V_p) \subset (\wedge V)_{p-1}$ and $d_2(V_p) \subset (\wedge V)_{p-2}$.

We are therefore motivated to define the graded Lie algebra of derivations $\mathrm{Der}_{q,s}$ as the set of derivations θ of $\wedge V$ that decrease the lower degree by q and increase the usual degree by s:

$$\begin{cases} \theta(V_p^r) \subset (\wedge V)_{p-q}^{r+s}, \\ \theta(ab) = \theta(a) \cdot b + (-1)^{s \cdot |a|} a \cdot \theta(b). \end{cases}$$

The commutator with d and the commutator bracket give $\mathrm{Der}_{*,*}$ the structure of a differential graded Lie algebra $(\mathrm{Der}_{*,*}, \mathcal{D})$. The differential $\mathcal{D}$ is given by

$$\mathcal{D}(\theta) = d \circ \theta - (-1)^s \theta \circ d,$$

so that (recalling $d(V_p) \subset (\wedge V)_{p-1}$) we have

$$\mathcal{D} \colon \mathrm{Der}_{q,s} \to \mathrm{Der}_{q+1,s+1}.$$

Note that, for each $\theta \in \mathrm{Der}_{q,s}$, $q > 0$, it is certainly true that $\theta^{q+1}(x) = 0$ for $x \in V_q$. Therefore the formula

$$e^\theta = \sum_{n \geq 0} \frac{\theta^n}{n!} = \mathrm{id} + \theta + \frac{\theta^2}{2} + \cdots$$

gives a well defined automorphism of $\wedge V$.

Now we can construct a sequence of obstructions to the formality of a cdga (A, d_A). Let $(\wedge V, D)$ be a filtered model of (A, d_A). Since $D^2 = 0$, we have $dd_2 + d_2 d = 0$ since it is the only part of D^2 that decreases the lower degree by 3. The derivation d_2 is thus a cycle in $\mathrm{Der}_{2,1}$. If there is an isomorphism $\varphi \colon (\wedge V, D) \to (\wedge V, d)$ of the form $\varphi = \mathrm{id} + \varphi_1 + \cdots$, then

$$d_2 = d\varphi_1 - \varphi_1 d = \mathcal{D}(\varphi_1).$$

2 : Minimal models

Therefore the first obstruction to formality is the class of d_2,

$$[d_2] \in H_{2,1}(\text{Der}_{*,*}).$$

On the other hand, if the class of d_2 is zero, then there is a derivation φ_1 such that $d_2 = d\varphi_1 - \varphi_1 d$. We then can form the automorphism $e^{-\varphi_1}$ and consider the new Sullivan model $(\wedge V, D')$ with $D' = e^{\varphi_1} D e^{-\varphi_1}$. By construction, we have an isomorphism

$$e^{-\varphi_1} : (\wedge V, D') \to (\wedge V, D),$$

but the main advantage of D' is that $D' - d$ decreases the lower gradation by at least three. We can see this by writing

$$D' = (\text{id} + \varphi_1 + \ldots)(d + d_2 + \ldots)(\text{id} - \varphi_1 + \ldots)$$
$$= d + (\varphi_1 d - d\varphi_1 + d_2) + \text{ terms decreasing lower degree by at least } 3$$
$$= d + \text{ terms decreasing lower degree by at least } 3$$

since $\varphi_1 d - d\varphi_1 + d_2 = 0$ by assumption. We can therefore write $D' = d + d_3 + d_4 + \cdots$ with $d_q(V_p) \subset (\wedge V)_{p-q}$. We then see that the second obstruction is the class of d_3 in $H_{3,1}(\text{Der}_{*,*})$. Of course, this process leads to an inductive construction.

Therefore, suppose by induction that (A, d_A) admits a filtered model of the form $(\wedge V, D_r)$ with $D_r = d + d_r + d_{r+1} + \cdots$ with $r \geq 3$. The element d_r is a cycle in $\text{Der}_{r,1}$, and defines the $(r-1)$st obstruction $[d_r] \in H_{r,1}(\text{Der}_{*,*})$. Suppose we have an isomorphism $\varphi^{(r)} : (\wedge V, D_r) \to (\wedge V, d)$ of the form $\varphi^{(r)} = \text{id} + \varphi_1 + \varphi_2 + \cdots$. Then $\varphi_1 d = d\varphi_1$ since $d_2 = 0$. Thus, $e^{-\varphi_1}$ is an automorphism of $(\wedge V, d)$ and we obtain the composition $e^{-\varphi_1} \circ \varphi^{(r)} = \text{id} + \varphi_2 + \cdots$. Hence, we can suppose that $\varphi_1 = 0$. By the same process, we can suppose that $\varphi_2 = \varphi_3 = \ldots = \varphi_{r-2} = 0$; that is, $\varphi^{(r)} = \text{id} + \varphi^{(r)}_{r-1} + \cdots$, and $d_r = d\varphi^{(r)}_{r-1} - \varphi^{(r)}_{r-1} d$. Hence, $[d_r]$ must be equal to zero in $H^*(\text{Der}_{*,*}, D)$. Conversely, if we suppose that $[d_r] = 0$, we can then replace $(\wedge V, D_r)$ by $(\wedge V, D_{r+1})$ with $D_{r+1} = e^{\varphi^{(r)}_{r-1}} D_r e^{-\varphi^{(r)}_{r-1}}$. As above, we also obtain an isomorphism $e^{-\varphi^{(r)}_{r-1}} : (\wedge V, D_{r+1}) \to (\wedge V, D_r)$.

If all the obstructions $[d_i]$ vanish, then we have a sequence of isomorphisms

$$\cdots \longrightarrow (\wedge V, D_{r+1}) \xrightarrow{e^{-\varphi^{(r)}_{r-1}}} (\wedge V, D_r) \longrightarrow \cdots \longrightarrow (\wedge V, D),$$

where $D_r - d$ decreases the lower gradation by at least r, and where $e^{-\varphi^{(r)}_{r-1}} - \text{id}$ decreases the gradation by at least $r - 1$. In other words, each stage of the construction creates a cdga which is closer to $(\wedge V, d)$ than the previous stage. Therefore, the composition

$$\cdots \circ e^{-\varphi_{r-1}^{(r)}} \circ e^{-\varphi_{r-2}^{(r-1)}} \circ \cdots \circ e^{-\varphi_1}$$

is a well-defined isomorphism between $(\wedge V, d)$ and $(\wedge V, D)$. We therefore have the

Theorem 2.97 *Starting from the filtered model $(\wedge V, D)$ for (A, d_A), there is a sequence of obstructions in $H^*(\mathrm{Der}_{*,*}, \mathcal{D})$, $[d_2], [d_3], \ldots, [d_r], \ldots$ such that the following conditions are equivalent:*

- $[d_2] = [d_3] = \ldots = [d_r] = 0$;
- $(\wedge V, D) \cong (\wedge V, D')$ *with* $(D' - d)(V_q) \subset (\wedge V)_{q-r}$.

In particular, all the obstructions vanish if and only if the cdga (A, d_A) is formal.

The following proposition will significantly simplify future computations of the obstructions $[d_r]$.

Proposition 2.98 *Let $(\wedge V, D)$ be the filtered model of a simply connected space X with finite Betti numbers such that $H^{>n}(X; \mathbb{Q}) = 0$. Suppose $D = d + d_q + d_{q+1} + \cdots$ and suppose there exists a derivation θ defined on $\wedge V_{\leq q}^{\leq n}$ such that $\theta(V_r^s) \subset (\wedge V)_{r-q+1}^{s}$ and $d_q = d\theta - \theta d$ on $V_q^{<n}$. Then the obstruction $[d_q]$ is zero.*

Proof First, define $\theta = 0$ on $V_{\leq q}^{\geq n}$. We clearly have $d_q = d\theta - \theta d = 0$ on $V_{<q}$. Now, by induction on r, for $r \geq n$ we extend θ on V_q^r such that $d_q = d\theta - \theta d$. By hypothesis, this has been done for $r < n$. Suppose this has also been done for degrees $\leq r$ and let x be an element of a basis of V_q^{r+1}. Then $d_q(x) + \theta d(x)$ is a cocycle in $(\wedge V)_0^{r+2}$. By hypothesis, we have $H^{>n}(X; \mathbb{Q}) = 0$, so there is some element y such that $d(y) = d_q(x) + \theta d(x)$. We define a linear map $\theta \colon V_q^{r+1} \to (\wedge V)_0^{r+1}$ by putting $\theta(x) = y$. This proves that θ can be defined on V_q with $d_q = d\theta - \theta d$.

Suppose now that we have defined θ on $V_{<s}$ and on $V_s^{<r}$ such that $d_q = d\theta - \theta d$ and θ decreases the lower gradation by exactly $q - 1$. Then, for each element x of a basis of V_s^r, the element $d_q(x) + \theta d(x)$ is a d-cocycle of positive lower degree in $\wedge V$. By Theorem 2.93 (2), there is an element y such that $dy = d_q(x) + \theta d(x)$. We extend θ linearly by putting $\theta(x) = y$.

In this way, we can define θ on V with $d_q = d\theta - \theta d$. This shows that the obstruction $[d_q]$ is zero. $\square$

Proposition 2.99 *Let X be a $(p-1)$-connected space, $p \geq 2$, of dimension $\leq 3p - 2$. Then X is formal.*

Proof Let $(\wedge V, d)$ denote the bigraded model of $H^*(X; \mathbb{Q})$. By the connectivity hypothesis, $V_0^q = 0$ for $q < p$. By the minimality property of the model, $V_1^q = 0$ for $q < 2p - 1$ and $V_2^q = 0$ for $q < 3p - 2$. We show that all the obstructions are zero. For the first one, $[d_2]$, we define θ to be zero on V_1. The hypotheses of Proposition 2.98 are satisfied, and so $[d_2] = 0$. By Proposition 2.98, the other obstructions are zero because $V_q^{< 3p-2} = 0$ for $q \geq 3$. $\square$

For a cdga (A, d_A), we can extend scalars to obtain a cdga $(A \otimes \Bbbk, d_A \otimes 1_{\Bbbk})$ over any field $\Bbbk$ of characteristic zero. We can also find a minimal model $(\wedge V_{\Bbbk}, d) \to (A \otimes \Bbbk, d_A \otimes 1_{\Bbbk})$ over $\Bbbk$. Then, formality over the field $\Bbbk$ simply means the following.

Definition 2.100 *We say that a space X is $\Bbbk$-formal if its $\Bbbk$-minimal model is quasi-isomorphic to $(H^*(X; \Bbbk), 0)$.*

The first question that comes to mind now is whether descent phenomena occur. That is, could a space "become" formal under a field extension. The following application of the framework of obstruction theory answers this question and will be very useful in Chapter 4. (Note that we have already discussed this result in Remark 2.86.)

Proposition 2.101 *A space X is $\Bbbk$-formal if and only if X is formal.*

Proof The obstruction theory enunciated in Theorem 2.97 can be defined over the field $\Bbbk$. The associated complex of derivations $\mathrm{Der}_{*,*}^{(\Bbbk)}$ is obtained by extension of the scalars from $\mathbb{Q}$ to $\Bbbk$ of the complex $\mathrm{Der}_{*,*}$ described above. Since the obstructions are linear, the obstructions $[d_q]$ are zero over the field $\Bbbk$ if and only if they are zero over $\mathbb{Q}$. $\square$

2.8 Semifree models

Semifree models for differential modules are introduced in this section because they will be very useful tools in Chapters 7 and 8.

Definition 2.102 *Let (A, d) be a differential graded algebra. A differential A-module (M, d) is a complex equipped with a structure of A-module such that $d(a \cdot m) = da \cdot m + (-1)^{|a|} a \cdot dm$.*

Definition 2.103 *Let (A, d) be a differential graded algebra. A differential A-module (M, d) is called semifree if M is equipped with a filtration*

$$0 = M(-1) \subset M(0) \subset \cdots \subset M = \cup_{p \geq 0} M(p)$$

such that $d(M(p)) \subset M(p-1)$ and such that as an A-module, $M(p)/M(p-1)$ is isomorphic to $(A,d) \otimes (V(p), 0)$. We write $M = A \otimes V$.

Proposition 2.104 ([87]) *Let (A,d) be a differential graded algebra and (N,d) be a differential A-module. There then exists a semifree A-module (M,d) and a quasi-isomorphism of A-modules $\varphi \colon (M,d) \to (N,d)$.*

Definition 2.105 *The differential A-module (M,d) is called a* semifree model *of (N,d).*

Definition 2.106 *Let (A,d) be a differential graded algebra and suppose (M,d) is a semifree A-module. Two morphisms of differential A-modules, $f, g \colon (M,d) \to (N,d)$ are* homotopic, $f \simeq g$, *if there is a morphism of A-modules $H \colon M \to N$ of degree -1 such that $f - g = dH + Hd$. We denote by $[(M,d), (N,d)]$ the vector space of homotopy classes from (M,d) to (N,d).*

Semifree modules enjoy properties very similar to those of minimal models. The next proposition contains some of these that will be useful in later chapters.

Proposition 2.107 ([87])

1. *Let $\varphi \colon (B,d) \to (A,d)$ be a quasi-isomorphism of differential graded algebras and let (M,d) be a semifree A-module, $(M,d) = (A \otimes V, d)$. Then there is a semifree B-module $(N,d) = (B \otimes V, d)$ and a quasi-isomorphism $\psi \colon (B \otimes V, d) \to (A \otimes V, d)$ such that $\psi(b \otimes v) - \varphi(b) \otimes v \in A^{>n} \otimes V$, for $b \in B^n$.*
2. *Let $\varphi \colon (B,d) \to (A,d)$ be a quasi-isomorphism of differential graded algebras and (M,d) be a semifree B-module. Then $A \otimes_B M$ is a semifree A-module and the map $\varphi \otimes \mathrm{id} \colon M = B \otimes_B M \to A \otimes_B M$ is a quasi-isomorphism.*
3. *(Lifting property) Given a commutative diagram of differential A-modules*

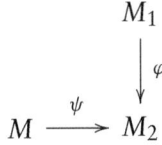

   where M is semifree and φ is a quasi-isomorphism, then there is a unique, up to homotopy, morphism of differential A-modules $\theta \colon M \to M_1$ such that $\psi \simeq \varphi \circ \theta$.

Part (3) of Proposition 2.107 can be rephrased in the form

2 : Minimal models

Proposition 2.108 *If (M,d) is a semifree A-module and $f\colon (N,d) \to (N',d)$ is a quasi-isomorphism of differential A-modules, then the composition with f induces an isomorphism $[(M,d),(N,d)] \to [(M,d),(N',d)]$.*

We denote by $\operatorname{Hom}_A^p(M,N)$ the vector space of A-morphisms of degree p from M into N. The differential $D\colon \operatorname{Hom}_A^p(M,N) \to \operatorname{Hom}_A^{p+1}(M,N)$ defined by $D(f) = df - (-1)^{|f|}fd$ makes $\operatorname{Hom}_A(M,N) = \oplus_q \operatorname{Hom}_A^q(M,N)$ into a cochain complex. By construction, $H^0(\operatorname{Hom}_A(M,N)) = [(M,d),(N,d)]$.

Let (M,d) and (N,d) be two differential A-modules, and let $(P,d) \to (M,d)$ be a semifree model for (M,d). Then, by Proposition 2.108, the cohomology $H^q(\operatorname{Hom}_A(P,N))$ is independent of the choice of semifree model (P,d) and is denoted by $\operatorname{Ext}_A^q(M,N)$.

Semifree modules are very useful for the study of fibrations. Let $F \to E \xrightarrow{p} B$ be a fibration and denote by $C^*(-)$ the singular cochains functor with coefficients in $\mathbb{Q}$. Then the morphism $C^*(p)$ makes the cochain algebra $C^*(E;\mathbb{Q})$ a module over the cochain algebra $C^*(B;\mathbb{Q})$. We then have:

Proposition 2.109 ([87]) *With the above structure, $C^*(E;\mathbb{Q})$ admits a semifree model of the form $(C^*(B;\mathbb{Q}) \otimes H^*(F;\mathbb{Q}), D)$. In the same way, when B is simply connected, a semifree model for the minimal model of E is given by $(\wedge V \otimes H^*(F;\mathbb{Q}), D)$, where $(\wedge V, D)$ is the minimal model of B.*

Exercises for Chapter 2

Exercise 2.1 Show that the correspondence $f \mapsto H^*(f)$ induces a bijection
$$[(\wedge V, 0)(A,d)] \xrightarrow{\cong} \operatorname{Hom}(V, H^*(A,d)).$$

Exercise 2.2 Let $(\wedge V, d)$ be a minimal model. Show that the correspondence $f \mapsto Q(f)$ induces a bijection
$$[(\wedge V, d), \mathcal{M}_{S^n}] \xrightarrow{\cong} \operatorname{Hom}(V^n, \mathbb{Q}).$$

Exercise 2.3 In [70, Theorem 4.1], a criterion for formality was given that sometimes makes it easy to say that a minimal cdga is not formal.

Proposition A minimal cdga $(\wedge V, d)$ is formal if and only if V decomposes as a direct sum $V = C \oplus N$ with $d(C) = 0$ and d injective on N such that every closed element in the ideal generated by N is exact.

Prove this result. Hints: Suppose that the property holds. This says that all cohomology of $(\wedge V, d)$ comes from the subalgebra $\wedge C$ where $C = \oplus C^i$ consists of all generators which are cocycles. Define a linear map $\psi\colon \oplus_i V^i \to H^*(\wedge V, d)$ by
$$\begin{cases} \psi(c) = [c] & \text{for } c \in \oplus_i C^i \\ \psi(n) = 0 & \text{for } n \in \oplus_i N^i. \end{cases}$$

Extend ψ multiplicatively to $\wedge V$ and show that the extension is a cdga homomorphism. Then show that, since the cohomology is generated by C, ψ induces the identity on cohomology.

For the other direction, suppose that $(\wedge V, d)$ is formal. Let $\psi \colon (\wedge V, d) \to H^*(\wedge V, d)$ be such that ψ^* is the identity. Let C be the kernel of ψ restricted to V. See [257], for example, for a proof.

Exercise 2.4 Let X be a simply connected CW complex of finite type. Suppose that there is an integer $n > 0$ so that $\pi_r(X) \otimes \mathbb{Q} = 0$ for $r > n$ and $H^p(X; \mathbb{Q}) = 0$ for $n < p \leq 2n$. Show that $H^p(X; \mathbb{Q}) = 0$ for $p > n$. In particular, X must be rationally elliptic of dimension n. Hint: if $H^q(X; \mathbb{Q}) \neq 0$ for some $q > 2n$, add generators to the minimal model (in degrees greater than or equal to $2n$) to kill *all* cohomology above degree $2n$. Show by Theorem 2.74 (3) that this is a contradiction.

Exercise 2.5 Suppose X is a finite CW complex that is $(r-1)$-connected with dim $X \leq 4r - 4$ and $H^{2q+1}(X; \mathbb{Q}) = 0$ for $q \geq 0$. Show that X is formal. (Note that X is *not* assumed to be a manifold.)

Exercise 2.6 Characterize the compact homogeneous spaces that are formal in terms of their minimal models. (Hint: see Theorem 2.77).

Exercise 2.7 Let $f \colon M_1 \to M_2$ be a quasi-isomorphism of semifree A-modules. Show that there is a morphism of differential A-modules $g \colon M_2 \to M_1$ such that $g \circ f$ and $f \circ g$ are homotopic to the identity.

3 Manifolds

A smooth compact manifold has many properties that make it distinct from an ordinary topological space. From the topological viewpoint, the existence of Poincaré duality in (co)homology is crucial to almost any result about the manifold. From the geometric viewpoint, the existence of a Riemannian metric allows the manifold to be studied using analytic techniques. In subsequent chapters, we shall see how these two points of view mix together to yield interesting results in both directions: topology applied to geometry and geometry applied to topology.

In this chapter, we show how minimal models of manifolds reflect the special properties of manifolds. In particular, we see how Poincaré duality plays a huge role in almost all aspects of the rational homotopy of manifolds, but especially in the geometric realization of algebraic data *and* in the problem of determining whether a manifold is formal. Also, restricting manifolds to satisfy certain properties constrains their minimal models as well, and we will see this clearly in the case of nilmanifolds and biquotients. Now let's describe exactly what is in the chapter.

In Section 3.1, we study how manifolds are linked to minimal models through Poincaré duality. In particular, we state the Barge–Sullivan realization criteria which tell us which minimal models contain manifolds inside their rational homotopy types. We also describe a model for connected sums, understand their rational homotopy groups and show which simply connected 4-manifolds are rationally elliptic. We end this section by giving proofs of the theorems of Miller and Stasheff, which state two important criteria for formality, using the obstruction theory of Subsection 2.7.2.

In Section 3.2, we consider a particular class of manifolds (which we will meet throughout the book) called nilmanifolds. The minimal model of a nilmanifold is very special indeed and this means that we can understand many general properties of nilmanifolds.

The construction of models can easily be extended to the case of cdga's equipped with the action of a finite group and we develop this in Section 3.3. This then leads to an understanding of the cohomology of the classifying space of a compact connected Lie group. For instance, we prove the well-known isomorphism $H^*(BG;\mathbb{Q}) \cong H^*(BT;\mathbb{Q})^{W(G)}$ between the cohomology of BG and the invariant subalgebra of the cohomology of the classifying space of a maximal torus T of G under the action of the Weyl group $W(G)$ of G.

Section 3.4 prepares the way for one of the main objects of Chapter 6, the biquotients $G/\!/H$ of a compact connected Lie group G by a closed connected subgroup H. We define biquotients here and give their principal properties. We also show how Sullivan's theory allows the study of biquotients through a special minimal model obtained from a description of $G/\!/H$ as a certain type of pullback.

Finally, in Section 3.5, we describe the canonical model for a Riemannian manifold in terms of Hodge theory. While this type of model is not well-known, its potential for application in geometry is apparent from Theorem 3.54, which says that the real homotopy theory of an isometry of a closed simply connected Riemannian manifold is determined by the induced homomorphism on real cohomology.

3.1 Minimal models and manifolds

In order to apply the algebra of minimal models to geometry, we have to understand how algebraic data can be realized geometrically. We have already seen in Subsection 2.6.1 that a minimal model may be realized by a space. While it is useful to have spatial realizations of models, for geometry it is essential to have realizations as closed manifolds. This, of course, leads to many restrictions on the original algebra. Let's consider this now.

3.1.1 Sullivan–Barge classification

The most fundamental algebraic aspect of compact manifolds is the fact that their cohomology satisfies Poincaré duality. We want to be able to realize minimal models by manifolds, so we need to build in the Poincaré duality requirement.

Definition 3.1 *A cga H is a $\Bbbk$-Poincaré duality algebra of dimension n if each H^q is of finite $\Bbbk$-dimension, $H^n = \Bbbk\omega$, $H^{>n} = 0$ and the multiplication induces a nondegenerate bilinear pairing $H^q \otimes H^{n-q} \to H^n \cong \Bbbk$ for $0 \leq q \leq n$.*

Let $(\wedge V, d)$ be a minimal model, with $H^1(\wedge V, d) = 0$ and such that $H^*(\wedge V, d)$ is a Poincaré duality algebra of dimension n. Also suppose that cohomology classes p_i in degree $4i$ for $4i < n$ have been chosen. Let p denote the collection $\{p_i\}$. A realization of the pair $((\wedge V, d), p)$ is a manifold whose minimal model is $(\wedge V, d)$ and whose Pontryagin classes are the p_i. The theorem of Sullivan and Barge (see [21], [246]) gives necessary and sufficient conditions for the realization of this data. Note first that in case the dimension is $4k$, we have a quadratic form on H^{2k} whose signature is related to the Pontryagin numbers by the Hirzebruch signature formula (see [141]).

Theorem 3.2 ([246]) *Let $(\wedge V, d)$ be a Sullivan model whose cohomology satisfies Poincaré duality with a fundamental class in dimension n and $V^1 = 0$. We also choose cohomology classes $p = \{p_i\} \in H^{4*}(\wedge V, d)$.*

1. *If n is not of the form $4k$, then there is a compact simply connected manifold that realizes the pair $((\wedge V, d), p)$.*
2. *If $n = 4k$, and the signature is zero, there is a compact simply connected manifold that realizes the pair $((\wedge V, d), p)$ if and only if the quadratic form on H^{2k} is equivalent over $\mathbb{Q}$ to a quadratic form $\sum \pm x_i^2$.*
3. *If $n = 4k$ and the signature is nonzero, then there exists a compact simply connected manifold realizing the pair $((\wedge V, d), p)$ if and only if the quadratic form on H^{2k} is equivalent over $\mathbb{Q}$ to a quadratic form $\sum \pm x_i^2$ and the Pontryagin numbers are numbers satisfying the congruence of a cobordism [244].*

In other words, the conditions that are necessary for the realization of algebraic data by a closed manifold are also sufficient. To see if a quadratic form has the form $\sum \pm x_i^2$ over $\mathbb{Q}$ is not easy in general. Let's look at an elementary example to illustrate this point. First, the rational quadratic form $2x^2 + 2y^2$ can be written $(x+y)^2 + (x-y)^2$, and therefore has the form desired. On the other hand, the quadratic form $x^2 + 2y^2$ cannot be written in the form $\pm z^2 \pm t^2$ by a standard computation. So we have, in particular, that the graded algebra $\wedge(x, y)/(2x^2 - y^2, xy)$, with $|x| = |y| = 2$, is not the cohomology of a manifold. The beginning reader will check that the quadratic form associated to this algebra is $x^2 + 2y^2$.

3.1.2 The rational homotopy groups of a manifold

Let M be a simply connected n-dimensional compact manifold, p a point in M, and D an n-dimensional disk centered at p in M. The spaces $M' = M \backslash \{p\}$ and $M'' = M \backslash \operatorname{int} D$ have the same homotopy type. Denote by $\varphi \colon S^{n-1} \to M''$ the inclusion of the boundary of D. Then M is homeomorphic to the space obtained by attaching an n-dimensional cell to M'' along the map φ.

Theorem 3.3 ([87, Section 37]) *Let M be a simply connected manifold such that the cohomology algebra requires at least two generators. Then,*

1. *the inclusion* $i \colon M' \hookrightarrow M$ *induces a surjective map*

$$i_\# \colon \pi_*(\Omega M') \otimes \mathbb{Q} \to \pi_*(\Omega M) \otimes \mathbb{Q};$$

2. *the kernel of* $i_\#$ *is a free graded Lie algebra;*
3. *if we filter the Lie algebra* $\pi_*(\Omega M') \otimes \mathbb{Q}$ *by the powers of the ideal generated by* $[\varphi]$, *we obtain a filtered Lie algebra whose graded associated Lie algebra is the free product of Lie algebras*

$$\mathbb{L}([\varphi]) \coprod (\pi_*(\Omega M) \otimes \mathbb{Q});$$

4. *if* φ *is an indecomposable element in the Lie algebra* $\pi_*(\Omega M') \otimes \mathbb{Q}$, *then we have an isomorphism of graded Lie algebras*

$$\pi_*(\Omega M') \otimes \mathbb{Q} \cong \mathbb{L}([\varphi]) \coprod (\pi_*(\Omega M) \otimes \mathbb{Q}).$$

Corollary 3.4 (The rational homotopy of a connected sum) *Suppose M and N are n-dimensional manifolds whose cohomologies require at least two generators. Then there is a filtration on* $\pi_*(\Omega(M\#N)) \otimes \mathbb{Q}$ *such that the associated graded Lie algebra is isomorphic to* $(\pi_*(\Omega M) \otimes \mathbb{Q}) \coprod (\pi_*(\Omega N) \otimes \mathbb{Q}) \coprod \mathbb{L}(x)$ *for some element x in degree* $n-1$. *In particular, we have an isomorphism of graded vector spaces*

$$\pi_*(\Omega(M\#N)) \otimes \mathbb{Q} \cong (\pi_*(\Omega M) \otimes \mathbb{Q}) \coprod (\pi_*(\Omega N) \otimes \mathbb{Q}) \coprod \mathbb{L}(x).$$

Proof Denote by ψ_1 and ψ_2 the attaching maps of the top cells in M and N. By Theorem 3.3, the inclusion $i \colon M' \vee N' \to M \vee N$ induces a surjective map on the rational homotopy groups (see Example 2.58). The connected sum $M\#N$ is obtained from $M' \vee N'$ by attaching a cell along $\psi_1 + \psi_2$. Therefore we have a factorization of i as

$$M' \vee N' \xrightarrow{j} M\#N \xrightarrow{q} M \vee N,$$

where j is the canonical injection and q pinches the tube $S^{n-1} \times [0,1]$ connecting the two components to a point. This shows that the pinching map q induces a surjective map on the rational homotopy groups. Filtering $\pi_*(\Omega(M'\vee N')) \otimes \mathbb{Q}$ by the ideal generated by $[\psi_1]$ and $[\psi_2]$, we get a filtered Lie algebra whose associated graded Lie algebra is

$$(\pi_*(\Omega M) \otimes \mathbb{Q}) \coprod (\pi_*(\Omega N) \otimes \mathbb{Q}) \coprod \mathbb{L}([\psi_1]) \coprod \mathbb{L}([\psi_2]).$$

Therefore filtering $\pi_*(\Omega(M\#N)) \otimes \mathbb{Q}$ by the ideal generated by $[\psi_1]$, we get a graded Lie algebra isomorphic to $(\pi_*(\Omega M) \otimes \mathbb{Q}) \coprod (\pi_*(\Omega N) \otimes \mathbb{Q}) \coprod \mathbb{L}(x)$, with $x = [\psi_1]$. $\square$

Remark 3.5 Note that, under the hypothesis of Corollary 3.4, the connected sum $M\#N$ is a rational hyperbolic space.

Example 3.6 (Model of a connected sum) Let M^n and N^n be simply connected compact n-manifolds. Denote by q the pinching map $M\#N \to M \vee N$ and by $\varphi \colon \mathcal{M}_{M \vee N} \to \mathcal{M}_{M\#N}$ the minimal model of q. Denote also by $\omega_M \in \mathcal{M}_{M \vee N}$ and $\omega_N \in \mathcal{M}_{M \vee N}$ cocycles representing the fundamental classes of M and N. Since $[\varphi^*(\omega_M)] = [\varphi^*(\omega_N)]$ is the fundamental class of $M\#N$, we can introduce a new generator x and define an extension of φ,

$$\varphi \colon (\mathcal{M}_{M \vee N} \otimes \wedge x, d) \to \mathcal{M}_{M\#N},$$

by putting $dx = \omega_M - \omega_N$. Since $H^{\le n}(\varphi)$ is an isomorphism and we know that $H^{>n}(M\#N; \mathbb{Q}) = 0$, to obtain a quasi-isomorphism, we have only to inductively introduce new generators x_i in degrees $\ge n$ so that $H^{>n}(\mathcal{M}_{M \vee N} \otimes \wedge(x, x_i), d) = 0$.

Recall that in Example 2.47 we have given a process to construct the minimal model of $M \vee N$. That computation combined with the present process gives a procedure to derive a Sullivan model of $M\#N$ from the minimal models of M and N. When the algebras $H^*(M; \mathbb{Q})$ and $H^*(N; \mathbb{Q})$ are not generated by only one element, then the map $\pi_*(q) \otimes \mathbb{Q}$ is surjective, which implies that φ is injective. The relative minimal model we have constructed above is then the minimal model of $M\#N$.

Example 3.7 The minimal model of a connected sum $M\#N$ depends on the choice of the fundamental classes ω_N and ω_M of N and M. If we reverse the orientation of N the fundamental class becomes $-\omega_N$, and we have $dx = \omega_M + \omega_N$. This change of orientation can change the rational homotopy type. This is the case when $M = N = \mathbb{CP}(2)$.

Write $H^*(M; \mathbb{Q}) = \mathbb{Q}[x]/x^3$ and $H^*(N; \mathbb{Q}) = \mathbb{Q}[y]/y^3$. If the fundamental classes are x^2 and y^2, the minimal model for $M\#N = \mathbb{CP}(2)\#\mathbb{CP}(2)$ is $(\wedge(x, y, z, t), d)$, $dx = dy = 0$, $dt = xy$, $dz = x^2 - y^2$. By Exercise 2.4, we can see that no other generators are required. Hence, $\mathbb{CP}(2)\#\mathbb{CP}(2)$ is rationally elliptic.

If we reverse the orientation in N, we get $(\wedge(x, y, z', t'), d)$, $dx = dy = 0$, $dt' = xy$, $dz' = x^2 + y^2$ and this is a model of $\mathbb{CP}(2)\#\overline{\mathbb{CP}}(2)$, the blow-up of $\mathbb{CP}(2)$ at a point (see Subsection 8.2.1). It is quite easy to see that there is no isomorphism between the two cdga's (see Exercise 3.2). Therefore, the two manifolds do not have the same rational homotopy type.

Example 3.8 (Rationally elliptic 4-manifolds) Rational homotopy conditions imposed on manifolds often restrict possibilities greatly. For instance, which rational homotopy types of closed simply connected 4-manifolds are rationally elliptic? Of course, the key properties that we shall use are contained in Theorem 2.75. In particular, there is the dimension formula (6) as well as the refinements (7) and (8) of Theorem 2.75. In what follows, for each i, we shall refer to the i-th rational homotopy group simply by π_i.

The first thing to notice is that, by (8), $0 \leq \dim \pi_5 \leq 1$ and $0 \leq \dim \pi_7 \leq 1$ and they cannot be non-zero simultaneously. Moreover, again by (8), if $\dim \pi_7 = 1$ or $\dim \pi_5 = 1$, then $\pi_3 = 0$. Similarly, by (7), if $\pi_4 \neq 0$, then $\pi_2 = 0$ and, if $\pi_4 = 0$, then $\dim \pi_2 \leq 2$. Finally, by (7), $\dim \pi_4 = 0$ or $\dim \pi_4 = 1$. This means we can work with each case separately.

So suppose $\pi_4 = \mathbb{Q}$. Then the dimension formula (6) restricts the possibilities for homotopy. For instance, if $\pi_5 = \mathbb{Q}$, then the formula becomes $4 = 5 - 4 + 1 = 2$, which is not true. Thus, in this case we cannot have a degree five generator. The other two cases are when $\pi_7 = \mathbb{Q}$ and when $\pi_5 = 0 = \pi_7$. The respective dimension formulas are $4 = 7 - 4 + 1$, which is true, and $4 = 3 \dim \pi_3 - 4 + 1 = 3 \dim \pi_3 - 3$, which is false since 4 is not divisible by 3. The true case gives a model with a degree 4 generator and a degree 7 generator. The differential is then forced since we need finite cohomology. The model is therefore $(\wedge(x_4, y_7), dy = x^2)$ and we recognize this as a model for S^4.

Now suppose $\pi_4 = 0$. Then, since $\dim \pi_2 \leq 2$, the reader can show that the only dimension formula possibilities are $4 = 5 - \dim \pi_2$ and $4 = 3 \dim \pi_3 - \dim \pi_2$. The first, with $\dim \pi_2 = 1$, gives a model $(\wedge(x_2, y_5), dy = x^3)$ and this is a model for $\mathbb{CP}(2)$. The other can only hold when $\dim \pi_3 = 2$ and $\dim \pi_2 = 2$. The possible models are:

- $(\wedge(x_1, x_2, y_1, y_2), dy_1 = x_1^2, dy_2 = x_2^2)$; a model for $S^2 \times S^2$.
- $(\wedge(x_1, x_2, y_1, y_2), dy_1 = x_1 x_2, dy_2 = x_1^2 - x_2^2)$; a model for $\mathbb{CP}(2) \# \mathbb{CP}(2)$.
- $(\wedge(x_1, x_2, y_1, y_2), dy_1 = x_1 x_2, dy_2 = x_1^2 + x_2^2)$; a model for $\mathbb{CP}(2) \# \overline{\mathbb{CP}}(2)$.

Therefore, the rational homotopy types of simply connected rationally elliptic 4-manifolds are given by S^4, $S^2 \times S^2$, $\mathbb{CP}(2)$, $\mathbb{CP}(2) \# \mathbb{CP}(2)$ and $\mathbb{CP}(2) \# \overline{\mathbb{CP}}(2)$. See Exercise 3.1 for the case of elliptic 5-manifolds.

3.1.3 Poincaré duality models

A *Poincaré duality model* for a compact simply connected n-dimensional manifold M is a cdga (A, d) that satisfies the following properties.

1. There are quasi-isomorphisms $(A, d) \xleftarrow{\simeq} \mathcal{M}_M \xrightarrow{\simeq} A_{PL}(M)$.
2. $A^p = 0$ for $p > n$, $A^0 = \mathbb{Q}$, $A^1 = 0$, each A^q is finite dimensional and $A^n = \mathbb{Q}\omega$, for $\omega \in A^n$.
3. The map $\varphi \colon A^p \to \text{Hom}(A^{n-p}, \mathbb{Q})$ given by $\varphi(a)(b) = \lambda$ if $ab = \lambda \omega$, is an isomorphism.

When (A,d) is a Poincaré duality model, then its dual $\mathrm{Hom}((A,d),\mathbb{Q})$ is a free (A,d)-module of rank one generated by the homomorphism that takes the value 1 on ω.

When $\dim V < \infty$ and $V = V^{\mathrm{odd}}$, then $(\wedge V, d)$ is a Poincaré duality algebra. A result of Lambrechts and Stanley generalizes this fact (see [169]).

Theorem 3.9 *Every compact simply connected manifold admits a Poincaré duality model.*

3.1.4 Formality of manifolds

Since we wish to study aspects of the geometry of manifolds using algebraic models, our first step might be to see if the particular manifold of interest is formal (see Section 2.7). Of course, there *are* some important classes of manifolds that are formal. For instance, spheres, Lie groups and Kähler manifolds (see Theorem 4.43) are formal spaces. The first important result on the formality of manifolds is due to Miller [192]. We give here a proof that is inspired by Miller's, but which uses differential graded algebra models instead of Lie models. Our proof makes use of the obstruction theory for formality presented in Subsection 2.7.2. A very different and interesting proof can also be found in [97]. After Miller's theorem, we give a powerful theorem of Stasheff [241] that says that an n-dimensional manifold M is formal if its $(n-1)$-skeleton is. The original complicated proof of this result also used Lie models (which are generally good for situations where cell-attaching occurs), but here we stay within the world of differential graded algebras. Let's begin now with Miller.

Proposition 3.10 ([192]) *Let M be a $(p-1)$-connected compact manifold, $p \geq 2$, of dimension $m \leq 4p-2$. Then M is formal.*

Proof To prove the theorem we will use the obstruction theory developed in Proposition 2.98. Let $(\wedge V, d)$ denote the *bigraded model* of $H = H^*(M;\mathbb{Q})$. By the connectivity hypothesis, $V_0^q = 0$ for $q < p$, and this implies $V_r^q = 0$ for $q < (r+1)p - r$. By Poincaré duality, we have $H^q = 0$ for $m-p < q < m$ and $H^m = \mathbb{Q} \cdot \omega \cong \mathbb{Q}$.

Before we look at the obstructions we make two observations. First of all, we notice that $\wedge^+ V_{\geq 1}$ does not contain any non-zero cocycle. To see this, suppose that $\wedge^+ V_{\geq 1}$ has nonzero cocycles. We denote by (x_r) an ordered basis of $V_{\geq 1}$ such that $d(x_r) \in \wedge V_0 \otimes \wedge(x_1, \ldots, x_{r-1})$, and we denote by a a nonzero cocycle in $\wedge(x_1, \ldots, x_r)$, where r is the smallest possible for a cocycle. We write $a = a_n x_r^n + \ldots a_1 x_r + a_0$ with

$a_n \neq 0$ and $a_i \in \wedge(x_1, \ldots, x_{r-1})$. Then $d(a_n) = 0$ and, by our minimality condition, a_n is a constant. From $da = 0$, we then deduce that $n\, a_n d(x_r) + d(a_{n-1}) = 0$. This shows that by making a change of generators we can suppose $dx_r = 0$, which is impossible. Therefore, no cocycles exist in $\wedge^+ V_{\geq 1}$.

Poincaré duality is crucial in the proof. We will explicitly use the following property of Poincaré duality algebras: every linear map $f: W \otimes H^{m-r} \to H^m$ factors as a composition

$$W \otimes H^{m-r} \xrightarrow{\varphi \otimes 1} H^r \otimes H^{m-r} \xrightarrow{\text{mult}} H^m.$$

Given f, for every element $w \in W$, we obtain by adjunction a linear map $f_w: H^{m-r} \to H^m$ defined by $f_w(x) = f(w \otimes x)$. By Poincaré duality, there is then an element $x_w \in H^r$ such that $f_w(y) = x_w \cdot y$ for any y. We define $\varphi(w) = x_w$ linearly and obtain a map $\varphi: W \to H^r$ such that $f(w \otimes x) = \varphi(w) \cdot x$.

Now we can consider the obstructions to the formality. By Proposition 2.98, the only obstructions are $[d_2: V_2^{m-1} \to (\wedge V_0)^m]$ and $[d_3: V_3^{m-1} \to (\wedge V_0)^m]$. The other obstructions are zero because $V_q^{<m} = 0$ for $q \geq 4$. Denote by $p: \wedge V_0 \to H$ the natural projection. For degree reasons, the composition

$$\tilde{d}: V_2^{m-1} \xrightarrow{d} V_1 \otimes (\wedge V_0) \xrightarrow{1 \otimes p} V_1 \otimes H$$

is injective. More precisely, in degree $m = 4p - 2$ we must have $d(V_2^{m-1}) \subseteq V_1 \cdot V_0$ and $p: V_0 \to H$ is injective. There is thus a linear map $\theta: V_1 \otimes H \to H^m$ defined by $\theta|_{\text{Im}\,\tilde{d}} = p \circ d_2$ and $\theta|_C = 0$ for any complement C to $\text{Im}\,\tilde{d}$. By definition, we have $\theta \circ \tilde{d} = p \circ d_2$. This is expressed in the following diagram

$$\begin{array}{ccc} V_2^{m-1} & \xrightarrow{d_2} & (\wedge V_0)^m \\ {\scriptstyle \tilde{d}}\downarrow & & \downarrow{\scriptstyle p} \\ V_1 \otimes H & \xrightarrow{\theta} & H^m \end{array}$$

By the property of Poincaré duality algebras given above, there is then a linear map $\varphi: V_1 \to H$ such that $\theta(x \otimes h) = \varphi(x) \cdot h$. We now choose a linear map $\tilde{\varphi}_1: V_1 \to \wedge V_0$ such that $p \circ \tilde{\varphi}_1 = \varphi$. We define $\tilde{\varphi}_1$ to be zero on V_0 and extend it as a derivation φ_1 on $\wedge V_1 \otimes \wedge V_0$. Note that we have a commutative diagram (with the top row being φ_1 and where μ and $\bar{\mu}$ are the multiplications)

$$
\begin{array}{ccc}
V_1 \otimes \wedge V_0 & \xrightarrow{\tilde{\varphi}_1 \otimes 1} \wedge V_0 \otimes \wedge V_0 & \xrightarrow{\bar{\mu}} \wedge V_0 \\
{\scriptstyle p \otimes p} \downarrow & & \downarrow {\scriptstyle p} \\
H \otimes H & \xrightarrow{\mu} & H
\end{array}
$$

Hence, $p\varphi_1 = p\bar{\mu}(\tilde{\varphi}_1 \otimes 1) = \mu(p \otimes p)(\tilde{\varphi}_1 \otimes 1)$.

The equality $\theta(x \otimes h) = \varphi(x) \cdot h = \mu(\varphi(x) \otimes h)$ also provides a commutative diagram

$$
\begin{array}{ccccc}
V_1 \otimes \wedge V_0 & \xrightarrow{\varphi \otimes 1} & H \otimes \wedge V_0 & \xrightarrow{1 \otimes p} & H \otimes H \xrightarrow{\mu} H \\
{\scriptstyle 1 \otimes p} \downarrow & & & \nearrow {\scriptstyle \theta} & \\
V_1 \otimes H & & & &
\end{array}
$$

Hence, $\theta(1 \otimes p) = \mu(1 \otimes p)(\varphi \otimes 1)$.

Using the relations expressed by these diagrams, consider, for each element $z \in V_2$, the element $d_2(z) - \varphi_1(dz)$. This element is a coboundary as can be seen from the following calculation.

$$
\begin{aligned}
p(d_2(z) - \varphi_1(dz)) &= pd_2(z) - p\varphi_1(dz) \\
&= \theta\tilde{d}(z) - \mu(p \otimes p)(\tilde{\varphi}_1 \otimes 1)(dz) \\
&= \theta(1 \otimes p)(dz) - \mu(p\tilde{\varphi}_1 \otimes p)(dz) \\
&= \mu(1 \otimes p)(\varphi \otimes 1)(dz) - \mu(\varphi \otimes p)(dz) \\
&= \mu(\varphi \otimes p)(dz) - \mu(\varphi \otimes p)(dz) \\
&= 0.
\end{aligned}
$$

Let $dw = d_2(z) - \varphi_1(dz)$ and define $\varphi_1(z) = w$. We can thus define φ_1 on V_2 so that $d_2 = d\varphi_1 + \varphi_1 d$ and the first obstruction is zero.

For $[d_3]$, we have only to consider the case $m = 4p - 2$. The composition

$$V_3^{4p-3} \xrightarrow{d} (V_2^{3p-2} \otimes V_0^p) \oplus (\wedge^2 V_1^{2p-1}) \xrightarrow{\text{proj}} V_2^{3p-2} \otimes V_0^p$$

is injective because there is always a nonzero part in $V_2^{3p-2} \otimes V_0^p$. The same construction as above gives a map $\varphi_2 \colon V_r \to (\wedge V)_{r-2}$ such that $d\varphi_2 + \varphi_2 d = d_3$. Thus, the second obstruction vanishes as well and the manifold is formal. □

Now let's turn to Stasheff's theorem. Before we can prove the result, we need some preliminaries on the homology of bigraded models (see Theorem 2.93). Let $(\wedge V, d)$ be the bigraded model of a commutative graded algebra H satisfying Poincaré duality in dimension m. We write $R_n = H_n^m(\wedge V_{\leq n}, d)$ for $n \geq 1$. Then denote by $\pi_1 \colon (\wedge V_{\leq n})_n \to V_n \otimes \wedge V_0$ the natural projection with kernel $(\wedge^{\geq 2} V_{\geq 1}) \otimes \wedge V_0$ and by $\pi_2 \colon V_n \otimes \wedge V_0 \to V_n \otimes H$ the projection that associates to $\alpha \otimes \beta$ the tensor product $\alpha \otimes [\beta]$. The composition $\pi_2 \circ \pi_1$ vanishes on $[d(\wedge V_{\leq n})]_n$ and the restriction to the cocycles induces a linear map

$$\theta \colon R_n \to V_n \otimes H.$$

Lemma 3.11 *The map θ is injective.*

Proof We proceed by induction on n. We suppose the result is true for $q < n$, and we write $\wedge V_{\leq n} = \wedge(x_1, \ldots, x_\ell)$ with $dx_i \in \wedge(x_1, \ldots, x_{i-1})$. Now let α be an element of R_n and $a \in \wedge(x_1, \ldots, x_r)$ be a cocycle with $[a] = \alpha$ such that r is minimal among all representatives of α. Then,

$$a = x_r^p u_p + x_r^{p-1} u_{p-1} + \ldots + x_r u_1 + u_0,$$

with $u_i \in \wedge(x_1, \ldots, x_{r-1})$. Since a is a cocycle, $du_p = 0$. If $u_p = dv$, then, replacing a by $a' = a - (-1)^{|x_r|} d(x_r^p v)$, we get a new cocycle in the class of α such that $a' = x_r^{p-1} u'_{p-1} + \ldots + u'_0$. We can therefore suppose $[u_p] \neq 0$. When $p > 1$, $da = 0$ implies that $p(dx_r) u_p + du_{p-1} = 0$ as well. The element $p x_r u_p + u_{p-1}$ is therefore a cocycle in $(\wedge V)_q$ for some q, $0 < q < n$, and is a coboundary in $(\wedge V)_{\leq n}$:

$$p x_r u_p + u_{p-1} = dv.$$

We then replace a by $a'' = a - d\left(\frac{1}{p} x_r^{p-1} v\right)$ and we obtain a new representative of α of the form

$$a'' = x_r^{p-1} u''_{p-1} + \ldots + u''_0.$$

In conclusion, we can always suppose $p = 1$ and that a representative of α has the form $a = u_1 x_r + u_0$, with $u_1 \in \wedge V_0$, $[u_1] \neq 0$, and $x_r \in V_n$. $\square$

We can now state and prove Stasheff's theorem.

Theorem 3.12 *Let M^m be a simply connected compact manifold. If $M \setminus \{*\}$ is formal, then M is formal.*

Proof Let $(\wedge V, d)$ be the bigraded model of $H = H^*(M; \mathbb{Q})$ with associated filtered model $(\wedge V, D)$. Because $M \setminus \{*\}$ is formal, we can assume that $D = d$ on $V^{<m-1}$. We will now inductively construct derivations $\varphi_i \colon V_* \to (\wedge V)_{*-i}$, $i \geq 1$, such that, denoting $D_1 = D$ and $D_r = D_{r-1} - (d\varphi_{r-1} - \varphi_{r-1}d)$, we have $D_r = d$ on $V_{\leq r} + V^{<m-1}$. This will imply the result by the obstruction theory of Subsection 2.7.2.

Now suppose $\varphi_1, \ldots, \varphi_{r-2}$ have been constructed and write $D = D_{r-1}$. Since $D = D_{r-1} = d$ on $V_{<r}$, we can suppose that $(D - d)(V_p) \subset (\wedge V)_{\leq p-r}$. In particular, $(D - d)(V_r) \subset \wedge V_0$. Since $D = d$ on $(\wedge V)_{<r}$, we have $R_{r-1} = H^m_{r-1}((\wedge V)_{<r}, d) = H^m_{r-1}((\wedge V)_{<r}, D)$ and the injection $(\wedge V_{<r}, D) \to (\wedge V, D)$ induces in cohomology a map $\psi \colon R_{r-1} \to H^m(\wedge V, d) = \mathbb{Q}\omega$. At this point we use Poincaré duality just as we did in the proof of Miller's theorem. Namely, for each graded vector space W and each degree 0 map $g \colon W \otimes H \to H^m = \mathbb{Q}\omega$, there is a degree 0 map $f \colon W \to H$ such that $g(w \otimes h) = f(w) \cdot h$. Since $\theta \colon R_{r-1} \to V_{r-1} \otimes H$ is an injection (by Lemma 3.11), there is a linear map $\widetilde{\varphi}_{r-1} \colon V_{r-1} \to H$ such that $\psi = \mu(\widetilde{\varphi}_{r-1} \otimes 1) \circ \theta$, where $\mu \colon H \otimes H \to H$ is the multiplication. This is expressed by the following diagram

$$\begin{array}{ccc} R_{r-1} & \xrightarrow{\psi} & H \\ \downarrow{\theta} & & \uparrow{\mu} \\ V_{r-1} \otimes H & \xrightarrow{\widetilde{\varphi}_{r-1} \otimes 1} & H \otimes H \end{array}$$

We then lift $\widetilde{\varphi}_{r-1}$ into a linear map $\varphi_{r-1} \colon V_{r-1} \to \wedge V_0$ such that $[\varphi_{r-1}(v)] = \widetilde{\varphi}_{r-1}(v)$ for $v \in V_{r-1}$. Now we extend φ_{r-1} to a derivation on $\wedge V_{\leq r-1}$ by putting $\varphi_{r-1}(V_q) = 0$ if $q < r - 1$.

Now let u be a d-cocycle in $(\wedge V)_{r-1}$. Then we have

$$[\varphi_{r-1}(u)] = \mu(\widetilde{\varphi}_{r-1} \otimes 1)\theta([u]) = \psi([u]).$$

To see this, write $u = \sum v_i \otimes u_i + u'$, with $v_i \in V_{r-1}$, $u_i \in \wedge V_0$ and $u' \in (\wedge^{\geq 2} V_{\geq 1}) \otimes \wedge V_0$. Then, since $\varphi(V_q) = 0$ for $q < r - 1$ and u' is a nontrivial product, we have

$$[\varphi_{r-1}(u)] = \left[\varphi_{r-1}\left(\sum v_i \otimes u_i\right)\right] = \sum [\widetilde{\varphi}_{r-1}(v_i)] \cdot [u_i]$$
$$= \mu(\widetilde{\varphi}_{r-1} \otimes 1)\theta([u]) = \psi([u]).$$

Now let $z \in V_r^{m-1}$. By our description of D, we see that $Dz = dz + d_r z$. Hence, with respect to D-cohomology, we have $\psi([dz]) = [dz] = -[d_r z]$ in $H^m(\wedge V, D)$. Note that dz is a D-cocycle because $D = d$ on $\wedge V_{r-1}$. By the calculation $[\varphi_{r-1}(u)] = \psi([u])$ above, we get

$$[d_r z + \varphi_{r-1}(dz)] = [d_r z] + [\varphi_{r-1}(dz)] = -\psi([dz]) + \psi([dz]) = 0$$

and $d_r z + \varphi_{r-1}(dz)$ is a coboundary. Hence, there is a $u \in (\wedge V)_1$ such that $du = d_r z + \varphi_{r-1}(dz)$. Define φ_{r-1} linearly on V_r^{m-1} by $\varphi_{r-1}(z) = u$ to obtain $d_r z = d\varphi_{r-1}(z) - \varphi_{r-1}(dz)$ on V_r^{m-1}.

Now let t be in V_r^q for some $q < m - 1$. If $[\varphi_{r-1}dt] \neq 0$, then there is a cocycle a such that $[\varphi_{r-1}dt] \cdot [a] = \omega \in H^m$. But this is impossible because (recalling that $D = d$ on V^{m-1} and φ_{r-1} is a derivation)

$$[\varphi_{r-1}dt] \cdot [a] = [\varphi_{r-1}d(ta)] = \psi([dt\,a]) = 0.$$

Therefore, $[\varphi_{r-1}dt] = 0$ and there is an element $u \in (\wedge V)_1$ such that $du = \varphi_{r-1}(dt)$. We define $\varphi_{r-1}(t) = u$ on a basis of $V_r^{<m-1}$. We can continue and define φ_{r-1} on $V_{>r}^{\leq m-1}$ such that $d_r = d\varphi_{r-1} - \varphi_{r-1}d$. Then by putting $D_r = e^{\varphi_{r-1}}De^{-\varphi_{r-1}}$, we obtain a derivation satisfying our inductive hypothesis $D_r = d$ on $V^{<m-1} + V_{\leq r}$. □

Now let's use Stasheff's theorem to give another proof of Miller's theorem. While the proof of Miller above showed the power of the obstruction theory developed earlier, it still lacked a good intuitive interpretation. Stasheff's result shows us exactly how the hypotheses of Miller's theorem produce a formal $(4p - 3)$-skeleton.

Proof 2 of Miller's Theorem 3.10. Let $(\wedge V, d)$ denote the bigraded model of $H = H^*(M; \mathbb{Q})$. By the connectivity hypothesis, $V_0^q = 0$ for $q < p$, and this implies $V_r^q = 0$ for $q < (r+1)p - r$. By Poincaré duality, we have $H^q(M\setminus\{*\}; \mathbb{Q}) = 0$ for $q > 3p - 2$. Therefore, by Proposition 2.99, $M\setminus\{*\}$ is formal, and by Stasheff's Theorem 3.12, M is formal. □

The results above are the most fundamental formality results for manifolds. Once we have these basic results, however, we can ask how various geometric constructions behave with respect to formality. In particular, we can see that formality is preserved by certain constructions. The following result exemplifies this fact.

Theorem 3.13 *The connected sum of two compact simply connected formal n-dimensional manifolds is formal.*

Proof Let M and N be nilpotent compact n-dimensional manifolds, and $M\#N$ be the connected sum. With the notation of Example 3.6, we have a quasi-isomorphism

$$\varphi: (\mathcal{M}_{M\vee N} \otimes \wedge(x, x_i), d) \to \mathcal{M}_{M\#N},$$

where $dx = \omega_M - \omega_N$, and the x_i are in degrees $\geq n$.

On the other hand we have a surjective quasi-isomorphism (see Example 2.47)

$$\sigma: \mathcal{M}_{M \vee N} \to \mathcal{M}_M \oplus_{\mathbb{Q}} \mathcal{M}_N.$$

We obtain a quasi-isomorphism by tensoring with $\wedge(x, x_i), d$:

$$\sigma': (\mathcal{M}_{M \vee N} \otimes \wedge(x, x_i), d) \to (\mathcal{M}_M \oplus_{\mathbb{Q}} \mathcal{M}_N \otimes \wedge(x, x_i), D),$$

with $D(x) = \sigma d(x)$ and $D(x_i) = (\sigma \otimes 1)d(x_i)$.

By using the formality of M and N, we finally obtain quasi-isomorphisms

$$(\mathcal{M}_M \oplus_{\mathbb{Q}} \mathcal{M}_N \otimes \wedge(x, x_i), D) \xrightarrow{\cong} (H^*(M; \mathbb{Q}) \oplus_{\mathbb{Q}} H^*(N; \mathbb{Q}) \otimes \wedge(x, x_i), \widetilde{D})$$

$$\xrightarrow{\cong} (H^*(M; \mathbb{Q}) \oplus_{\mathbb{Q}} H^*(N; \mathbb{Q})/(\omega_M - \omega_N), 0)$$

where $\widetilde{D} = \theta \circ D$ and $\theta: \mathcal{M}_M \oplus \mathcal{M}_N \to H^*(M; \mathbb{Q}) \oplus H^*(N; \mathbb{Q})$ is a quasi-isomorphism obtained from the formality of M and N. □

3.2 Nilmanifolds

In [183], Malcev studied nilpotent, simply connected Lie groups N acting transitively and properly on a compact manifold M. Recall that any connected, nilpotent, locally compact group acting properly on a manifold is a Lie group. Directly from the definition, we see that M is isomorphic to the quotient of N by the co-compact discrete subgroup Γ stabilizing a point. We call $M = N/\Gamma$ a *nilmanifold* and sometimes write the pair (N, Γ) to denote $M = N/\Gamma$.

Let (N, Γ) and (N', Γ') be two nilmanifolds. Malcev observed that any isomorphism between Γ and Γ' can be extended to a homeomorphism between N and N'. Thus, a nilmanifold is determined by its fundamental group. We therefore arrive at

Question 3.14 *What are the conditions for an abstract group Γ to be the fundamental group of a nilmanifold? Note that this is equivalent to asking when a group Γ is a uniform, co-compact discrete subgroup of a Lie group.*

Malcev provided the answer by giving a necessary and sufficient condition: *Γ is the fundamental group of a nilmanifold if and only if Γ is nilpotent, finitely generated and torsion-free.*

A simply connected nilpotent Lie group is diffeomorphic to a Euclidean space, so a nilmanifold has a fundamental group that is a finitely generated torsionfree nilpotent group and has higher order homotopy groups which are trivial. Nilmanifolds then provide prime examples of $K(\pi, 1)$-manifolds;

3.2 Nilmanifolds

that is, compact manifolds with the fundamental group as the only nontrivial homotopy group. Clearly, any nilmanifold is orientable. Examples are given by any torus $T^n = \mathbb{R}^n/\mathbb{Z}^n$ and the Heisenberg manifold formed by the quotient of the Lie group of matrices of the form

$$\begin{pmatrix} 1 & a & b \\ 0 & 1 & c \\ 0 & 0 & 1 \end{pmatrix},$$

with a, b and c real numbers, by the subgroup of the corresponding matrices with integer entries.

3.2.1 Relations with Lie algebras

Instead of starting with the discrete group Γ, we may start with the Lie group N and ask:

Question 3.15 *What are the conditions on a connected, simply connected, nilpotent Lie group N that ensure the existence of a uniform, co-compact subgroup?*

This was also answered by Malcev. A necessary and sufficient condition is that *the Lie algebra of N has rational structure constants relative to some chosen basis*. To any nilmanifold, we can associate a rational nilpotent Lie algebra $\mathfrak{n}$ which naturally elicits the following

Question 3.16 *What is the topological invariant corresponding to this rational Lie algebra?*

To answer this question, Malcev gave the following necessary and sufficient condition: (N,Γ_1) *and* (N,Γ_2) *have isomorphic rational Lie algebras if and only if* $N/(\Gamma_1 \cap \Gamma_2)$ *finitely covers* N/Γ_1 *and* N/Γ_2. We also note that, in [183], Malcev gives an interesting example of two nonisomorphic rational Lie algebras which become isomorphic over the reals. An example of this type of descent phenomena for the rational and real homotopy type of manifolds is given in Example 6.15.

With all of this in mind, let $\mathfrak{n}$ be a nilpotent Lie algebra with the property that there exists a basis in $\mathfrak{n}$, $e_1, e_2, \ldots, e_n$, such that the structure constants c_{ij}^k arising in brackets

$$[e_i, e_j] = \sum_k c_{ij}^k e_k$$

are rational numbers for all i, j, k. In fact, Malcev showed that, corresponding to $\mathfrak{n}$, there is a simply connected nilpotent Lie group N which admits a lattice (i.e. a discrete co-compact subgroup) Γ so that N/Γ is a compact nilmanifold.

Example 3.17 Consider the nilpotent Lie group of upper triangular matrices having 1's along the diagonal, $\mathbb{U}_n(\mathbb{R})$.

$$\mathbb{U}_n(\mathbb{R}) = \left\{ \begin{pmatrix} 1 & x_{12} & x_{13} & \cdots & x_{1n} \\ 0 & 1 & x_{23} & \cdots & x_{2n} \\ \cdots & \cdots & \cdots & \cdots & \cdots \\ 0 & \cdots & \cdots & 0 & 1 \end{pmatrix} \;\Big|\; x_{ij} \in \mathbb{R} \right\}.$$

Let $\mathbb{U}_n(\mathbb{Z}) \subset \mathbb{U}_n(\mathbb{R})$ denote the set of matrices having integral entries. Then, $\mathbb{U}_n(\mathbb{Z})$ is a lattice and the quotient

$$M = \mathbb{U}_n(\mathbb{R})/\mathbb{U}_n(\mathbb{Z})$$

is a nilmanifold. The group $\mathbb{U}_3(\mathbb{R})$ is called the *Heisenberg* group and the resulting nilmanifold is called the Heisenberg (nil)manifold. For other examples, see [257, Chapter 2].

Let $\mathfrak{g}$ denote a Lie algebra with basis $\{X_1, \ldots, X_s\}$. Then the dual of $\mathfrak{g}$, $\mathfrak{g}^*$, has basis $\{x_1, \ldots, x_s\}$ and there is a differential δ on the exterior algebra $\wedge \mathfrak{g}^*$ given by defining it to be dual to the bracket on degree 1 elements,

$$\delta x_k(X_i, X_j) = -x_k([X_i, X_j]),$$

and then extending δ to be a graded derivation. Now, $[X_i, X_j] = \sum c_{ij}^l X_l$, where c_{ij}^l are the structure constants of $\mathfrak{g}$, so duality then gives

$$\delta x_k(X_i, X_j) = -c_{ij}^k$$

and the differential has the form (on generators)

$$\delta x_k = -\sum_{i<j} c_{ij}^k x_i \wedge x_j.$$

We note that the Jacobi identity in the Lie algebra is equivalent to the condition $\delta^2 = 0$. Therefore, we obtain a cdga $(\wedge \mathfrak{g}^*, \delta)$ associated to the Lie algebra $\mathfrak{g}$. The cdga $(\wedge \mathfrak{g}^*, \delta)$ is the cochain algebra on the Lie algebra $L = \mathfrak{g}$ (see Subsection 2.6.2), and the differential δ is a particular case of the differential defined in Subsection 2.6.2 because $\mathfrak{g}^*$ is concentrated in degree one.

Theorem 3.18 (Model of a nilmanifold I) *If N/Γ is a nilmanifold, then the complex $(\wedge \mathfrak{n}^*, \delta)$ associated to $\mathfrak{n}$, is isomorphic to the minimal model of N/Γ.*

Proof The proof follows from Nomizu's theorem (see [209]) which says that the natural inclusion of invariant de Rham forms on the nilpotent Lie group into de Rham forms on the nilmanifold

$$A_{DR}^{inv}(N) \to A_{DR}(N/\Gamma)$$

is a quasi-isomorphism. Here, observe that the Lie group is not compact as in Theorem 1.30. (Also, note that the notation $\Omega_L(N)$ for invariant forms was used in Definition 1.25.) From the definition of the Lie algebra associated to a Lie group, we have

$$(\wedge \mathfrak{n}^*, \delta) \cong (A_{DR}^{inv}(N), d).$$

Hence, the composition

$$(\wedge \mathfrak{n}^*, \delta) \to A_{DR}^{inv}(N) \to A_{DR}(N/\Gamma)$$

displays $(\wedge \mathfrak{n}^*, \delta)$ as a *Sullivan model* for N/Γ. It remains to prove that the complex of $\mathfrak{n}$ is a minimal cdga. This follows from the general dual relationship between nilpotent Lie algebras and cdga's. Consider the dual basis $x_1, \ldots, x_n$ to the basis $X_1, \ldots, X_n$ of $\mathfrak{n}$ (ordered by central series extensions of $\mathfrak{n}$). Then, as above, the differential δ is defined on generators by

$$\delta x_k = \sum -c_{ij}^k x_i \wedge x_j$$

where $[X_i, X_j] = \sum c_{ij}^k X_k$. Since the X_i are ordered according to the way the nilpotent Lie algebra $\mathfrak{n}$ is built from central series extensions, the differential must be decomposable in terms of earlier generators. □

Example 3.19
Consider the Lie algebra $\mathfrak{n}$ of dimension $2m+2$ having basis

$$\{X_1, \ldots, X_m, Y_1, \ldots, Y_m, Z, W\}$$

with bracket structure given by

$$[X_i, Y_i] = -Z \quad \text{for all } i = 1, \ldots, m$$

and all other brackets zero. The associated minimal model is given by

$$(\wedge(x_1, \ldots, x_m, y_1, \ldots, y_m, z, w), d)$$

with $dx_i = dy_i = dw = 0$ and

$$dz = \sum_{i=1}^{m} x_i \wedge y_i.$$

For instance, if $m = 1$, then $\mathfrak{n} = \langle U, V, Y, T \rangle$ with only nonzero bracket $[U, Y] = -V$. The associated minimal model is $\wedge(u, v, y, t)$ with only nonzero differential $dv = uy$. Note that, in this case, the element $\omega = ut + vy$ has $\omega^2 = 2utvy \neq 0$. We will see in Subsection 4.6.4 that this implies that the corresponding nilmanifold, called the Kodaira–Thurston manifold, is a *symplectic* manifold.

Perhaps the most important rational homotopy property of nilmanifolds is that they are rarely formal spaces.

Proposition 3.20 *Any formal nilmanifold M^n has the rational homotopy type of a torus T^n.*

Proof Suppose that M is formal. Let $(\wedge V, d) = (\wedge(x_1, \ldots, x_n), d)$ be the minimal model of M and suppose $\varphi \colon (\wedge V, d) \to (H^*(M; \mathbb{Q}), 0)$ is a formality quasi-isomorphism. Now, each generator x_i has degree 1, so $\wedge V$ is an exterior algebra with $\dim (\wedge V)^n = 1$. A basis element for this top dimension is the product of all generators $\mu = x_1 \cdots x_n$. Since φ^* is an isomorphism, $\varphi^*(\mu) \neq 0$. Therefore $\varphi(\mu) \neq 0$ and this implies that $\varphi(x_i) \neq 0$ for each $i = 1, \ldots, n$. In fact, φ must be injective. For, suppose that $y = c \cdot x_{i_1} \cdots x_{i_j} + \ldots$ with $|y| = j$ and $\varphi(y) = 0$. Then there is a complementary set of generators $x_{i_{j+1}}, \ldots, x_{i_{n-j}}$ such that

$$x_{i_{j+1}} \cdots x_{i_{n-j}} \cdot y = K \cdot \mu$$

for some $K \in \mathbb{Q}$ with $K \neq 0$, and we have

$$\varphi(\mu) = \frac{1}{K} \cdot \varphi(x_{i_{j+1}} \cdots x_{i_{n-j}}) \cdot \varphi(y) = 0,$$

and this is a contradiction.

Now, φ is an injective cdga morphism to a cdga with differential equal to zero, so the differential d in $(\wedge V, d)$ must be zero as well. Therefore, M has the rational homotopy type of a torus. $\square$

Remark 3.21

1. It is also instructive to see how Proposition 3.20 fits with our approach to formality via the bigraded model. Of course, the key feature is again the fact that the top cohomology class is represented by the product of all generators of the model. Let $\varphi \colon (\wedge V, d) \to (H^*(M; \mathbb{Q}), 0)$ be the bigraded model of the cohomology of M. Because we assume that M is formal, this bigraded model is, in fact, the minimal model of M. By hypothesis, $V = V^1$ is finite dimensional. Denote by $x_1, \ldots, x_r$ a bigraded basis of V^1. The element obtained by multiplying all generators, $\mu = x_1 \cdot x_2 \cdots x_r$, is a cocycle in $(\wedge V)^r$ and a basis of $(\wedge V)^r$.

Indeed, since the model is an exterior algebra and since the nilmanifold is orientable, the element μ is a nontrivial top class and we must have $r = n$. Also, since $\varphi(V_p) = 0$ if $p > 0$, we must have $\mu \in (\wedge V)_0$. Because the lower grading is multiplicative, this implies that *all* the elements x_i belong to V_0. Hence, the differential d is zero and $(\wedge V, 0)$ is the minimal model of a torus T^n.

2. Another way to say this is that, unless each generator is a cocycle (in which case, the nilmanifold is a torus), then the top degree element is a cocycle, representing a nontrivial cohomology class, in the ideal generated by the noncocycle generators of the minimal model. By Exercise 2.3, we again obtain the fact that nontoral nilmanifolds are never formal.

3. In fact, a formal nilmanifold $M = K(\pi, 1)$ must be diffeomorphic to a torus by the following argument. Suppose M has the rational homotopy type of an m-torus. Then, since $\pi_1(M) = \pi$ is finitely generated nilpotent, it rationalizes to $(\oplus_m \mathbb{Z})_\mathbb{Q} = (\oplus_m \mathbb{Q})$ and, since it is torsion-free, the rationalization $\pi \to \oplus \mathbb{Q}$ is injective. Hence π is finitely generated, torsion-free and abelian, so we have $\pi = \oplus \mathbb{Z}$. By Mostow's classification of solvmanifolds, M is diffeomorphic to a torus.

3.2.2 Relations with principal bundles

Let $M = N/\Gamma$ be the quotient of the nilpotent Lie group N by the discrete co-compact subgroup Γ. It is well known that the exponential map $\exp\colon \mathfrak{n} = T_e N \to N$ is a global diffeomorphism, and this in fact gives a diffeomorphism from N^n to $\mathbb{R}^n$. Since $N \to M$ is a covering because Γ acts on N by translations, M is a $K(\Gamma, 1)$ with Γ a finitely generated torsion-free nilpotent group. On the algebraic side, there is a refinement of the *upper central series* of Γ,

$$\Gamma \supseteq \Gamma_2 \supseteq \Gamma_3 \supseteq \cdots \supseteq \Gamma_n \supseteq 1,$$

with each $\Gamma_i/\Gamma_{i+1} \cong \mathbb{Z}$. The length of this series is invariant and is called the *rank* of Γ. So, for Γ above, rank $(\Gamma) = n$. This description implies that any $u \in \Gamma$ has a decomposition $u = u_1^{x_1} \cdots u_n^{x_n}$, where $\langle u_n \rangle = \Gamma_n, \cdots \langle u_i \rangle = \Gamma_i/\Gamma_{i+1}$. The set $\{u_1, \cdots u_n\}$ is called a *Malcev basis* for Γ and, using this basis, the multiplication in Γ takes the form

$$u_1^{x_1} \cdots u_n^{x_n} u_1^{y_1} \cdots u_n^{y_n} = u_1^{\rho_1(x,y)} \cdots u_n^{\rho_n(x,y)}$$

where

$$\rho_i(x, y) = x_i + y_i + \tau_i(x_1, \ldots x_{i-1}, y_1, \ldots y_{i-1}).$$

For instance, for the group $N = U_n(\mathbb{R})$ with $\Gamma = U_n(\mathbb{Z})$, a Malcev basis is given by $\{u_{ij} \mid 1 \leq i < j \leq n\}$ where $u_{ij} = I + e_{ij}$ and e_{ij} denotes the matrix

3 : Manifolds

with all zeros except for a 1 in the ij-th position. We then have

$$\rho_{ij}(x,y) = x_{ij} + y_{ij} + \sum_{i<k<j} x_{ik} y_{kj}.$$

Now, consider the central extension $\Gamma_n \to \Gamma \to \overline{\Gamma}$. The cocycle for the extension is $\tau_n \colon \overline{\Gamma} \times \overline{\Gamma} \to \mathbb{Z}$. Of course $\overline{\Gamma}$ is also finitely generated torsion-free with refined upper central series,

$$\overline{\Gamma} = \frac{\Gamma}{\Gamma_n} \supseteq \frac{\Gamma_2}{\Gamma_n} \supseteq \cdots \supseteq \frac{\Gamma_{n-1}}{\Gamma_n} \supseteq \frac{\Gamma_n}{\Gamma_n} = 1.$$

Hence, rank $(\overline{\Gamma}) = n - 1$ and

$$\overline{\rho}_i(x,y) = \rho_i((x,0),(y,0)) = x_i + y_i + \tau_i(x_1, \ldots x_{i-1}, y_1, \ldots y_{i-1})$$

for $i < n$. The cocycle τ_n gives an extension cohomology class $[\tau_n] \in H^2(\overline{\Gamma}; \mathbb{Z}) \cong H^2(K(\overline{\Gamma}, 1); \mathbb{Z}) \cong [K(\overline{\Gamma}, 1), K(\mathbb{Z}, 2)] = [K(\overline{\Gamma}, 1), BS^1]$, so we obtain a principal circle bundle over $K(\overline{\Gamma}, 1)$: $S^1 \to K(\Gamma, 1) \to K(\overline{\Gamma}, 1) \xrightarrow{\tau_n} \mathbb{CP}(\infty)$. We iterate this procedure modeled on the algebraic decomposition of Γ to obtain an iterated sequence of principal S^1-bundles classified by extension classes $[\tau_i] \in H^2(\overline{\Gamma}_i; \mathbb{Z})$ (where the coefficients are untwisted since the extension is central and $\overline{\Gamma}_i$ arises at the ith stage of the construction).

This sequence of bundles produces a nilmanifold because the sequence of extensions gives a torsion-free nilpotent group. On the other hand, the decomposition of $M = K(\Gamma, 1)$ into a tower of principal S^1-bundles classified by the τ_i is precisely the right information allowing us to construct the minimal model of M from relative Sullivan cdga's with the twisting of the differential corresponding to the τ_i. Alternatively, for the more homotopically-minded reader, we can see that the sequence of principal S^1-bundles is precisely the (refined) Postnikov tower for M (see Subsection 2.5.4) with the τ_i being the k-invariants. Therefore, we have

Theorem 3.22 (Model of a nilmanifold II) *The minimal model of a nilmanifold $M = N/\Gamma$ of dimension n has the form*

$$\mathcal{M}_M = (\wedge(x_1, \ldots x_n), d) \quad \text{with} \quad |x_i| = 1,$$

and $dx_i = \tau_i$, the extension cocycle for the ith stage of the upper central series of Γ.

Example 3.23 Take generators u and y for $H^1(T^2; \mathbb{Z})$ corresponding to the torus's circle factors and note that the fundamental class of T^2 is the cup product uy. Since a map $T^2 \to BS^1 = K(\mathbb{Z}, 2)$ is characterized by its effect on cohomology, we can use uy to classify a principal circle bundle

$S^1 \to H \to T^2$. In fact, the total space is just the Heisenberg manifold $H = \mathbb{U}_3(\mathbb{R})/\mathbb{U}_3(\mathbb{Z})$ from Example 3.17. Now let $KT = H \times S^1$ and note that we now have a principal T^2 bundle, $T^2 \to KT \to T^2$ with classifying map $T^2 \to BT^2 = K(\mathbb{Z} \oplus \mathbb{Z}, 2)$ given by the map $(uy, *)$. In this way we get one factor of the the total space being trivial. The relative minimal model is then given by the relative cdga

$$(\wedge(u, y), 0) \to (\wedge(u, y, v, t), D) \to (\wedge(v, t), 0)$$

with $D(u) = 0$, $D(y) = 0$, $D(v) = uy$ and $D(t) = 0$. Here the differential is determined by the classifying map. Indeed, principal bundles are always easy to model because the classifying map defines the differential.

Remark 3.24 We have given two descriptions above for the minimal model of a nilmanifold. The connection between them rests on the fact that the duals of the lower central series quotients of a Lie algebra together with the dual to the Lie bracket define a cdga (see Exercise 1.7 or [246] for instance). This duality is reflected by the relative models associated to the tower of principal fibrations above.

3.3 Finite group actions

In geometry, we often have occasion to view a manifold through the symmetries it admits. In order to use algebraic models in this context, we need to know how to transport the action of a symmetry group to the model. Here we will consider the case of finite transformation groups and in Chapter 7 we will consider connected groups.

3.3.1 An equivariant model for Γ-spaces

Let Γ be a finite group. A Γ-cdga is a cdga on which the group Γ acts by a homomorphism $\Gamma \to \mathrm{aut}_{cdga}(A, d_A)$.

Definition 3.25 A Γ-cdga (A, d_A) is called *minimal* if $(A, d_A) = (\wedge V, d)$ with

1. $d(V) \subset \wedge^{\geq 2}(V)$.
2. Each V^n is a Γ-module (i.e. this gives a Γ-structure to $\wedge V$).
3. d is Γ-equivariant: $d(ga) = gd(a)$.
4. V admits a filtration by sub Γ-spaces

$$0 \subset V(0) \subset V(1) \subset \cdots \subset V(n) \subset \cdots V = \cup_n V(v),$$

with $d(V(n)) \subset (\wedge V(n-1))$.

Generalizing the nonequivariant case, we have

3 : Manifolds

Theorem 3.26 (see [111], [215], [163], and [258])
Let (A, d_A) be a Γ-cdga. Suppose that $H^0(A, d_A) = \mathbb{Q}$ and $H^1(A, d_A) = 0$. Then there exists a Γ-minimal algebra $(\wedge V, d)$ and a Γ-equivariant quasi-isomorphism $\varphi \colon (\wedge V, d) \to (A, d_A)$. The Γ-minimal algebra $(\wedge V, d)$ is called the Γ-minimal model of the Γ-cdga (A, d_A), and it is unique up to Γ-isomorphism.

We give here the main lines of the proof. First recall that if V is a Γ-module and W is a sub Γ-module, then W admits a Γ-complement S in V. To construct S, we first choose a complement T of W and we denote by $\pi \colon V \to W$ the projection with kernel T. We can now make π equivariant by putting

$$\pi'(x) = \sum_{g \in \Gamma} g^{-1} \pi(gx).$$

The kernel S of π' is a Γ-complement of W in V.

This is very useful. Suppose that $p \colon E \to E'$ is a surjective Γ-module morphism. We denote by T a Γ-complement of $\operatorname{Ker} p$ in E, and we note that $p \colon T \to E'$ is a Γ-equivariant isomorphism. Hence, p admits an equivariant section.

Finally suppose that $W \subset V$ is a sub-Γ-module with Γ-complement S. The projection $V \to V/S$ is a Γ-equivariant projection $V \to W$.

We now construct the minimal model $\varphi \colon (\wedge V, d) \to (A, d_A)$ by induction on the degree of V. We put $V^1 = 0$, $V^2 = H^2(A, d_A)$ and we let π denote the canonical projection from the cocycles to the cohomology, $\pi \colon Z^2(A, d_A) \to H^2(A, d_A)$. We also denote by ρ a Γ-equivariant section of π. Then we define $d(V^2) = 0$ and $\varphi_2 = \varphi_{|V^2} = \rho$.

Consider the inductive step now. Suppose we have constructed a Γ-minimal algebra $(\wedge V^{<k}, d)$ and a Γ-equivariant morphism of cdga's

$$\varphi_{k-1} \colon (\wedge V^{<k}, d) \to (A, d_A)$$

such that $H^r(\varphi_{k-1})$ is an isomorphism for $r \le k-1$ and an injection for $r = k$. We define $V^k = W \oplus Z$ where $W = \operatorname{Coker} H^k(\varphi_{k-1})$ and $sZ = \operatorname{Ker} H^{k+1}(\varphi_{k-1})$. We then define the extension

$$\varphi_k \colon (\wedge V^{\le k}, d) \to (A, d_A),$$

in the following way. First of all, $d(W) = 0$, and $(\varphi_k)_{|W} = \sigma$, where σ is an equivariant section of the projection $Z^k(A, d_A) \to H^k(A, d_A) \to \operatorname{Coker} H^k(\varphi_{k-1})$.

We now choose an equivariant projection, $\pi \colon H^{k+1}(\wedge V) \to \operatorname{Ker} H^{k+1}(\varphi_{k-1})$ and we denote by σ' an equivariant section of the composition $Z^{k+1}(\wedge V) \to H^{k+1}(\wedge V) \xrightarrow{\pi} \operatorname{Ker} H^{k+1}(\varphi_{k-1})$. We put $d(z) = \sigma'(sz)$.

Note that $B = \sigma's(Z)$ is a Γ-module, and $\varphi_{k-1}(B) \subset \operatorname{Im} d_A$. We denote by π' an equivariant projection $\operatorname{Im} d_A \to \varphi_{k-1}(B)$ and by $\tau \colon \varphi_{k-1}(B) \to A^k$ an equivariant section of the projection $A^k \xrightarrow{d_A} \operatorname{Im} d_A \xrightarrow{\pi'} \varphi_{k-1}(B)$. We finally define the restriction of φ to Z by $\varphi(z) = \tau \varphi_{k-1}\sigma'(sz)$.

In [40], Bredon proves the following result about the projection $\pi \colon X \to X/\Gamma$.

Theorem 3.27 ([40, Theorem 2.4]) *If Γ is a finite group acting on a space X that is a manifold or a CW complex, then the projection π induces an isomorphism*

$$\pi^* \colon H^*(X/\Gamma; \mathbb{Q}) \to H^*(X; \mathbb{Q})^\Gamma,$$

where $H^*(X; \mathbb{Q})^\Gamma$ denotes the cohomology invariant under the induced action of Γ.

What is very interesting about the invariant cohomology $H^*(X; \mathbb{Q})^\Gamma$ is that each equivariant quasi-isomorphism induces an isomorphism on invariant cohomology. Moreover, the invariant part of the cohomology of a Γ-complex is the cohomology of the invariant part of the complex. This is the content of the following theorem.

Theorem 3.28 *The following properties hold:*

1. *If (A, d_A) is a Γ-cdga, then A^Γ is a subcomplex, and the injection of A^Γ into A induces an isomorphism $H^*(A^\Gamma) \cong (H^*(A))^\Gamma$.*
2. *If $f \colon (A, d_A) \to (B, d_B)$ is an equivariant quasi-isomorphism, then the induced map $(A^\Gamma, d_A) \to (B^\Gamma, d_B)$ is also a quasi-isomorphism.*

Proof (1) Note first that the injection $A^\Gamma \hookrightarrow A$ induces an injection in cohomology. If $a \in A^\Gamma$ such that $a = d_A(b)$, then

$$a = \left(\frac{1}{|\Gamma|} \sum_{g \in \Gamma} ga\right) = \left(\frac{1}{|\Gamma|} \sum_{g \in \Gamma} gd_A(b)\right) = d_A\left(\frac{1}{|\Gamma|} \sum_{g \in \Gamma} gb\right).$$

Therefore, a is exact by an invariant element.

Now, if $[a] \in H^*(A)^\Gamma$, then

$$[a] = \left[\frac{1}{|\Gamma|} \sum_{g \in \Gamma} ga\right].$$

Since $\frac{1}{|\Gamma|} \sum_{g \in \Gamma} ga \in A^\Gamma$, $[a] \in H^*(A^\Gamma)$.

(2) Suppose $f: (A, d_A) \to (B, d_B)$ is an equivariant quasi-isomorphism. We then have the following commutative diagram.

$$\begin{array}{ccc} H(A)^\Gamma & \xrightarrow{f'} & H(B)^\Gamma \\ \downarrow & & \downarrow \\ H(A) & \xrightarrow[\cong]{H^*(f)} & H(B) \end{array}$$

Since the vertical arrows are monomorphisms, f' is injective. Let $x \in H(B^\Gamma) = H(B)^\Gamma$. By surjectivity of $H^*(f)$, there is some element $y \in H^*(A)$ such that $H^*(f)(y) = x$. Since $H^*(f)$ is an equivariant isomorphism, $H^*(f)(\bar{y}) = x$ where $\bar{y} = 1/|\Gamma| \Sigma_g gy \in H^*(A)^\Gamma$. This proves the surjectivity of f'. □

Corollary 3.29 *Let Γ be a finite group acting on a CW complex X, and let $(\wedge V, d)$ be its Γ-minimal model. Then $H^*(X/\Gamma; \mathbb{Q}) \cong H^*((\wedge V)^\Gamma, d)$.*

Proof We have

$$H^*((\wedge V)^\Gamma, d) \cong (H^*(\wedge V, d))^\Gamma \cong H^*(X; \mathbb{Q})^\Gamma \cong H^*(X/\Gamma; \mathbb{Q}).$$

□

Note that the computation of the integer cohomology of X/Γ is more difficult because a homotopy equivalence, $f: X \to Y$, which is equivariant does not necessary give a homotopy equivalence $X/\Gamma \to Y/\Gamma$. Consider, for instance, the $\mathbb{Z}/2$-equivariant map $S^\infty \to \{*\}$. This is a homotopy equivalence, but the quotient map $\mathbb{R}P(\infty) \to \{*\}$ is not a homotopy equivalence.

Remark 3.30
(1) In fact, if X/Γ is simply connected, then its minimal model is the minimal model of $(\wedge V, d)^\Gamma$. By [40, Corollary II.6.3], if the action has a connected orbit, then $\pi_1(X) \to \pi_1(X/\Gamma)$ is a surjection. If X is simply connected and, for instance, there is a fixed point, then certainly the orbit of the fixed point is connected, so $\pi_1(X/\Gamma) = 0$ as well. This is then a case when the minimal model of X/Γ can be identified as the minimal model of $(\wedge V, d)^\Gamma$.

(2) Suppose a finite group Γ acts on the space X. If the Γ-equivariant minimal model of X, $(\wedge V, d)$, is equivariantly isomorphic to the Γ-equivariant minimal model of $H^*(X; \mathbb{Q})$, then we say that (X, Γ) is Γ-*formal*. Using the results of [215], it can be shown that a formal Γ-space is Γ-formal. That is, if X is a formal space with an action of a finite group Γ, then the equivariant minimal model can be constructed from the action of Γ

on $H^*(X;\mathbb{Q})$. Moreover, in this situation, we can show that the minimal model of X/Γ is the minimal model of $H^*(X/\Gamma;\mathbb{Q})$, so that X/Γ is formal. To see this, let $\phi\colon (\wedge W, d) \to (\wedge V, d)^\Gamma$ be the minimal model of $(\wedge V, d)^\Gamma$. By Corollary 3.29, we know that $(\wedge W, d)$ is the minimal model of X/Γ. Now consider the commutative diagram below, where the right square comes from the inclusion of invariant elements and the formality quasi-isomorphism θ, and the left square comes from lifting the composition $(\wedge W, d) \xrightarrow{\phi} (\wedge V, d)^\Gamma \xrightarrow{\theta^\Gamma} H^*(X;\mathbb{Q})^\Gamma$ via the isomorphism $H^*(X/\Gamma;\mathbb{Q}) \cong H^*(X;\mathbb{Q})^\Gamma$.

$$\begin{array}{ccccc}
(\wedge W, d) & \xrightarrow{\phi} & (\wedge V, d)^\Gamma & \longrightarrow & (\wedge V, d) \\
\downarrow & & \downarrow \theta^\Gamma & & \downarrow \theta \\
H^*(X/\Gamma;\mathbb{Q}) & \xrightarrow{\cong} & H^*(X;\mathbb{Q})^\Gamma & \longrightarrow & H^*(X;\mathbb{Q})
\end{array}$$

By Theorem 3.28, since θ is a quasi-isomorphism, so is θ^Γ. But then the lift $(\wedge W, d) \to H^*(X/\Gamma;\mathbb{Q})$ is also a quasi-isomorphism. Hence, X/Γ is formal if X is.

(3) There is a more complicated notion of equivariant model that is also truer to the spirit of modern equivariant homotopy theory (see [98], [258] for example). There are various types of formality which pertain to these models, even in the case of toral actions. In particular, symplectic geometers now refer to an action whose Borel fibration has collapsing spectral sequence as *equivariantly formal* or *TNCZ formal*. An interesting comparison of these varying notions of formality is given in [235].

3.3.2 Weyl group and cohomology of *BG*

Let G be a compact connected Lie group. We prove that the cohomology of the classifying space BG can be recovered from the action of the Weyl group on the cohomology of a maximal torus. We begin by considering the Euler characteristic of the quotient of the group by a maximal torus.

Proposition 3.31 *Let T be a maximal torus of a compact Lie group G. Then the Euler characteristic of G/T is strictly positive, $\chi(G/T) > 0$. In fact, this Euler characteristic is equal to the number of elements of the Weyl group, $\chi(G/T) = |W(G)|$.*

Proof Consider the action of the torus T on G/T. The fixed points set is determined by

$$(G/T)^T = \left\{ g \in G \mid g^{-1}Tg \subseteq T \right\}/T,$$

which means $(G/T)^T = N(T)/T = W(G)$ by definition. The result follows now from $\chi((G/T)^T) = \chi(G/T)$ (see Proposition 3.32 below). □

In the proof of the proposition we have used the following standard proposition whose proof will be given in Chapter 7 (see Theorem 7.33).

Proposition 3.32 *Let T be a torus acting smoothly on a compact manifold M with fixed point set M^T. Then M^T has the same Euler characteristic as M:*

$$\chi(M^T) = \chi(M).$$

Now, as an application of the previous results, we obtain a characterization theorem for maximal tori.

Theorem 3.33 *Let T be a torus contained in a compact Lie group G. Then the following conditions are equivalent.*

1. *T is a maximal torus.*
2. *The Euler characteristic of G/T is positive: $\chi(G/T) > 0$.*
3. *The cohomology of G/T is concentrated in even degrees.*
4. *The rank of T is equal to the number of generators of the algebra $H^*(G;\mathbb{Q})$.*

Proof By Proposition 3.31, if T is maximal, then $\chi(G/T) > 0$. When T is not a maximal torus, T injects into a maximal one T' and we have a fiber bundle

$$T'/T \to G/T \to G/T'.$$

Since $\chi(T'/T) = 0$ and since the Euler characteristic of the total space of a fibration is the product of the Euler characteristics of the base and the fiber, we have $\chi(G/T) = 0$.

The conditions (1) and (2) are therefore equivalent. Recall now that, by Theorem 2.71, a minimal model of G/T has the form

$$(\wedge Q \otimes \wedge P, d),$$

where $Q = Q^2$, $P = P^{odd}$, $d(Q) = 0$, and $d(P) \subset \wedge Q$. Moreover $\dim Q = \operatorname{rank} T$, and $\dim P = \operatorname{rank} G$. The equivalence of the properties (2), (3) and (4) is then a particular case of Theorem 2.75 (5) (also see Theorem B.18). □

In [75], in order to study homotopy-theoretic analogues of Lie groups, Dwyer and Wilkerson observed that the result above can be used to give a definition of maximal tori: namely, a torus T included in a compact Lie group is *maximal* if the Euler characteristic of G/T is strictly positive.

Recall, from Example 1.79, that if $G \to EG \to BG$ is the universal bundle for G, we can choose $T \to EG \to BT = EG/T$ as the universal bundle for T. More precisely, BT is the set of orbits of EG under the right action of T induced by the action of G. We define a right action of $W(G)$ on EG/T by $[x] \cdot [n] = [xn]$, if $x \in EG$ and $[n] \in W(G) = N(T)/T$. On $BG = EG/G$ this action becomes trivial: $[x] \cdot [n] = [xn] = [x]$. In summary, we have an action of the Weyl group and the map $B\iota \colon BT \to BG$ is equivariant. Now we can give the main result of this section.

Theorem 3.34 *Let T be a maximal torus of a compact connected Lie group G. Then the canonical morphism $H^*(BG; \mathbb{Q}) \to H^*(BT; \mathbb{Q})$ is injective and $H^*(BG; \mathbb{Q})$ can be identified with the invariant set of $H^*(BT; \mathbb{Q})$ under the action of the Weyl group:*

$$H^*(BG; \mathbb{Q}) \cong H^*(BT; \mathbb{Q})^{W(G)}.$$

Proof Consider the Serre spectral sequence of the fibration $G/T \to BT \to BG$. The second page of it, $E_2^{p,q} = H^p(BG; \mathbb{Q}) \otimes H^q(G/T; \mathbb{Q})$, is entirely concentrated in even degrees by Theorems 1.81 and 3.33. Therefore, the spectral sequence collapses and $H^*(BT; \mathbb{Q}) \cong H^*(G/T; \mathbb{Q}) \otimes H^*(BG; \mathbb{Q})$. Thus $H^*(BG; \mathbb{Q}) \to H^*(BT; \mathbb{Q})$ is an injection.

For the second part, we observe that if $F \xrightarrow{j} E \xrightarrow{p} B$ is a fibration with an action of a group Γ such that the maps j and p are Γ-equivariant, then the Serre spectral sequence $E_r^{p,q}$ induces a spectral sequence of the invariants $\left(E_r^{p,q}\right)^{\Gamma}$. Let's apply this to the fibration $G/T \to BT \to BG$ with the given actions of the Weyl group $W(G)$. We then get an isomorphism

$$H^*(BT; \mathbb{Q})^{W(G)} \cong H^*(BG; \mathbb{Q})^{W(G)} \otimes H^*(G/T; \mathbb{Q})^{W(G)}.$$

As we saw above, the action of $W(G)$ on $H^*(BG; \mathbb{Q})$ is trivial, so $H^*(BG; \mathbb{Q}) = H^*(BG; \mathbb{Q})^{W(G)}$. From Lemma 3.35 below, we deduce that $H^*(BG; \mathbb{Q}) = H^*(BT; \mathbb{Q})^{W(G)}$. $\square$

Lemma 3.35

$$H^*(G/T; \mathbb{Q})^{W(G)} = H^0(G/T; \mathbb{Q})^{W(G)} \cong \mathbb{Q}.$$

Proof From the covering

$$W(G) = N(T)/T \longrightarrow G/T \longrightarrow G/N(T),$$

we get:

$$\chi(G/T) = \chi(N(T)/T) \cdot \chi(G/N(T))$$
$$= |W(G)| \cdot \chi(G/N(T)).$$

Since $\chi(G/T) = |W(G)|$ by Proposition 3.31, we obtain $\chi(G/N(T)) = 1$. On the other hand, we know that $H^*(G/N(T);\mathbb{Q}) = H^*(G/T;\mathbb{Q})^{W(G)}$ by Theorem 3.27. Therefore, the vector space $H^*(G/N(T);\mathbb{Q})$ is evenly graded with an Euler characteristic equal to 1. This can only happen if $H^+(G/N(T);\mathbb{Q}) = 0$. □

Remark 3.36 The Serre spectral sequence of $G/T \to BT \to BG$ gives an isomorphism $H^*(BT) \cong H^*(G/T) \otimes H^*(BG)$ which is not an isomorphism of algebras even in elementary cases. For instance, if $G = \mathrm{SU}(2) \cong S^3$, then we have $T = S^1$ and $G/T = \mathrm{SU}(2)/S^1 = S^2$ so that we obtain the Hopf fibration. The spectral sequence isomorphism above is then

$$H^*(S^2;\mathbb{Q}) \otimes H^*(\mathrm{BSO}(3);\mathbb{Q}) = \mathbb{Q}[x_2]/(x_2^2) \otimes \mathbb{Q}[x_4] \cong \mathbb{Q}[y_2]$$
$$= H^*(BT;\mathbb{Q})$$

where the indices on generators denote the degrees of the generators. This is clearly *not* an isomorphism of algebras.

Example 3.37 Let T be a maximal torus of $\mathrm{Sp}(n)$. Here, we determine the canonical homomorphism $H^*(\mathrm{BSp}(n);\mathbb{Q}) \to H^*(BT,\mathbb{Q})$ that is described in Theorem 3.34. First recall from Corollary 1.86 and Section 1.5 that:

- the rational cohomology of $\mathrm{BSp}(n)$ is given by $H^*(\mathrm{BSp}(n);\mathbb{Q}) = \mathbb{Q}[q_1, q_2, \ldots, q_n]$ with $q_i \in H^{4i}(\mathrm{BSp}(n);\mathbb{Q})$;
- the maximal torus T of $\mathrm{Sp}(n)$ is the product of n circles, corresponding to the set of diagonal matrices, and its classifying space has rational cohomology $H^*(BT;\mathbb{Q}) = \mathbb{Q}[t_1, \ldots, t_n]$;
- the Weyl group of $\mathrm{Sp}(n)$ has $2^n n!$ elements acting by permutation of coordinates possibly composed with reverses of orientations. This implies that the set of invariants $H^*(BT;\mathbb{Q})^{W(G)}$ is generated by the symmetric polynomials in the square of the t_i's.

Therefore, the canonical isomorphism $H^*(\mathrm{BSp}(n);\mathbb{Q}) \cong H^*(BT;\mathbb{Q})^{W(G)}$ is obtained by sending q_k onto the kth symmetric polynomial in the t_i^2, $i = 1, \ldots, n$. For instance, the canonical inclusion $j \colon H^*(\mathrm{BSp}(3);\mathbb{Q}) = \mathbb{Q}[q_1, q_2, q_3] \hookrightarrow H^*(BT;\mathbb{Q}) = \mathbb{Q}[t_1, t_2, t_3]$ is defined by $j(q_1) = t_1^2 + t_2^2 + t_3^2$, $j(q_2) = t_1^2 t_2^2 + t_1^2 t_3^2 + t_2^2 t_3^2$, $j(q_3) = t_1^2 t_2^2 t_3^2$.

The minimal model of the injection of BT into $\mathrm{BSp}(3)$ is given by

$$j \colon (\wedge(q_1, q_2, q_3), 0) \to (\wedge(t_1, t_2, t_3), 0),$$
$$|q_1| = 4, |q_2| = 8, |q_3| = 12,$$

with

$$j(q_1) = t_1^2 + t_2^2 + t_3^2, \; j(q_2) = t_1^2 t_2^2 + t_1^2 t_3^2 + t_2^2 t_3^2, \; j(q_3) = t_1^2 t_2^2 t_3^2.$$

Therefore a Sullivan model of $Sp(3)/T$ is given by

$$(\wedge(t_1,t_2,t_3,y_3,y_7,y_{11}),d)$$

with $|y_i| = i$ and

$$d(y_3) = t_1^2 + t_2^2 + t_3^2, \quad d(y_7) = t_1^2 t_2^2 + t_1^2 t_3^2 + t_2^2 t_3^2, \quad d(y_{11}) = t_1^2 t_2^2 t_3^2.$$

This model is minimal and is therefore the minimal model of $Sp(3)/T$.

Example 3.38 Let $j \colon H = Sp(1) \times Sp(1) \hookrightarrow G = Sp(3)$, where H is viewed as a subgroup of G through the inclusion $Sp(1) \times Sp(1) \times 1 \subset Sp(3)$. A description of $H^*(Bj; \mathbb{Q})$ can be derived in a similar way to that of the maximal torus inclusion of Example 3.37, and this leads to a calculation of the model of Bj:

$$j^* \colon (\wedge(q_1,q_2,q_3),0) \to (\wedge(y_4,z_4),0)$$

with

$$|q_1| = 4, \quad |q_2| = 8, \quad |q_3| = 12, \quad |y_4| = |z_4| = 4,$$

and

$$j^*(q_1) = y_4 + z_4, \quad j^*(q_2) = y_4 z_4, \quad j^*(q_3) = 0.$$

We deduce from Theorem 2.71 the following model of $Sp(3)/(Sp(1) \times Sp(1))$:

$$(\wedge(y_4,z_4,y_3,y_7,y_{11}),d)$$

with $|y_i| = i$, $d(y_4) = d(z_4) = 0$, $d(y_3) = y_4 + z_4$, $d(y_7) = y_4 z_4$, $d(y_{11}) = 0$. The cancellation of the acyclic ideal generated by y_3 and $y_4 + z_4$ gives a minimal model

$$(\wedge(v_4,y_7,y_{11}),d),$$

with $d(v_4) = 0$, $d(y_7) = v_4^2$, $d(y_{11}) = 0$. As a consequence, the homogeneous space $Sp(3)/(Sp(1) \times Sp(1))$ has the rational homotopy type of $S^4 \times S^{11}$.

Example 3.39 We consider $K = Sp(1)$ as a subgroup of $G = Sp(3)$, through the diagonal inclusion

$$Sp(1) \xrightarrow{\Delta} Sp(1) \times Sp(1) \times Sp(1) \subset Sp(3) \qquad c \mapsto \begin{pmatrix} c & 0 & 0 \\ 0 & c & 0 \\ 0 & 0 & c \end{pmatrix},$$

and we are looking for a model of the homogeneous space $Sp(3)/Sp(1)$.

3: Manifolds

We first note that the injections j_i, $i = 1,\ldots,4$, of Sp(1) into Sp(3) given by

$$c \xmapsto{j_1} \begin{pmatrix} c & 0 & 0 \\ 0 & 1 & 0 \\ 0 & 0 & 1 \end{pmatrix}, \quad c \xmapsto{j_2} \begin{pmatrix} 0 & c & 0 \\ -1 & 0 & 0 \\ 0 & 0 & 1 \end{pmatrix}, \quad c \xmapsto{j_3} \begin{pmatrix} 1 & 0 & 0 \\ 0 & c & 0 \\ 0 & 0 & 1 \end{pmatrix}, \quad c \xmapsto{j_4} \begin{pmatrix} 1 & 0 & 0 \\ 0 & 1 & 0 \\ 0 & 0 & c \end{pmatrix}$$

are homotopic, the homotopies between j_1 and j_2, and between j_2 and j_3 being given by the following morphisms

$$t \mapsto \begin{pmatrix} c & 0 & 0 \\ 0 & 1 & 0 \\ 0 & 0 & 1 \end{pmatrix} \cdot \begin{pmatrix} \cos t & \sin t & 0 \\ -\sin t & \cos t & 0 \\ 0 & 0 & 1 \end{pmatrix}, \quad t \in [0, \pi/2],$$

$$t \mapsto \begin{pmatrix} \cos t & \sin t & 0 \\ -\sin t & \cos t & 0 \\ 0 & 0 & 1 \end{pmatrix} \cdot \begin{pmatrix} 0 & c & 0 \\ -1 & 0 & 0 \\ 0 & 0 & 1 \end{pmatrix}, \quad t \in [0, 3\pi/2].$$

The injection $j_1 \colon \text{Sp}(1) \to \text{Sp}(3)$ is the fiber of the principal fiber bundle $\text{Sp}(3) \to \text{Sp}(3)/\text{Sp}(1) = V_{3,2}(\mathbb{H})$. From the determination of the model of Sp(n) in Example 2.39, we see that the relative minimal model of this fiber bundle is

$$(\wedge(y_7, y_{11}), 0) \to (\wedge(y_7, y_{11}, y_3), 0) \xrightarrow{\rho} (\wedge y_3, 0).$$

For degree reasons, there is no choice for the morphism ρ: namely, we must have $\rho(y_3) = y_3$ and $\rho(y_7) = \rho(y_{11}) = 0$.

The multiplication μ on Sp(3) is associative with unit. For degree reasons the model of the multiplication is given by

$$\nu \colon (\wedge(y_3, y_7, y_{11}), 0) \to (\wedge(y_3, y_7, y_{11}), 0) \otimes (\wedge(y_3, y_7, y_{11}), 0),$$

with $\nu(y_3) = y_3 \otimes 1 + 1 \otimes y_3$, $\nu(y_7) = y_7 \otimes 1 + 1 \otimes y_7$ and $\nu(y_{11}) = y_{11} \otimes 1 + 1 \otimes y_{11}$.

We now observe that the morphism $\Delta \colon \text{Sp}(1) \to \text{Sp}(3)$ can be viewed as the composition

$$\text{Sp}(1) \xrightarrow{\sigma} \text{Sp}(1)^3 \xrightarrow{j_1 \times j_3 \times j_4} \text{Sp}(3)^3 \xrightarrow{\mu} \text{Sp}(3),$$

where σ is the diagonal map. We know the models of each of these maps. We then deduce that the model of Δ is

$$(\wedge(y_3, y_7, y_{11}), 0) \to (\wedge y_3, 0), \quad y_3 \mapsto 3y_3.$$

In particular, $\pi_3(\Delta) \otimes \mathbb{Q}$ is an isomorphism. It then follows from the long homotopy exact sequence of the fiber bundle $Sp(1) \xrightarrow{\Delta} Sp(3) \to Sp(3)/Sp(1)$ that the minimal model of $Sp(3)/Sp(1)$ is $(\wedge(y_7, y_{11}), d)$ for some differential d. For degree reasons the only possibility is $d = 0$. Therefore, the homogeneous space $Sp(3)/Sp(1)$ has the rational homotopy type of $S^7 \times S^{11}$.

Example 3.40 Let T be a maximal torus of $U(n)$ and $W(U(n))$ its Weyl group (see Section 1.5). With the same method as in Example 3.37, we prove that the inclusion

$$H^*(BU(n); \mathbb{Q}) = \mathbb{Q}[c_1, \ldots, c_n] \longrightarrow H^*(BT; \mathbb{Q}) = \mathbb{Q}[t_1, \ldots, t_n]$$

sends the class c_i onto the ith elementary symmetric polynomial in the t_i. See [198] or [199, Theorem 5.5, page 136].

Example 3.41 Let T be a maximal torus of $SO(2m+1)$ and $W(SO(2m+1))$ its Weyl group (see Section 1.5). The inclusion

$$H^*(BSO(2m+1); \mathbb{Q}) = \mathbb{Q}[p_1, \ldots, p_m] \longrightarrow H^*(BT; \mathbb{Q}) = \mathbb{Q}[t_1, \ldots, t_m]$$

sends the class p_i onto the ith elementary symmetric polynomial in the t_i^2. See [198] or [199, Theorem 5.16, page 144].

Example 3.42 Let T be a maximal torus of $SO(2m)$ and $W(SO(2m))$ its Weyl group (see Section 1.5). The inclusion

$$H^*(SO(2m); \mathbb{Q}) = \mathbb{Q}[p_1, \ldots, p_{m-1}, \chi] \longrightarrow H^*(BT; \mathbb{Q}) = \mathbb{Q}[t_1, \ldots, t_m]$$

sends the class p_i onto the ith elementary symmetric polynomial in the t_i^2 and χ on $t_1 \ldots t_m$. See [198] or [199, Theorem 5.16, page 144].

3.4 Biquotients

3.4.1 Definitions and properties

Let G be a compact Lie group and H be a closed subgroup of G. We introduced the homogeneous space G/H as the manifold of left classes, $G/H = \{xH \mid x \in G\}$, obtained as the quotient of G by the right action of H (i.e. $x \sim y$ if and only if $x^{-1}y \in H$). If K is a closed subgroup of G, we could also consider the manifold $K\backslash G$ of right classes, $K\backslash G = \{Kx \mid x \in G\}$, obtained as the quotient of G by the left action of K (i.e. $x \sim y$ if and only if $yx^{-1} \in K$). When K acts on the quotient G/H, we obtain a new space from

3 : Manifolds

the data (G, K, H). More precisely, since we are interested in manifolds, we consider the following situation.

Definition 3.43 *Let H and K be closed connected subgroups of a compact connected Lie group G such that K acts freely on G/H. The quotient, denoted $K\backslash G/H$, is a closed manifold whose elements are denoted by KxH, $x \in G$. Any closed manifold diffeomorphic to $K\backslash G/H$ is called* a biquotient *of G.*

As we will see in Chapter 6, biquotients are important examples of Riemannian manifolds with non-negative sectional curvature. In fact, they are the only known examples of manifolds with positive sectional curvature. In Subsection 3.4.2, we will present the construction of a Sullivan model of a biquotient. In order to do that, we need to consider biquotients as they are presented in [81] and [238].

First, the definition of biquotient is more symmetric in H and K than it appears. From

$$Kx.h = Kx \Leftrightarrow h \in x^{-1}Kx \text{ and } k.xH = xH \Leftrightarrow k \in xHx^{-1}$$

we see the following.

Proposition 3.44 *The following properties are equivalent:*

- *the group H acts freely on $K\backslash G$;*
- *the only pair $(k, h) \in K \times H$ where k is conjugate to h is (e, e);*
- *the group K acts freely on G/H.*

Remark 3.45 Let T_H and T_K be maximal tori of H and K respectively. Then (G, H, K) satisfies the second property above if and only if (G, T_H, T_K) does. Since the biquotient is unchanged if we replace H and K by conjugate subgroups, one can always suppose that T_H and T_K are contained in a maximal torus of G.

Proposition 3.44 gives us two principal bundles whose base is the biquotient:

$$H \to K\backslash G \to K\backslash G/H \text{ and } K \to G/H \to K\backslash G/H.$$

In the first bundle, H acts on the right of $K\backslash G$ and, in the second, K acts on the left of G/H. We are looking now for a third principal bundle with $K\backslash G/H$ as base. Observe that G acts on the right of $K\backslash G$ by $Kx.g = Kxg$ and on the left of G/H by $g.xH = gxH$. We therefore have a right action of G on the product $K\backslash G \times G/H$, defined by $(Kx, yH).g = (Kxg, g^{-1}yH)$. It is now easy to see that G acts freely on $K\backslash G \times G/H$ if and only if the properties of Proposition 3.44 are satisfied. In this case, the canonical map

$K\backslash G \times G/H \to K\backslash G/H$, $(Kx, yH) \mapsto KxyH$, induces a diffeomorphism between $K\backslash G \times_G G/H = (K\backslash G \times G/H)/G$ and $K\backslash G/H$. Hence,

$$G \to K\backslash G \times G/H \to K\backslash G/H$$

is a principal bundle and from the three principal bundles, we can deduce the following result.

Theorem 3.46 ([238]) *Let G be a compact connected Lie group such that H and K are closed connected subgroups of G satisfying the equivalent properties of Proposition 3.44. Then we have the following homotopy pullback*

$$\begin{array}{ccc} K\backslash G/H & \longrightarrow & BH \\ \downarrow & & \downarrow \\ BK & \longrightarrow & BG \end{array}$$

The maps $BK \to BG$, $BH \to BG$ are induced by the canonical inclusions $K \hookrightarrow G$, $H \hookrightarrow G$. The composition $K\backslash G/H \to BG$ is a classifying map for the principal G-bundle $K\backslash G \times G/H \to K\backslash G/H$.

Proof Consider the following morphisms of principal bundles defined above.

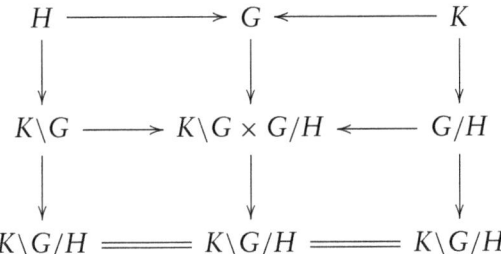

The classifying maps of these bundles give a commutative diagram

$$\begin{array}{ccc} K\backslash G/H = K\backslash G/H = K\backslash G/H \\ \downarrow \qquad\qquad \downarrow \qquad\qquad \downarrow \\ BH \longrightarrow BG \longleftarrow BK \end{array}$$

which is the commutative diagram of the statement. Observe now that G/H is the common fiber of $K\backslash G/H \to BK$ and $BH \to BG$. Therefore, we have a homotopy pullback. Finally, the description of maps follows directly from the construction of the square. □

Remark 3.47 In [82], Eschenburg defined a biquotient as the base space of a principal bundle with a homogeneous space as total space. He then reduced the situation to our Definition 3.43 by considering actions of subgroups of $G \times G$ on G. For completeness, we now develop this viewpoint following Totaro's presentation in [255].

Let G be a compact connected Lie group with center $Z(G)$. We let $G \times G$ act on the left of G by $(x,y).g = xgy^{-1}$. We embed $Z(G)$ diagonally in $G \times G$ and denote by $\Delta(Z(G))$ the subgroup image in $G \times G$. This subgroup acts trivially on G, so we get an action of $(G \times G)/\Delta(Z(G))$ on G. Observe that this action is transitive with $G/Z(G)$ as the stabilizer at e.

Now consider a second compact Lie group L with a homomorphism $L \to (G \times G)/\Delta(Z(G))$. With the action described above, this homomorphism makes L act on G and *we suppose that this action is free*. From the observations above, we see that the quotient $G//L$ of G by the action of L can be written as a biquotient of $(G \times G)/\Delta(Z(G))$ by $G/Z(G)$ and the image of L in $(G \times G)/\Delta(Z(G))$:

$$G//L \cong (G/Z(G))\backslash(G \times G)/\Delta(Z(G)))/L.$$

We thus recover our initial presentation of a biquotient.

If U is a closed subgroup of $G \times G$, we are in the framework just described. We may observe that the action of U on G is free if and only if the only pair consisting of conjugate elements $(x,y) \in U$ is (e,e). In this case, the quotient manifold, denoted by $G//U$, and called *the biquotient of G by U*, is diffeomorphic to $U\backslash(G \times G)/\Delta(G)$. Here, $G//U$ is the quotient of the U-action on G given by $(u_1, u_2)g = u_1 g u_2^{-1}$ for $(u_1, u_2) \in U$ and $g \in G$.

Note that the case $U = K \times \{1\}$ gives the homogeneous space $K\backslash G$.

Remark 3.48 Let G be a compact connected Lie group and H and K be closed connected subgroups of G giving raise to a biquotient $K\backslash G/H$. If the rank of G is equal to the sum of the ranks of H and K, then the same argument as in Exercise 1.14 implies that $K\backslash G/H$ is simply connected.

In this direction, we also quote an observation of Totaro [255, Lemma 3.3]: with the notation above, if a biquotient is simply connected, then it can be written as a biquotient of a simply connected group G by a connected group H acting on G by a homomorphism $H \to (G \times G)/\Delta(Z(G))$. In the case of a 2-connected biquotient this homomorphism can be lifted to $H \to G \times G$.

Remark 3.49 Let U be a subgroup of $G \times G$. In [151], Kapovitch and Ziller studied the biquotients $G//U$ such that the algebra $H^*(G//U; \mathbb{Q})$ is generated by one element. They obtained a complete classification and, as a consequence, proved that the only biquotient which can be an exotic sphere is diffeomorphic to the Gromoll–Meyer sphere $\text{Sp}(2)//S^3$.

This last result is also proved independently by Totaro in [255]. In [54], Cheeger constructed closed manifolds of non-negative sectional curvature as the connected sum of two rank-one symmetric spaces. The purpose of [255] is to determine the Cheeger manifolds which are diffeomorphic to biquotients. Totaro obtains a complete answer and, among other results, shows that there are only finitely many diffeomorphism classes of 2-connected biquotient manifolds of a given dimension. The proof uses Friedlander's and Halperin's result on elliptic spaces (see [101]).

3.4.2 Models of biquotients

Now that we understand the basic properties of biquotients, let's give algebraic models for them. Throughout this section, we suppose that the biquotients we are considering are simply connected. The main tool we shall use is Theorem 3.46.

Theorem 3.50 ([150]) *Let H and K be closed connected subgroups of a compact connected Lie group G defining a biquotient $K\backslash G/H$. We denote by $\iota_H\colon H \to G$, $\iota_K\colon K \to G$ the canonical inclusions and by $B\iota_H\colon BH \to BG$, $B\iota_K\colon BK \to BG$ the induced maps on classifying spaces. Let $\wedge V = H^*(BG;\mathbb{Q})$, $\wedge W_H = H^*(BH;\mathbb{Q})$, $\wedge W_K = H^*(BK;\mathbb{Q})$ be the cohomology algebras of BG, BH and BK respectively. We denote by sV a copy of the vector space V shifted by one degree, $|sv| = |v| - 1$ if $v \in V$, and define a differential d on $\wedge W_H \otimes \wedge W_K \otimes \wedge(sV)$ by $dw = 0$ if $w \in W_H \oplus W_K$ and $d(sv) = H^*(B\iota_H)(v) - H^*(B\iota_K)(v)$ if $sv \in sV$.*

Then the cdga $(\wedge(W_H \oplus W_K) \otimes \wedge(sV), d)$ is a model for the biquotient $K\backslash G/H$. In particular, $H^*(K\backslash G/H;\mathbb{Q}) = H(\wedge(W_H \oplus W_K) \otimes \wedge(sV), d)$.

Proof Using Theorem 3.46, we construct a biquotient as a homotopy pullback. A model of the homomorphisms between the classifying spaces is given by

$$(\wedge W_H, 0) \xleftarrow{\varphi_H} (\wedge V, 0) \xrightarrow{\varphi_K} (\wedge W_K, 0)$$

where $\varphi_H = H^*(B\iota_H)$ and $\varphi_K = H^*(B\iota_K)$. We first have to transform one of these two maps into a relative model, such as

$$(\wedge W_H, 0) \xleftarrow{\varphi_H} (\wedge V, 0) \xrightarrow{\overline{\varphi_K}} (\wedge W_K \otimes \wedge V \otimes \wedge(sV), D)$$

where $\overline{\varphi}_K$ is the canonical inclusion and $D_{|W_K} = D_{|V} = 0$ and $Dsv = v - \varphi_K(v)$ if $sv \in sV$. From Theorem 2.70, we know that a model of the

3 : Manifolds

homotopy pullback is given by

$$(\wedge W_H, 0) \otimes_{(\wedge V, 0)} (\wedge W_K \otimes \wedge V \otimes \wedge(sV), D)$$
$$\cong (\wedge W_H \otimes \wedge W_K \otimes \wedge(sV), d)$$

with $d_{|W_H} = d_{|W_K} = 0$ and $dsv = \varphi_H(v) - \varphi_K(v)$ if $sv \in sV$. □

Corollary 3.51 *The minimal model of a simply connected biquotient is pure.*

Example 3.52 Consider the biquotient $Sp(1)\backslash Sp(3)/Sp(1) \times Sp(1)$ introduced by Eschenburg (see [82, page 164]). This space is obtained from the canonical inclusion of $K = Sp(1)$ in $Sp(3)$ through the diagonal map (see Example 3.39) and the inclusion $H = Sp(1) \times Sp(1) \times 1 \subset Sp(3)$ (see Example 3.38). We have already determined the following canonical homomorphisms $\varphi_H \colon H^*(BG; \mathbb{Q}) = \wedge(q_1, q_2, q_3) \to H^*(BH; \mathbb{Q}) = \wedge(\bar{q}_1, \bar{q}_1')$ and $\varphi_K \colon H^*(BG; \mathbb{Q}) \to H^*(BK; \mathbb{Q}) = \wedge(\tilde{q}_1)$. Theorem 3.50 gives as a model of the biquotient,

$$\left(\wedge(\bar{q}_1, \bar{q}_1', \tilde{q}_1) \otimes \wedge(y_3, y_7, y_{11}), d\right)$$

with

$$\begin{aligned}
d\bar{q}_1 &= d\bar{q}_1' = d\tilde{q}_1 = 0, \\
dy_3 &= \varphi_H(q_1) - \varphi_K(q_1) = \bar{q}_1 + \bar{q}_1' - 3\tilde{q}_1, \\
dy_7 &= \varphi_H(q_2) - \varphi_K(q_2) = \bar{q}_1 \bar{q}_1' - 3\tilde{q}_1^2, \\
dy_{11} &= \varphi_H(q_3) - \varphi_K(q_3) = -\tilde{q}_1^3.
\end{aligned}$$

Since the differential dy_3 is linear, we may simplify this model by quotienting out a contractible ideal generated by y_3 and dy_3. We then obtain the minimal model of $Sp(1)\backslash Sp(3)/Sp(1) \times Sp(1)$:

$$\left(\wedge(\bar{q}_1', \tilde{q}_1) \otimes \Lambda(y_7, y_{11}), d\right)$$

with

$$\begin{aligned}
d\bar{q}_1 &= d\tilde{q}_1 = 0, \\
dy_7 &= \bar{q}_1'(3\tilde{q}_1 - \bar{q}_1') - 3\tilde{q}_1^2 = -\bar{q}_1'^2 + 3\bar{q}_1' \tilde{q}_1 - 3\tilde{q}_1^2, \\
dy_{11} &= -\tilde{q}_1^3.
\end{aligned}$$

This cdga is pure and Theorem B.18 implies:

$$H^*(Sp(1)\backslash Sp(3)/Sp(1) \times Sp(1); \mathbb{Q}) = \wedge(\bar{q}_1, \tilde{q}_1)/I,$$

where I is the ideal generated by dy_7 and dy_{11}. Changing the generators by $a = -\tilde{q}_1$ and $b = \bar{q}_1 - \tilde{q}_1$ gives:

$$H^*(\text{Sp}(1)\backslash\text{Sp}(3)/\text{Sp}(1) \times \text{Sp}(1); \mathbb{Q}) = \wedge(a,b)/(a^3, b^2 + ab + a^2).$$

This algebra is isomorphic to $H^*(\text{Sp}(3)/\text{Sp}(1) \times \text{Sp}(1) \times \text{Sp}(1); \mathbb{Q})$ (see Exercise 3.3). Since these spaces are formal, they have the same *rational homotopy type*. But, by considering their Pontryagin classes, Singhof proved that they do not have the same *homotopy* type (see [238, Theorem 4.2 and Example 4.4]).

Example 3.53 Consider the biquotient $\text{Sp}(2)//S^3$, obtained from the canonical inclusion $S^3 \times 1 \subset \text{Sp}(2)$ and the diagonal inclusion $S^3 \to \text{Sp}(2)$, $g \mapsto (g,g)$. With the same technique as above, it is easy to see that $\text{Sp}(2)//S^3$ has the rational homotopy type of S^7. In fact, it is homeomorphic to S^7 (as Gromoll and Meyer proved in [118]) and has non-negative curvature since it is a biquotient. Since any homogeneous space homeomorphic to a sphere is diffeomorphic to a sphere (as Borel observed), $\text{Sp}(2)//S^3$ cannot be obtained as a homogeneous space. This proves that the class of biquotients is larger than the class of homogeneous spaces.

3.5 The canonical model of a Riemannian manifold

In Section A.4, it is shown that the de Rham p-forms on a closed smooth manifold M have a direct sum Hodge decomposition for each p,

$$A_{DR}^p(M) = \mathcal{H}^p \oplus \text{Im}(\Delta) = \mathcal{H}^p \oplus \text{Im}(d) \oplus \text{Im}(\delta),$$

which, in fact, is an orthogonal decomposition with respect to the inner product on forms arising from a metric on the manifold and defined by

$$(\alpha, \beta) = \int_M \alpha \wedge *\beta.$$

Here, $\delta \colon A_{DR}^{p+1}(M) \to A_{DR}^p(M)$ is adjoint to the exterior derivative d with respect to the inner product and $\Delta = \delta d + d \delta$. The subspace $\mathcal{H}^p$ denotes the *harmonic p-forms*; namely, the forms α such that $\delta \alpha = 0$ and $d \alpha = 0$. Furthermore, each α has a unique decomposition

$$\alpha = \mathcal{H}(\alpha) + \alpha_d + \alpha_\delta,$$

where $\mathcal{H}(\alpha) \in \mathcal{H}^p$, $\alpha_d \in \text{Im}(d)$ and $\alpha_\delta \in \text{Im}(\delta)$. Moreover, $\mathcal{H}^p$ and $\text{Im}(d)$ give all closed forms in $A_{DR}^p(M)$ and, from the discussion following Corollary A.11, we see that α is exact precisely when $\mathcal{H}(\alpha) = 0$, $\alpha_\delta = 0$ and

$\alpha = d\beta_\delta$ for a *unique* element $\beta_\delta \in \text{Im}(\delta)$. In particular, we see that, for each k, $H^k(M) \cong \mathcal{H}^k$ canonically (once a metric is fixed). This rigid metric structure on a closed manifold gives an indication that our standard minimal models can be improved in this case. In particular, recall that Proposition 2.26 gives a *homotopy* commutative diagram lifting a morphism of cdga's to minimal models. The lack of *exact* commutativity presents various problems when translating from geometry to algebra and back again, so a remedy for this deficiency is highly desired. A model built on harmonic forms using the Hodge decomposition provides this remedy, so we will now describe this *canonical model* (see [245]).

To begin, we note that we can introduce an inner product to any cga of the form $\wedge(V^1 \oplus V^2 \oplus \ldots)$. Suppose each vector space V^k (where V^i is the set of homogeneous elements of degree i) has an inner product which we denote by $\langle -, - \rangle_k$. Then, for $\lambda = \lambda_1 \wedge \ldots \wedge \lambda_p$, $\beta = \beta_1 \wedge \ldots \wedge \beta_p \in \wedge^p(V^{2k+1})$, we define

$$\langle \lambda, \beta \rangle = \det(\langle \lambda_i, \beta_j \rangle_{2k+1}).$$

Monomials of different lengths are declared orthogonal. By extending bilinearly, this defines an inner product on the exterior algebra $\wedge(V^{2k+1})$. Note that this definition only works because both exterior multiplication and the determinant are alternating. So what can be done for V^{2k} since $\wedge(V^{2k})$ is a symmetric algebra? Well, the determinant must be replaced by a non-alternating version called the *permanent* of a matrix. For an $n \times n$-matrix A, the permanent is defined to be

$$\text{perm}(A) = \sum_{\sigma \in S_n} a_{1\sigma(1)} \cdots a_{n\sigma(n)}.$$

Note that this differs from the definition of the determinant only in the missing sign factor $(-1)^{\text{sgn}\,\sigma}$. The key property of the permanent is that switching columns has no effect. Therefore, for $\lambda = \lambda_1 \wedge \ldots \wedge \lambda_p$, $\beta = \beta_1 \wedge \ldots \wedge \beta_p \in \wedge^p(V^{2k})$, we define

$$\langle \lambda, \beta \rangle = \text{perm}(\langle \lambda_i, \beta_j \rangle_{2k}).$$

By extending bilinearly, this defines an inner product on the symmetric algebra $\wedge(V^{2k})$. Putting the two definitions together provides an inner product on all of $\wedge(V^1 \oplus \ldots)$.

Now let's construct the canonical model with inner product for a closed simply connected Riemannian manifold M. Let $(A_{DR}(M), d)$ denote the de Rham algebra of M with a Hodge decomposition associated to a given Riemannian metric. Begin constructing the model by taking the symmetric algebra freely generated by degree 2 harmonic forms, $(\wedge(\mathcal{H}^2), d = 0)$, and mapping the harmonic forms identically to $A_{DR}(M)$. Clearly, this induces an isomorphism in degree 2 cohomology and, since $H^3(\wedge(\mathcal{H}^2), d=0) = 0$, an injection in degree 3 cohomology.

Inductively assume that $\wedge(n-1) = \wedge(V^{\le n-1})$ has been constructed with inner product and canonical morphism $\rho_{n-1} \colon \wedge(n-1) \to A_{DR}(M)$ with ρ_{n-1}^* an isomorphism in degrees through $n-1$ and an injection in degree n.

The injection $\rho_{n-1}^* \colon H^n(\wedge(n-1)) \to H^n(M) \cong \mathcal{H}^n$ induces an orthogonal decomposition

$$\mathcal{H}^n = \operatorname{Im} \rho_{n-1}^* \oplus U^n,$$

where the Hodge inner product on $\mathcal{H}^n$ provides the orthogonal complement U^n. Now let $L = \wedge(n-1) \otimes \wedge(U^n)$ with inner product defined as above and $\rho_n \colon L \to A_{DR}(M)$ preliminarily defined by

$$\rho_n|_{\wedge(n-1)} = \rho_{n-1}, \quad \text{and} \quad \rho_n|_{U^n} = U^n \hookrightarrow \mathcal{H}^n.$$

By the definition of U^n, we see that ρ_n^* is an isomorphism through degree n. However, to carry out the inductive step, we must also define ρ_n^* so as to be injective in degree $n+1$. With this in mind, suppose $\rho_n^*([\alpha]) = 0$ for $\alpha \in L^{n+1}$. Then there is some $\beta \in A_{DR}^n(M)$ with $d\beta = \rho_n(\alpha)$. Now, using the Hodge decomposition,

$$\rho_n(\alpha) = h_{\rho_n(\alpha)} + dx_{\rho_n(\alpha)} + \delta y_{\rho_n(\alpha)} = d\beta,$$

and the discussion above, we see that $h_{\rho_n(\alpha)} = 0$, $y_{\rho_n(\alpha)} = 0$ and a canonical choice for β is $\beta = x_{\rho_n(\alpha)} \in \operatorname{Im}(\delta)$. Thus, if we take a basis $\alpha_1, \ldots, \alpha_r$ for $\operatorname{Ker}(\rho_n^*)^{n+1}$, there are canonical choices $x_{\rho_n(\alpha_1)}, \ldots, x_{\rho_n(\alpha_r)} \in \operatorname{Im}(\delta)$ with $dx_{\rho_n(\alpha_i)} = \rho_n(\alpha_i)$. Let $W^n \cong \operatorname{Ker}(\rho_n^*)^{n+1}$ with basis $\{\bar\alpha_1, \ldots, \bar\alpha_r\}$, each $\bar\alpha_i$ in degree n. Form

$$\wedge(n) = L \otimes \wedge(W^n) \quad \text{with} \quad d|_L = d, \quad d|_W = \operatorname{id}_{\operatorname{Ker}(\rho_n^*)^{n+1}}.$$

Also, using the basis $\{\bar\alpha_1, \ldots, \bar\alpha_r\}$, define $\rho_n|_W$ by $\rho_n(\bar\alpha_i) = x_{\rho_n(\alpha_i)}$. Then ρ_n is defined on all of $\wedge(n)$ and, clearly now, ρ_n^* is an isomorphism through degree n and an injection in degree $n+1$. Also, the inner product extends since $H^{n+1}(L)$ has an inner product inherited from the orthogonal decomposition of cocycles $Z^{n+1}(L) = B^{n+1}(L) \oplus H^{n+1}(L)$ and so therefore does $\operatorname{Ker}(\rho_n^*)^{n+1}$ from $H^{n+1}(L) = \operatorname{Ker}(\rho_n^*)^{n+1} \oplus \operatorname{Im}(\rho_n^*)^{n+1}$. Finally, note that $\rho_n(U) \subseteq \mathcal{H}$ and $\rho_n(W) \subseteq \operatorname{Im}(\delta)$ by uniqueness of the Hodge decomposition.

Hence, $(\wedge(n), d)$ is the nth stage of a (canonical) minimal model for M. If we continue in this way, we obtain the canonical model $(\mathcal{CM}_M, d) \xrightarrow{\rho} (A_{DR}(M), d)$. Note first that this canonical model is an $\mathbb{R}$-minimal model and that it requires a Riemannian metric to produce the Hodge decomposition. Secondly, note that the construction produces the following geometrically important result (compare Proposition 2.26).

3 : Manifolds

Theorem 3.54 *Let $f: M \to M$ be an isometry of a closed simply connected Riemannian manifold M. Then:*

1. *The induced morphism $A_{DR}(f): A_{DR}(M) \to A_{DR}(M)$ has a lift to the canonical model, $f^C: CM_M \to CM_M$, and this makes the following diagram strictly commutative*

$$\begin{array}{ccc} CM_M & \xrightarrow{f^C} & CM_M \\ \rho \downarrow & & \downarrow \rho \\ A_{DR}(M) & \xrightarrow{A_{DR}(f)} & A_{DR}(M) \end{array}$$

2. *The real homotopy theory of f is determined by f^*, the induced homomorphism on cohomology.*

Proof Because f is an isometry, $A_{DR}(f)$ preserves the Hodge decomposition. In particular, $A_{DR}(f)(\mathcal{H}^2) = \mathcal{H}^2$ and this defines $f^C|: \wedge(\mathcal{H}^2) \to \wedge(\mathcal{H}^2)$ which obviously makes the restricted diagram commute. Inductively, assume $f^C: \wedge(n-1) \to \wedge(n-1)$ exists with $\rho f^C = A_{DR}(f)\rho$ and which preserves the inner product on $\wedge(n-1)$. Now using the notation above, $\wedge(n) = \wedge(n-1) \otimes \wedge(U^n) \otimes \wedge(W^n)$, so if we can define f^C correctly on U and W, we shall be done. Of course, we have an orthogonal decomposition $H^n(M) = \mathcal{H}^n = \text{Im}\,\rho^* \oplus U$, so because $f^* = A_{DR}(f)|_{\mathcal{H}}$ is an isometry, we have $f^C|_U = f^*|_U = A_{DR}(f)|_U$. Clearly, even with this extended definition of f^C, we have $\rho f^C = A_{DR}(f)\rho$ and the inner product is preserved. Now, we know that

$$dA_{DR}(f)(\rho(\overline{\alpha}_i)) = dA_{DR}(f)(x_{\rho(\alpha_i)})$$
$$= A_{DR}(f)(dx_{\rho(\alpha_i)})$$
$$= A_{DR}(f)(\rho(\alpha_i))$$
$$= \rho(f^C(\alpha_i)),$$

since $\alpha_i \in \wedge(n-1) \otimes \wedge(U^n)$. Also, because $\rho(W) \subseteq \text{Im}(\delta)$ and $A_{DR}(f)$ preserves the Hodge decomposition, we see that $A_{DR}(f)(\rho(\overline{\alpha}_i))$ is the unique element in $\text{Im}(\delta)$ with $dA_{DR}(f)(\rho(\overline{\alpha}_i)) = \rho(f^C(\alpha_i))$. By definition of W^n, there is a unique element $w_i \in W^n$ with $dw_i = f^C(\alpha_i)$. Define $f^C(\overline{\alpha}_i)) = w_i$ and do this for each i to obtain $f^C|_{W^n}$. Hence, $f^C: \wedge(n) \to \wedge(n)$ is defined with $\rho f^C = A_{DR}(f)\rho$. By induction, we have therefore defined the required f^C.

3.5 The canonical model of a Riemannian manifold

The construction of f^C shows that it is completely determined by $A_{DR}(f)|_{\mathcal{H}}$ and this is precisely f^* since $\mathcal{H} = H^*(M)$. Hence, the real homotopy type of f, represented by the morphism of models f^C, is determined by the induced homomorphism on cohomology. □

Exercises for Chapter 3

Exercise 3.1 Show that the only rational homotopy types of simply connected closed rationally elliptic 5-manifolds are S^5 and $S^2 \times S^3$.

Exercise 3.2 Show that $\mathbb{CP}(2)\#\mathbb{CP}(2)$ and $\mathbb{CP}(2)\#\overline{\mathbb{CP}}(2)$ have different rational homotopy types. Hint: suppose $\phi \colon H^*(\mathbb{CP}(2)\#\mathbb{CP}(2);\mathbb{Q}) \to H^*(\mathbb{CP}(2)\#\overline{\mathbb{CP}}(2);\mathbb{Q})$ is a rational equivalence. It must have the form: $\phi(x_1) = ax_1 + bx_2$, $\phi(x_2) = cx_1 + dx_2$. Now use $\phi(x_1 x_2) = 0$, $\phi(x_1^2) = \phi(x_2^2)$ to derive a contradiction.

Exercise 3.3 Prove that $H^*(\mathrm{Sp}(3)/\mathrm{Sp}(1) \times \mathrm{Sp}(1) \times \mathrm{Sp}(1);\mathbb{Q}) = \mathbb{Q}[a,b]/(a^3, b^2 + ab + a^2)$, with $|a| = |b| = 2$. Hint: Use the techniques developed in Example 3.39 and Example 3.38.

Exercise 3.4 Show that a model for the biquotient $S^1\backslash \mathrm{Sp}(n)/\mathrm{SU}(n)$ is given by:

$$\wedge(t, \sigma_2, \ldots, \sigma_n, y_3, \ldots, y_{4n-1}), d),$$

where $|t| = 2$, $|\sigma_i| = 2i$, $|y_i| = i$, $dt = d\sigma_i = 0$, $dy_3 = 2\sigma_2 - t^2$ and $dy_{4i-1} = 2\sigma_{2i} + \sum_{r+s=2i}(-1)^s \sigma_r \sigma_s$. Hint: [150, Example 6].

Exercise 3.5 (1) Show that the minimal model of $\mathrm{Sp}(5)/\mathrm{SU}(5)$ is

$$\bigl(\wedge(\alpha_6, \alpha_{10}, \alpha_{11}, \alpha_{15}, \alpha_{19}), d\bigr),$$

with $|\alpha_i| = i$, $d\alpha_6 = d\alpha_{10} = 0$, $d\alpha_{11} = \alpha_6^2$, $d\alpha_{15} = \alpha_6 \alpha_{10}$, $d\alpha_{19} = \alpha_{10}^2$.
(2) From this model compute the cohomology algebra of $\mathrm{Sp}(5)/\mathrm{SU}(5)$ and prove

$$H^*(\mathrm{Sp}(5)/\mathrm{SU}(5);\mathbb{Q}) = \wedge (c_6, c_{10}, a, b)/I,$$

where I is the ideal generated by $(\alpha_6^2, \alpha_{10}^2, \alpha_6\alpha_{10}, \alpha_6 a, \alpha_{10} a + \alpha_6 b, \alpha_{10} b, ab)$, where c_6, c_{10}, a and b denote the classes associated respectively to α_6, α_{10}, $\alpha_6 \alpha_{15} - \alpha_{11}\alpha_{10}$, $\alpha_6 \alpha_{19} - \alpha_{10}\alpha_{15}$.
(3) Show that the space $\mathrm{Sp}(5)/\mathrm{SU}(5)$ is not formal.
(4) Construct the bigraded model and the filtered model of $\mathrm{Sp}(5)/\mathrm{SU}(5)$ in degrees ≤ 32.

Exercise 3.6 Let $X = S^3 \times S^3 \times S^3 \times S^3$. We define a $\mathbb{Z}_4$-action on X by the shift map $T(u_1, u_2, u_3, u_4) = (u_2, u_3, u_4, u_1)$, for $u_i \in S^3$. Prove that the orbit space $X/\mathbb{Z}_4$ has the rational homotopy type of $S^3 \times S^6$. Hint: Use Corollary 3.29 and Remark 3.30.

3: Manifolds

Exercise 3.7 Show that a formal rationally elliptic space with $\pi_{even} \otimes \mathbb{Q} = 0$ must be a product of odd spheres. Hint: imitate the proof of Proposition 3.20.

Exercise 3.8 Prove the following result.
  Theorem. Any compact Riemannian symmetric space is formal.
  Hints: Consider the linear map

$$\rho: H^*(M) \to A_{DR}(M), \qquad \rho([u]) = h_u,$$

where h_u is a *unique* harmonic representative (see Theorem A.10). It is well known that, for a Riemannian *symmetric space* (see Exercise 1.6), harmonic forms are identified with left-invariant forms under the action of the isometry group of the manifold (also see Remark 1.51). But the product in the de Rham algebra of invariant forms is again an invariant form in $A_{DR}(M)$ and therefore ρ is a cdga-morphism inducing an isomorphism in cohomology.

4 Complex and symplectic manifolds

Just as a complex analytic function has special properties compared to an ordinary smooth function, so does the de Rham algebra of a complex manifold display additional structure to that derived from the manifold's smooth structure alone. In particular, the splitting of the de Rham algebra into (p, q)-parts endows a complex manifold with two differentials whose interactions are extremely powerful in many regards, but especially in discovering rational homotopy properties of the manifold.

In this chapter, we study models of the de Rham *and* Dolbeault algebras of a complex manifold M. We present the particular topological properties of Kähler manifolds M, and, in particular, we prove that they are *formal*. The main tool for the link between Dolbeault and de Rham algebras is a perturbation theorem which allows the construction of a model of a filtered cdga and starts from a model of any stage of the spectral sequence associated to the filtration. Examples of this are given which come from the Borel spectral sequence of a principal holomorphic bundle or from the Frölicher spectral sequence of a complex manifold. The use of models is particularly well-suited to the study of these spectral sequences and, in this context, we revisit some examples, due to Pittie, in which the Frölicher spectral sequence collapses at level 3. We relate these examples to the notion of Dolbeault formality of a complex manifold.

Properties of symplectic manifolds have also been discovered using algebraic models (e.g. see [165], [178], [257]). In the last part of this chapter, we describe some of the implications of models for symplectic topology. For a compact symplectic manifold, it has been shown by Mathieu [185] and Merkulov [190] that the hard Lefschetz property is related to the existence of symplectically harmonic forms in each cohomology class and with a more technical property called the $d\delta$-lemma. Since this last property is revealed as the key to the proof of formality of Kähler manifolds, we may wonder if it has the same effect in the symplectic case. This question is answered negatively by Cavalcanti [52] and we will take it up in Chapter 8.

Section 4.1 contains definitions and examples of *complex and almost complex manifolds*. Recall that a complex structure on a real vector space gives a bigradation on the complexification, V^c, of V which extends to the exterior algebra $\wedge V^c$. We first describe the bigradation of the algebra A^c of complex-valued smooth differential forms on an almost complex manifold. In the case of a complex manifold, the de Rham differential on A^c can be decomposed as $d = \partial + \bar{\partial}$, with ∂ of bidegree $(1,0)$ and $\bar{\partial}$ of bidegree $(0,1)$. This property is a necessary and sufficient condition for an almost complex manifold to be a complex manifold. The almost complex manifold is then called *integrable* and, in this case, the derivations ∂ and $\bar{\partial}$ are differentials ($\partial^2 = \bar{\partial}^2 = 0$) which also commute ($\partial\bar{\partial} + \bar{\partial}\partial = 0$). The complex $(A^c, \bar{\partial})$ is called the *Dolbeault complex* or the *Dolbeault algebra* of the complex manifold M.

In Section 4.2, we recall the definition of *Kähler manifolds*. The *Kähler form* of a Kähler manifold is the skew-symmetric part of a hermitian product. It has bidegree $(1,1)$ if we consider it as a form on each complexified tangent vector space. One purpose of this section is simply to describe certain properties of the de Rham and Dolbeault algebras of compact Kähler manifolds. For instance, the fact that the differentials d, ∂ and $\bar{\partial}$ induce the same Laplacian is a necessary and sufficient condition for the existence of a Kähler metric. This also entails a technical result, called the $\partial\bar{\partial}$-lemma (see Lemma 4.24), that, in turn, implies the *formality of the rational homotopy type of a Kähler manifold*. This is stated and proved in Theorem 4.43. We present other topological properties of Kähler manifolds as well involving the vector space of homology (see Proposition 4.33) or the algebra of cohomology (see Theorem 4.35). In this section, we also show how we can put a complex manifold structure on the total space of a principal bundle with base a Kähler manifold and with structure group an even dimensional torus, see Subsection 4.2.2. These manifolds are called *Calabi–Eckmann manifolds*. The complex structures of the *Hopf surface*, or the *Kodaira–Thurston manifold* or of any even dimensional compact connected Lie group are particular cases of this situation. A theorem of Blanchard [29] implies the non-Kählerness of these manifolds.

Section 4.3 is dedicated to understanding the *Dolbeault model of a complex manifold*. We define this model, prove its existence and introduce the notion of *Dolbeault formality*. In the case of a Kähler manifold, the de Rham and the Dolbeault complexes coincide. Thus, the Dolbeault model of a Kähler manifold comes about from the usual methods. In a more general situation however, the Dolbeault model is not so easy to construct. In order to do this, we introduce the notion of deformation; namely, to any filtered cdga $(A, d, \mathcal{F})$ (see Definition 4.55), we associate a spectral sequence (E_r, d_r), each of whose stages is a commutative differential bigraded algebra (*cdba*). In [133], it is proved that a model of $(A, d, \mathcal{F})$ can be recovered from a model of the cdba (E_r, d_r) for any $r \geq 0$ (see Theorem 4.56). This is used here for the Dolbeault model of the total space of a holomorphic principal bundle by way of the *Borel spectral sequence*. A more geometrical computation involving Chern–Weil theory is also presented for the Calabi–Eckmann manifolds, $T^{2n} \longrightarrow E \longrightarrow B$. As examples, we examine the product of two odd spheres and of the total space of a principal fibration having base the product of two complex Grassmann manifolds of 2-planes in $\mathbb{C}^4$.

Indeed, Theorem 4.56 provides a general point of view for many different circumstances. For instance, it gives the filtered model of Chapter 2 from the bigraded model and it also comes into play in Section 4.4 as follows. The complex valued de Rham algebra of a complex manifold is a bicomplex and it can be filtered by the first degree. From this filtration, we obtain a spectral sequence, called the *Frölicher spectral sequence*, such that $(E_0^{p,q}, d_0)$ is the Dolbeault complex $(A^{p,q}, \bar{\partial})$. From Theorem 4.56 and this spectral sequence, we see that a de Rham model can be constructed as a perturbation of the Dolbeault model. This perturbation is the right object for the study of the degeneracy of the Frölicher spectral sequence. First, we observe that we always have $E_1 \cong E_\infty$ in the case of a Kähler manifold. At first glance, since this spectral sequence is built from a bigraded model, it appears to degenerate at level 2. It is now well-known that this does not, in general, happen and we end this section with an example where $E_2 \not\cong E_3$ and $E_3 \cong E_\infty$. The context, due to Pittie [228], is that of an even dimensional Lie group viewed as the total space of the holomorphic principal bundle $T \longrightarrow G \longrightarrow G/T$ built from the inclusion of a maximal torus T. More particularly, Pittie proves the non-degeneracy of the Frölicher spectral sequence at stage 2 for $G = SO(9)$. In fact, if G is an even dimensional Lie group, with a complex structure coming from the holomorphic principal bundle, we can prove that the associated Frölicher spectral sequence collapses at stage 2 if and only if G is Dolbeault formal (see [249] and Theorem 4.74). We illustrate this situation in the case $G = SO(9)$.

Section 4.5 is devoted to properties of *symplectic manifolds*. We recall the definition and the most basic examples: $\mathbb{R}^{2n}$, the cotangent bundle and Kähler manifolds. In Subsection 4.5.3, we study what kind of properties and tools of complex manifolds carry over to the symplectic world. We consider the hard Lefschetz property, the $d\delta$-lemma and Brylinski's symplectically harmonic forms (see [44], or also Libermann [173]). In particular, we present Mathieu's result disproving a well-known conjecture of Brylinski and Merkulov's result linking several of these different notions. The relations among all these notions magnify the need for a more general viewpoint that includes the symplectic and complex settings.

This is the generalized complex geometry of Hitchin that we mention briefly.

When the properties we are looking for are purely homological, symplectic manifolds may be replaced by *c-symplectic manifolds*; that is, manifolds which mimic symplectic manifolds cohomologically. We consider these types of manifolds in Section 4.6 and study several examples. In particular, we demonstrate that (c-)symplectic homogeneous spaces and (c-)symplectic biquotients must have maximal rank. For the important test class of nilmanifolds, the two notions, symplectic and c-symplectic, coincide. Moreover, we show that the only symplectic nilmanifolds of Lefschetz type are tori (up to diffeomorphism). This property, combined with Mathieu's theorem, then provides counterexamples to the Brylinski conjecture. In Subsection 4.6.5, we pose and discuss a question on the boundary of symplectic geometry and homotopy theory; namely, which symplectic manifolds are nilpotent spaces?

In an Appendix (Section 4.7), we recall the basics of complex (and symplectic) linear algebra: in particular, how the structure of a complex vector space can be expressed in terms of real objects. We end with a discussion of hermitian products which provides background for the discussion of Kähler metrics in Section 4.2.

4.1 Complex and almost complex manifolds

Let's first recall some well-known definitions and properties concerning complex manifolds. References for these notions are [155], [264]. Complex structures in linear algebra are recalled in Section 4.7 and this serves also as a reference for the notation.

4.1.1 Complex manifolds

Definition 4.1 *A complex manifold M, of (complex) dimension n, is a manifold which admits an open cover $(U_j)_{j \in I}$ and coordinate maps $\varphi_j \colon U_j \to \mathbb{C}^n$, such that $\varphi_j \circ \varphi_k^{-1}$ is holomorphic on $\varphi_k(U_j \cap U_k) \subset \mathbb{C}^n$, for any j and k. The open sets U_j are called* charts *of M.*

Let $f \colon U \subset \mathbb{C}^n \to \mathbb{C}^n$ be defined on an open set U of $\mathbb{C}^n$. We set $\mathbf{x} = (x_1, \ldots, x_n)$, $\mathbf{y} = (y_1, \ldots, y_n)$ with $\mathbf{z} = \mathbf{x} + i\mathbf{y}$ and we recall that $f(\mathbf{x} + i\mathbf{y}) = (P_1(\mathbf{x}, \mathbf{y}), +iQ_1(\mathbf{x}, \mathbf{y}), \ldots)$ is a *holomorphic map* if the real components of f, P_k and Q_k, are differentiable as functions $\mathbb{R}^{2n} \to \mathbb{R}$ and satisfy the Cauchy–Riemann equations:

$$\frac{\partial P_k}{\partial x_j} = \frac{\partial Q_k}{\partial y_j}, \quad \frac{\partial P_k}{\partial y_j} = -\frac{\partial Q_k}{\partial x_j}.$$

4.1 Complex and almost complex manifolds

The first obvious example of a complex manifold is given by $\mathbb{C}^n$ or any open set of $\mathbb{C}^n$. Complex projective space is an example of a compact complex manifold as we see next.

Example 4.2 Recall that $\mathbb{C}P^n$ can be defined as the quotient of $\mathbb{C}^{n+1} - \{0\}$ by the equivalence relation

$$(z_1, \ldots, z_{n+1}) \sim (\lambda z_1, \ldots, \lambda z_{n+1}),$$

where $\lambda \in \mathbb{C} - \{0\}$. Denote by $[z_1, \ldots, z_n]$ the equivalence class of $(z_1, \ldots, z_n) \in \mathbb{C}^{n+1}$. We cover $\mathbb{C}P^n$ by the open sets $U_j = \{[z_1, \ldots, z_{n+1}] \mid z_j \neq 0\}$ with coordinate maps

$$\varphi_j : U_j \to \mathbb{C}^n; \; \varphi_j([z_1, \ldots, z_{n+1}]) = \left(\frac{z_1}{z_j}, \ldots, \hat{z}_j, \ldots, \frac{z_{n+1}}{z_j}\right),$$

where the hat over a coordinate indicates that it has been deleted. On $\varphi_k(U_j \cap U_k)$, we have the holomorphic map

$$\varphi_j \circ \varphi_k^{-1}(z_1, \ldots, z_n) = \varphi_j[z_1, \ldots, z_{k-1}, 1, \ldots, z_n] = \left(\frac{z_1}{z_j}, \ldots, \frac{z_{k-1}}{z_j}, \frac{1}{z_j}, \ldots, \frac{z_n}{z_j}\right).$$

Thus, $\mathbb{C}P(n)$ is a complex manifold of complex dimension n.

Example 4.3 Let $f : M \to N$ be a covering space. If the base N is a complex manifold, then the map f induces a complex manifold structure on M. Indeed, if $\varphi_j : U_j \to \mathbb{C}^n$ is a chart in N, then $\varphi_j \circ f$ is a chart on each component of $f^{-1}(U_j)$.

If we suppose now that M is a complex manifold and also that the covering group acts holomorphically, then we have a complex manifold structure on N. For instance, if $\mathbb{Z}^{2n}$ is a discrete lattice of $\mathbb{C}^n$, we obtain a complex manifold structure on the quotient $\mathbb{C}^n/\mathbb{Z}^{2n}$. We call it a *complex torus*.

A complex manifold M of complex dimension n is also a smooth manifold of dimension $2n$. We denote by $T_{\mathbb{R}}(M)$, (or $T(M)$ if there is no confusion), the tangent bundle of M. The tangent space at one point, x, is denoted by $T_x(M)$. For instance, if $M = \mathbb{C}^n$, then the tangent space $T(\mathbb{C}^n) \cong \mathbb{R}^{2n}$ is generated by $\left\{\frac{\partial}{\partial x_j}; \frac{\partial}{\partial y_j}\right\}$. The real vector space $T(\mathbb{C}^n)$ has a complex structure given by $J\left(\frac{\partial}{\partial x_j}\right) = \frac{\partial}{\partial y_j}$ and $J\left(\frac{\partial}{\partial y_j}\right) = -\frac{\partial}{\partial x_j}$.

Proposition 4.4 Let $f : U \subset \mathbb{C}^n \to \mathbb{C}^n$ be smooth on the open set U as a map from $\mathbb{R}^{2n}$ to $\mathbb{R}^{2n}$. The map f is holomorphic if and only if the induced map $f_* : T(\mathbb{R}^{2n}) \to T(\mathbb{R}^{2n})$ is compatible with the complex structures in the sense that $f_* \circ J = J \circ f_*$.

Proof As done above, denote by (P_k, Q_k) the real components of f. The derivation law gives:

$$f_*\left(\frac{\partial}{\partial x_j}\right) = \sum_k \frac{\partial P_k}{\partial x_j} \frac{\partial}{\partial x_k} + \sum_k \frac{\partial Q_k}{\partial x_j} \frac{\partial}{\partial y_k}$$

$$f_*\left(\frac{\partial}{\partial y_j}\right) = \sum_k \frac{\partial P_k}{\partial y_j} \frac{\partial}{\partial x_k} + \sum_k \frac{\partial Q_k}{\partial y_j} \frac{\partial}{\partial y_k}.$$

It is now easy to see that f is holomorphic if and only if $f_* \circ J = J \circ f_*$. □

The notions of bundle, principal bundle (see Section 1.10) and vector bundle carry over from the smooth case to the holomorphic case and give the definitions of *holomorphic bundle, principal holomorphic bundle* and *holomorphic vector bundle* respectively.

For instance, the tangent vector bundle of a complex manifold M is a holomorphic vector bundle. Let U_j be a chart of M and $x \in U_j$. The complex structure on $T(\mathbb{C}^n) \cong \mathbb{R}^{2n}$ can be transferred to $T_x(M)$ by φ_{j*}, where $\varphi_j: U_j \to \mathbb{C}^n$ is a coordinate map. More precisely, the complex structure on $T_x(M)$, also denoted by J, is defined locally by $J = \varphi_{j*}^{-1} \circ J \circ \varphi_{j*}$. Observe that this structure does not depend on the choice of U_j and φ_j because, on $T(\mathbb{C}^n) \cong \mathbb{R}^{2n}$, we have $\varphi_{j*}^{-1} \circ \varphi_{k*} \circ J \circ \varphi_{k*}^{-1} \circ \varphi_{j*} = J$, due to the holomorphicity of $\varphi_j^{-1} \circ \varphi_k$. Therefore, the next result follows immediately.

Proposition 4.5 *Let M be a complex manifold with real tangent bundle $T(M)$. There exists a bundle map $J: T(M) \to T(M)$ such that $J_x: T_x(M) \to T_x(M)$ is a complex structure on $T_x(M)$.*

4.1.2 Almost complex manifolds

Almost complex manifolds are exactly those manifolds whose tangent bundle satisfies the conclusion of Proposition 4.5.

Definition 4.6 *Let M be a smooth manifold of dimension $2n$. An almost complex structure on M is a bundle map $J: T(M) \to T(M)$, such that each J_x is a complex structure on $T_x(M)$. The couple (M, J) is called an* almost complex manifold. *If (M, J) comes from a complex structure on M, we say that the almost complex structure is* complex.

Remark 4.7 Suppose that an almost complex manifold (M, J) comes from two complex manifold structures on M. Then, by Proposition 4.4, the identity map is a holomorphic map between the two complex manifolds. Thus, if an almost complex structure (M, J) is complex, this complex structure

4.1 Complex and almost complex manifolds

on M is unique. The existence of almost complex manifolds that are not complex will be discussed in Subsection 4.1.4.

Proposition 4.4 leads naturally to the next notion.

Definition 4.8 *A map $f: (M, J) \to (M', J')$ between almost complex manifolds is said to be* almost complex *if it is a smooth map such that $J' \circ f_* = f_* \circ J$.*

The existence of an almost complex structure on a manifold M is not very common. Let's look first at the case of spheres. Evidently, the dimension of the manifold must be even and we have only to consider even dimensional spheres.

Example 4.9 (Even dimensional spheres) Since $S^2 = \mathbb{CP}(1)$, we already know that a complex structure exists on S^2. Nonetheless, let us now take a new look at S^2, starting with the canonical inclusion of $\mathbb{R}$ in the quaternionic field $\mathbb{H}$. The orthogonal subspace E of $\mathbb{R}$ in $\mathbb{H}$ gives a decomposition $\mathbb{H} = \mathbb{R} \oplus E$ and a unique way of writing $q \in \mathbb{H}$ as $q = a + \vec{X}$ with $a \in \mathbb{R}$ and $\vec{X} \in E \cong \mathbb{R}^3$. In this context, the quaternionic product becomes:

$$(a + \vec{X})(b + \vec{Y}) = (ab - \langle \vec{X}, \vec{Y} \rangle) + (a\vec{Y} + b\vec{X} + \vec{X} \wedge \vec{Y}),$$

where $\langle -, - \rangle$ is the scalar product and $\wedge$ the vector (or cross) product in $\mathbb{R}^3$. On E, the subspace of *pure* quaternionic numbers, the previous law reduces to:

$$\vec{X}\vec{Y} = -\langle \vec{X}, \vec{Y} \rangle + \vec{X} \wedge \vec{Y}.$$

This formula shows that the vector product on $E \cong \mathbb{R}^3$ can be directly deduced from the quaternionic product of $\mathbb{H}$. Let's recall now the central role played by this vector product in the construction of the usual almost complex structure on S^2.

The tangent subspace $T_{\vec{X}}(S^2)$ at $\vec{X} \in S^2$ is identified with the subspace $\{\vec{Y} \in \mathbb{R}^3 \mid \langle \vec{X}, \vec{Y} \rangle = 0\}$. We define $J_{\vec{X}}: T_{\vec{X}}(S^2) \to T_{\vec{X}}(S^2)$ by $J_{\vec{X}}(\vec{Y}) = \vec{X} \wedge \vec{Y}$. This is in fact the almost complex structure of S^2 coming from its classical complex structure.

The interest of this viewpoint is that exactly the same procedure works for the field $\mathbb{K}$ of Cayley numbers. The choice of an orthogonal subspace F of $\mathbb{R}$ in $\mathbb{K}$ gives a decomposition $\mathbb{K} = \mathbb{R} \oplus F$ and the Cayley product on $\mathbb{K}$ induces a vector product $\wedge$ on $F \cong \mathbb{R}^7$ by:

$$\vec{X} \wedge \vec{Y} = \vec{X}\vec{Y} + \langle \vec{X}, \vec{Y} \rangle.$$

Since the tangent subspace $T_{\vec{X}}(S^6)$ at $\vec{X} \in S^6$ is identified with the subspace $\left\{\vec{Y} \in \mathbb{R}^7 \mid \langle \vec{X}, \vec{Y} \rangle = 0\right\}$, we define $J_{\vec{X}} \colon T_{\vec{X}}(S^6) \to T_{\vec{X}}(S^6)$ by $J_{\vec{X}}(\vec{Y}) = \vec{X} \wedge \vec{Y}$. This gives an almost complex structure on S^6. The previous argument works also for any smooth oriented six-dimensional manifold embedded in $\mathbb{R}^7$, see [49] or [155, Example 2.6]. For instance, since the map

$$S^4 \times S^2 \longrightarrow \mathbb{R}^7, \quad (\mathbf{x}, y_1, y_2, y_3) \longmapsto \left(\left(1 + \tfrac{y_1}{2}\right)\mathbf{x}, \tfrac{y_2}{2}, \tfrac{y_3}{2}\right)$$

is an embedding, we see that $S^4 \times S^2$ is an almost complex manifold.

The constructions of almost complex structures on S^2 and S^6 above required some type of product law on an ambient space $\mathbb{R}^k$. If the existence of such a law is a necessary path to an almost complex structure, then we should expect that very few even dimensional spheres can have almost complex structures. In fact, that is the case, for by using reduced Steenrod powers, Borel and Serre proved [35] that *S^2 and S^6 are the only spheres that admit almost complex structures*. A consequence of this nonexistence of almost complex structures on spheres is that, in general, a connected sum of almost complex manifolds does not have an almost complex structure compatible with the structures on the summands [16]. See Exercise 4.6 for a stronger statement for 4-manifolds.

We have already mentioned that, for obvious dimensional reasons, there is no almost complex structure on an odd dimensional sphere. However, if we take a product of two such spheres this obstruction vanishes. Indeed, we can put an almost complex structure (and even a complex structure) on any product $S^{2k+1} \times S^{2l+1}$ as we will show in Subsection 4.2.2. (The case $S^1 \times S^1$ is, of course, the complex torus of Example 4.3.) Therefore, we may wonder if it is possible to find almost complex structures on a product of two even dimensional spheres. The answer is more or less no: in [67], it is proven that the product $S^{2k} \times S^{2l}$, with $l \leq k$, admits an almost complex structure if and only if $l = 1$ and $1 \leq k \leq 3$ or $k = l = 3$.

4.1.3 Differential forms on an almost complex manifold

We now use the algebraic recollections of complex linear algebra in Section 4.7 to describe differential forms on an almost complex manifold (M, J) of dimension $2n$. By definition, the *real tangent space* at $x \in M$, $T_x(M) = \mathbb{R}\left\{\frac{\partial}{\partial x_j}, \frac{\partial}{\partial y_j}\right\}$, has a complex structure J_x. The *complex tangent*

4.1 Complex and almost complex manifolds

space at x is the complexification

$$T_x^c(M) = T_x(M) \otimes_{\mathbb{R}} \mathbb{C} = \mathbb{C}\left\{\frac{\partial}{\partial x_j}, \frac{\partial}{\partial y_j}\right\},$$

and this complex vector space has a decomposition into J-eigenspaces, $T_x^c(M) = T_x^{1,0} \oplus T_x^{0,1}$, which defines two smooth sub-bundles $T^{1,0}$, $T^{0,1}$ of $T^c(M)$.

By taking duals, the almost complex structure on $T(M)$ induces an almost complex structure on the real cotangent bundle $T(M)^*$. This implies a decomposition of the complexified space as $T_x^c(M)^* = T_x^{*\,1,0} \oplus T_x^{*\,0,1}$, where $T^{*\,1,0}$ and $T^{*\,0,1}$ are the $\mathbb{C}$-dual bundles of $T^{1,0}$ and $T^{0,1}$ respectively. This decomposition of the cotangent bundle gives a *bigradation on the algebra of differential forms with values in* $\mathbb{C}$,

$$A_{DR}^c(M) = A_{DR}(M) \otimes_{\mathbb{R}} \mathbb{C} = \oplus_{p,q} A^{p,q}(M).$$

More precisely, let $(\omega_1, \ldots, \omega_n)$ be a local frame for $A^{1,0}$. Then $(\overline{\omega}_1, \ldots, \overline{\omega}_n)$ is a local frame for $A^{0,1}$ and an element ω of $A^{p,q}(M)$ can be written locally as:

$$\omega = \sum_{J,K} \varphi_{J,K}\, \omega_{j_1} \wedge \ldots \omega_{j_p} \wedge \overline{\omega}_{k_1} \wedge \ldots \overline{\omega}_{k_q},$$

where $\varphi_{J,K}\colon U_j \to \mathbb{C}$ is a smooth complex function for any multi-indices $J = (j_1, \ldots, j_p)$, $K = (k_1, \ldots, k_q)$.

We now want to decompose the differential $d\colon A_{DR}^c(M)^r \to A_{DR}^c(M)^{r+1}$ in accordance with this bigradation. In order to do that, observe first that $A^{p,q}$ is generated by elements of $A^{0,0}$, $A^{1,0}$ and $A^{0,1}$. Secondly, we know the behavior of the differential on these subspaces:

$$dA^{0,0} \subset A^{0,1} + A^{1,0},\ dA^{0,1} \subset A^{2,0} + A^{1,1} + A^{0,2},\ dA^{1,0} \subset A^{2,0} + A^{1,1} + A^{0,2}.$$

In short, we have $dA^{p,q} \subset A^{p+2,q-1} + A^{p+1,q} + A^{p,q+1} + A^{p-1,q+2}$. We single out two components of d which correspond to the bidegrees $(1, 0)$ and $(0, 1)$. They are respectively denoted by

$$\partial\colon A^{p,q} \to A^{p+1,q} \quad \text{and} \quad \overline{\partial}\colon A^{p,q} \to A^{p,q+1}.$$

Example 4.10 Let M be a complex manifold with local coordinates $(z_1, \ldots, z_n)$ and $z_j = x_j + iy_j$. The complex structure on the real tangent space $T_x(M)$ is given by $J\left(\frac{\partial}{\partial x_j}\right) = \frac{\partial}{\partial y_j}$ and $J\left(\frac{\partial}{\partial y_j}\right) = -\frac{\partial}{\partial x_j}$. The decomposition $T_x^c(M) = T_x^{1,0} \oplus T_x^{0,1}$ of the complexification $T_x^c(M)$ is given locally

4: Complex and symplectic manifolds

by $T_{|U}^{1,0} = \mathbb{C}\left\{\frac{\partial}{\partial z_j}\right\}$ and $T_{|U}^{0,1} = \mathbb{C}\left\{\frac{\partial}{\partial \bar{z}_j}\right\}$, with $\frac{\partial}{\partial z_j} = \frac{1}{2}\left(\frac{\partial}{\partial x_j} - i\frac{\partial}{\partial y_j}\right)$ and $\frac{\partial}{\partial \bar{z}_j} = \frac{1}{2}\left(\frac{\partial}{\partial x_j} + i\frac{\partial}{\partial y_j}\right)$. The complexified cotangent complex of M decomposes as $T_{|U}^{*\,1,0} = \mathbb{C}\{dz_j\}$ and $T_{|U}^{*\,0,1} = \mathbb{C}\{d\bar{z}_j\}$, where $(dz_j, d\bar{z}_j)$ is the basis dual to $\left(\frac{\partial}{\partial z_j}, \frac{\partial}{\partial \bar{z}_j}\right)$. This induces a bigradation on the algebra of $\mathbb{C}$-differential forms given by:

$$\omega = \sum_{J,K} \varphi_{J,K}\, dz_{j_1} \wedge \ldots dz_{j_p} \wedge d\bar{z}_{k_1} \wedge \ldots d\bar{z}_{k_q} \in A^{p,q}(M).$$

The differential of ω can be computed as:

$$d\omega = \sum_l \sum_{J,K} \frac{\partial \varphi_{J,K}}{\partial z_l} dz_l \wedge dz_{j_1} \wedge \ldots dz_{j_p} \wedge d\bar{z}_{k_1} \wedge \ldots d\bar{z}_{k_q}$$

$$+ \sum_l \sum_{J,K} \frac{\partial \varphi_{J,K}}{\partial \bar{z}_l} d\bar{z}_l \wedge dz_{j_1} \wedge \ldots dz_{j_p} \wedge d\bar{z}_{k_1} \wedge \ldots d\bar{z}_{k_q},$$

and as a consequence, we see that *if an almost complex manifold is complex, then* $d = \partial + \bar{\partial}$.

4.1.4 Integrability of almost complex manifolds

Of course, the first question to ask about almost complex manifolds is how to know if they are complex. From Subsection 4.1.3, we know that the equality between d and the sum $\partial + \bar{\partial}$ is a necessary condition for having a complex structure which induces (M, J). But is this condition sufficient to guarantee the existence of an inducing complex structure? This leads to the next definition and the subsequent fundamental theorem of complex manifold theory.

Definition 4.11 *An almost complex manifold, (M, J), is integrable if $d = \partial + \bar{\partial}$.*

Theorem 4.12 *An almost complex manifold is complex if and only if it is integrable.*

We simply mention that the real analytic case is due to Ehresmann [78] (also see Eckmann and Frölicher [76]), and the smooth case to Newlander and Nirenberg [208]. Since $A^{p,q}$ is generated by $A^{0,0}$, $A^{1,0}$ and $A^{0,1}$, Theorem 4.12 also takes the following form.

4.1 Complex and almost complex manifolds

Corollary 4.13 *An almost complex manifold (M, J) is complex if and only if $d\, A^{p,q} \subset A^{p+1,q} + A^{p,q+1}$. This is equivalent to*

$$d\, A^{0,1} \subset A^{1,1} + A^{0,2} \text{ and } d\, A^{1,0} \subset A^{2,0} + A^{1,1}.$$

Note the existence of equivalent criteria (see [155, Theorem 2.8 of Chapter IX]) such as:

- $T^{1,0}$ is stable under the Jacobi bracket (or the same requirement for $T^{0,1}$);
- the vanishing of the Nijenhuis tensor defined by

$$N(U, V) = 2\left([JU, JV] - [U, V] - J[U, JV] - J[JU, V]\right),$$

where U and V are any tangent vector fields.

In the case of an integrable manifold, the equalities $d = \partial + \bar{\partial}$ and $d^2 = 0$ imply $\partial^2 = \bar{\partial}^2 = \partial\bar{\partial} + \bar{\partial}\partial = 0$. In particular, $\bar{\partial}$ is a differential which allows the definition of cohomology groups.

Definition 4.14 (Definition of Dolbeault Cohomology) *If (M, J) is an integrable almost complex manifold (i.e. induced from a complex manifold), the complex $(A^{p,q}(M), \bar{\partial})$ is called the* Dolbeault complex *of (M, J). Its cohomology is denoted by $H^{p,q}_{\bar{\partial}}$ and called the* Dolbeault cohomology *of (M, J).*

Definition 4.15 *A* holomorphic form *is a form ω such that $\omega \in A^{p,0}$ and $\bar{\partial}\omega = 0$. Such a form gives a Dolbeault cohomology class in $H^{p,0}_{\bar{\partial}}$.*

If $f: M \to N$ is a holomorphic map between complex manifolds, then the derivative map f_* respects the almost complex structure, see Proposition 4.4. Dually, on forms we have: $f^* A^{p,q}(N) \subset A^{p,q}(M)$ and $\bar{\partial} \circ f^* = f^* \circ \bar{\partial}$. Therefore a holomorphic map f induces a homomorphism between the Dolbeault cohomology groups.

Example 4.16 (Nilmanifolds) An almost complex structure J may be defined on a nilmanifold by first defining it on the Lie algebra of the associated nilpotent Lie group (see [134]). For instance, consider the Lie algebra $\mathfrak{n}$ of dimension $2m + 2$ having basis $\{X_1, \ldots, X_m, Y_1, \ldots, Y_m, Z, W\}$ with bracket structure given by $[X_i, Y_i] = -Z$ for all $i = 1, \ldots, m$ and all other brackets zero. Let

$$J(X_i) = Y_i, \quad J(Y_i) = -X_i, \quad J(Z) = W, \quad J(W) = -Z.$$

This defines a left invariant almost complex structure on the nilpotent Lie group associated to $\mathfrak{n}$ which then descends to the nilmanifold level. The

Nijenhuis tensor of the almost complex structure is then

$$N(U,V) = 2\left([JU,JV] - [U,V] - J[U,JV] - J[JU,V]\right)$$

and this is easily computed to be zero for X, Y, Z, W. Hence, by Theorem 4.12 and the discussion above, the almost complex structure J is integrable and *the nilmanifold has the structure of a complex manifold*. Note that in case $m = 1$ we get *the Kodaira–Thurston manifold* (which is symplectic as we will see later).

Remark 4.17 Let's return now to the examples S^2 and S^6 developed in Example 4.9. While the Riemann sphere $S^2 = \mathbb{CP}(1)$ is manifestly a complex manifold, in [79], Ehresmann and Liebermann show that the almost complex structure on S^6 induced by Cayley number multiplication is not integrable, see also [102] and [34]. More generally, in [49], Calabi proves the nonintegrability of all the almost complex manifold structures on closed hypersurfaces of $\mathbb{R}^7$ which are obtained from Cayley multiplication. By contrast, there exist open hypersurfaces of $\mathbb{R}^7$ which *are* integrable, see [49, Theorem 6]. Finally, we note that it is still unknown whether S^6 has a complex manifold structure.

Remark 4.18 In [78], Ehresmann proves that the Stiefel–Whitney characteristic 3-class is an obstruction to the existence of an almost complex structure. This can be seen as follows. The existence of an almost complex structure on a manifold M^{2n} gives a lifting of the classifying map of the tangent bundle (i.e. a reduction of the structure group from $O(2n)$ to $U(n) \subset O(2n)$),

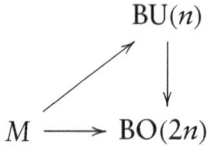

Since the cohomology of $BU(n)$ is evenly graded for all coefficients (see Corollary 1.86 for the particular case of coefficients in $\mathbb{Q}$), all odd dimensional Stiefel-Whitney classes of M must vanish.

4.2 Kähler manifolds

4.2.1 Definitions and properties

Definition 4.19 *A hermitian metric on an almost complex manifold (M, J) is a Riemannian metric, h, invariant with respect to the complex structure;*

4.2 Kähler manifolds

that is,

$$h(JX, JY) = h(X, Y).$$

A hermitian metric defines a hermitian inner product on $T_x(M)$ with respect to the complex structure defined by J_x. As a consequence of Remark 4.105, hermitian metrics on (M, J) always exist if M is paracompact.

Definition 4.20 *Let (M, J) be an almost complex manifold endowed with a hermitian metric h. The* fundamental 2-form ω *on (M, J), associated to h, is defined by:*

$$\omega(X, Y) = h(X, JY).$$

The fundamental 2-form ω is a real differential form of degree 2. Considered as a form with values in $\mathbb{C}$, we have $\omega \in A^{1,1}(M)$, see Subsection 4.7.3. From the nondegeneracy of the metric h, we see that ω is of rank $2n$ and that ω^n is a volume form on M. Therefore, any almost complex manifold M is orientable as a smooth manifold.

Definition 4.21 *The metric h is said to be* Kähler *if the associated fundamental 2-form ω is closed for the de Rham differential, $d\omega = 0$. A* Kähler manifold *is a complex manifold admitting a Kähler metric. The fundamental 2-form associated to a Kähler metric is called a* Kähler form.

We can see directly from the definition that $\mathbb{C}^n$ is Kähler and that the product of two Kähler manifolds is Kähler. Also, if N is a complex submanifold of a Kähler manifold M (i.e. the canonical injection is holomorphic), then N is also Kähler. Let's now give some examples of compact Kähler manifolds, see [155, pages 159–165] for details.

Example 4.22 (1) A compact Riemann surface M is Kähler since $d\omega$ is in $A_{DR}^3(M) = 0$.

(2) Any complex Grassmann manifold is Kähler. In particular, any complex projective space is Kähler. Furthermore, because the Kähler form of $\mathbb{C}P(n)$ pulls back via the inclusion of a smooth projective variety, we see that any such variety is Kähler.

(3) The complex tori are Kähler.

In Subsection 4.2.3, we will give some necessary conditions for the existence of a Kähler metric on a compact complex manifold. As a consequence, we will be able to identify certain complex manifolds as ones without Kähler structures.

Suppose now that M is a *compact* complex manifold. From a hermitian metric h on M, one can define as usual a star operator, $*$, which gives Laplace operators Δ_d, Δ_∂, $\Delta_{\bar\partial}$ for the respective differentials d, ∂, $\bar\partial$. Let's recall briefly how these Laplace operators are constructed (also see Section A.4 of Appendix A). The star operator $*\colon A^{p,q} \to A^{n-p,n-q}$ sends $\eta = \sum_{I,J} \eta_{I,J}\, \varphi_I \wedge \bar\varphi_J$ to $*\eta = 2^{p+q-n} \sum_{I,J} \varepsilon_{I,J} \bar\eta_{I,J} \varphi_{I^0} \wedge \bar\varphi_{J^0}$, where I^0 and J^0 are the complements of I and J and $\varepsilon_{I,J}$ is the sign of the permutation

$$(1,\ldots,n,1,\ldots,n) \mapsto (i_1,\ldots,i_p,j_1,\ldots,j_q,i_1^0,\ldots,i_{n-p}^0,j_1^0,\ldots,j_{n-q}^0),$$

see [264, page 161] for more details. We now define three operators, d^*, ∂^* and $\bar\partial^*$ by $d^* = *d*\colon A^m \to A^{m-1}$, $\partial^* = *\partial*\colon A^{p,q} \to A^{p-1,q}$ and $\bar\partial^* = *\bar\partial*\colon A^{p,q} \to A^{p,q-1}$. The associated Laplace operators are $\Delta_d = dd^* + d^*d$, $\Delta_\partial = \partial\partial^* + \partial^*\partial$ and $\Delta_{\bar\partial} = \bar\partial\bar\partial^* + \bar\partial^*\bar\partial$.

Let δ be any one of the differentials d, ∂ or $\bar\partial$. The harmonic forms for the Laplace operator Δ_δ are $\mathcal{H}_\delta^{p,q} = (\operatorname{Ker}\Delta_\delta) \cap A^{p,q}$ and there exists a Green's operator, G_δ, uniquely determined by $G_\delta(\mathcal{H}_\delta^{p,q}) = 0$, $\operatorname{Id} = \mathcal{H}_\delta + \Delta_\delta\, G_\delta$, $\delta\, G_\delta = G_\delta\, \delta$ and $\delta^*\, G_\delta = G_\delta\, \delta^*$, (see [264, page 147] or [115, page 84] for more details). These Laplace operators lead to a characterization of Kähler manifolds.

Theorem 4.23 ([71, Section 5]) *Let M be a compact complex manifold. Then M is Kähler if and only if either of the following equivalent properties is satisfied:*

1. *parallel transport preserves the almost complex structure;*
2. *the different Laplace operators coincide, $\Delta_d = 2\Delta_\partial = 2\Delta_{\bar\partial}$.*

We do not give the proof here and instead refer the reader to [71]. Recall also that certain commutation relations hold in a Kähler manifold. For instance, we will often use $[\bar\partial, \partial^*] = [\partial, \bar\partial^*] = 0$, (see [71] or [264, page 193]). The next result is a key tool for the study of compact Kähler manifolds.

Lemma 4.24 ($\partial\bar\partial$-lemma) *Let M be a compact Kähler manifold and let $\alpha \in A^c(M)$ such that $\partial\alpha = \bar\partial\alpha = 0$.*

1. *If $\alpha = \partial\gamma$ for some γ, then there exists β such that $\alpha = \partial\bar\partial(\beta)$.*
2. *If $\alpha = \bar\partial\gamma$ for some γ, then there exists β such that $\alpha = \bar\partial\partial(\beta)$.*

Proof We prove the first assertion, the argument for the second one being similar. Let α be as in the statement. The relation $\operatorname{Id} = \mathcal{H}_\delta + \Delta_\delta\, G_\delta$ implies $\alpha = \mathcal{H}_{\bar\partial}(\alpha) + \bar\partial\bar\partial^*\, G_{\bar\partial}(\alpha) = \mathcal{H}_\partial(\alpha) + \partial\partial^*\, G_\partial(\alpha)$. Since $\alpha = \partial\gamma$, the ∂-cohomology class associated to α is trivial and, therefore, $\mathcal{H}_\partial(\alpha) = 0$. Since the manifold M is Kähler, the equality $\Delta_\partial = \Delta_{\bar\partial}$ of the two Laplace operators implies $\mathcal{H}_{\bar{\partial}}(\alpha) = 0$ and $\alpha = \bar{\partial}\bar{\partial}^* G_{\bar{\partial}}(\alpha)$. Therefore, one has

$$\begin{aligned}\alpha &= \partial\,\partial^* G_\partial(\alpha) \\ &= \partial\,\partial^* G_\partial(\bar{\partial}\bar{\partial}^* G_{\bar{\partial}}(\alpha)) \\ &= \partial\,\partial^* \bar{\partial} G_\partial(\bar{\partial}^* G_{\bar{\partial}}(\alpha)) \\ &= -\partial\,\bar{\partial}\,(\partial^* G_\partial(\bar{\partial}^* G_{\bar{\partial}}(\alpha))),\end{aligned}$$

by using the relation $[\bar{\partial}, \partial^*] = 0$. It is then sufficient to set $\beta = -\partial^* G_\partial(\bar{\partial}^* G_{\bar{\partial}}(\alpha))$. □

4.2.2 Examples: Calabi–Eckmann manifolds

Now let's enlarge our herbarium of complex manifolds and then, in the next sections, test their Kählerness, see Remark 4.73. For instance, the product $S^{2p+1} \times S^1$ admits a complex manifold structure which is called a Hopf manifold, see [143] and Example 4.30. As with some other classical constructions of complex manifolds, these structures appear as the result of a general process involving a principal bundle. We describe this general process in this section and give some examples of this situation. This construction will be explicitly used in Section 4.3.4 for the determination of a Dolbeault model of these manifolds.

Proposition 4.25 *Let $G \longrightarrow E \xrightarrow{f} B$ be a principal fiber bundle. Suppose that B is an almost complex manifold and G a Lie group of even real dimension. Then there exist almost complex structures on E which are liftings of the structure on B.*

Proof Denote by $J: T(B) \to T(B)$ the structure map on the tangent space of B. Since the induced map $f_*: T(E) \to T(B)$ is surjective, we have a bundle defined by $V = \mathrm{Ker}\, f_*$ which is called the *vertical sub-bundle*. In the case of a principal bundle, this vertical sub-bundle is trivial [113, page 236]. On the other hand, we can choose a connection in the principal fiber bundle that induces a decomposition of the tangent bundle $T(E)$ as $T(E) = H_E \oplus V_E$. The map $f_*: T(E) \to T(B)$ restricts to a bundle isomorphism, $H_E \xrightarrow{\cong} T(B)$. We now build the complex structure J on the tangent bundle $T(E)$ as follows:

- on the horizontal sub-bundle, J comes from a transfer of the structure on B through the isomorphism $H_E \xrightarrow{\cong} T(B)$;
- since the vertical sub-bundle V is trivial, any complex structure on $T_e(G)$ fits. □

4 : Complex and symplectic manifolds

The next result provides an explicit way for obtaining *integrable* almost complex structures on the total space of a T^{2n}-principal bundle.

Proposition 4.26 *Let T^{2n} be an even dimensional torus and let $T^{2n} \to E \xrightarrow{f} B$ be a principal bundle endowed with a connection $\tilde{\omega} \in A^1_{DR}(E; \mathbb{R}^{2n})$. We suppose that B is a complex manifold and that the curvature $\omega \in A^2_{DR}(B; \mathbb{R}^{2n})$ of the connection form has a complexification of bidegree $(1,1)$. Then there exists a complex manifold structure on E and the principal bundle f is holomorphic.*

Proof We write the proof in the case $T = S^1 \times S^1$, the general case being similar. On $T_{(x,y)}(S^1 \times S^1) \cong T_x(S^1) \oplus T_y(S^1)$, we define a complex structure by $J(a, 0) = (0, a)$ and $J(0, b) = (-b, 0)$. With the process described in Proposition 4.25, we get an almost complex structure on the total space E.

Let $(A^c(E), d)$ be the complex de Rham algebra of E. Its subspace of forms of degree one is generated by $f^*\left(A^c(B)^1\right)$, $\tilde{\omega}_1$ and $\tilde{\omega}_2$, where $\tilde{\omega}_1$ and $\tilde{\omega}_2$ are the components of $\tilde{\omega} \in A^c_{DR}(E; \mathbb{R}^2)$. Note that $\tilde{\omega}_2 = J\tilde{\omega}_1$. Recall from Subsection 4.1.3 that this de Rham algebra has a bigradation. On $f^*\left(A^c(B)^1\right)$, the differential d is of the form $\partial + \bar{\partial}$ because B is complex. From basic linear algebra (see Section 4.7), we see that the elements $\tilde{\omega}_1 - i\tilde{\omega}_2$ and $\tilde{\omega}_1 + i\tilde{\omega}_2$ are of bidegree $(1,0)$ and $(0,1)$ respectively. Since $d\tilde{\omega} = f^*(\omega)$, we therefore have $d(\tilde{\omega}_1 - i\tilde{\omega}_2) \in A^{1,1}(E)$ and $d(\tilde{\omega}_1 + i\tilde{\omega}_2) \in A^{1,1}(E)$. The integrability of the almost complex structure is now a consequence of Corollary 4.13. □

Definition 4.27 *With the complex structure of Proposition 4.26, the total space E is called a* Calabi–Eckmann manifold.

We now want to create concrete examples of principal bundles as in Proposition 4.26. Recall (see [264, Chapter VI]) that a *Hodge manifold* B is a Kähler manifold such that the cohomology class induced by the Kähler form $\omega \in A^{1,1}(B)$ is in the image of $H^*(B;\mathbb{Z}) \to H^*(B;\mathbb{R}) \to H^*(B;\mathbb{C})$. We still denote the class by $[\omega] \in H^2(B;\mathbb{Z})$ or $[\omega] \in H^2(B;\mathbb{R})$. Because $H^2(B;\mathbb{Z}) \cong [B, BS^1]$, we thus have a principal bundle

$$S^1 \longrightarrow E \xrightarrow{f} B$$

whose Euler class is $[\omega] \in H^2(B;\mathbb{R})$, see Remark 1.88. This implies the existence of a form $\tilde{\omega} \in A^1_{DR}(E)$ such that $d\tilde{\omega} = f^*(\omega)$ and $\oint_{S^1} \tilde{\omega} = 1$, where $\oint_{S^1}$ denotes integration along the fiber. The form $\tilde{\omega}$ is a connection for the principal bundle f and ω is its curvature. The next result, which considers a product of two bundles of this type, is a direct consequence of Proposition 4.26.

Corollary 4.28 *Let B and B' be two Hodge manifolds with respective Kähler forms ω and ω'. In the product of bundles*

$$S^1 \times S^1 \longrightarrow E \times E' \xrightarrow{f \times f'} B \times B',$$

the total space $E \times E'$ admits a complex manifold structure.

Observe also that, as in Example 2.72, we can construct a relative model of the fibration f by

$$\varphi \colon \left(A_{DR}(B) \otimes \wedge \tilde{\omega}, \, d\right) \longrightarrow A_{DR}(E),$$

with $d(\xi \otimes 1) = d_B(\xi) \otimes 1$, $d(1 \otimes \tilde{\omega}) = \omega \otimes 1$, $\varphi(\xi \otimes 1) = f^*(\xi)$ and $\varphi(1 \otimes \tilde{\omega}) = \tilde{\omega}$.

Remark 4.29 Let $F \to E \to B$ be a holomorphic bundle where $\pi_1(B)$ acts trivially on $H^1(F)$. Blanchard [29] proved that the complex manifold E is Kähler if and only if

- there exists a Kähler form on F which represents a cohomology class invariant with respect to the action of $\pi_1(B)$;
- the manifold B is Kähler;
- the transgression $H^1(F) \to H^2(B)$ is zero (i.e. the Betti numbers satisfy $b_1(E) = b_1(F) + b_1(B)$).

In the previous construction, the transgression $H^1(F) \to H^2(B)$ is never zero, so these Calabi–Eckmann manifolds cannot be Kähler. We will come back to this point when we discuss the Frölicher spectral sequence in Remark 4.73.

Example 4.30 (Hopf manifold) If we apply the construction above to complex projective spaces, we obtain a principal bundle

$$S^1 \times S^1 \longrightarrow S^{2n+1} \times S^{2m+1} \longrightarrow \mathbb{CP}(n) \times \mathbb{CP}(m),$$

whose total space we denote by $M_{n,m}$. An application of Corollary 4.28 implies that $S^{2n+1} \times S^{2m+1}$ has the structure of a complex manifold. In the case $n = 0$ and $m = 1$, this is the Hopf surface, see [143] or [264, page 200].

Example 4.31 (Kodaira–Thurston manifold) KT is the manifold obtained as the quotient $\mathbb{R}^4 / \Gamma$ of $\mathbb{R}^4$ by the discrete affine group generated by unit translations along the x_1, x_2, x_3 axes together with the transformation

$(x_1, x_2, x_3, x_4) \mapsto (x_1 + x_2, x_2, x_3, x_4 + 1)$. The manifold KT can also be expressed as a principal bundle

$$S^1 \times S^1 \longrightarrow KT \longrightarrow S^1 \times S^1$$

and therefore admits the structure of a complex manifold. In fact, KT is the product of the Heisenberg manifold (see Section 3.2, Example 3.23 and Example 4.92) with a circle. A minimal model of KT is given by $(\wedge(u, v, y, t), d)$, with all generators of degree 1 and the differential given by $du = dy = dt = 0$, $dv = uy$. Clearly, we see from this model that $H^1(\wedge(u, v, y, t), d)$ is of dimension 3. Now see Example 4.34.

Example 4.32 Let T be a maximal torus of a Lie group G. The quotient G/T is a Kähler manifold, [34] and, if G is an even dimensional, compact connected Lie group, we have a principal fibration $T \longrightarrow G \longrightarrow G/T$ which fits the requirements of Proposition 4.26. We thus re-discover an observation of Samelson: *any even dimensional, compact connected Lie group G admits the structure of a complex manifold.* Moreover, left (or right) translations are holomorphic but note that this does not imply that G is a complex Lie group.

4.2.3 Topology of compact Kähler manifolds

The existence of a Kähler metric on a compact manifold M imposes strong constraints on the homotopy type of M. Here we give the most important ones with a particular emphasis on rational homotopy constraints. We begin with the *vector space of homology*.

Proposition 4.33 *Let M be a compact Kähler manifold.*

- *The even Betti numbers of M are nonzero.*
- *The odd Betti numbers of M are even.*

Proof Because the Kähler form ω gives a volume form ω^n representing the fundamental cohomology class, all wedge products ω^j, $j \leq n$ represent non-trivial cohomology classes. Hence, any even Betti number must be nonzero.

Recall, from Theorem 4.23, that we have an equality between Laplace operators: $\Delta_d = 2\Delta_{\bar{\partial}} = 2\Delta_{\partial}$. Therefore, we have the following properties.

- Δ_d respects the bidegree. This implies $H^r(M; \mathbb{C}) \cong \oplus_{p+q=r} H^{p,q}(M)$.
- Δ_{∂} and $\Delta_{\bar{\partial}}$ are real operators. This implies $H^{p,q}(M) \cong \overline{H^{q,p}(M)}$. (Remember that complex conjugation gives an isomorphism between the spaces of (p, q) and (q, p) forms, see Subsection 4.7.2.)

From these two observations, we obtain

$$\dim H^{2r+1}(M;\mathbb{C}) = 2\left(\sum_{p=0}^{r} \dim H^{p,2r+1-p}(M)\right),$$

and we see that odd Betti numbers are even. □

Example 4.34 From the first property, we see that the complex manifolds $S^{2n+1} \times S^{2m+1}$ constructed in Example 4.30 are not Kähler. Using the second property, we see that the Kodaira–Thurston manifold, KT, presented in Example 4.31 is not Kähler. (Again we note that KT is a symplectic manifold, so it is symplectic non-Kähler. We will come back to this in Section 4.5.)

Now consider the *algebra of cohomology*.

Theorem 4.35 (Hard Lefschetz theorem) *Let M be a compact Kähler manifold of real dimension $2n$. Then, the Lefschetz map, defined by*

$$L^k : H^{n-k}(M;\mathbb{C}) \longrightarrow H^{n+k}(M;\mathbb{C}),$$
$$\eta \mapsto \omega^k \wedge \eta$$

is an isomorphism for all $k \geq 0$, where ω denotes the Kähler form of M as well as its associated cohomology class.

An algebra H is said to have the *hard Lefschetz property* if it satisfies the condition of Theorem 4.35 for an element $\omega \in H^2$; namely, multiplication by ω^k induces isomorphisms $H^{n-k} \to H^{n+k}$ for each k. (This entails the fact that ω^n is in the top degree of the algebra.) Theorem 4.35 is a consequence of properties of representations of $\mathfrak{sl}(2,\mathbb{C})$ and the proof is beyond the scope of this book. We refer the reader to [115, page 122], [264, Chapter 5]. This result imposes strict restrictions on the cohomology algebra as shown by the

Theorem 4.36 (Blanchard [29]) *Let M be a compact manifold whose cohomology algebra satisfies the hard Lefschetz property. If θ is a derivation of negative degree on $H^*(M;\mathbb{C})$ such that the restriction of θ to $H^1(M;\mathbb{C})$ is zero, then $\theta(x) = 0$ for any $x \in H^*(M;\mathbb{C})$.*

Proof Let $\omega \in H^2(M;\mathbb{C})$ be the class for which the hard Lefschetz property holds and let θ be an algebra derivation of negative degree on $H^*(M;\mathbb{C})$. Recall that ω^n is fundamental cohomology class of M and denote $H^*(M;\mathbb{C})$ by H^*.

4 : Complex and symplectic manifolds

We first claim that $\theta(\omega) = 0$. If not, we have two possibilities:

- θ is a derivation of degree -2 with $\theta(\omega) = \lambda \in \mathbb{C}$. From $\omega^{n+1} = 0$, we obtain $0 = \theta(\omega^{n+1}) = (n+1)\lambda\omega^n$, which implies $\lambda = 0$.
- θ is a derivation of degree -1 with $\theta(\omega) = \alpha \in H^1(M; \mathbb{C})$. Let (α_j) be a basis of $H^1(M; \mathbb{C})$. From $\alpha_j\omega^n = 0$ and the hypothesis that the restriction of θ to $H^1(M; \mathbb{C})$ is zero, we obtain $\theta(\alpha_j\omega^n) = n\alpha\alpha_j\omega^{n-1} = 0$, and therefore α has cup-product 0 with any $\alpha_j\omega^{n-1}$. Now we know that $(\alpha_j\omega^{n-1})$ is a basis of H^{2n-1} by the hard Lefschetz property, so Poincaré duality implies $\alpha = 0$.

Now let r be the integer such that $\theta = 0$ on $H^{\leq r}$ and $\theta(x) \neq 0$ for some $x \in H^{r+1}$, $r < 2n$. We claim first that θ is zero on $H^{\geq 2n-r}$. If $y \in H^{2n-p}$ with $2n - p \geq 2n - r$, then, by hypothesis, there exists $z \in H^p$ such that $y = z\omega^{n-p}$. The claim results from $\theta(y) = \theta(z)\omega^{n-p} + z\theta(\omega^{n-p}) = 0$ by the choice of r and $\theta(\omega) = 0$.

Denote by $-s$ the degree of θ, so $\theta(x) \in H^{r+1-s}$. Let $z \in H^{2n-r-1+s}$. From $xz = 0$, we deduce $\theta(xz) = \theta(x)z = 0$. Since this is true for any z, we have $\theta(x) = 0$ by Poincaré duality. □

Remark 4.37 The hard Lefschetz property by itself implies that odd degree Betti numbers are even (see Exercise 4.3), so the results of Proposition 4.33 hold for the manifolds in Theorem 4.36 as well as for Kähler manifolds.

Theorem 4.36 is related to the collapsing of spectral sequences. Before we see this, recall the following.

Definition 4.38 *The fibration* $F \longrightarrow E \xrightarrow{f} B$ *is called* orientable *if the fundamental group $\pi_1(B)$ acts trivially on the cohomology $H^*(F; \mathbb{Q})$.*

Definition 4.39 *An orientable fibration* $F \xrightarrow{\iota} E \xrightarrow{f} B$ *is called* TNCZ *(for totally non-cohomologous to zero) if one of the following equivalent conditions is satisfied.*

1. *The Serre spectral sequence of f collapses at level 2. This means there is an isomorphism of vector spaces:*

$$H^n(E; \mathbb{Q}) \cong \bigoplus_{p+q=n} E_2^{p,q} = \bigoplus_{p+q=n} H^p(B; \mathbb{Q}) \otimes H^q(F; \mathbb{Q}).$$

2. *The induced map $H^*(\iota; \mathbb{Q}) \colon H^*(E; \mathbb{Q}) \to H^*(F; \mathbb{Q})$ is surjective.*

Observe that

- there are obviously fibrations which are not TNCZ, such as, for instance, the Hopf fibration $S^3 \longrightarrow S^7 \longrightarrow S^4$;
- by combining Theorem 2.71 and Theorem B.18, we see that the universal bundle $G/H \longrightarrow BH \longrightarrow BG$ is TNCZ if H is a closed subgroup of *maximal rank* of a compact connected Lie group G;
- some authors use the term *c-split* to mean TNCZ.

Proposition 4.40 *Let $F \longrightarrow E \xrightarrow{f} B$ be an orientable fibration satisfying the following properties:*

- *the vector space $H^*(F; \mathbb{Q})$ is finite in each dimension;*
- *any derivation θ of negative degree on the algebra $H^*(F; \mathbb{Q})$ whose restriction to $H^1(F; \mathbb{Q})$ vanishes, is identically zero;*
- *the transgression $H^1(F; \mathbb{Q}) \to H^2(B; \mathbb{Q})$ is zero.*

Then the fibration $f: E \to B$ is TNCZ.

Proof If the fibration is not TNCZ, there is an integer $k \geq 2$ such that $d_k : H^s(B; \mathbb{Q}) \otimes H^t(F; \mathbb{Q}) \to H^{s+k}(B; \mathbb{Q}) \otimes H^{t-k+1}(F; \mathbb{Q})$ is the first nonzero differential in the Serre spectral sequence. The fact that d_k is zero on $H^s(B; \mathbb{Q}) \otimes 1$ implies the existence of an algebra generator $x \in H^t(F; \mathbb{Q})$ with $d_k(1 \otimes x) \neq 0$. Let (w_i) be a basis of $H^{s+k}(B; \mathbb{Q})$. We decompose $d_k(1 \otimes x) = \sum_j w_j \otimes \theta_j(x)$ and extend the θ_j to all of $H^*(F; \mathbb{Q})$ as algebra derivations. Observe first that the θ_j are of negative degree by construction. Now, if $x \in H^1(F; \mathbb{Q})$, then we can only have $\theta_j(x) \neq 0$ if $k = 2$. In this case, $d_2(1 \otimes x)$ is precisely the transgression of x and we have assumed that it vanishes. Therefore, by hypothesis, the θ_j are zero and we get a contradiction. □

Theorem 4.36 and Proposition 4.40 immediately imply the next result due to Blanchard.

Corollary 4.41 ([29, Théorème II.1.2.]) *Let $F \longrightarrow E \xrightarrow{f} B$ be an orientable fibration such that the fiber F is a compact connected manifold whose cohomology satisfies the hard Lefschetz property. If the transgression $H^1(F; \mathbb{Q}) \to H^2(B; \mathbb{Q})$ vanishes, then the fibration $f: E \to B$ is TNCZ.*

Remark 4.42 In [188], D. McDuff constructed a blow-up of $\mathbb{C}P(5)$ along an embedding of the Kodaira–Thurston manifold $KT \hookrightarrow \mathbb{C}P(5)$. This gave the first known example of a compact *simply connected* symplectic non-Kähler manifold. More generally, if $\mathbb{C}P(n)$ is blown up along a submanifold M, then McDuff proved that if the hard Lefschetz property fails for M, it also fails for the blow-up (see [188, Proposition 2.5] or [257, Chapter 4] for

4 : Complex and symplectic manifolds

details). More recently, G. Cavalcanti [52] has made a systematic study of the behavior of the hard Lefschetz property under taking blow-ups and, in particular, has shown that there are compact simply connected symplectic manifolds (created as blow-ups) which satisfy Hard Lefschetz, but which are nonformal.

We now turn to the impact on *the de Rham algebra $A_{DR}(M)$* of a Kähler structure on M. First recall that we have a form $\omega \in A^2_{DR}(M)$ whose power ω^n is a volume form. This gives M a symplectic structure and we will discuss this in Section 4.5. Secondly, we have the following constraint on the *rational homotopy type* of M.

Theorem 4.43 ([71, Section 5]) *Any compact Kähler manifold is a formal space.*

Proof Recall that in any complex manifold M the complex valued de Rham algebra, $(A^{*,*}(M), d)$, is bigraded and that the d, ∂ and $\bar{\partial}$ are differentials, related by $d = \partial + \bar{\partial}$ such that $\partial^2 = \bar{\partial}^2 = \partial\bar{\partial} + \bar{\partial}\partial = 0$. We denote by $Z^{*,*}_\partial(M)$ and $H^{*,*}_\partial(M)$ the space of cocycles and the cohomology with respect to the differential ∂. We put the differential induced by $\bar{\partial}$ on $Z^{*,*}_\partial(M)$ and $H^{*,*}_\partial(M)$ and consider the following diagram of cdga's:

$$(H^{*,*}_\partial(M), \bar{\partial}) \xleftarrow{\rho} (Z^{*,*}_\partial(M), \bar{\partial}) \xrightarrow{j} (A^{*,*}(M), d).$$

The proof follows from the next steps, each of which uses the $\partial\bar{\partial}$-lemma stated in Lemma 4.24.

- $H(j)$ *is onto.* If $[\alpha] \in H(A(M), d)$, we have $\partial(\partial\alpha) = 0$, $\bar{\partial}(\partial\alpha) = \bar{\partial}(d\alpha - \bar{\partial}\alpha) = -\bar{\partial}^2(\alpha) = 0$. Then there exists β such that $\partial\alpha = \partial\bar{\partial}\beta$. Set $\gamma = \alpha - d\beta$, one has $\partial\gamma = \partial\alpha - \partial d\beta = \partial\alpha - \partial\bar{\partial}\beta = 0$ and $\bar{\partial}\gamma = \bar{\partial}\alpha - \bar{\partial}d\beta = -\bar{\partial}\alpha - \bar{\partial}\partial\beta = -(\partial\alpha - \partial\bar{\partial}\beta) = 0$. Therefore, γ is a ∂-cocycle and induces a $\bar{\partial}$-cohomology class such that $H(j)([\gamma]) = [\alpha]$.
- $H(j)$ *is injective.* Let $\alpha \in Z_\partial(M)$ such that $\alpha = d\beta$. From $\partial(\partial\beta) = 0$ and $\bar{\partial}(\partial\beta) = -\partial(\bar{\partial}\beta) = -\partial(d\beta - \partial\beta) = -\partial\alpha = 0$, we deduce the existence of γ such that $\partial\beta = \partial\bar{\partial}\gamma = -\bar{\partial}\partial\gamma$. Thus $\alpha = \partial\beta + \bar{\partial}\beta = -\bar{\partial}\partial\gamma + \bar{\partial}\beta = \bar{\partial}(\beta - \partial\gamma)$ is a $\bar{\partial}$-coboundary.
- *The induced differential by $\bar{\partial}$ on $H_\partial(M)$ is zero.* Let α be such that $\partial\alpha = 0$. From $\bar{\partial}(\bar{\partial}\alpha) = 0$ and $\partial(\bar{\partial}\alpha) = -\bar{\partial}(\partial\alpha) = 0$, we deduce the existence of γ such that $\bar{\partial}\alpha = \bar{\partial}\partial\gamma = -\partial\bar{\partial}\gamma$. Therefore $\bar{\partial}\alpha$ is in the image of ∂ and this implies that $[\bar{\partial}\alpha]$ is 0 in $H_\partial(M)$.
- $H(\rho)$ *is onto.* Let α be such that $\partial\alpha = 0$. From $\bar{\partial}(\bar{\partial}\alpha) = 0$ and $\partial(\bar{\partial}\alpha) = -\bar{\partial}(\partial\alpha) = 0$, we deduce the existence of β such that $\bar{\partial}\alpha = \bar{\partial}\partial\beta$. If we set $\gamma = \alpha - \partial\beta$, we have $\partial\gamma = \bar{\partial}\gamma = 0$ and $H(\rho)[\gamma] = [\alpha]$.

- $H(\rho)$ *is injective.* Let α be such that $\partial\alpha = \bar\partial\alpha = 0$ and $\alpha = \partial\beta$. Then there exists γ such that $\alpha = \partial\bar\partial\gamma = -\bar\partial\partial\gamma$ and $[\alpha]$ is zero in $H(Z_\partial, \bar\partial)$.

These steps show that $(A_{DR}^{*,*}(M), d)$ is linked by quasi-isomorphisms to a cdga with differential zero, $(H_\partial^{*,*}(M), 0)$ and this is the definition of formality; see Definition 2.85 and Remark 2.86. Observe also, from Proposition 2.101 that formality over $\mathbb{C}$ implies formality over $\mathbb{Q}$. $\square$

Corollary 4.44 *Except for tori, nilmanifolds are never Kähler manifolds.*

Proof This is a immediate consequence of Theorem 4.43 and Proposition 3.20. $\square$

In [64], [96], Cordero, Fernández and Gray showed the existence of Massey products in the cohomology of certain nilmanifolds. Hasegawa [134] proved that non-toral nilmanifolds are not formal directly (see Proposition 3.20) and discussed almost complex structures on them. In [25], C. Benson and C. Gordon used the hard Lefschetz theorem to prove that nontoral nilmanifolds cannot satisfy even that condition (see Theorem 4.98).

Example 4.45 The space $\mathbb{C}P^4 \# (S^3 \times S^5) \# (S^3 \times S^5)$ has the following properties:

- it is a formal space;
- its cohomology algebra satisfies Poincaré duality and the odd Betti numbers are even;
- there exists a closed form of degree 2, ω, such that ω^4 induces a fundamental class;
- it does not satisfy the hard Lefschetz theorem.

Example 4.46 There exist examples of nonformal spaces such that their cohomology satisfies Poincaré duality and the hard Lefschetz theorem, [177]. For instance, take spaces with the same cohomology algebra as

$$\left\{\left\{\left\{\mathbb{C}P(2)\#(S^2 \times S^2)\right\} \times S^2\right\} \#(S^3 \times S^3)\right\} \times S^2,$$

but with a different rational homotopy type.

Remark 4.47 Theorem 4.43 implies the vanishing of Massey products with coefficients in $\mathbb{R}$. The analogous result is not true with coefficients in $\mathbb{Z}_p$, as shown by an example of T. Ekedahl in [80].

Remark 4.48 If G is a finite group acting holomorphically on a Kähler manifold (with nonempty, connected and simply connected fixed point sets), there is a notion of *equivariant formality*. One can prove that *a compact G-Kähler manifold* is equivariantly formal. This is due to T. Lambre [163] in the case $G = \mathbb{Z}_p$ and to B. Fine and G. Triantafillou [98] in the general case.

The purpose of the next conjecture, due to S. Halperin, is to put conditions on a space F such that any orientable fibration with fiber F is TNCZ. Before stating the conjecture, recall from Definition 2.73 that a rationally elliptic space X is a nilpotent space with finite dimensional rational cohomology such that $\sum_{p\geq 2} \operatorname{rank} \pi_p(X) < \infty$. Such spaces are known to have non-negative Euler characteristics and their rational cohomology algebras satisfy Poincaré duality. Moreover, for such spaces, the following conditions are equivalent (see Theorem B.18 and [87, Proposition 32.10]).

- The Euler characteristic is positive; that is, $\chi(X) > 0$.
- The rational cohomology is evenly graded; that is, $H^{2p+1}(X; \mathbb{Q}) = 0$.
- The homotopy Euler characteristic is zero; that is, $\sum_{q>0} \operatorname{rank} \pi_{2q}(X) = \sum_{q>0} \operatorname{rank} \pi_{2q+1}(X)$.

Conjecture 4.49 (**The Halperin conjecture**) *Let X be a simply connected rationally elliptic space with positive Euler characteristic. Then, any orientable fibration whose fiber is X is TNCZ.*

From Proposition 4.40, it is easy to see that Conjecture 4.49 is equivalent to the nonexistence of negative derivations on the cohomology of X. Theorem 4.36 of Blanchard thus implies that the Halperin conjecture is true if X is a compact simply connected Kähler manifold. Shiga and Tezuka have also proved the conjecture for the quotient of a compact connected Lie group by a closed subgroup of maximal rank by using the action of the Weyl group on the cohomology as in Subsection 3.3.2 (see [237]).

We will not go further into this topic, but instead refer the reader to [86, page 516] for a more complete list of references, see also Section 9.7. We simply mention one tantalizing link to geometry. Let $\mathcal{H}$ denote the class of simply connected, finite CW-complexes whose rational cohomology has no nonzero derivations of negative degree. In [23] (see Chapter 6, in particular Theorem 6.40), Belegradek and Kapovitch relate $\mathcal{H}$ to concrete properties of curvature. They also motivate the question of whether an extension of the Shiga–Tezuka result to the class of biquotients is possible, see [23, Section 13.2].

4.3 The Dolbeault model of a complex manifold

Let M be a complex manifold. Recall that the de Rham algebra of complex-valued smooth forms on M admits a bigradation, $A_{DR}^c(M) = \oplus_{p,q} A^{p,q}(M)$,

4.3 The Dolbeault model of a complex manifold

and that there exist two differentials, $\partial: A^{p,q}(M) \to A^{p+1,q}(M)$ and $\bar\partial: A^{p,q}(M) \to A^{p,q+1}(M)$. In this section, we study the Dolbeault complex $(A^{*,*}(M), \bar\partial)$ and its relation with the de Rham complex $(A^{*,*}(M), d)$ when M is Kähler.

4.3.1 Definition and existence

If $V = \oplus_{p,q} V^{p,q}$ is a bigraded vector space, the expression *degree of an element* $v \in V^{p,q}$ will always mean the *total degree* $p+q$. When we consider the commutative free graded algebra on V, we will refer also to the total degree of V:

$$V = \bigoplus_n V^n = \bigoplus_n \left(\bigoplus_{p+q=n} V^{p,q} \right).$$

The bigradation on V is naturally extended as a bigradation of algebras on $\wedge V$. The homogeneous elements of bidegree (p,q) are the $\omega = \sum \omega_{i_1} \ldots \omega_{i_r}$ such that $\omega_{i_j} \in V^{p_j, q_j}$ with $p = \sum_{j=1}^r p_j$ and $q = \sum_{j=1}^r q_j$.

Theorem 4.50 ([207]) *Let M be a complex manifold with no nonconstant holomorphic functions (e.g. a connected compact complex manifold). Then there exists a model of the Dolbeault complex, provided with a bigradation,*

$$\varphi: \left(\wedge V^{*,*}, \bar\partial\right) \longrightarrow \left(A^{*,*}, \bar\partial\right),$$

such that $\varphi(V^{p,q}) \subset A^{p,q}$ and $\bar\partial V^{p,q} \subset (\wedge V)^{p,q+1}$.

We will give a proof in a more general setting (see Theorem 4.53), but, before that, Theorem 4.50 elicits the next notion of formality.

Definition 4.51 A complex manifold M is *Dolbeault formal* if there exist two morphisms of differential graded algebras φ and ψ,

$$(H^{*,*}_{\bar\partial}(M), 0) \xleftarrow{\psi} (\wedge V^{*,*}, \bar\partial) \xrightarrow{\varphi} (A^{*,*}(M), \bar\partial),$$

such that φ and ψ are of bidegree $(0,0)$ and induce isomorphisms in cohomology.

The hypothesis on holomorphic functions assumed in Theorem 4.50 corresponds to the cohomological connectedness required for the construction of a minimal model (see Theorem 2.24). This theorem is a particular case of a model for a bigraded object. We briefly discuss this framework since it gives a unified point of view for various situations.

4: Complex and symplectic manifolds

Definition 4.52 *Let $r \geq 0$ be fixed. An r-bigraded cdga (A, d) is a pair (A, d) consisting of an algebra A and a differential d satisfying:*

- $A = \bigoplus_{p,q} A^{p,q}$ for $p \in \mathbb{Z}$, $q \in \mathbb{Z}$,
- $A^{p,q}.A^{p',q'} \subset A^{p+p',q+q'}$,
- $xy = (-1)^{(p+q)(p'+q')} yx$ and $d(xy) = (dx)y + (-1)^{p+q} x(dy)$ if $x \in A^{p,q}$ and $y \in A^{p',q'}$,
- $dA^{p,q} \subset A^{p+r,q-r+1}$.

A morphism of r-bigraded cdga's is a morphism of cdga's of bidegree $(0,0)$.

Theorem 4.53 *Let $f : (A, d) \to (A', d')$ be a morphism of r-bigraded cdga's with $H^1(f)$ injective. Then there exists a factorization of f into r-bigraded morphisms*

$$(A, d) \hookrightarrow (A \otimes \wedge Z, \delta) \xrightarrow{\varphi} (A', d')$$

where φ induces an isomorphism in cohomology and the differential $\bar{\delta}$, induced by δ on the quotient of $A \otimes \wedge Z$ by the ideal generated by A, is decomposable.

An absolute version of the result devolves to the particular case where the domain (A, d) is the field of reference with trivial differential and trivial bigradation.

Proof The construction of the minimal model in Chapter 2 (see Theorem 2.24) is done in two stages: a *cokernel* stage which introduces a new space of generators V_1 and a *kernel* stage introducing V_2 as a space of new generators. To prove Theorem 4.53, we mimic this approach with regard to the total degree by giving the following bidegree to the generators:

$$V_1^{p,q} = \left(\operatorname{Coker} H^{n+1}\right)^{p,q} \text{ and } V_2^{p-r,q+r-1} = \left(\operatorname{Ker} H^{n+1}\right)^{p,q}.$$

The remainder of the proof is straightforward. □

Definition 4.54 *A decomposition as in Theorem 4.53 is called a relative minimal r-bigraded model of f or an r-bigraded minimal model of (A', d') in the absolute case.*

The r-bigraded cdga's arise naturally.

- The bigraded model (see Theorem 2.93) is a 1-bigraded model of $(H, 0)$ where the elements of $(\wedge V)_p^n$ are of bidegree $(-p, n + p)$.
- The differential of the Dolbeault cdga verifies $\bar{\partial} A^{p,q} \subset A^{p,q+1}$, so this is a 0-bigraded cdga.

More generally any r-stage of a spectral sequence of a filtered cdga is an r-bigraded cdga, see Appendix B. Let's now recall this notion of filtered cdga.

Definition 4.55 *A filtered cdga $(A, d, \mathcal{F})$ is a cdga (A, d) together with a filtration $\mathcal{F}^p A$ on A such that:*

- $\mathcal{F}^{p+1} A \subset \mathcal{F}^p A$; $\mathcal{F}^p A . \mathcal{F}^q A \subset \mathcal{F}^{p+q} A$, $d\mathcal{F}^p A \subset \mathcal{F}^p A$;
- $A = \bigcup_p \mathcal{F}^p A$;
- *for any integer n, there exists an integer $p(n)$ such that $\mathcal{F}^{p(n)} A^n = \{0\}$.*

A morphism of filtered cdga's is a morphism of cdga's that respects the filtrations.

To any r-bigraded cdga (B, d), we associate a filtered cdga $(B, d, \mathcal{F})$ by $\mathcal{F}^p B = \oplus_{j \geq p} B^{j,*}$. Reciprocally, if $f : (A, d, \mathcal{F}) \to (A', d', \mathcal{F})$ is a morphism of filtered cdga's, we denote by $E_r(f) : (E_r^{p,q}(A), d_r) \to (E_r^{p,q}(A'), d_r')$ the induced morphism between the associated spectral sequences. This morphism $E_r(f)$ is a morphism of r-bigraded cdga's for any $r \geq 0$. The surprise is that we can get a model of the filtered algebra built from a model of its r-stage as we describe now.

Theorem 4.56 ([133]) *Let $f : (A, d, \mathcal{F}) \to (A', d', \mathcal{F})$ be a morphism of filtered cdga's and $r \geq 0$ be an integer. We consider a relative r-bigraded model of $E_r(f)$,*

$$(E_r(A), d_r) \longrightarrow (E_r(A) \otimes \wedge V, D_r) \xrightarrow{\varphi} (E_r(A'), d_r').$$

Then there exists a factorization of f into morphisms of filtered cdga's

$$(A, d, \mathcal{F}) \longrightarrow (A \otimes \wedge V, D, \mathcal{F}) \xrightarrow{\psi} (A', d', \mathcal{F})$$

such that $E_r(\psi) = \varphi$.

An absolute version of *this* result also devolves to the particular case where the domain $(A, d, \mathcal{F})$ is the field of reference with trivial differential and trivial filtration.

Definition 4.57 *A decomposition as in Theorem 4.56 is called a relative r-filtered model of f or an r-filtered model of $(A', d', \mathcal{F})$ in the absolute case.*

Let (A, d) be a cdga and consider the filtration associated to the degree of A. Then the first stage of the associated spectral sequence is $E_1(f) = H(f)$. The relative 1-filtered model of f is exactly the filtered model of Definition 2.96.

4 : Complex and symplectic manifolds

The *proof of Theorem 4.56* consists of a careful study of derivations in a spectral sequence. We refer the reader to [133] where the framework is more general than here. We will use Theorem 4.56 in:

- Subsection 4.3.3 for the construction of a Dolbeault model of the total space of a holomorphic fibration, using the Borel spectral sequence, (see Proposition 4.62);
- Section 4.4 for the study of the Frölicher spectral sequence (see Theorem 4.70).

4.3.2 The Dolbeault model of a Kähler manifold

For Kähler manifolds, Dolbeault models are directly related to de Rham models. To see this, we introduce a notion of formality well-suited to this situation.

Definition 4.58 ([207]) *A complex manifold M is strictly formal if, for some 0-bigraded differential algebra $(\wedge Z^{*,*}, \delta)$, there exist quasi-isomorphisms of bidegree $(0,0)$,*

$$(A^{*,*}(M), \bar{\partial}) \xleftarrow{\simeq} (\wedge Z^{*,*}, \delta) \xrightarrow{\simeq} (A^{*,*}(M), d)$$

$$\downarrow \simeq$$

$$(H_{\bar{\partial}}^{*,*}(M), 0),$$

linking together the de Rham algebra, the Dolbeault algebra and the Dolbeault cohomology.

Observe that, if M is strictly formal, then M is de Rham formal and Dolbeault formal. Moreover, we have $H_{DR}(M) = H_{\bar{\partial}}(M)$.

Theorem 4.59 ([207]) *A compact Kähler manifold is strictly formal.*

Proof We show the existence of quasi-isomorphisms in the following diagram.

$$(A^{*,*}(M), \bar{\partial}) \xleftarrow{\psi} (Z_{\partial}^{*,*}(M), \bar{\partial}) \xrightarrow{j} (A^{*,*}(M), d)$$

$$\downarrow \rho$$

$$(H_{\partial}^{*,*}(M), 0)$$

4.3 The Dolbeault model of a complex manifold

Here, $Z_{\bar\partial}^{p,q}(M)$ is the subspace of $A^{p,q}(M)$, consisting of the $\bar\partial$-cocycles. From the proof of Theorem 4.43, we already know that $H(j)$ and $H(\rho)$ are isomorphisms, so let's start here with $H(\psi)$ using Lemma 4.24.

- $H(\psi)$ *is onto.* Let $\alpha \in A^{p,q}(M)$ be such that $\bar\partial \alpha = 0$. We denote the associated $\bar\partial$-cohomology class by $[\alpha]$. With the $\partial\bar\partial$-lemma applied to $\partial\alpha$, we obtain β such that $\partial\alpha = \partial\bar\partial\beta$. Thus, the $\bar\partial$-cohomology class $[\alpha]$ in $A^{p,q}(M)$ is hit by the $\bar\partial$-cohomology class associated to the ∂-cocycle $\alpha - \bar\partial\beta$; that is, $H(\psi)([\alpha - \bar\partial\beta]) = [\alpha]$.

- $H(\psi)$ *is injective.* Let $\alpha \in Z_\partial^{*,*}(M)$ such that $\alpha = \bar\partial(\beta)$. We have to show that α is the coboundary of an element of Z_∂. From the $\partial\bar\partial$-lemma applied to $\partial\beta$, we get γ such that $\partial\beta = \partial\bar\partial\gamma$. The element $\beta - \bar\partial\gamma$ belongs to $Z_\partial^{*,*}(M)$ and satisfies $\alpha = \bar\partial(\beta - \bar\partial\gamma)$. Therefore the $\bar\partial$-class of α is zero in $Z_\partial^{*,*}(M)$.

Note that we have shown that $H_\partial^{*,*}(M) \cong H_{\bar\partial}^{*,*}(M)$, so we can replace $H_\partial^{*,*}(M)$ in the diagram by $H_{\bar\partial}^{*,*}(M)$ to achieve strict formality. □

Example 4.60 By Theorem 4.59, since the manifold $\mathbb{C}P(n)$ is Kähler, its Dolbeault model coincides with the de Rham model except for the bigradation. Recall, from Example 2.44, that the minimal model of $\mathbb{C}P(n)$ is given by $(\wedge(a,b), \bar\partial)$, $\bar\partial b = a^{n+1}$, $\bar\partial a = 0$. The element a, representing the Kähler form, must be of bidegree $(1, 1)$. The bidegree of b must be chosen so that $\bar\partial$ is of bidegree $(0, 1)$. So b is of bidegree $(n + 1, n)$.

4.3.3 The Borel spectral sequence

In this section, we recall the Borel spectral sequence and how Borel used it to compute the Dolbeault cohomology of the Hopf manifold $S^1 \times S^{2n+1}$.

Theorem 4.61 ([141, Appendix]) *Let $F \to E \to B$ be a holomorphic fiber bundle where E, B, F are connected and F is compact. Suppose that every connected component of the structure group acts trivially on $H_{\bar\partial}(F)$. Then there exists a spectral sequence (E_r, d_r), $r \geq 0$, whose terms are trigraded by type (p, q) and the basic degree s such that:*

$$\left({}_sE_1^{p,q}, d_1\right) = \left(\sum_i A^{i,s-i}(B) \otimes H_{\bar\partial}^{p-i,q-s+i}(F), \bar\partial \otimes 1\right)$$

$${}_sE_2^{p,q} = \sum_i H_{\bar\partial}^{i,s-i}(B) \otimes H_{\bar\partial}^{p-i,q-s+i}(F),$$

which converges to $H_{\bar\partial}^{*,*}(E)$.

For instance, if F is Kähler, then the spectral sequence exists, see [141, Remark 2.2 of the Appendix]. We now use Theorem 4.61 to get a Dolbeault model of the total space E of a holomorphic bundle.

Proposition 4.62 *Let $F \longrightarrow E \xrightarrow{f} B$ be a holomorphic bundle between compact, connected, nilpotent complex manifolds, with fiber F a Kähler manifold and with a trivial action of the fundamental group of the base on the cohomology of the fiber. Then there exists a Dolbeault model of the total space*

$$(\wedge W \otimes \wedge V, \delta) \xrightarrow{\psi} (A^{*,*}(E), \overline{\partial}),$$

such that

- *$(\wedge W, \delta)$ is a Dolbeault model of B;*
- *with respect to the total degree, the quotient $(\wedge V, \overline{\delta})$ of $(\wedge W \otimes \wedge V, \delta)$ by the ideal generated by $\wedge W$ is a model of $(H^*(F; \mathbb{C}), 0)$.*

Proof Since the fiber F is a Kähler manifold, we have an isomorphism $H_{\overline{\partial}}(F) \cong H^*(F; \mathbb{C})$. Let $(\wedge V, \overline{\delta})$ be a 0-bigraded model of $(H_{\overline{\partial}}(F), 0)$. Theorem 4.56 gives a Dolbeault model of E by perturbing a model

$$(\wedge W, \delta) \otimes (\wedge V, \overline{\delta}) \xrightarrow{\sim} (_s E_1^{p,q}, d_1)$$

of the E_1-term of the Borel spectral sequence. This is the desired model. □

In the next section, we will give a more geometric argument, based on Chern–Weil theory, which is sufficient for most of our examples. Before that, we consider the Dolbeault model of the Hopf manifold.

Example 4.63 (Dolbeault cohomology of $S^1 \times S^{2n+1}$) Recall that there is a complex manifold structure on the product $S^1 \times S^{2n+1}$ (see Example 4.30) such that

$$S^1 \times S^1 \longrightarrow S^1 \times S^{2n+1} \longrightarrow \mathbb{CP}(n)$$

is a principal holomorphic bundle. This complex manifold is called a Hopf manifold and is denoted by $M_{0,n}$. We now want to construct a Dolbeault model of it.

Take a Dolbeault model of $\mathbb{CP}(n)$, $(\wedge(a, b), \overline{\partial})$, $\overline{\partial} b = a^{n+1}$, $\overline{\partial} a = 0$, with a and b of respective bidegrees $(1,1)$ and $(n+1,n)$, see Example 4.60. We give the fiber $S^1 \times S^1$ the structure of a complex torus. Its Dolbeault model is $(\wedge(\alpha, \beta), \overline{\partial})$, $\overline{\partial}(\alpha) = \overline{\partial}(\beta) = 0$, with α and β of respective bidegrees $(0,1)$

and $(1, 0)$. A model of the (E_1, d_1)-term of the Borel spectral sequence is given by:

$$(\wedge(a, b), \overline{\partial}) \otimes (\wedge(\alpha, \beta), \overline{\partial}) = (\wedge(a, b, \alpha, \beta), \overline{\partial}).$$

From Proposition 4.62, we know that a model of the Dolbeault complex of the Hopf manifold is a bigraded differential algebra $(\wedge(a, b, \alpha, \beta), \delta)$. The differential δ is related to $\overline{\partial}$ by $\delta = \overline{\partial} + \tau$ where τ increases the basic degree and is of bidegree $(0, 1)$; that is, $\tau \colon {}_s E_1^{p,q} \to {}_{s'} E_1^{p,q+1}$ with $s' > s$. The only possible choice is $\tau(\beta) = \lambda a$, where $\lambda \in \mathbb{C}$. By modifying generators, we have, in fact, only two possibilities: $\lambda = 0$ or $\lambda = 1$. At this point, only a deeper analysis of the complex structure allows a choice between these two possibilities. We shall not do that here, but instead recall that Borel proves $H^{1,0}(M_{0,n}) = 0$, see [141, Lemma 9.4]. Therefore we must have $\tau(\beta) = a$. Now an easy computation gives $H_{\overline{\partial}}(M_{0,n}) = \wedge(b, \alpha)$, with b and α of respective bidegrees $(n+1, n)$ and $(0, 1)$ (as in [141]).

In Subsection 4.3.4, we will reconsider this example and construct the Dolbeault model without using Borel's result on $H^{1,0}(M_{0,n})$, see Example 4.67.

4.3.4 The Dolbeault model of Calabi–Eckmann manifolds

Let's briefly recall some points of Chern–Weil theory for fibrations with toral fibers. Let $T^{2n} \longrightarrow E \xrightarrow{f} B$ be a real principal bundle with B simply connected. Denote by $\mathcal{T} \cong \mathbb{R}^{2n}$ the Lie algebra associated to T^{2n}, $\mathcal{T}^*$ the dual of $\mathcal{T}$ and by $V \cong \mathcal{T}^*$ a graded vector space concentrated in degree 1. The relative minimal model of a fibration (see Theorem 2.64) gives a morphism of real cdga's

$$\Phi \colon (A_{DR}(B) \otimes \wedge V, D) \longrightarrow (A_{DR}(E), d),$$

inducing an isomorphism in cohomology such that $\Phi(\alpha) = f^*(\alpha)$ and $D(\alpha) = d_B(\alpha)$ for $\alpha \in A_{DR}(B)$. We now want to determine, from geometrical data, the map Φ and the differential D on V. For that, we choose a connection $\omega \in A^1_{DR}(E; \mathcal{T})$ of curvature $R_B \in A^2_{DR}(B; \mathcal{T})$. Denoting by $\langle -, - \rangle$ the duality between $\mathcal{T}$ and $\mathcal{T}^*$, we set

$$\Phi(v) = \langle v, \omega \rangle \text{ and } Dv = \langle v, R_B \rangle,$$

for $v \in V$. We specify now by the exponent ^{c} the fact that we have complexified the objects as $\Phi \colon (A^c(B) \otimes \wedge V^c, D) \to (A^c(E), d)$, the maps Φ, D and d being extended by $\mathbb{C}$-linearity.

The complex structure on T^{2n} gives a decomposition $V^c \cong V^{0,1} \oplus V^{1,0}$ that we extend into an algebra bigradation on $\wedge V^c$ as before. The next

4: Complex and symplectic manifolds

result is contained in [14]. We give here a direct proof of a particular case sufficient for our purposes.

Lemma 4.64 *There exists a connection ω such that the above morphism, $\Phi: (A^c(B) \otimes \wedge V^c, D) \to (A^c(E), d)$, is compatible with the bidegree. Such a connection is said to be* compatible with the complex structures.

Proof Let (e_j) be a real basis of T such that $J(e_j) = e_{j+n}$, $J(e_{j+n}) = -e_j$ if $j \le n$ and denote by (e^j) the dual basis. The space T^c decomposes in $T^c = T^{1,0} \oplus T^{0,1}$ with basis $f_j = e_j - ie_{j+n}$ (resp. $\bar{f}_j = e_j + ie_{j+n}$) for $T^{1,0}$ (resp. $T^{0,1}$). The complex dual $T^{c,*}$ of T^c splits into $T^{c,*} = T^{*0,1} \oplus T^{*1,0}$ with bases $f^j = (e^j + ie^{j+n})/2$ for $T^{*0,1}$ and $\bar{f}^j = (e^j - ie^{j+n})/2$ for $T^{*1,0}$.

Let (U_α) be a family of trivializing open sets and let ρ_U be an associated partition of unity. Denote by t^j the coordinates of T^{2n} in $f^{-1}(U) \cong U \times T^{2n}$. From the choice of the basis (e_j), $t^j + it^{j+n}$ is a holomorphic coordinate in T^{2n}. Locally, we choose $\omega_U^j = dt^j$, and this gives the connection $\omega = \sum_U \rho_U \omega_U^j e_j$. A computation shows that $\langle a_j f^j, \omega \rangle = \sum_U \rho_U a_j d(t^j + it^{j+n})$ is of bidegree $(0,1)$ and $\langle a_j \bar{f}^j, \omega \rangle$ of bidegree $(1,0)$. Since the family is holomorphic, the result is established. □

Theorem 4.65 ([249]) *Let $T^{2n} \to E \xrightarrow{f} B$ be a holomorphic principal bundle with Chern–Weil homomorphism $\Phi: (A^c(B) \otimes \wedge V^c, D) \to (A^c(E), d)$ associated to a connection compatible with the complex structures. Then a Dolbeault model of E is given by*

$$\Phi: (A^c(B) \otimes \wedge(V^{0,1} \oplus V^{1,0}), \delta) \longrightarrow (A^c(E), \bar{\partial})$$

where

- $V^c = V^{0,1} \oplus V^{1,0}$ *is a decomposition associated to the complex structure of the torus;*
- $\delta \alpha = \bar{\partial} \alpha$ *if $\alpha \in A^c(B)$;*
- $\delta v = \bar{D} v$, *if $v \in V^c$, where $\bar{D}$ is the homogeneous part of bidegree $(0,1)$ in D.*

Proof From Lemma 4.64, we know that the Chern–Weil homomorphism Φ is of bidegree $(0,0)$. The differential D is compatible with the decreasing filtration associated to the first degree and Φ induces a morphism of cdga's at level 0 of the two associated spectral sequences (i.e. the Frölicher spectral sequences studied in Section 4.4.):

$$\Phi: (E_0, d_0) \cong (A^c(B) \otimes \wedge(V^c), \delta) \to (A^c(E), \bar{\partial}).$$

We are reduced to proving that we get an isomorphism in cohomology. For that, we filter the domain by

$$\mathcal{F}^p = \sum_{i \geq p} A^c(B)^i \otimes \wedge(V^c)$$

and the target by the classical basic degree. The morphism Φ induces a morphism between the associated spectral sequences. On the target, this is the Borel spectral sequence. Thus Φ induces an isomorphism at the second level of the two spectral sequences; $E_2' \cong H_{\bar\partial}(B) \otimes \wedge V^c$. □

Corollary 4.66 *With the notation of Theorem 4.65, and with B a simply connected Kähler manifold such that $H^2(B;\mathbb{C}) \cong H_{\bar\partial}^{1,1}(B)$, there exists a Dolbeault model of E, $(\wedge Y \otimes \wedge(V^{0,1} \oplus V^{1,0}), \delta)$, and a de Rham model of E, $(\wedge Y \otimes \wedge(V^{0,1} \oplus V^{1,0}), D)$, linked by*

- $(\wedge Y, \delta) \cong (\wedge Y, D)$ *is a Dolbeault model of B;*
- $\delta v = Dv$ *if* $v \in V^{1,0}$;
- $\delta v = 0$ *if* $v \in V^{0,1}$.

Proof In Theorem 4.65, we have to replace $A^c(B)$ by its minimal model $(\wedge Y, \delta)$. The conclusion comes from the hypothesis on $H^2(B;\mathbb{C})$. □

Example 4.67 Recall that a complex manifold structure, denoted by $M_{m,n}$, exists on the product $S^{2m+1} \times S^{2n+1}$ such that

$$S^1 \times S^1 \longrightarrow S^{2m+1} \times S^{2n+1} \longrightarrow \mathbb{CP}(m) \times \mathbb{CP}(n)$$

is a principal holomorphic bundle. Suppose $0 < n \leq m$. Directly from Corollary 4.66 and Example 4.60, we obtain the Dolbeault model of $M_{m,n}$ to be $\left(\wedge (a, b, a', b', \alpha, \beta), \bar\partial\right)$, with:

- $|a| = (1,1), |b| = (m+1, m), \bar\partial(a) = 0, \bar\partial(b) = a^{m+1}$;
- $|a'| = (1,1), |b'| = (n+1, n), \bar\partial(a') = 0, \bar\partial(b') = a'^{n+1}$;
- $|\alpha| = (1,0), |\beta| = (0,1), \bar\partial(\alpha) = a - ia', \bar\partial(\beta) = 0$.

A computation gives the Dolbeault cohomology algebra of $M_{m,n}$,

$$H_{\bar\partial}(M_{m,n}) \cong \left(\frac{\wedge a'}{a'^{n+1}}\right) \otimes \wedge (b, \beta),$$

already determined by A. Borel in [141].

Example 4.68 The Grassmannian of 2-planes in $\mathbb{C}^4$, $G_{4,2}(\mathbb{C}) = \dfrac{U(4)}{U(2) \times U(2)}$, is a Kähler manifold of dimension 8. As in Corollary 4.28, we construct a holomorphic principal bundle

$$S^1 \times S^1 \longrightarrow E \longrightarrow \frac{U(4)}{U(2) \times U(2)} \times \frac{U(4)}{U(2) \times U(2)}$$

and we want to determine a Dolbeault model of the total space E. From Theorem 2.71, we compute the minimal model of $G_{4,2}(\mathbb{C})$:

$$(\wedge V, d) = (\wedge (y_2, y_4) \otimes \wedge (x_5, x_7), \delta)$$

with $|y_k| = |x_k| = k$, $\delta y_2 = \delta y_4 = 0$, $\delta x_5 = y_2^3 - 2 y_4 y_2$ and $\delta x_7 = y_2^2 y_4 - y_4^2$. Since $G_{4,2}(\mathbb{C})$ is Kähler, a Dolbeault model is obtained from the de Rham model by adding a bigradation:

- the element y_2 is of bidegree $(1,1)$ because it corresponds to the Kähler form;
- y_4 is of bidegree $(2,2)$ because, in a Kähler manifold, one has $H^{p,q} \cong H^{q,p}$ and therefore there is no other possibility;
- the two other generators are given the right bidegree to obtain a differential of bidegree $(0,1)$: $|x_5| = (3,2)$ and $|x_7| = (4,3)$.

Corollary 4.66 now gives a Dolbeault model of the total space E:

$$(\wedge(y_2, y_4, x_5, x_7) \otimes \wedge(y_2', y_4', x_5', x_7') \otimes \wedge(\alpha, \beta), \overline{\partial})$$

with $|\alpha| = (1,0)$, $\overline{\partial}(\alpha) = y_2 - i y_2'$, $|\beta| = (0,1)$, $\overline{\partial}(\beta) = 0$. On the other generators, the differential has been described before. From this model, a computation gives the minimal Dolbeault model of the total space E:

$$(\wedge \beta \otimes \wedge (y_2', y_4, y_4', x_5, x_7, x_5', x_7'), \overline{\partial})$$

with $\overline{\partial}(\beta) = \overline{\partial}(y_2') = \overline{\partial}(y_4) = \overline{\partial}(y_4') = 0$, $\overline{\partial}(x_5) = -i y_2'^3 - 2 i y_4 y_2'$, $\overline{\partial}(x_5') = y_2'^3 - 2 y_4' y_2'$, $\overline{\partial}(x_7) = -y_2'^2 y_4 - y_4^2$, $\overline{\partial}(x_7') = y_2'^2 y_4' - y_4'^2$.

4.4 The Frölicher spectral sequence

4.4.1 Definition and properties

Let M be a complex manifold with complex de Rham algebra

$$A_{DR}^c(M) = \left(\bigoplus_{p,q} A^{p,q}(M), \partial + \overline{\partial} \right).$$

We filter this complex by the first degree:

$$\mathcal{F}^p A_{DR}^c(M) = \oplus_{j \geq p} A^{j,*}.$$

The filtration $\mathcal{F}^p$ is decreasing, $\mathcal{F}^p \supset \mathcal{F}^{p+1}$, and stable under the differential, $d\mathcal{F}^p \subset \mathcal{F}^p$.

Definition 4.69 *Let M be a complex manifold. The spectral sequence associated to the filtration $\mathcal{F}^p A^c_{DR}(M)$ is called the Frölicher spectral sequence. It satisfies $E_0^{p,q} = A^{p,q}(M)$, $d_0 = \bar\partial$, $E_1^{p,q} = H^{p,q}_{\bar\partial}(M)$ and converges to $H_{DR}(M)$.*

As a direct application of Theorem 4.56, we have the following result.

Theorem 4.70 *Let M be a complex manifold and $(\wedge V^{*,*}, \bar\partial) \xrightarrow{\varphi} (A^{*,*}(M), \bar\partial)$ be a Dolbeault model of M. Then there exist perturbations of the morphism φ and of the differential $\bar\partial$ which give a model of the de Rham complex,*

$$(\wedge V^{*,*}, \bar\partial + \tau) \xrightarrow{\psi} (A^{*,*}, d),$$

such that: $\tau\left(V^{p,q}\right) \subset (\wedge V)^{>p,*}$ and $\psi(v) - \varphi(v) \in A^{>p,*}$ if $v \in V^{p,*}$.

Theorem 4.59 directly characterizes the behavior of the Frölicher spectral sequence of a compact Kähler manifold.

Corollary 4.71 *The Frölicher spectral sequence of a compact Kähler manifold collapses at the E_1-term; that is, $E_1 \cong E_\infty$.*

Example 4.72 Recall that the Hopf manifold of Example 4.63 has Dolbeault model $(\wedge(a, b, \alpha, \beta), \bar\partial)$. A *model of the de Rham complex* of $M_{0,n}$ is obtained by a perturbation of the differential $\bar\partial$, which increases the first degree by at least one. But it is easy to see that $H_{DR}(M_{0,n}) = H_{DR}\left(S^1 \times S^{2n+1}\right) \cong H_{\bar\partial}(M_{0,n})$. The complex manifold $M_{0,n}$ is thus strictly formal and its de Rham model coincides with its Dolbeault model. Observe that $M_{0,n}$ is not Kähler because, for instance, there are no cohomology classes in degree 2.

Remark 4.73 From Corollary 4.66 and using its notation, the element $v \in V^{0,1}$ is a Dolbeault cocycle and not a de Rham cocycle. Therefore, since the Frölicher spectral sequence has $E_1^{p,q} = H^{p,q}_{\bar\partial}(M)$ and it converges to $H_{DR}(M)$, we see that it does not collapse at level 1 for any Calabi–Eckmann manifold built as in Corollary 4.28.

4.4.2 Pittie's examples

Let's now follow the approach of [227] and [228] in using even dimensional compact connected Lie groups G to find examples of compact complex manifolds whose Frölicher spectral sequence does not collapse at level 2. (Note that there is another family of examples coming from nilmanifolds

4: Complex and symplectic manifolds

that is described in [65].) Now let's give a characterization in terms of models for the degeneracy at level 2 of the Frölicher spectral sequence for G.

Theorem 4.74 ([249]) *Let G be a compact, connected, even dimensional Lie group and let T be a maximal torus of G. If $T \longrightarrow G \longrightarrow G/T$ is a principal holomorphic bundle, then the Frölicher spectral sequence of G degenerates at level 2 if and only if the complex manifold G is Dolbeault formal.*

In the course of the proof, we build a special model of the Dolbeault algebra of G, using Corollary 4.66. This model is the positive answer to Conjecture 5.12 of [227].

Proof As we said above, we use Corollary 4.66. First, we need a Dolbeault model of G/T. A de Rham model is given by $(\wedge P \otimes \wedge U, \delta)$, where P is the primitive space of G, the vector space U is concentrated in degree 2, $\delta(u) = 0$ if $u \in U$ and $\delta(v) \in \wedge U$ if $v \in P$. For the Dolbeault model, we need a bigradation of the generators: the elements of U are in bidegree $(1, 1)$ and the elements of P are bigraded such that the differential δ is of bidegree $(0, 1)$.

A de Rham model of G is obtained as a relative model of the fibration above; that is $(\wedge P \otimes \wedge U \otimes \wedge V, D)$, with V concentrated in degree 1, $D = \delta$ on $U \oplus P$ and D creating an isomorphism from V to U.

From the complex structure on V, we get a decomposition into eigenspaces $V = V^{0,1} \oplus V^{1,0}$. The Dolbeault model of G can thus be written as

$$(\wedge P \otimes \wedge U \otimes \wedge(V^{0,1} \oplus V^{1,0}), \overline{\partial}),$$

with $\overline{\partial} z = 0$ if $z \in U \oplus V^{0,1}$, $\overline{\partial} z = Dz$ if $z \in P \oplus V^{1,0}$.

We now want a Dolbeault model of G with a decomposable differential. For that, we decompose U as $U = D(V^{1,0}) \oplus D(V^{0,1})$. Denote $W = D(V^{0,1})$ and replace $\wedge(P \oplus U \oplus V^{1,0})$ by $\wedge(P \oplus W)$ in the de Rham model of G. We get a new de Rham model of G:

$$(\wedge(P \oplus W \oplus V^{0,1}), \delta_1 + \delta_2),$$

with $\delta_1 w = 0$ if $w \in W$, $\delta_1 x \in \wedge W$ if $x \in P$, $\delta_1 v = 0$ if $v \in V^{0,1}$, $\delta_2 x = 0$ if $x \in P \oplus W$, and δ_2 is an isomorphism from $V^{0,1}$ to W.

Denote the complex $(\wedge(P \oplus W), \delta_1)$ by (C, δ_1). The Frölicher spectral sequence of G can be written as

- $(E_0, d_0) = (C, \delta_1) \otimes (\wedge V^{0,1}, 0)$;
- $(E_1, d_1) = (H(C, \delta_1) \otimes \wedge V^{0,1}, \delta_2^\sharp)$;
- $E_2 = H(H(C, \delta_1) \otimes \wedge V^{0,1}, \delta_2^\sharp)$;

where $\delta_2^\sharp \colon V^{0,1} \to H(C, \delta_1)$ is induced by $\delta_2 \colon V^{0,1} \to C$.

First step. Suppose G is Dolbeault formal. Observe that the quasi-isomorphism $(C \otimes \wedge V^{0,1}, \delta_1) \xrightarrow{\sim} (H(C, \delta_1) \otimes \wedge V^{0,1}, 0)$ induces a quasi-isomorphism

$$(C \otimes \wedge V^{0,1}, \delta_1 + \delta_2) \xrightarrow{\sim} (H(C, \delta_1) \otimes \wedge V^{0,1}, \delta_2^\sharp),$$

which implies the collapsing of the spectral sequence at level 2. (For more details, see [114, Example, page 151].)

Second step. Suppose now that G is not Dolbeault formal and decompose (C, δ_1) into $(C, \delta_1) = (\wedge(\tilde{P} \oplus W), \delta_1) \otimes (\wedge \hat{P}, 0)$, where the canonical surjection $\rho \colon (\wedge(\tilde{P} \oplus W), \delta_1) \to (\wedge \hat{P}, 0)$ is such that $H(\rho) = 0$. We follow the argument of [114, Theorem VIII, page 83]: on $(\wedge(\tilde{P} \oplus W), \delta_1)$ we put the gradation by length of words in $\tilde{P}$. The differential δ_1 decreases this bigradation by exactly 1 and therefore induces a gradation (called the lower gradation) on the cohomology $H(\wedge(\tilde{P} \oplus W), \delta_1)$. From [114, Theorem VIII, page 83], we know that $(\wedge(\tilde{P} \oplus W), \delta_1)$ is formal if and only if $H_1(\wedge(\tilde{P} \oplus W), \delta_1) = 0$. Therefore, by hypothesis, there exists a nonzero element ω in the cohomology $H_1(\wedge(\tilde{P} \oplus W), \delta_1)$. We choose such an ω with the smallest degree and we show that ω a coboundary for the total differential that survives to stage 2. This will imply $E_2 \not\cong E_\infty$.

Suppose there exists ω' such that $\omega = \delta_2^\sharp \omega'$. Since the elements of P have a bidegree of type $(k, k-1)$ and elements of W have bidegree $(1, 1)$, the element ω of wedge length 1 in $\tilde{P}$ has bidegree equal to $(p, p-1)$. Thus ω' must have bidegree equal to $(p-1, p-1)$. We decompose ω' along the length of words in V as $\omega' = \omega_0 + \sum_i v_i \omega_i + \sum_{i,j} v_i v_j \omega_{i,j} + \cdots$, with $\omega_0, \omega_i, \omega_{i,j} \ldots$ in $H(\wedge(\tilde{P} \oplus W), \delta_1)$. From the definition of δ_2, we have $\delta_2^\sharp \omega_0 = 0$. Observe that the two equalities $|v_i| = (0, 1)$ and $|v_i \omega_i| = (p-1, p-1)$ imply that ω_i contains one, and only one, element in $\tilde{P}$; that is, $\omega_i \in H_1(\wedge(\tilde{P} \oplus W), \delta_1)$. Since the total degree of ω_i is $2p - 3$, which is strictly less than the total degree of ω, $2p - 2$, we have $\omega_i = 0$ by the choice of ω.

In conclusion, we get $\omega = \delta_2^\sharp(\sum_{i,j} v_i v_j \omega_{i,j} + \cdots)$ in the ideal generated by V. This implies $\omega = 0$, which is a contradiction to the choice of ω. Therefore ω gives a non-zero class in E_2.

We are now reduced to proving that ω is trivial for the total differential. Let v be a δ_1-cocycle representing ω. We write $v = \sum_i a_i \xi_i$, with $a_i \in \wedge W$ and $\xi_i \in \tilde{P}$. Because the cdga $(\wedge(W \otimes \wedge V^{0,1}, \delta_2)$ is acyclic, there exist b_i such that $\delta_2 b_i = a_i$. Since the element $\sum_i b_i \delta_1 \xi_i$ is a δ_2-cocycle in $\wedge W \otimes \wedge V^{0,1}$, there exists $c \in \wedge W \otimes \wedge V^{0,1}$ such that $\delta_2 c = \sum_i b_i \delta_1 \xi_i$. The conclusion then follows from $(\delta_1 + \delta_2)(c + \sum_i b_i \xi_i) = \sum_i a_i \xi_i$. $\square$

4 : Complex and symplectic manifolds

Example 4.75 A Dolbeault model of SO(9), viewed as the total space of the principal holomorphic fibration $T \longrightarrow SO(9) \longrightarrow SO(9)/T$ (see Example 4.32) is given by:

$$(\wedge(x_3, x_7, x_{11}, x_{15}) \otimes \wedge(a_1, a_2) \otimes \wedge(b_1, b_2), \bar{\partial}),$$

with x_i, a_i, b_i of degree i and $\bar{\partial}a_1 = \bar{\partial}a_2 = \bar{\partial}b_1 = \bar{\partial}b_2 = \bar{\partial}x_3 = 0$, $\bar{\partial}x_7 = (a_1^2 + a_2^2)^2$, $\bar{\partial}x_{11} = (a_1^2 + a_2^2)a_1^2 a_2^2$, $\bar{\partial}x_{15} = a_1^4 a_2^4$.

The cocycle $\big((a_1^2 + a_2^2)x_{11} - x_7 a_1^2 a_2^2\big)$ is a Massey product in the Dolbeault model of SO(9). Therefore, by Theorem 4.74, the Frölicher spectral sequence of SO(9) does not collapse at level 2.

4.5 Symplectic manifolds

A class of smooth manifolds that share some of the properties of complex manifolds are the symplectic manifolds. In this section, we compare and contrast these two types of manifolds.

4.5.1 Definition of symplectic manifold

Definition 4.76 *A manifold M^{2n} is* symplectic *if it possesses a nondegenerate 2-form ω which is closed (i.e. $d\omega = 0$). The symplectic manifold is then denoted (M^{2n}, ω).*

The nondegeneracy condition is equivalent to saying that ω^n is a true volume form (i.e. nonzero at each point) on M (see Section 4.7). Furthermore, the nondegeneracy of ω sets up an isomorphism between 1-forms and vector fields on M by assigning to a vector field X the 1-form $i_X \omega = \omega(X, -)$. Observe also that, if M is closed, the symplectic form ω cannot be exact (since a volume form is a representative of the fundamental cohomology class). The most important theorem about symplectic manifolds says that all symplectic manifolds locally look like $\mathbb{R}^{2n}$.

Theorem 4.77 (Darboux's theorem) *Around each point in a symplectic manifold (M^{2k}, ω) there are local coordinates $(x_1, \ldots, x_k, y_1, \ldots, y_k)$ such that*

$$\omega = \sum_{j=1}^{k} dx_j \wedge dy_j.$$

This result says that symplectic manifolds have no *local* distinguishing invariants. In other words, all symplectic manifolds look alike locally. In this sense then, symplectic *geometry* is a global subject. A general reference for symplectic geometry and topology is [189].

4.5.2 Examples of symplectic manifolds

We list our first examples of symplectic manifolds. Other ones appear in Section 4.6.

Example 4.78 For any manifold M^n, the cotangent bundle T^*M is a symplectic manifold with symplectic form defined as follows. Let $p\colon T^*M \to M$ be the cotangent projection and take the induced map on tangent bundles $Tp\colon T(T^*M) \to TM$. Then, for $\beta_m \in T^*M$ and $\Gamma \in T_{\beta_m}(T^*M)$, define a 1-form (the *Liouville form*) pointwise on T^*M by

$$\theta_{(m,\beta_m)}(\Gamma) = \beta_m(T_m p(\Gamma))$$

and then globally by

$$\theta\colon T^*M \to T^*(T^*M) \qquad \theta(m,\beta_m) = \theta_{(m,\beta_m)}.$$

In local coordinates $(q_1,\ldots,q_n,p_1,\ldots,p_n)$, where the q_j's come from M and the p_j's come from T^*M, we have $\theta = \sum_{j=1}^n p_j\, dq_j$ with exact 2-form

$$\omega = -d\theta = \sum_{j=1}^n dq_j \wedge dp_j.$$

Example 4.79 The first example specializes to the case $\mathbb{R}^{2n}$ with coordinates $(x_1,\ldots,x_n,y_1,\ldots,y_n)$ and symplectic form $\omega = \sum_{j=1}^n dx_j \wedge dy_j$. Then, since the symplectic form is invariant under translations $x_j \mapsto x_j + 2\pi$ and $y_j \mapsto y_j + 2\pi$, the form induces a symplectic form on the orbit space $T^{2n} = \mathbb{R}^{2n}/\mathbb{Z}^{2n}$.

Let $\mathbb{C}^n$ be the complex n-space with the hermitian inner product

$$\langle z, w \rangle = \sum_{j=1}^n z_j \bar{w}_j$$

where $z = (z_1,\ldots,z_n)$ and $w = (w_1,\ldots,w_n)$ are in $\mathbb{C}^n$. If we write $z_j = x_j + iy_j$ and $w_j = u_j + iv_j$, then

$$\langle z, w \rangle = \sum_{j=1}^n x_j u_j + y_j v_j + i \sum_{j=1}^n y_j u_j - x_j v_j.$$

The first term is the standard dot product in $\mathbb{R}^{2n}$. The second term is alternating, so is a 2-form on $\mathbb{R}^{2n}$. Define $\omega = \sum_{j=1}^n y_j u_j - x_j v_j$ to be this 2-form. Then ω is closed and nondegenerate and $\mathbb{C}^n$ is symplectic.

Example 4.80 Recall from Definition 4.21 that a Kähler manifold possesses a nondegenerate closed 2-form ω. Thus, by definition, any Kähler manifold is symplectic. Since compact Kähler manifolds constitute an important class of compact symplectic manifolds, a focus of research has been to understand which general properties of Kähler manifolds are shared by symplectic manifolds. In the next section, we will address this question by considering certain examples and theorems which focus, in particular, on the hard Lefschetz property. Later in Chapter 8, we will explore the question of the formality of (1-connected) symplectic manifolds.

4.5.3 Symplectic manifolds and the hard Lefschetz property

In this section, we examine whether certain tools and results about complex manifolds carry over to symplectic manifolds.

Let (M^{2n}, ω) be a symplectic manifold. Brylinski observed that, since the 2-form ω gives an adjoint, there is a star operator and a new differential on M (see [44]; also [157], [173] and [174]). (Compare with the usual Hodge star operator described in Section A.4.)

1. The *symplectic star operator*, $*: A_{DR}^k(M) \to A_{DR}^{2n-k}(M)$, is defined by $\beta \wedge *\alpha = \wedge^k(G)(\beta, \alpha) v_M$, where:
   - $v_M = \omega^n/n!$ is a volume form on M;
   - G is the tensor field dual to ω.

   (Locally, that means $G = \sum_{j=1}^{n} \dfrac{\partial}{\partial q_j} \wedge \dfrac{\partial}{\partial p_j}$ if $\omega = \sum_{j=1}^{n} dq_j \wedge dp_j$.)

   This operator satisfies $**\beta = \beta$.

2. The differential $\delta: A_{DR}^k(M) \to A_{DR}^{k-1}(M)$ is given by $\delta = (-1)^k * d *$.

Naturally, we want to compare the behavior of the symplectic star operator with that of the star operator coming from Hodge theory, see Section A.4. At first glance, it appears this behavior will be totally different because we have the relation $d\delta + \delta d = 0$ (see [157]) and this affords no opportunity to define a Laplacian. Nevertheless, we can make the following

Definition 4.81 *A form $\beta \in A_{DR}(M)$ is said to be* symplectically harmonic *if $d(\beta) = \delta(\beta) = 0$.*

Of course, we can continue our comparison with Hodge theory and ask if the symplectically harmonic forms play an analogous role to the usual harmonic forms. Here also, the answer is no at a first glance: in the Hodge situation, any de Rham cohomology class can be represented by a *unique* harmonic form, and that cannot be the case for symplectically harmonic forms because any 1-form β with $d\beta = 0$ is symplectically harmonic. Thus, *uniqueness* of representation cannot occur in general. Nevertheless, Brylinski asked if one could find, for any cohomology class in $H^*(M; \mathbb{C})$, a symplectically harmonic representative. Indeed, Brylinski proved that this is true for compact Kähler manifolds. (This comes from the fact that the symplectic and Hodge star operators coincide up to sign in this case, see [44, Theorem 2.4.1].) He also conjectured that the result is true for any compact symplectic manifold. In fact, however, this conjecture was disproved by O. Mathieu in [185] (see also [266]). Moreover, the failure of Brylinski's conjecture is (surprisingly!) connected with the hard Lefschetz property (Theorem 4.35).

Theorem 4.82 ([185]) *Let (M^{2n}, ω) be a symplectic manifold. Then the following two statements are equivalent:*

1. *Any cohomology class contains at least one symplectically harmonic form.*
2. *(M^{2n}, ω) satisfies the hard Lefschetz property.*

The proofs of Mathieu [185] and Yan [266] use representation theory. We refer the reader to the original papers for more details.

In Definition 4.96, we will introduce a weaker version of the hard Lefschetz property called Lefschetz type and prove that a nilmanifold of Lefschetz type is diffeomorphic to a torus (Theorem 4.98). Therefore, *any non-toral symplectic nilmanifold is a counterexample to Brylinski's conjecture*. As for simply connected examples, we will briefly discuss how the symplectic blow-up gives counterexamples in Remark 8.30.

Now recall the following three notions that occur in the study of Kähler manifolds. Let M be Kähler. Then

1. the $\partial\bar{\partial}$-Lemma holds for $A_{DR}(M)$, see Lemma 4.24;
2. the hard Lefschetz theorem holds for $H^*(M)$, see Theorem 4.35;
3. M is formal, see Theorem 4.43.

Since the property that each cohomology class may be represented by a symplectically harmonic form is equivalent to the hard Lefschetz property, we would like to compare the three statements above in the symplectic setting. A good reference is the nicely written monograph of G. Cavalcanti [52] on this subject.

In Theorem 4.43, we chose a proof of the formality of Kähler manifolds that used the differentials ∂ and $\bar{\partial}$ and that could also be adapted to the case of strict formality (see Theorem 4.59). But an alternative presentation with real objects already existed in [71]. There, the two differentials involved were the de Rham differential d and a differential d^c defined by $d^c = i(\bar{\partial} - \partial)$. Indeed, a $d\,d^c$-lemma was given analogous to Lemma 4.24 and that lemma was able to be cast into different forms (see [71, Lemma 5.15]). Adapted to the symplectic setting, we have the following.

4 : Complex and symplectic manifolds

Definition 4.83 *A symplectic manifold* (M^{2n}, ω) *satisfies the $d\delta$-lemma if*

$$\operatorname{Im} d \cap \operatorname{Ker} \delta = \operatorname{Im} \delta \cap \operatorname{Ker} d = \operatorname{Im} d\delta.$$

Merkulov established the equivalence of this property with the hard Lefschetz property.

Theorem 4.84 ([190]) *A compact symplectic manifold satisfies the hard Lefschetz property if and only if it satisfies the $d\delta$-lemma.*

For the proof we refer the reader to [190] or to [52, Theorem 5.4]. From the $d\delta$-lemma, we can prove, exactly as in Theorem 4.43, that we have a sequence of quasi-isomorphisms

$$(H_\delta(M), 0) \xleftarrow{\rho} (Z_\delta(M), d) \xrightarrow{j} (A_{DR}(M), d).$$

But, as noted by Cavalcanti (see [52, Remark, pages 14 and 83]), this does not imply the formality of M because the differential δ is not a derivation of algebras (see [175]), and, therefore, the vector space of cocycles Z_δ is not a graded algebra. So the question of the existence of an equivalence between the hard Lefschetz property and formality is not answered by the $d\delta$-lemma. We will come back to this question in Chapter 8.

In Section 4.4, we introduced the Frölicher spectral sequence associated to any complex manifold. Its degeneracy in the case of a Kähler manifold gives an obstruction to the existence of a Kähler metric. We can ask if such a spectral sequence exists in the symplectic setting and the answer is yes, but it always degenerates at the E_1-stage for any symplectic manifold, see [44].

Remark 4.85 Although the results appear different, this section shows that symplectic and complex manifolds have some common aspects. In fact, there is a more general notion that contains the two settings: *the general complex geometry* introduced by Hitchin [142] (see also Gualtieri [124]). The idea is to consider the direct sum of the tangent and cotangent bundles, $T(M) \oplus T^*(M)$, endowed with the classical scalar product

$$\langle X + \alpha, Y + \beta \rangle = \frac{1}{2}(\alpha(Y) + \beta(X)).$$

A general complex manifold is a complex structure on this direct sum, orthogonal with respect to this scalar product and satisfying an integrability condition. (See Subsection 4.7.6 for the linear setting.) We refer the reader to [124] and [142] for more details, emphasizing that most of [52] is written in this general framework.

4.5.4 Symplectic and complex manifolds

In the previous section, we looked at properties or tools shared by symplectic and complex manifolds. Now we wish to study manifolds endowed with both of these structures. For instance, we already noticed that a Kähler manifold is always symplectic, but we shall see that more can be said about this. First, let's note an important property of symplectic manifolds which provides the fundamental link to the complex world. The proof follows as in Subsection 4.7.5 with the Riemannian metric $g(-,-)$ and the symplectic form ω on the manifold globalizing the scalar product $\langle -, - \rangle$ and vector space symplectic form ω respectively.

Proposition 4.86 *A symplectic manifold has an almost complex structure.*

Although each symplectic manifold has an almost complex structure, there are many more almost complex manifolds than symplectic manifolds. The reason is that the condition that the 2-form ω be closed is very restrictive. For instance, as we have seen, the sphere S^6 is almost complex but is not symplectic because there are no nonexact closed 2-forms on S^6. The requirement that $d\omega = 0$ points out once again that being symplectic is a global condition (compare Theorem 4.77). In Subsection 4.7.5, we see that, given a scalar product, the properties of having complex and symplectic structures on a vector space are essentially equivalent. Indeed, some authors refer to *almost symplectic manifolds* as manifolds having a nonclosed nondegenerate 2-form and it can be shown that these manifolds have almost complex structures. On the other hand, the hermitian product on an almost complex manifold provides a nondegenerate 2-form as well, so it is not local properties that distinguish these notions, but rather the global property $d\omega = 0$.

4.6 Cohomologically symplectic manifolds

4.6.1 C-symplectic manifolds

From the point of view of algebraic models, it is often not necessary to use the full force of the symplectic structure. Because the form ω is closed and ω^n is a volume form in a symplectic manifold M^{2n}, then there are nonzero cohomology classes $[\omega^i] \in H^{2i}(M; \mathbb{R})$ for each $i = 1, \ldots, n$ *when M is compact*. (We often use the same symbol ω for a form and a cohomology class, the context making it clear which is intended.) Note that it is not the case that any manifold having a degree two cohomology class which cups to a top class is symplectic. For instance, $\mathbb{C}P(2)\#\mathbb{C}P(2)$ has several such classes, but it cannot be symplectic since it cannot have an almost complex structure (see Exercise 4.6 for a characteristic class argument). Sometimes however, the existence of a cohomology class in degree 2 with the property that it cups to a top class (over $\mathbb{Q}$ or $\mathbb{R}$) is sufficient to give interesting information. Of course, we can say the same thing about models, and to formalize this, we make the following

Definition 4.87 *A closed manifold* (M^{2n}, ω) *is* cohomologically symplectic *(or c-symplectic) if* $\omega \in H^2(M; \mathbb{R})$ *has the property that* $\omega^n \neq 0$. *Note that this implies that the manifold is orientable.*

A cdga (A, d_A) *is called* cohomologically symplectic *or c-symplectic if:*

- $H^*(A, d_A) = \oplus_{k=0}^{2n} H^k(A, d_A)$ satisfies Poincaré duality with top dimension $2n$;
- there exists an element $\omega \in A^2$ such that for $[\omega] \in H^2(A, d_A)$, $[\omega]^n \neq 0$.

4.6.2 Symplectic homogeneous spaces and biquotients

In Theorem 2.77, we saw that the algebraic assumption of *purity* can have a strong effect on the algebraic structure of a model. Here, we will see that the same algebraic assumption, in conjunction with a c-symplectic structure, profoundly constrains the rational homotopy types that can arise. Recall from Definition 2.76 that a pure minimal *cdga* may be written as

$$(\wedge V, d) = (\wedge(Q \oplus P), d)$$

with $Q = V^{\text{even}}$ denoting the subspace of V consisting of even generators, $P = V^{\text{odd}}$ denoting the subspace of odd generators, $d(Q) = 0$ and $d(P) \subset \wedge^{\geq 2} Q$. A compact manifold M^{2n} is called a *pure manifold* if it has a pure minimal model. If the minimal model of M, $(\wedge V, d) = (\wedge(Q \oplus P), d)$, has P and Q finite dimensional, then M is rationally elliptic as well.

Now, we can give a second *lower grading* to elements of $(\wedge(Q \oplus P), d)$ by defining the lower degree of an even generator $x \in Q$ to be zero, the lower degree of an odd degree generator $y \in P$ to be one and extending to products by requiring $\wedge_i \cdot \wedge_j = \wedge_{i+j}$ (also see Theorem B.18). Note that the hypothesis of pureness implies $d(\wedge_i) \subseteq \wedge_{i-1}$. Therefore, the lower grading descends to the level of cohomology.

Theorem 4.88 ([178]) *A simply connected c-symplectic pure elliptic cdga* $(\wedge V, d) = (\wedge(Q \oplus P), d)$ *has* $\dim Q = \dim P$. *Hence, it is formal.*

Proof We shall use the equivalent criteria of Theorem B.18. Namely, we shall prove that the cohomology of $(\wedge V, d)$ is evenly graded.

If ω denotes the symplectic element in $(\wedge(Q \oplus P), d)$, then $\omega \in Q_0^2$ since $(\wedge V, d)$ is simply connected. By the multiplicative property of the lower grading, the top class $\omega^n \in (\wedge V)_0^{2n}$.

Now suppose that $\eta \in (\wedge V)^{2j+1}$ is an odd degree cocycle. Then clearly, $\eta \in \text{Ideal}(P)$. By Poincaré duality, there exists a class $[\tau] \in H^{2n-2j-1}(\wedge V, d)$ with $[\eta] \cup [\tau] = [\omega^n]$. But, because $\eta \in \text{Ideal}(P)$, its lower grading is greater than or equal to one. Since multiplication by $[\tau]$ can never decrease the lower grading, the product $[\eta] \cup [\tau]$ must also have lower grading greater than or equal to one. This contradicts the fact that $\omega^n \in (\wedge V)_0^{2n}$.

Hence, $[\eta] = 0$ and η is exact. We therefore see that all odd cocycles are exact and the cohomology of $(\wedge V, d)$ is evenly graded. By the equivalence of the criteria in Theorem B.18, we have $\dim Q = \dim P$. Moreover, this implies that $(\wedge V, d)$ is formal as well. $\square$

We have the following immediate implication.

Corollary 4.89 *A simply connected, c-symplectic homogeneous space G/H is a maximal rank homogeneous space. That is, $\text{rank } H = \text{rank } G$.*

A simply connected c-symplectic biquotient $K\backslash G/H$ has $\text{rank } G = \text{rank } K + \text{rank } H$.

Proof Note that, by Theorem 2.71 and Corollary 3.51, homogeneous spaces and biquotients are pure manifolds. Furthermore, for the homogeneous space, the description of the pure model $(\wedge Q \oplus P, d)$ shows that Q may be identified with $\pi_*(BH) \otimes \mathbb{Q}$ and P may be identified with $\pi_*(BG) \otimes \mathbb{Q}$. From Theorem 1.81 and Theorem 3.33, we see that $\text{rank } G = \dim \pi_*(BG) \otimes \mathbb{Q}$ and $\text{rank } H = \dim \pi_*(BH) \otimes \mathbb{Q}$, so the equality $\text{rank } H = \text{rank } G$ follows immediately from Theorem 4.88.

For biquotients, by Theorem 3.50, Q may be identified with $(\pi_*(BH) \oplus \pi_*(BK)) \otimes \mathbb{Q}$ and P may be identified with $\pi_*(BG) \otimes \mathbb{Q}$. Then Theorem 4.88 gives

$$\text{rank } G = \text{rank } H + \text{rank } K.$$

$\square$

4.6.3 Symplectic fibrations

Purity can also be relativized. A fibration $F \to E \to B$ is said to be a *pure fibration* if its relative Sullivan model has the form,

$$(\wedge X, d_X) \to (\wedge X \otimes \wedge Y, D) \to (\wedge Y, d_Y)$$

with $D(Y^{\text{even}}) = 0$ and $D(Y^{\text{odd}}) \subset \wedge X \otimes \wedge Y^{\text{even}}$. As shown in [250], if $F = G/H$, where G is a compact connected Lie group, $H \subset G$ a closed connected subgroup and the fibration is the bundle associated to a principal bundle $G \to P \to B$ via the usual action of G on G/H, then the fibration is pure.

4: Complex and symplectic manifolds

Definition 4.90 *A bundle $F \to E \to B$ is symplectic if (F, ω_F) is symplectic and the structure group of the bundle acts by symplectomorphisms (i.e. diffeomorphisms which preserve the symplectic form) on F.*

A fundamental question is whether, for a symplectic bundle, there is a symplectic structure on E that restricts to the given symplectic structure on F. The main result in this direction is due to Thurston (see [189, Theorem 6.3]).

Theorem 4.91 *Let $(F, \omega_F) \to E \xrightarrow{p} (B, \omega_B)$ be a symplectic bundle with base and fiber compact symplectic manifolds. If there is some cohomology class $\alpha \in H^2(E; \mathbb{R})$ which restricts to the class $[\omega_F] \in H^2(F; \mathbb{R})$ by $i^*(\alpha) = [\omega_F]$, then for sufficiently large K, there exists a symplectic form ω_E on E with cohomology class $[\omega_E] = \alpha + K p^*([\omega_B])$.*

Example 4.92 The Kodaira–Thurston manifold KT (see Example 4.31 and Example 3.23) is a principal torus bundle over a torus classified by the map $T^2 \to BT^2 = BS^1 \times BS^1$ defined by $(uy, *)$, where $u, y \in H^1(T^2; \mathbb{Z})$ correspond to the torus's circle factors and the fundamental class of T^2 is the cup product uy. Since the torus acts symplectically on itself by translation, we see that this bundle is a symplectic bundle. The relative model of the bundle $T^2 \to KT \xrightarrow{p} T^2$ is

$$(\wedge(u,y), 0) \to (\wedge(u,y,v,t), D) \to (\wedge(v,t), 0)$$

with $D(u) = 0, D(y) = 0, D(v) = uy$ and $D(t) = 0$. Note that the fiber is a torus with symplectic cohomology class vt. If we are to apply Theorem 4.91, then we must find a cohomology class in $(\wedge(u,y,v,t), D)$ which restricts to vt. But it is easy to see that *no such class exists*. Therefore, the principal bundle structure of KT is not the one compatible with KT's symplectic structure.

Instead of the relative model above, consider

$$(\wedge(u,t), 0) \to (\wedge(u,y,v,t), D) \to (\wedge(v,y), 0)$$

with $D(u) = 0, D(y) = 0, D(v) = uy$ and $D(t) = 0$. This is *not* the model of a principal bundle because the differential is not defined by a classifying map on the base. But now we can see that the class $vy \in (\wedge(u,y,v,t), D)$ is both closed and maps to $vy \in (\wedge(v,y), 0)$ under restriction to the fiber. Thus, Thurston's criterion is satisfied and, for some K large enough, $vy + K p^*(ut)$ is a symplectic form on KT. Now, by Proposition 4.94, we know that $\omega = vy + ut$ is a symplectic form on KT, so $K = 1$ is good enough.

Now let's combine the geometry of Thurston's result with the algebra inherent in the definition of *pure fibration*. Indeed, the simplicity of the

4.6 Cohomologically symplectic manifolds

proof hints at a result with less restrictive algebraic assumptions. Note also that the second condition in the hypothesis holds for all simply connected fibers.

Proposition 4.93 ([153]) *Let* $(F, \omega_F) \to E \to B$ *be a pure symplectic fibration with F and B compact and with relative Sullivan model* $(\wedge X, d_X) \to (\wedge X \otimes \wedge Y, D) \to (\wedge Y, d_Y)$. *If there exists* $\alpha \in \wedge Y$ *with*

- $[\alpha] \mapsto [\omega_F]$ *under the induced cohomology homomorphism* $H^*(\wedge Y, d_Y) \to H^*(A_{DR}(F), d)$;
- $\alpha = \sum_j c_j y_j$ *for* $c_j \in \mathbb{R}$ *and* $y_j \in Y^2$;

then there exists a compatible symplectic structure on E.

Proof The hypothesis $\alpha = \sum_j c_j y_j$ implies that $D\alpha = 0$ since, by the assumption of purity, $D(Y^{\text{even}}) = 0$. Therefore, $\alpha \in \wedge X \otimes \wedge Y$ is a cocycle restricting to $\alpha \in \wedge Y$ and Thurston's result guarantees a compatible symplectic structure. □

Explicit examples of torus bundles illustrating the proposition may be found in [153].

4.6.4 Symplectic nilmanifolds

The question of exactly how c-symplectic differs from symplectic is an interesting one. For 4-manifolds, the existence of an almost complex structure is paramount (see Exercise 4.6). But for higher dimensions, sophisticated quantities such as Seiberg–Witten invariants have been necessary to address this problem. For nilmanifolds, the question is much easier because of the strong relationship between the Lie algebra of the nilpotent Lie group and the nilmanifold itself. Recall that Nomizu's theorem says that $H^*(\wedge \mathfrak{n}^*, \delta)) \cong H^*(N/\Gamma; \mathbb{Q})$, where N is a nilpotent Lie group, N/Γ is the associated nilmanifold and $\mathfrak{n}$ is the nilpotent Lie algebra of N.

Suppose now that N/Γ is a c-symplectic nilmanifold. Nomizu's theorem (see Theorem 3.18 and its proof) guarantees the existence of a cochain in $(\wedge \mathfrak{n}^*, \delta)$ representing the cohomology class which cups to a top class. Since the cohomology product cannot be zero, then the cochain (i.e. wedge) product of the cochain cannot be zero. This then defines a left invariant volume form on the nilpotent Lie group. Both the cochain (as a left invariant form) and the volume form then descend to the nilmanifold to provide a symplectic form wedging to a volume form. Therefore, we have

Proposition 4.94 *A nilmanifold is c-symplectic if and only if it is symplectic.*

Now, if a nilmanifold M^{2k} is symplectic, then there must be a degree 2 element of the minimal model which multiplies up to the top element

4: Complex and symplectic manifolds

$v = x_1 \cdots x_{2k}$. We can write this degree 2 element as

$$\omega = \sum a_{ij} x_i x_j.$$

Since $\omega^k = v$, the sum in the expression for ω *must contain all the degree 1 generators*. Of course, ω must be closed as well.

Example 4.95 What must a four-dimensional nilmanifold look like rationally? By what we have said, the minimal model must have four generators and two of those must be cocycles. Thus, there are three rational homotopy types to deal with (up to isomorphism).

1. **The torus T^4.** The minimal model is $(\wedge(u,v,y,t), d = 0)$ and a symplectic element is given by $\omega = uv + yt$.
2. **The Kodaira–Thurston manifold KT.** This manifold is obtained by taking the product of the Heisenberg manifold $M = U_3(\mathbb{R})/U_3(\mathbb{Z})$ and the circle S^1. The minimal model is given by

    $(\wedge(u,v,y,t), d)$ with $du = 0,\ dy = 0,\ dv = uy,\ dt = 0$

    where the first three generators come from M and t comes from the circle. A symplectic element is then given by $\omega = vy + ut$ having $\omega^2 = 2vyut \neq 0$. Note that the degree 1 cohomology of KT is generated by the classes of u, y and t. Hence, the first Betti number is three and, by Proposition 4.33, KT cannot be Kähler.
3. Take the minimal model $(\wedge(u,v,y,t), d)$ with $du = 0$, $dv = 0$, $dy = uv$ and $dt = uy$. Recall that the corresponding finitely generated torsion-free nilpotent group may be realized as a nilmanifold. Then a symplectic element is given by $\omega = ut + vy$.
4. By the above, all four-dimensional nilmanifolds admit symplectic structures. By contrast, the six-dimensional nilmanifold $U_4(\mathbb{R})/U_4(\mathbb{Z})$ is not symplectic because no *closed* element ω in its minimal model can be found whose cube does not vanish.

We have seen above that nilmanifolds can be symplectic, but can they be Kähler? By Section 4.2, we know that Kähler manifolds are formal, but by Proposition 3.20, the only formal nilmanifolds are tori. So we can ask if weaker conditions derived from Kählerness can still hold for nilmanifolds. For instance, we make the

Definition 4.96 A symplectic manifold (M^{2n}, ω) has Lefschetz type *if the multiplication by* $[\omega]^{n-1}$,

$$\text{Lef} = [\omega]^{n-1} \cup : H^1(M;\mathbb{R}) \to H^{2n-1}(M;\mathbb{R}),$$

is an isomorphism.

4.6 Cohomologically symplectic manifolds

Example 4.97 If X is a nonsimply connected Kähler manifold and Y is a simply connected symplectic manifold with some odd odd-degree Betti number (and these exist by [187]), then $X \times Y$ is not Kähler, but *does* have Lefschetz type.

We also observe that the six-dimensional manifold constructed in [95] does not satisfy the hard Lefschetz property, but is of Lefschetz type.

So what can we say about Lefschetz type nilmanifolds? Surprisingly, we have the same restriction as for the Kähler hypothesis. Benson and Gordon ([25]) were the first to establish the fact that tori are the only nilmanifolds satisfying the hard Lefschetz property. Here we present a proof of this fact that uses properties of the model derived from symplecticness as well as the existence of a particular derivation reminiscent of Blanchard's Theorem 4.36.

Theorem 4.98 ([178]) *A symplectic nilmanifold M of Lefschetz type is diffeomorphic to a torus.*

Proof Write the minimal model for M as $(\wedge(x_1, \ldots, x_{2n}), d)$ with symplectic element given by

$$\omega = \sum_{i,j < 2n} a_{ij} x_i x_j + z x_{2n}$$

where we have taken out all terms involving the last generator x_{2n}. Also, from the discussion following Proposition 4.94, we see that all generators $x_1, \ldots, x_{2n}$ appear in the expression for ω. Assume that $d \neq 0$. First note that, because of the stage-by-stage construction of the minimal model, the last generator x_{2n} cannot appear in any differential dx_i, $i = 1, \ldots, 2n$. Hence, since $d\omega = 0$, the only way to have the term $dz \cdot x_{2n}$ cancelled is for z to be a cocycle, $dz = 0$. Now define a derivation λ of degree -1 by

$$\begin{cases} \lambda(x_i) = 0 & \text{for } i < 2n \\ \lambda(x_{2n}) = 1. \end{cases}$$

Extend λ freely to $(\wedge(x_1, \ldots, x_{2n}), d)$ as a derivation of algebras. The effect of λ on ω follows from the definition and the derivation property:

$$\lambda(\omega) = z.$$

The derivation λ obeys the relation $\lambda d = -d\lambda$ because d is decomposable, so a derivation of cohomology is induced with $\lambda([\omega]) = [z] \neq 0$ since z is a nonzero degree 1 cocycle.

Now, a basis for $H^1(M)$ consists precisely of the generators $x_1, \ldots, x_s$ with $dx_i = 0$, $i = 1, \ldots, s$. By the definition of λ then, $\lambda(H^1(M)) = 0$. (Note that this only holds when $dx_{2n} \neq 0$.) Let $[\alpha]$ be any element of $H^1(M)$ and consider $[\alpha] \cup [\omega^n]$. Since this class is above the top degree, $[\alpha] \cup [\omega^n] = 0$. Applying λ, we obtain

$$0 = \lambda([\alpha] \cup [\omega^n]) = \lambda([\alpha]) \cup [\omega^n] - [\alpha] \cup \lambda([\omega^n]) = -[\alpha] \cup \lambda([\omega^n]).$$

Now $[\alpha] \cup \lambda([\omega^n]) = 0$ for any $[\alpha] \in H^1(M)$, where $\lambda([\omega^n]) \in H^{2n-1}(M)$. By Poincaré duality (i.e. the nondegeneracy of the bilinear form), this can only be true if $\lambda([\omega^n]) = 0$. But then

$$0 = \lambda([\omega^n]) = n\,\lambda([\omega]) \cup [\omega^{n-1}] = n\operatorname{Lef}(\lambda([\omega]))$$

and the hypothesis of Lefschetz type implies that this can happen only when $\lambda([\omega]) = 0$. This contradicts the fact that $\lambda([\omega]) = [z] \neq 0$. Thus, a Lefschetz type nilmanifold must have $d = 0$ and therefore have the rational homotopy type of a torus. Now, the same argument as in Remark 3.21 shows that M is diffeomorphic to a torus. □

4.6.5 Homotopy of nilpotent symplectic manifolds

R. Gompf showed in [108] that any finitely presented group can be realized as the fundamental group of certain symplectic 4-manifolds. More recently, in [146] certain restrictions were found on the groups that can arise as fundamental groups of symplectic manifolds where the symplectic cohomology class annihilates the image of the Hurewicz homomorphism. These are the so-called symplectically aspherical manifolds. A fair amount is known about the homotopy theory of symplectically aspherical manifolds (see [179] as well as the references above), but these manifolds are very special. These are just first steps in understanding the homotopy theory of symplectic manifolds and certain classes of symplectic manifolds. We have seen throughout the first three chapters that minimal models best reflect the homotopy theory of spaces when the spaces are nilpotent. Therefore, in order for minimal models to be truly applicable in symplectic geometry, we will need to know the answer to the following

Question 4.99 *How can nilpotent symplectic manifolds be recognized? If a symplectic manifold is a nilpotent space, what special homotopical properties are apparent? Conversely, what nilpotent spaces have symplectic or c-symplectic structures.*

This is not a question directly related to models, of course, but rather a question which connects geometry and homotopy theory in a fundamental way. Although we don't know of any general results in this direction, here is a result that gives a slight indication of how the action of the fundamental group on higher homotopy may be recognized in the symplectic world. While the proposition holds in general, it pays to think of ω as the symplectic (or c-symplectic) class. Now, for a symplectic manifold (M, ω), the condition of symplectic asphericity mentioned above is equivalent to the condition that $\omega = f^*(\widetilde{\omega})$, where $f\colon M \to K(\pi_1 M, 1)$ classifies the universal cover and $\widetilde{\omega}$ is some class in $H^2(K(\pi_1 M, 1); \mathbb{R})$ (see [179]). Hence, $p^*(\omega) = 0$, where $p\colon \widetilde{M} \to M$ is the universal cover. On the other hand, the result below applies to the generic case of symplectic manifolds that are not symplectically aspherical. (Also note that the proposition is a special case of that found in [162].)

Proposition 4.100 ([162]) *Suppose that M is a path connected space with $a \in H^1(M; \mathbb{Q})$ and $\omega \in H^2(M; \mathbb{Q})$ obeying $a \cup \omega = 0$ and $p^*(\omega) \neq 0$, where $p\colon \widetilde{M} \to M$ is the universal cover. Then the action of $\pi_1(M)$ on $\pi_2(M)$ is nontrivial.*

Proof First, we can take multiples of a and ω so that they are integral. We therefore assume this. Now note that the condition $p^*(\omega) \neq 0$ is equivalent to saying that $\omega|_{\pi_2(M)} \neq 0$, where $\omega \in H^2(M) \to \operatorname{Hom}(H_2(M; \mathbb{Z}), \mathbb{Z})$ is considered dual to homology and operating on the image of Hurewicz in $H_2(M; \mathbb{Z})$. The condition $\omega|_{\pi_2(M)} \neq 0$ removes the universal cover from consideration and focuses on M and ω. So now take $\gamma \in \pi_2(M)$ such that $\omega(h(\gamma)) \neq 0$ and $\alpha \in \pi_1(M)$ such that $a(h(\alpha)) \neq 0$. Classically, we know that the deviation of the action $\alpha \cdot \gamma$ from being trivial is detected by the Whitehead product $[\alpha, \gamma]$:

$$\alpha \cdot \gamma - \gamma = [\alpha, \gamma].$$

Thus, to show that the action of $\pi_1(M)$ on $\pi_2(M)$ is nontrivial, it is sufficient to show that the Whitehead product $[\alpha, \gamma]$ is nonzero.

The cohomology classes a and ω give a map $a \times \omega\colon M \to K(\mathbb{Z}, 1) \times K(\mathbb{Z}, 2)$ which, composed with $\iota_1 \cup \iota_2\colon K(\mathbb{Z}, 1) \times K(\mathbb{Z}, 2) \to K(\mathbb{Z}, 3)$ yields $(a \times \omega)^*(\iota_1 \cup \iota_2)^*(\iota_3) = a \cup \omega = 0$. Here, ι_j is the fundamental cohomology class of $K(\mathbb{Z}, j)$. The equality $a \cup \omega = 0$ then shows that there is a lifting ϕ in the following diagram (where the right square is a pullback)

$$\begin{array}{ccc} & E & \longrightarrow PK(\mathbb{Z}, 3) \\ {}^{\phi}\nearrow & \downarrow & \downarrow \\ M \xrightarrow{a \times \omega} & K(\mathbb{Z}, 1) \times K(\mathbb{Z}, 2) \xrightarrow{\iota_1 \cup \iota_2} & K(\mathbb{Z}, 3) \end{array}$$

Now, the minimal model of E is apparent: $\mathcal{M}_E = (\wedge(x, y, z), d)$ with $|x| = 1$, $|y| = 2$, $|z| = 2$ and only nonzero differential $dz = xy$. By Theorem 2.56, the quadratic part of the differential, d_1, corresponds to the bracket of homotopy Lie algebra elements. In fact, the isomorphism $s\colon \pi_q(\Omega X)\otimes \mathbb{Q} \to \pi_{q+1}(X)\otimes \mathbb{Q}$ also identifies the bracket with the classical Whitehead product, up to sign. Therefore, Whitehead products may be read off from d_1. Thus, we see that the Whitehead product in $\pi_2(E) \otimes \mathbb{Q}$ is nonzero, $[\hat{\imath}_1, \hat{\imath}_2]\otimes \mathbb{Q} \neq 0$, where $\hat{\imath}_1 \in \pi_1(K(\mathbb{Z},1))$ and $\hat{\imath}_2 \in \pi_2(K(\mathbb{Z},2))$ are the generators of the respective homotopy groups. But then any integral multiple of $[\hat{\imath}_1, \hat{\imath}_2]$ is also nonzero as well. Now, since $a(h(\alpha)) \neq 0$ and $\omega(h(\gamma)) \neq 0$, the lift ϕ can be used to push the Whitehead product $[\alpha, \gamma] \in \pi_2(M)$ forward to an integral multiple of $[\hat{\imath}_1, \hat{\imath}_2]$. Since the latter is nontrivial, so is the former and we are done. □

The preceding discussion illustrates an important point. Minimal models may be useful even in the non-nilpotent situation. Minimal models exist for any path connected space and may be used as intermediate steps in a rational homotopy analysis as long as the analysis never relies on identifying their homotopy properties with their algebraic properties. For other approaches to viewing the action of the fundamental group on higher homotopy geometrically, see Exercise 4.7 and Subsection 9.6.6.

4.7 Appendix: Complex and symplectic linear algebra

Almost complex manifolds or symplectic manifolds are globalizations of complex or symplectic vector spaces. The bigradation on differential forms in the complex case or Theorem 4.77 in the symplectic setting are expressions of this philosophy. Here we will remind the reader of some basic facts about symplectic and complex linear algebra.

4.7.1 Complex structure on a real vector space

From the canonical inclusion $\mathbb{R} \hookrightarrow \mathbb{C}$, any $\mathbb{C}$-vector space V inherits the structure of an $\mathbb{R}$-vector space. Moreover, multiplication by the complex number i induces an $\mathbb{R}$-linear map, $J\colon V \to V$, such that $J^2 = -1$. In fact, this data is sufficient for recovering the complex structure of V and allows the description of a $\mathbb{C}$-vector space structure with real objects.

Definition 4.101 *Let V be a real vector space. A* complex structure *on V is an $\mathbb{R}$-linear map, $J\colon V \to V$, such that $J^2 = -1$. The associated complex vector space is denoted by (V, J).*

Observe that a basis $(v_1, \ldots, v_n)$ of the $\mathbb{C}$-vector space (V, J) corresponds to the basis $(v_1, \ldots, v_n, Jv_1, \ldots, Jv_n)$ of the $\mathbb{R}$-vector space V. Moreover, if (V, J) and (V, J') are two complex vector spaces, an $\mathbb{R}$-linear map $f\colon V \to V'$ is $\mathbb{C}$-linear if and only if we have $f \circ J = J' \circ f$. Finally, if (V, J) is a complex vector space, one defines a complex structure on the dual vector space, $V^* = \text{Hom}_\mathbb{R}(V, \mathbb{R})$, by $\langle Jv^*, v \rangle = \langle v^*, Jv \rangle$, where $\langle \, , \, \rangle$ denotes the duality between V and V^*.

Example 4.102 First, we describe the usual complex structure on $V = \mathbb{R}^{2n} \cong \mathbb{C}^n$. An element $(x_1, \ldots, x_n, y_1, \ldots, y_n)$ of $\mathbb{R}^{2n}$ is identified with $(z_1, \ldots, z_n) \in \mathbb{C}^n$, by $z_j = x_j + i y_j$. The corresponding $\mathbb{R}$-linear map, $J: V \to V$, is given by $J(x_1, \ldots, x_n, y_1, \ldots, y_n) = (-y_1, \ldots, -y_n, x_1, \ldots, x_n)$. Relative to the canonical basis, the matrix associated to J is

$$J_0 = \begin{pmatrix} 0 & -I_n \\ I_n & 0 \end{pmatrix}.$$

Thus, the complex linear group, $\text{Gl}(n, \mathbb{C})$, may be identified with the subgroup of the real linear group $\text{Gl}(2n, \mathbb{R})$ consisting of matrices which commute with J_0:

$$\text{Gl}(n, \mathbb{C}) = \left\{ S \in \text{Gl}(2n, \mathbb{R}) \mid S J_0 S^{-1} = J_0 \right\}.$$

Moreover, the complex linear group embeds in the real linear group by

$$\text{Gl}(n, \mathbb{C}) \to \text{Gl}(2n, \mathbb{R})$$

$$A + iB \mapsto \begin{pmatrix} A & -B \\ B & A \end{pmatrix}.$$

Observe that the set of complex structures on $\mathbb{R}^{2n}$ can be identified with the homogeneous space $\dfrac{\text{Gl}(2n, \mathbb{R})}{\text{Gl}(n, \mathbb{C})}$. The class associated with $S \in \text{Gl}(2n, \mathbb{R})$ corresponds to the complex structure $J = S J_0 S^{-1}$.

4.7.2 Complexification of a complex structure

Let (V, J) be a complex vector space and let $V^c = V \otimes_\mathbb{R} \mathbb{C}$ denote the *complexification* of V. We emphasize that two complex structures are at play here: the letter i denotes the element of $\mathbb{C}$ as usual while J is the complex structure on V. An element of V^c can be written $z = x + iy$, with $x \in V$ and $y \in V$. Therefore a complex conjugation can be defined on V^c by $\bar{z} = x - iy$.

We now extend the $\mathbb{R}$-linear map $J: V \to V$ to a $\mathbb{C}$-linear map $J: V^c \to V^c$. This $\mathbb{C}$-linear map has two eigenvalues, i and $-i$, with corresponding eigenspaces $V^{1,0}$ and $V^{0,1}$ characterized by:

$$V^{1,0} = \{ z \in V^c \mid Jz = iz \} = \{ v - iJv \mid v \in V \},$$
$$V^{0,1} = \{ z \in V^c \mid Jz = -iz \} = \{ v + iJv \mid v \in V \}.$$

This gives a decomposition $V^c = V^{1,0} \oplus V^{0,1}$ and complex conjugation induces an isomorphism $V^{1,0} \cong V^{0,1}$.

Conversely, let V be a real vector space. Any decomposition of the complexification $V^c = V \otimes_{\mathbb{R}} \mathbb{C}$ into $V^c = V_1 \oplus V_2$ such that $\overline{V}_1 \cong V_2$, provides a complex structure on V with $V^{1,0} \cong V_1$ and $V^{0,1} \cong V_2$. Indeed, observe that a complex basis of V_1, $(a_1 - i b_1, \ldots, a_n - i b_n)$, $a_i \in V$, $b_i \in V$, produces a $\mathbb{C}$-basis of V, $(a_1, \ldots, a_n)$. The complex structure on V is determined by $J(a_i) = b_i$. If we carry out the same constructions for the complex structure on the dual, $V^{*c} = \text{Hom}(V^c, \mathbb{C})$, we obtain:

1. $V^{*c} = V^{*1,0} \oplus V^{*0,1}$;
2. $V^{*1,0} = \{v^* \in V^{*c} \,|\, \langle v^*, v\rangle = 0, \forall v \in V^{0,1}\} = (V^{0,1})^{\perp}$;
3. $V^{*0,1} = (V^{1,0})^{\perp}$.

Denote by $\wedge V^c$ the (complex) exterior algebra on V^c. We define a bigradation on $\wedge V^c$ by:

$$\wedge^{p,q} V^c := \wedge^p V^{1,0} \otimes \wedge^q V^{0,1}.$$

Observe directly from the definitions that:

1. $\wedge V^c = \wedge V^{1,0} \otimes \wedge V^{0,1} = \sum_{p,q} \wedge^{p,q} V^c$;
2. complex conjugation induces a real isomorphism $\wedge^{p,q} V^c \cong \wedge^{q,p} V^c$;
3. if $(e_1, \ldots, e_n)$ is a basis of $V^{1,0}$, then $(\bar{e}_1, \ldots, \bar{e}_n)$ is a basis of $V^{0,1}$ and $\wedge^{p,q} V^c$ admits the vectors $e_{j_1} \wedge \ldots e_{j_p} \wedge \bar{e}_{k_1} \wedge \ldots \bar{e}_{k_q}$ as a basis.

4.7.3 Hermitian products

The purpose of this section is to recall the notion of *hermitian product* and to see how hermitian products are expressed in terms of real structures. First, recall the classical definition.

Definition 4.103 *Let (V, J) be a complex vector space. A hermitian product on (V, J) is a map $h: V \times V \to \mathbb{C}$, such that:*

1. *for any $v' \in V$, the map $V \to \mathbb{C}$, $v \mapsto h(v, v')$, is $\mathbb{C}$-linear;*
2. *for any $(v, v') \in V \times V$, we have $h(v, v') = \overline{h(v', v)}$.*

Let h be a hermitian product on (V, J). We define $S: V \otimes V \to V$ and $A: V \otimes V \to V$ by $h(v, v') = S(v, v') + i A(v, v')$, with $S(v, v')$ and $A(v, v')$ real. We check easily that:

1. S is $\mathbb{R}$-bilinear and symmetric; A is $\mathbb{R}$-bilinear and skew symmetric;
2. $S(v, v') = -A(v, Jv')$; $A(v, v') = S(v, Jv')$;
3. $S(v, v') = S(Jv, Jv')$; $A(v, v') = A(Jv, Jv')$.

Given an $\mathbb{R}$-bilinear form S such that $S(v, v') = S(Jv, Jv')$, we define a skew symmetric form A by $A(v, v') = S(v, Jv')$. Using the fact that $h = S + iA$ is a hermitian product, we can define the notion of *inner product* on a complex vector space as follows.

Definition 4.104 *An* inner product *on the complex vector space* (V,J) *is an* $\mathbb{R}$-*bilinear map* $S\colon V \times V \to \mathbb{R}$ *such that, for any* $(v, v') \in V \times V$:

1. $S(Jv, Jv') = S(v, v') = S(v', v)$;
2. $S(v, v) \geq 0$;
3. $S(v, v) = 0$ *if and only if* $v = 0$.

Observe that $S(v, Jv)$ is always 0. Thus, in the case of a finite dimensional vector space V, one can find an orthonormal real basis for S, of the form $(v_1, \ldots, v_n, Jv_1, \ldots, Jv_n)$.

Remark 4.105 Let (V, J) be a complex vector space with $\langle\,,\,\rangle \colon V \times V \to \mathbb{R}$ a scalar product on the underlying real vector space. We define an inner product, S, on (V, J) by $S(v, v') = \langle v, v' \rangle + \langle Jv, Jv' \rangle$. Therefore, $h(v, v') = S(v, v') + i S(v, Jv')$ defines a hermitian product on (V, J).

Example 4.106 The usual complex structure J_0 on $V = \mathbb{R}^{2n} \cong \mathbb{C}^n$ was described in Example 4.102. Recall first the usual euclidian product $S_0 \colon \mathbb{R}^{2n} \times \mathbb{R}^{2n} \to \mathbb{R}$,

$$S_0\left((x_1,\ldots,x_n,y_1,\ldots,y_n),(x'_1,\ldots,x'_n,y'_1,\ldots,y'_n)\right) = \sum_1^n x_j x'_j + \sum_1^n y_j y'_j,$$

and its associated skew bilinear form $A_0 \colon \mathbb{R}^{2n} \times \mathbb{R}^{2n} \to \mathbb{R}$,

$$A_0\left((x_1,\ldots,x_n,y_1,\ldots,y_n),(x'_1,\ldots,x'_n,y'_1,\ldots,y'_n)\right) = \sum_1^n x'_j y_j - \sum_1^n x_j y'_j.$$

Together, they induce the usual hermitian product on $\mathbb{C}^n$, defined by:

$$h_0((z_1,\ldots,z_n),(z'_1,\ldots,z'_n))$$

$$= (S_0 + i A_0)\left((z_1,\ldots,z_n),(z'_1,\ldots,z'_n)\right) = \sum_1^n z_j \bar{z}'_j,$$

where we have used the usual identification $z = x + iy$.

Observe that the set of hermitian products on $\mathbb{R}^{2n}$ may be identified with the homogeneous space $\dfrac{\mathrm{Gl}(n, \mathbb{C})}{\mathrm{U}(n)}$. The class associated with $S \in \mathrm{Gl}(n, \mathbb{C})$ corresponds to the hermitian product h defined by $h(v, v') = h_0(Sv, Sv')$.

If S is an inner product on the complex vector space (V, J), we extend S to the complexification V^c as a $\mathbb{C}$-bilinear form, also denoted by S. We check easily that $S(z, z') = 0$, for any $z \in V^{1,0}$ and $z' \in V^{1,0}$.

4 : Complex and symplectic manifolds

By definition, the skew symmetric form, associated to S and defined by $A(v,v') = S(v,Jv')$, is an element of $\wedge^2 V^*$. Since $\wedge^2 V^*$ is a subspace of $\wedge^2 V^{*c}$, one can see A as an element of $\wedge^2 V^{*c}$. The properties of S written above and the links between (V,J) and (V^*,J) imply:

$$A \in \wedge^{1,1} V^{*c}.$$

4.7.4 Symplectic linear algebra

Symplectic linear algebra deals with the consequences of having a nondegenerate skew-symmetric bilinear form on a vector space. We have already seen that skew-symmetric bilinear forms arise in complex linear algebra, so this hints at a connection between these two areas which comes to fruition when we globalize symplectic linear algebra to symplectic geometry. A reference with many more details (and that then moves on to symplectic topology) is [189].

Definition 4.107 *A symplectic vector space* (V, ω) *consists of a finite dimensional real vector space V and a nondegenerate bilinear form* $\omega: V \times V \to \mathbb{R}$ *which is skew-symmetric.*

Skew-symmetric means that $\omega(v,w) = -\omega(w,v)$ and nondegenerate means that $\omega(v,w) = 0$ for all w implies $v = 0$. Such an ω is called a *symplectic form*. The key result about symplectic vector spaces is the following local version of Darboux's Theorem 4.77.

Proposition 4.108 *Let* (V, ω) *be a symplectic vector space. Then there exists a basis* $e_1, \ldots, e_n, f_1, \ldots, f_n$ *with the property that*

$$\omega(e_j, f_k) = \delta_{jk}; \ \omega(e_j, e_k) = 0; \ \omega(f_j, f_k) = 0,$$

where δ_{jk} is the Kronecker delta. In particular, $\dim V = 2n$ *is even.*

We shall not give the proof of this result here since it is standard (see [189] for example), but we emphasize that it is the nondegeneracy of ω that is the key point. This property both starts an inductive construction of the basis (by choosing any $e_1 \neq 0$ and being assured of the existence of f_1 with $\omega(e_1, f_1) = 1$) and allows the induction to continue (because then the subspace perpendicular to e_1 and f_1 is the complete annihilator in the dual). Nondegeneracy of ω also allows us to identify the volume form of the vector space and this property carries over to the manifold setting. Because it is essential for understanding symplectic manifolds, we provide a short proof.

Proposition 4.109 *A skew-symmetric bilinear form ω on a vector space V of dimension $2n$ is nondegenerate if and only if* $\omega^n = \omega \wedge \ldots \wedge \omega$ *(n-times) is a volume form on V.*

Proof Suppose ω is not nondegenerate. Then there is some $v \in V$ such that $\omega(v, u) = 0$ for all $u \in V$. Choose a basis for V with v as a member. Then, to test ω^n as a volume form, we apply it to all elements of the basis. But this means $\omega(v, -)$ appears in every monomial and, therefore, $\omega^n(v, v_2, \ldots, v_{2n}) = 0$. Hence, it is not a volume form.

Now suppose ω is nondegenerate. By Proposition 4.108, there is a symplectic basis $e_1, \ldots, e_n, f_1, \ldots, f_n$ and

$$\omega^n(e_1, \ldots, e_n, f_1, \ldots, f_n) = \prod_1^n \omega(e_j, f_j) = 1.$$

Hence, ω^n is a volume form. □

4.7.5 Symplectic and complex linear algebra

We have looked at complex linear algebra and symplectic linear algebra, but how are they related? In the presence of a scalar product, they are virtually the same. Let V be a real, finite dimensional, vector space endowed with a scalar product $\langle \, , \, \rangle : V \times V \to \mathbb{R}$.

Let ω be a symplectic form on V. We define an isomorphism A of V by $\omega(v, v') = \langle v, A(v') \rangle$. Note that the skew-symmetry and nondegeneracy of ω imply that $A^t = -A$. As usual, see [199, page 34], the endomorphism AA^t is diagonalizable with positive eigenvalues $\{\lambda_j\}$. Thus, we have an orthogonal matrix B such that $AA^t = B D(\lambda_j) B^{-1}$. From this, we get the so-called polar decomposition of A, $A = UJ$, where $U = B D(\sqrt{\lambda_j}) B^{-1}$ is symmetric, J is orthogonal and $JU = UJ$. Now, from

$$J^t = A^t \left(U^{-1}\right)^t = -A \left(U^{-1}\right)^t = -A U^{-1} = -U J U^{-1} = -J,$$

we deduce $J^{-1} = J^t = -J$ and, therefore, $J^2 = -1$. Hence, J is a complex structure.

Thus, *if (V, ω) is a symplectic vector space, we can define a complex structure on V.* Observe also that the complex structure J depends on the choice of the scalar product $\langle \, , \, \rangle$ on V.

Conversely, consider a complex structure J on V. We define an inner product on V by

$$S(v, v') = \langle v, v' \rangle + \langle Jv, Jv' \rangle.$$

This product satisfies $S(Jv, Jv') = S(v, v')$ and from $S(v, Jv') = \langle v, Jv' \rangle - \langle Jv, v' \rangle = -S(Jv, v')$, we see that $J^* = -J$ where the adjoint is relative to S. We set $\omega(v, v') = S(Jv, v')$ and check that

$$\omega(v', v) = S(Jv', v) = S(J^2 v', Jv) = -S(Jv, v') = -\omega(v, v').$$

Since the form ω is nondegenerate, (V, ω) is a symplectic vector space.

4.7.6 Generalized complex structure

Let's now unite complex and symplectic linear algebra. First, note that a symplectic form ω can be viewed as linear map $\omega \colon V \to V^*$ such that $\omega^t = -\omega$. On the other hand, a complex structure is a linear map $J \colon V \to V$ such that $J^2 = -1$. Now consider the direct sum $W = V \oplus V^*$ together with the *evaluation scalar product*:

$$\langle v + \alpha, v' + \alpha' \rangle = \frac{1}{2}(\alpha(v') + \alpha'(v)).$$

We define (see [142]) a *generalized complex structure* on V to be a complex structure on W, orthogonal with respect to this scalar product. As we did before, we decompose $W \otimes \mathbb{C}$ into eigenspaces as $W \otimes \mathbb{C} = W^{0,1} \oplus W^{1,0}$.

If (V, J) is a complex vector space, we set $W^{0,1} = V^{0,1} \oplus V^{*1,0}$ and this gives a generalized complex structure on V of associated matrix:

$$\begin{pmatrix} -J & 0 \\ 0 & J^* \end{pmatrix}.$$

If $\omega \colon V \to V^*$ is a symplectic form on V, we set $W^{0,1} = \{X - i\omega(X) \mid X \in V\}$ and this gives a generalized complex structure on V with associated matrix:

$$\begin{pmatrix} 0 & -\omega^t \\ \omega & 0 \end{pmatrix}.$$

Thus, complex and symplectic linear algebras are particular cases of this more general notion.

Exercises for Chapter 4

Exercise 4.1 Generalize Example 4.2 and prove that the complex Grassmann manifold, $G_{n+k,k}(\mathbb{C})$, of k-plans in $\mathbb{C}^{n+k}$ is a complex manifold. Hint: [155, page 133].

Exercise 4.2 Show that any oriented surface admits a complex structure. Hint: [107, page 149].

Exercise 4.3 Let H be a commutative graded algebra with an element ω of degree 2, satisfying the hard Lefschetz property with respect to ω and Poincaré duality with respect to ω^n. Let $\alpha \in H^{2k+1}$ and $\beta \in H^{2k+1}$.
(1) Show the existence of an integer j such that $\beta\omega^j$ is in the complementary degree of α.
(2) Define $\langle \alpha, \beta \rangle$ by $\alpha\beta\omega^j = \langle \alpha, \beta \rangle \omega^n$. Show that $\langle -, - \rangle$ defines a symplectic structure on H^{2k+1}.

(3) Prove that H^{2k+1} is even dimensional (see Proposition 4.108).

Exercise 4.4 Give the details of the determination of the Dolbeault model of SO(9) in Example 4.75. Hint: [228].

Exercise 4.5 Let $S^1 \longrightarrow E \xrightarrow{p} B$ be a principal bundle such that B is a compact irreducible hermitian symmetric space and the Euler class of p represented by the Kähler form of B.

(1) Let PH be the primitive cohomology of $H(B)$, see [264, Chapter 5] or [115, page 122]. Prove that if PH is generated, as algebra, by at most one element, then the space E is formal. Give an example where PH is generated by two elements and E is still formal.

(2) Let $(\wedge Y \otimes \wedge Z, D)$ be a cdga such that
- Y is a vector space of dimension n, concentrated in even degree,
- Z is a vector space of dimension $n + 1$, concentrated in odd degree,
- $D(Z) \subset \wedge Y$, $D(Y) = 0$,
- the canonical projection $\rho \colon (\wedge Y \otimes \wedge Z, D) \to (Z, 0)$ induces 0 in cohomology.

Prove that $(\wedge Y \otimes \wedge Z, D)$ is not formal. (Observe that we have a pure model as in Theorem 2.77.)

(3) Show that the space E is not formal if and only if
- $B = U(p + q)/U(p) \times U(q)$, with $p \geq 3$ and $q \geq 3$,
- or $B = SO(2n)/U(n) \cong Sp(n - 1)/U(n - 1)$, with $n > 5$,
- or $B = E_7/E_6 \times SO(2)$.

Hint: Find the list of compact irreducible Hermitian symmetric spaces in [136, page 518], determine the minimal models and use the first questions.

Exercise 4.6 Show that if M^4 and N^4 are almost complex 4-manifolds, then $M \# N$ cannot have an almost complex structure. Then show that $\mathbb{C}P(2)\#\mathbb{C}P(2)$ is c-symplectic, but not symplectic.

Hints: (1) First show that, if P is a closed 4-manifold with almost complex structure, then

$$1 - b_1 + b_+ = \frac{\chi + \sigma}{2}$$

is even, where b_1 is the first Betti number, b_+ is the number of positive diagonal entries for the signature form, χ is the Euler characteristic of P and σ is the signature of P. Note that $\chi = 1 - b_1 + b_2 - b_3 + 1$ and $\sigma = b_+ - b_-$ with $b_2 = b_+ + b_-$.

Now, Borel and Hirzebruch showed that, if x is any cohomology class with $x \equiv w_2$ mod 2, where w_2 is the Stiefel–Whitney class, then $x^2 \equiv \sigma$ mod 8. The first Chern class has $p_2(c_1) = w_2$, so we calculate the Pontryagin class $p_1(TM)$ to be $p_1(TM) = -c_1(TM \otimes \mathbb{C}) = 1 + 2c_2 - c_1^2$. Apply the Hirzebruch signature formula to get $2(\chi + \sigma) = c_1^2 - \sigma = 8\ell$.

(2) Now use this fact and compute $1 - b_1 + b_+$ for the connected sum (see, for instance, [257, Theorem 3.6]).

Exercise 4.7 Prove Proposition 4.100 using Steenrod's functional cup product. The definition and properties may be found in [265]. Hint: the key property is that, for the Whitehead product map $[\iota_p, \iota_q]\colon S^{p+q-1} \to S^p \vee S^q$ and u_p, u_q, u_{p+q-1} the respective generators of $H^p(S^p)$, $H^q(S^q)$ and $H^{p+q-1}(S^{p+q-1})$,

$$u_p \cup_{[\iota_p, \iota_q]} u_q = -u_{p+q-1}.$$

Now let $\alpha \in \pi_p(X)$ and $\beta \in \pi_q(X)$ and represent the Whitehead product $[\alpha, \beta]$ by

$$F\colon S^{p+q-1} \xrightarrow{[\iota_p, \iota_q]} S^p \vee S^q \xrightarrow{h} X$$

where $h|_{S^p} = \alpha$ and $h|_{S^q} = \beta$. The related functional cup product is

$$u \cup_F v \in H^{p+q-1}(S^{p+q-1}).$$

Also, $u \cup_F v \subset h^*(u) \cup_{[\iota_p, \iota_q]} h^*(v)$.

If $u \in H^p(X)$ is dual to an element $h(\alpha)$, $\alpha \in \pi_p(X)$, and $v \in H^q(X)$ is dual to an element $h(\beta)$, $\beta \in \pi_q(X)$, then $h^*(u)(\bar{u}_p) = u(h_*(\bar{u}_p)) = u(h(\alpha)) = 1$ and similarly $h^*(v)(\bar{u}_q) = 1$ for the dual homology generators $\bar{u}_p$ and $\bar{u}_q$. Thus, $h^*(u) = u_p$, $h^*(v) = u_q$ and $u \cup_F v = h^*(u) \cup_{[\iota_p, \iota_q]} h^*(v) = u_p \cup_{[\iota_p, \iota_q]} u_q = -u_{p+q-1}$. Since $u_{p+q-1} \neq 0$, $F = [\alpha, \beta] \neq 0$ as well.

Now modify the situation above to treat the case of Proposition 4.100.

5 Geodesics

Let M be a smooth Riemannian manifold. The Riemannian structure of the manifold is reflected in the form its geodesics take. For instance, we *see* the symmetry of the sphere S^2 (with the Riemannian structure induced from $\mathbb{R}^3$) in the great circles that are its geodesics. Moreover, while the differential geometric structure of a physical system is detected by geodesics (e.g. in general relativity), the actual physical nature of such a system is seen in its motions. A beautiful result of Jacobi (see [216, Chapter 8] for instance) says that the Riemannian metric on the manifold of interest may be modified in a simple way so that the geometric and physical viewpoints agree; namely, the motions of the system are along geodesics with respect to the new metric. Therefore, in some sense, the study of geodesics exemplifies the paradigm expressing the relationship between mathematics and physics. Of course, the motions that are most important in physics are the periodic ones, so we begin by studying the geometric counterpart, closed geodesics.

A geodesic $c(t)$ is *closed* if there is a real number T such that $c(t+T) = c(t)$ and $\dot{c}(t + T) = \dot{c}(t)$ for all $t \in \mathbb{R}$. When you have a closed geodesic, then rotation along the geodesic re-parametrizes the geodesic to produce other ones. Therefore it is natural to consider only the closed geodesics $c(t)$ and $c'(t)$ with distinct images. They are then said to be *geometrically distinct*.

The first natural question about closed geodesics concerns their very existence on any particular compact Riemannian manifold M. In 1898, Hadamard proved that each nontrivial conjugacy class of $\pi_1(M)$ contains a closed geodesic that is the shortest closed curve representing an element in the conjugacy class. In 1929, Lusternik and Schnirelmann proved that, on a surface homeomorphic to S^2, *but with any metric*, there exist at least three simple closed geodesics. Finally, in 1951, Lusternik and Fet [181] proved that each compact Riemannian manifold contains at least one closed geodesic.

The second question is then: How many geometrically distinct closed geodesics can we find on a compact manifold? Over the years, computations suggested the following question.

Question 5.1 (The closed geodesic problem) *Does every compact Riemannian manifold M of dimension at least two admit infinitely many geometrically distinct geodesics?*

The existence theory of geodesics and closed geodesics on a compact Riemannian manifold M is based on the calculus of variations for the energy function on path spaces. Using Morse theory in an infinite dimensional setting, Gromoll and Meyer reduced the closed geodesic problem to the computation of the cohomology of the free loop space $LM = M^{S^1}$. Indeed Gromoll and Meyer proved that the manifold M admits infinitely many geometrically distinct closed geodesics if the Betti numbers of LM are unbounded. Now it should be clear how rational homotopy enters the picture; namely, because it gives a process for computing the minimal model of LM, and as a consequence, the Betti numbers of LM. Using this approach, D. Sullivan and M. Vigué-Poirrier proved that, if the cohomology algebra $H^*(M; \mathbb{Q})$ is not singly generated, then M admits infinitely many geometrically closed geodesics. We explain all this in detail in the first three sections of this chapter. Note that the Sullivan and Vigué-Poirrier theorem misses the fundamental case of the 2-sphere S^2. It was not until the early 1990s that this case was settled affirmatively by the combined work of V. Bangert [19], J. Franks [99] and N. Hingston [140].

When we know that there exist infinitely many geometrically distinct closed geodesics, the natural second question is the following.

Question 5.2 (Asymptotic behavior problem) *What is the asymptotic behavior (with respect to T) of the sequence n_T consisting of the number of geometrically distinct closed geodesics of length $\leq T$?*

Once again, the sequence of Betti numbers of LM provides a solution. Indeed, M. Gromov proved that for a family of generic metrics called bumpy metrics, there are integers $a > 0$ and $b > 0$ such that

$$n_T \geq a \cdot \max_{p \leq bT} \dim H^p(LM; \mathbb{Q}).$$

Moreover, Gromov conjectured that

Conjecture 5.3 *For rationally hyperbolic manifolds, the sequence n_T has exponential growth.*

This conjecture remains essentially open. It has been proved only for some families of manifolds such as connected sums of manifolds with nonsingly generated cohomology (see Theorem 5.34). These results will be described in more detail in Section 5.6.

In Section 5.7, we briefly describe an important generalization of Gromov's work. Instead of looking at closed geodesics and the function n_T, we consider geodesics of length $\leq T$ connecting a point x to a point y. The behavior of the function analogous to n_T is related to the entropy of the geodesic flow and to the cohomology of certain path spaces. Here, rational homotopy appears as an important tool for formulating results properly.

In the 1970s, K. Grove initiated a series of works on A-invariant geodesics. Here we fix an isometry A, and a geodesic $\gamma(t)$ is called A-invariant if for some T, $\gamma(t+T) = A(\gamma(t))$ for all $t \in \mathbb{R}$. The main question is then

Question 5.4 (*A-invariant geodesic problem*) *What conditions must be imposed on a manifold M and on an isometry A to guarantee the existence of a nontrivial A-invariant geodesic?*

Here, "nontrivial" means a geodesic that is not a constant map at a fixed point of the isometry A. Sections 5.4 and 5.5 are devoted to a study of A-invariant geodesics. These sections contain the classical results of Grove, Grove–Halperin and Grove–Halperin–Vigué. We show, in particular, the existence of infinitely many nontrivial A-invariant geodesics on a rationally hyperbolic manifold, and the existence of at least one nontrivial A-invariant geodesic on a rationally elliptic manifold M when the rank of some homotopy group $\pi_k(M)$ is odd.

Finally, it is an inescapable fact that the cohomology of the free loop space is related to (and computed by) Hochschild cohomology. Since we need this relationship in certain proofs, we have recalled the definitions and basic properties on the bar construction and Hochschild homology and cohomology in Section 5.9.

5.1 The closed geodesic problem

Let's now begin our discussion of the closed geodesic problem. We first formulate it as a conjecture.

Conjecture 5.5 *Each compact Riemannian manifold of dimension at least two admits infinitely many geometrically distinct closed geodesics.*

The conjecture above is true when $\pi_1(M)$ has infinitely many conjugacy classes by the result of Hadamard. Using a different approach, Bangert and Hingston [20] proved that the conjecture is also true when $\mathbb{Z}$ is a subgroup of finite index in $\pi_1(M)$. In later work, the combined work of V. Bangert [19], J. Franks [99] and N. Hingston [140] proved that the conjecture is true for the sphere S^2 with any metric.

Example 5.6 Let M and N be n-dimensional compact manifolds with $n > 2$, and let $M\#N$ be their connected sum. Suppose $\pi_1(M) \neq 0$ and $|\pi_1(N)| \geq 4$. Let $x \in \pi_1(M)$, $x \neq 1$, and let $y_1, y_2, z \neq 1 \in \pi_1(N)$ be distinct nontrivial elements. Then the elements $(xy_i xy_j)^n xz$ belong to different conjugacy classes in $\pi_1(M) * \pi_1(N)$. This shows that $\pi_1(M\#N) = \pi_1(M) * \pi_1(N)$ contains infinitely many conjugacy classes. It follows from Hadamard's result that $M\#N$ admits infinitely many geometrically distinct closed geodesics.

Let us denote by $LM = M^{S^1}$ the space of continuous maps from S^1 to M with the compact open topology and by $\mathcal{H}^1(S^1, M)$ the Hilbert manifold consisting of absolutely continuous maps from S^1 into M with square integrable derivative. By a result of Klingenberg [154] the inclusion $i: \mathcal{H}^1(S^1, M) \hookrightarrow LM$ is a homotopy equivalence (see [120]). Because of this equivalence, we shall abandon the notation $\mathcal{H}^1(S^1, M)$ and simply use LM even in situations where we obviously are referring to $\mathcal{H}^1(S^1, M)$.

We can therefore consider the energy of a closed curve, $E: LM \to \mathbb{R}$, defined by

$$E(c(t)) = \frac{1}{2} \int_0^1 \langle c'(t), c'(t) \rangle \, dt.$$

Recall now that if W is a Hilbert manifold, and $f: W \to \mathbb{R}$ is a smooth map, then $m \in W$ is a critical point if $\nabla(f)(m) = 0$. In this case, $\nabla(E) = 0$ produces the Euler–Lagrange equations for the energy functional E. Of course, these Euler–Lagrange equations are simply the geodesic equations (see [216, page 441]), so we have the

Theorem 5.7 ([154]) *The critical points of the energy function of a manifold are the constant maps and the closed geodesics.*

In order to appreciate the significance of this result, let's see why every compact manifold must have at least one closed geodesic. The approach we describe is due to the combined efforts of many mathematicians including Birkhoff, Morse and Bott. Define

$$P_n(M) = \{(x_1, \ldots, x_n) \in M^n \mid d(x_1, x_2)^2 + \cdots + d(x_n, x_1)^2 \leq \epsilon\}$$

where $d(x, y)$ denotes the distance from x to y on M and $0 \leq \epsilon \leq \epsilon(M)$. Here, $\epsilon(M)$ is the particular number (called the injectivity radius) associated to M such that, for any x and y in M with $d(x, y) \leq \epsilon(M)$, there exists a unique minimal geodesic joining x and y. Such an $\epsilon(M)$ always exists for a compact manifold. Clearly, $P_n(M)$ is compact and every point of $P_n(M)$ represents a closed geodesic n-gon with corners at the x_i. We have an inclusion $j: P_n(M) \hookrightarrow LM$ (after parametrizing the n-gons with respect to arclength). Moreover, a theorem of Bott says that, for fixed r and all $k \leq r$, there exists n_r such that, for $n \geq n_r$,

$$j_\#\colon \pi_k(P_n(M)) \xrightarrow{\cong} \pi_k(LM).$$

From the point of view of homotopy theory then, the $P_n(M)$ approximate LM better and better as n goes to infinity. Therefore, for homotopy questions, we have an effective finite dimensional reduction.

Now let's look at the situation from the viewpoint of critical point theory. We can define a type of energy function $E\colon P_n(M) \to \mathbb{R}$ by

$$E(x_1,\ldots,x_n) = \sum_{i=1}^{n} d(x_i, x_{i+1})^2$$

where $x_{n+1} = x_1$. This discretizes the energy functional E described above. It can be shown that a critical point of E gives rise to a polygon without corners with sides of equal length; that is, a closed geodesic. Note that the section $s\colon M \to P_n(M) \subset LM$ embeds M as the "trivial" critical points which are absolute minima for the energy functional E having $E = 0$. Of course, we want to find critical points besides these trivial ones and this is the content of the theorem of Lusternik and Fet.

Theorem 5.8 ([181]) *On any compact simply connected smooth manifold, there exists a nontrivial closed geodesic.*

Proof Suppose no critical point of E exists with $E > 0$. By [66, Theorem 1.17], there is a deformation of $P_n(M)$ into an open neighborhood of M which may be taken as close to M as we desire. In particular, $P_n(M)$ deforms into a tubular neighborhood of M, which itself deformation retracts onto M. If H is the total deformation of $P_n(M)$ into M and $h = H_1$, then we have $s \circ h \simeq \mathrm{id}_{P_n(M)}$, where s is the inclusion $M \hookrightarrow P_n(M)$. In particular, this implies that $h_\#\colon \pi_*(P_n(M)) \to \pi_*(M)$ is *injective*.

Now suppose the first nonzero homotopy group of M occurs in degree r. Because $\pi_k(\Omega M) = \pi_{k+1}(M)$, the splitting $\pi_k(LM) \cong \pi_k(\Omega M) \oplus \pi_k(M)$ says that the first nonzero homotopy group of LM occurs in degree $r-1$. If we choose n large enough, then this will also be true for $P_n(M)$. Thus, we may assume that $\pi_t(P_n(M)) = 0$ for $t < r-1$ and $\pi_{r-1}(P_n(M)) \neq 0$. The injectivity of $h_\#$ then implies that $\pi_{r-1}(M) \neq 0$ and this is a contradiction to the assumption that the first nonzero homotopy group occurs in degree r. Hence, our original supposition is incorrect and a non-trivial critical point of E exists in $P_n(M)$. Thus, there exists a nonconstant closed geodesic on M. □

Of course, it is a long way from proving the existence of a single closed geodesic to showing that a manifold admits an infinite number of closed geodesics. The crucial step was taken by Gromoll and Meyer in 1969, so let's turn to that now.

The circle acts on LM by rotation along the curves

$$A\colon S^1 \times LM \to LM \qquad A(t,c)(s) = c(t+s \bmod \mathbb{Z}).$$

The orbit of a curve, $S^1(c)$, is a submanifold of LM. Clearly, two geodesics c and c' are geometrically distinct if $c' \notin S^1(c)$. Note that if c is a critical point of E, then the same is true for all the points in $S^1(c)$.

Relations between critical points of the energy function and the homology of LM can then be established by adapting Morse theory to the infinite dimensional case. This gives the following theorem due to Gromoll and Meyer.

Theorem 5.9 ([117]) *Let M be a compact simply connected manifold. If the sequence of Betti numbers of LM is not bounded, then M admits infinitely many geometrically distinct closed geodesics.*

And this then leads to the main question in the area for which algebraic models are relevant.

Question 5.10 *When is the sequence of Betti numbers of the free loop space an unbounded sequence?*

This question will be answered in Proposition 5.14.

5.2 A model for the free loop space

The Gromoll–Meyer theorem reduces the closed geodesic problem to the problem of the computation of the cohomology of the free loop space. In this section we give a procedure to compute the minimal model for the free loop space LX of a simply connected space X from its minimal model $(\wedge V, d)$.

Theorem 5.11 *Let X be a simply connected space with minimal model $(\wedge V, d)$. Then a model for the free loop space, LX, is given by*

$$(\wedge V \otimes \wedge sV, \delta)$$

with $\delta(v) = dv$ and $\delta(sv) = -sd(v)$ where s is the derivation defined by $s(v) = sv$.

5.2 A model for the free loop space

Proof The space LX is the pullback in the following diagram of fibrations

$$\begin{array}{ccc} LX & \longrightarrow & X^{[0,1]} \\ {\scriptstyle p_0}\downarrow & & \downarrow {\scriptstyle (p_0,p_1)} \\ X & \xrightarrow{\Delta} & X \times X \end{array}$$

where $p_j(c) = c(j)$. Denote by $i\colon X \to X^{[0,1]}$ the map that associates to a point x the constant path at x. The map i is a homotopy equivalence making the following diagram commutative

$$\begin{array}{ccc} X & \xrightarrow{i} & X^{[0,1]} \\ {\scriptstyle \Delta}\searrow & & \downarrow {\scriptstyle (p_0,p_1)} \\ & X \times X & \end{array}$$

This implies that (p_0, p_1) and Δ have the same relative minimal model.

Denote by $(\wedge V_1, d)$ and $(\wedge V_2, d)$ two copies of $(\wedge V, d)$. By Example 2.48, a minimal model of Δ is given by the multiplication

$$\mu \colon (\wedge V_1, d) \otimes (\wedge V_2, d) \to (\wedge V, d).$$

By Example 2.48 again, a relative minimal model of μ is given by the following commutative diagram

$$\begin{array}{ccc} (\wedge V_1, d) \otimes (\wedge V_2, d) & \xrightarrow{\mu} & (\wedge V, d) \\ & \searrow & \uparrow {\scriptstyle \varphi} \\ & & (\wedge V_1 \otimes \wedge V_2 \otimes \wedge sV, D) \end{array}$$

where φ is a quasi-isomorphism with $\varphi(sv) = 0$, and where $D(sv) - v_2 + v_1$ is a decomposable element in $\wedge V_1 \otimes \wedge V_2 \otimes \wedge sV$ (for, recall that a decomposable element in $\wedge T$ is an element in $\wedge^{\geq 2} T$).

By Theorem 2.70 together with the standard pullback description of LX, we see that a model for LX is given by the tensor product

$$(\wedge V, d) \otimes_{(\wedge V_1 \otimes \wedge V_2)} (\wedge V_1 \otimes \wedge V_2 \otimes \wedge sV, D).$$

The proof of the theorem will follow from the construction of an isomorphism

$$(\wedge V, d) \otimes_{(\wedge V_1 \otimes \wedge V_2)} (\wedge V_1 \otimes \wedge V_2 \otimes \wedge sV, D) \to (\wedge V \otimes \wedge sV, \delta).$$

5: Geodesics

For this purpose, we give a more explicit construction of the cochain algebra $(\wedge V_1 \otimes \wedge V_2 \otimes \wedge s V, D)$. Consider the model $(\wedge V \otimes \wedge \bar{V} \otimes \wedge \hat{V}, D)$ constructed in Chapter 2 for the definition of left homotopies (Definition 2.18) with differential

$$D(v) = dv, D(\hat{v}) = 0, D(\bar{v}) = \hat{v},$$

and $\bar{V} = sV$. This model is equipped with a derivation s satisfying

$$s(v) = \bar{v}, s(\bar{v}) = 0, s(\hat{v}) = 0.$$

The commutator $\theta = sD + Ds$ is a derivation of degree 0 and the morphism

$$e^\theta : (\wedge V, d) \to (\wedge V \otimes \wedge \bar{V} \otimes \wedge \hat{V}, D)$$

is a well-defined morphism of cdga's. By definition,

$$e^\theta(v) = v + \hat{v} + \text{decomposable}.$$

We then consider the isomorphism of algebras

$$\psi : \wedge V_1 \otimes \wedge V_2 \otimes \wedge s V \longrightarrow \wedge V \otimes \wedge \bar{V} \otimes \wedge \hat{V},$$

defined by $\psi(v_1) = v$, $\psi(v_2) = e^\theta(v)$, and $\psi(sv) = \bar{v}$. We make ψ an isomorphism of cdga's by defining D on $\wedge V_1 \otimes \wedge V_2 \otimes \wedge s V$ by the formula

$$D = \psi^{-1} D \psi.$$

This differential D then inherits the good properties of D on $\wedge V_1 \otimes \wedge V_2 \otimes \wedge s V$; that is, $D = d$ on V_1 and on V_2 and $D(sv) - v_2 + v_1$ is a decomposable element.

We now define a morphism of cdga's

$$q : (\wedge V \otimes \wedge \bar{V} \otimes \wedge \hat{V}, D) \to (\wedge V \otimes \wedge s V, \delta)$$

by $q(v) = v$, $q(\bar{v}) = sv$ and $q(\hat{v}) = -sd(v)$. A simple computation shows that $q(sD + Ds) = 0$. We then form the composition $q \circ \psi : (\wedge V_1 \otimes \wedge V_2 \otimes \wedge s V, D) \to (\wedge V \otimes \wedge s V, \delta)$. By the observation that $q(sD + Ds) = 0$, we have $q \circ \psi(v_1) = v$, $q \circ \psi(v_2) = v$ and $q \circ \psi(sv) = sv$. This induces the promised isomorphism

$$(\wedge V, d) \otimes_{(\wedge V_1 \otimes \wedge V_2)} (\wedge V_1 \otimes \wedge V_2 \otimes \wedge s V, D) \xrightarrow{1 \otimes (q \circ \psi)} (\wedge V \otimes \wedge s V, \delta).$$

$\square$

For the sake of simplicity, in the minimal model of the free loop space, $(\wedge V \otimes \wedge s V, \delta)$, we will write $\bar{x}$ instead of sx for an element of sV.

Example 5.12 Let M be the sphere S^2. A minimal model for S^2 has the form $(\wedge(x,y), d)$ with $|x| = 2$, $|y| = 3$, $dx = 0$ and $dy = x^2$. A minimal model for LM is thus given by the cochain algebra $(\wedge(x,y,\bar{x},\bar{y}), \delta)$, with $|\bar{x}| = 1$, $|\bar{y}| = 2$,
$$\delta x = 0, \ \delta y = x^2, \ \delta(\bar{x}) = 0, \ \delta(\bar{y}) = -2x\bar{x}.$$
We can compute the cohomology to be
$$H^*(LM; \mathbb{Q}) = \langle \bar{x}, x, \bar{x}\bar{y}, x\bar{y} - 2\bar{x}y, \bar{x}\bar{y}^2, \ldots \rangle.$$
Note that $\dim H^i(LM; \mathbb{Q}) = 1$ for all i and that all cup products are zero. Nevertheless, the free loop space of S^2 is not formal because there are many Massey products such as $x\bar{y} - 2\bar{x}y$.

5.3 A solution to the closed geodesic problem

In the historical development of rational homotopy theory, one of the first major applications of the theory was the theorem of Vigué-Poirrier and Sullivan on closed geodesics [261].

Theorem 5.13 *If M is a compact simply connected Riemannian manifold whose rational cohomology algebra requires at least two generators, then M has infinitely many geometrically distinct closed geodesics.*

This theorem is in fact a direct consequence of Theorem 5.14 below combined with Theorem 5.9.

Proposition 5.14 ([261]) *Let M be a simply connected space whose rational cohomology is finite dimensional. The following conditions are equivalent.*

1. *The sequence of rational Betti numbers of LM is unbounded.*
2. *The cohomology algebra $H^*(M; \mathbb{Q})$ requires at least two generators.*
3. *The dimension of $\pi_{odd}(M) \otimes \mathbb{Q}$ is at least two.*

Proof If the cohomology $H^*(M)$ is singly generated, then either $H^*(M) = \wedge x$, with x in odd degree, or else $H^*(M) = \mathbb{Q}[y]/y^n$. In the first case, the minimal models of M and LM are $(\wedge x, 0)$ and $(\wedge(x,\bar{x}), 0)$. The Betti numbers of LM are either 0 or 1, so they are bounded. In the second case, the minimal model of M is $(\wedge(y,z), d)$ with $dy = 0$ and $dz = y^n$. The minimal model for LM is thus $(\wedge(y,z,\bar{y},\bar{z}), \delta)$ with $\bar{y}$ in odd degree, $\bar{z}$ in even degree, $\delta(\bar{y}) = 0$ and $\delta(\bar{z}) = -ny^{n-1}\bar{y}$. Since the ideal generated by z and y^n is acyclic, the complex is quasi-isomorphic to the quotient complex

$$\left(\wedge(y,\bar{y})/y^n \otimes \wedge \bar{z}, \bar{\delta} \right), \qquad \bar{\delta}(\bar{z}) = -ny^{n-1}\bar{y}.$$

A standard computation shows that the reduced cohomology of LM is therefore

$$\wedge^+(y,\bar{y})/(y^n,\bar{y}y^{n-1}) \otimes \wedge \bar{z}.$$

Once again the Betti numbers are bounded. Observe also that in both cases, the dimension of $\pi_{\text{odd}}(M) \otimes \mathbb{Q}$ is one. This shows (1) $\Rightarrow$ (2) and (3) $\Rightarrow$ (2).

Suppose now that the cohomology requires at least two generators. We first prove that the dimension of $\pi_{\text{odd}}(M) \otimes \mathbb{Q}$ is at least two, i.e. (2) $\Rightarrow$ (3). If there is no odd generator, the differential is zero for degree reasons, and the cohomology is infinite, which is impossible. Suppose there is only one odd generator y. If $dy = 0$, then the first even generator is a cocycle, and no power of it is a coboundary because there is no other odd generator, and so the cohomology is infinite, which is impossible. Therefore dy is nonzero. We now order the generators by degree and we obtain

$$(\wedge(x_1,\ldots,x_n,y,x_{n+1},\ldots),d),$$

with $|x_i|$ even. Since dy is not a zero divisor in $\wedge(x_1,\ldots,x_n)$, the cohomology of $(\wedge(x_1,\ldots,x_n,y),d)$ is $\wedge(x_1,\ldots,x_n)/(dy)$ and therefore, the global cohomology is

$$\wedge(x_1,\ldots,x_n)/(dy) \otimes \wedge(x_{n+1},\ldots).$$

If $n \geq 2$ the dimension of $\wedge(x_1,\ldots,x_n)/(dy)$ is infinite and the cohomology is generated by the x_j. If $n = 1$, we must have another generator x_2 of even degree because the cohomology requires at least two generators. Now $d(x_2) = 0$ because there is no cocycle in $\mathbb{Q}y \cdot \wedge(x_1)$. It follows that the cohomology is infinite. This proves that we have at least two odd generators.

We finish by proving (2) $\Rightarrow$ (1). By our discussion above, we can assume that the model for M contains at least two odd generators. Therefore, let y_1 and y_2 be the first two odd generators and denote the generators by increasing degrees: $x_1,\ldots,x_n,y_1,x_{n+1},\ldots,x_r,y_2,\ldots$ The differential d is given on the first generators by polynomials P_1, Q_i and P_2:

$$dx_1 = 0, \ dx_2 = 0, \ \ldots, \ dx_n = 0,$$
$$dy_1 = P_1(x_1,\ldots x_n),$$
$$dx_{n+1} = y_1 Q_1(x_1,\ldots,x_n), \ \ldots, \ dx_r = y_1 Q_r(x_1,\ldots,x_{r-1}),$$
$$dy_2 = P_2(x_1,\ldots,x_r)$$

The ideal J generated by $x_1,\ldots x_r$ is then a differential ideal in the minimal model of the free loop space $(\wedge V \otimes \wedge sV, \delta)$. The quotient $(A, \delta) = (\wedge V \otimes \wedge sV, \delta)/J$ is the minimal cdga $(A, \delta) = (\wedge(y_1, y_2, \ldots) \otimes \wedge \bar{V}, \delta)$, where $\bar{V} = sV$ as usual. In this quotient, the elements

$$\overline{x_1} \cdots \overline{x_r}\, \overline{y_1}^p\, \overline{y_2}^q, \ p, q \geq 1$$

are cocycles that induce linearly independent classes in cohomology. It follows that the Betti numbers of the cohomology of (A, δ) are unbounded (see the loop space analogue in Exercise 5.5). We now apply Lemma 5.16 inductively (beginning with x_r) to deduce that the Betti numbers of $(\wedge V \otimes \wedge s V, \delta)$ are also unbounded. □

Lemma 5.15 *Let (A, d) be a connected cdga, y an odd dimensional generator, and $(A, d) \to (A \otimes \wedge y, d)$ a relative minimal model. If the Betti numbers of $(A \otimes \wedge y, d)$ are unbounded, then the same is true for (A, d).*

Proof We filter the cdga $A \otimes \wedge y$ by the ideals $A^{\geq p} \otimes \wedge y$. The E_2-term of the associated spectral sequence is $H^*(A, d) \otimes \wedge y$. Since the spectral sequence converges to the cohomology of $(A \otimes \wedge y, d)$, we deduce that

$$\dim H^q(A \otimes \wedge y, d) \leq \dim H^q(A, d) + \dim H^{q-m}(A, d),$$

where m denotes the degree of y. □

Lemma 5.16 *Let $(\wedge V, d)$ be a minimal model and x be a closed even dimensional generator of $\wedge V$. If the Betti numbers of the quotient cdga $(\wedge V/(x), \bar{d})$ are unbounded, then the Betti numbers of $(\wedge V, d)$ are also unbounded.*

Proof The quotient is quasi-isomorphic to the Sullivan model $(\wedge V \otimes \wedge y, d)$ where $dy = x$. The result follows then directly from Lemma 5.15. □

Example 5.17 Note also that S^2 does not fall under the hypotheses of Proposition 5.14. However, as we mentioned in the introduction, the combined work of V. Bangert [19], J. Franks [99] and N. Hingston [140] led to a proof of the closed geodesic problem for S^2 with any metric.

5.4 A-invariant closed geodesics

The symmetry of a Riemannian manifold is reflected by its isometry group. Therefore, we can try to mix together the geometric content engendered by knowledge of the geodesics on a manifold and the symmetry information underlying the existence of an isometry. To see how to accomplish this, let A denote an isometry of a compact simply connected Riemannian manifold M.

Definition 5.18 *A geodesic $\gamma(t)$ is called A-invariant if there exists some $T \in \mathbb{R}$ such that $\gamma(t + T) = A(\gamma(t))$ for all $t \in \mathbb{R}$.*

Example 5.19 Let M be the sphere S^2 with the usual metric. Denote by A the antipodal map; then the great circles are A-invariant geodesics. When

5 : Geodesics

M is a flat torus and A is the involution $A(x, y) = (y, x)$ then the A-invariant geodesics are the lines in the square $[0, 1]^2$ that are parallel to the ascending diagonal.

Let's denote the interval $[0, 1]$ by I. The natural space to consider for the study of A-invariant geodesics is the space M_A^I of paths $c: I \to M$ satisfying $c(1) = A(c(0))$. Observe that M_A^I is the pullback

$$\begin{array}{ccc} M_A^I & \longrightarrow & M^I \\ \downarrow & & \downarrow {\scriptstyle (p_0, p_1)} \\ M & \xrightarrow{(\mathrm{id}, A)} & M \times M \end{array}$$

and note that, if $A = \mathrm{id}$, we recover the free loop space LM.

Just as for loop spaces, the space of paths M^I has the same homotopy type as the Hilbert manifold $\mathcal{H}^1(I, M)$ consisting of absolutely continuous curves $\sigma: I \to M$ with square integrable derivative. We can then take the pullback of the diagram

$$\begin{array}{ccc} \mathcal{H}_A^1(I, M) & \longrightarrow & \mathcal{H}^1(I, M) \\ p \downarrow & & \downarrow p \\ G(A) & \longrightarrow & M \times M \end{array}$$

where $p(c) = (c(0), c(1))$ and $G(A) \subset M \times M$ is the graph of the isometry A. Observe that $G(A)$ is homeomorphic to M. Since p is a submersion, the Hilbert submanifold $\mathcal{H}_A^1(I, M)$ has the homotopy type of M_A^I.

In [120], K. Grove extended the classical theory of closed geodesics to the A-equivariant situation and proved the following.

Theorem 5.20 ([120]) *The critical points of the energy function $\mathcal{H}_A^1(I, M) \to \mathbb{R}$ are the closed A-invariant geodesics and the constant maps at fixed points.*

Note that the set of fixed points for A, Fix A, is a disjoint union of totally geodesic submanifolds because, when $A(x) = x$, A_* is the identity when restricted to the tangent space of the submanifold, and $\exp_x \circ A_{*x} = A \circ \exp_x$. The starting point of the study of A-invariant geodesics is the strong relation between Fix A and M_A^I:

Theorem 5.21 ([120]) *If Fix $A = \emptyset$, then M has an A-invariant geodesic. If Fix $A \neq \emptyset$, and M has no nontrivial A-invariant geodesic, then the natural injection Fix $A \to M_A^I$ is a homotopy equivalence.*

5.4 A-invariant closed geodesics

We use Theorem 5.21 to obtain a link between the existence of A-invariant geodesics on M and algebraic invariants of M such as homotopy groups or the Euler characteristic. In order to obtain explicit criteria for the existence of A-invariant geodesics, we must study certain properties of the fibration

$$\Omega M \to M_A^I \to G(A).$$

If we replace A by an isometry that is homotopic to it, then the homotopy type of the fibration does not change. This is interesting because the isometries form a compact Lie group whose elements of finite order are dense [205]. Therefore, A is homotopic to an element of finite order. *We can thus suppose without loss of generality that A has finite order, say $A^k = 1$, for some integer k.*

Let's first recall the isomorphism $\theta: \pi_q(M) \xrightarrow{\cong} \pi_{q+1}(\Omega M)$. Denote by x_0 the base point of M, $\bar{x}_0$ the constant loop at the base point and PM the subspace of M^I consisting of paths c starting at x_0. Each continuous map $c: (I^q, \partial I^q) \to (M, x_0)$ lifts into a map $\tilde{c}: I^q \to PM$ defined by $\tilde{c}(t_1, \cdots, t_q)(t) = c(t_1, \cdots, t_{q-1}, tt_q)$. By restriction to $I^{q-1} \times \{1\}$, this gives the isomorphism $\theta: \pi_q(M) \to \pi_{q-1}(\Omega M)$. The inverse map θ^{-1} is defined by

$$\theta^{-1}(c)(t_1, \cdots, t_q) = c(t_1, \cdots, t_{q-1})(t_q).$$

Now, in the fibration $\Omega M \to M^I \xrightarrow{p} M \times M$, $p(c) = (c(0), c(1))$, each continuous map $c = (c_1, c_2): (I^q, \partial I^q) \to (M \times M, (x_0, x_0))$ lifts into the map

$$\overline{\tilde{c}_1} \cdot \tilde{c}_2: I^q \to M^I,$$

where $\cdot$ denotes composition of paths and $\bar{c}_1$ the path inverse to c_1. Let δ_M denote the connecting map in the homotopy long exact sequence for the fibration p, $\delta_M(c) = \overline{\tilde{c}_1} \cdot \tilde{c}_2$ restricted to $I^{q-1} \times \{1\}$. This implies that the composition

$$\pi_q(M) \oplus \pi_q(M) \xrightarrow{\delta_M} \pi_{q-1}(\Omega M) \xrightarrow{\theta^{-1}} \pi_q(M)$$

maps (a, b) to $b - a$ when $q > 1$ and to $b \cdot a^{-1}$ when $q = 1$.

Now denote by $\varphi: M \to G(A)$ the homeomorphism defined by $\varphi(m) = (m, A(m))$. We then have:

Lemma 5.22 *Let $\delta: \pi_q(G(A)) \to \pi_{q-1}(\Omega M)$ be the connecting map of the fibration $\Omega M \to M_A^I \to G(A)$. Then the composition*

$$\rho: \pi_q(M) \xrightarrow{\pi_q(\varphi)} \pi_q(G(A)) \xrightarrow{\delta} \pi_{q-1}(\Omega M) \xrightarrow{\theta^{-1}} \pi_q(M)$$

satisfies $\rho(x) = \pi_1(A)(x) \cdot x^{-1}$ for $q = 1$ and $\rho(x) = \pi_q(A)(x) - x$ when $q \geq 2$.

Proof The proof follows directly from the above computations and the commutativity of the following pullback diagram (where $G(A) \to M \times M$ is the natural inclusion)

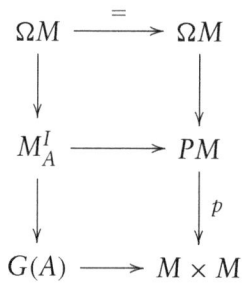

$\square$

Theorem 5.23 ([121]) *Let M be a compact connected Riemannian manifold. Then M has a nontrivial A-invariant geodesic if one of the following is satisfied.*

1. $\pi_q(M_A^I) \neq 0$ for some $q \geq 1$.
2. $\pi_q(A) - \mathrm{id}$ is not an isomorphism for some $q \geq 2$.
3. $\pi_1(A)(x) = x$ for some $x \in \pi_1(M)$.
4. rank $\pi_k(M)$ is odd for some $k \geq 2$.
5. $\pi_1(M)$ is finite and A^{p^m} is homotopic to the identity for some prime p and some integer m.
6. Fix A is not a connected set and M is simply connected.

Proof Suppose there is no nontrivial A-invariant geodesic. Then, by Theorem 5.21, Fix $A \neq \emptyset$. Furthermore, Fix A has no component of dimension at least one because, otherwise, this component would contain a nontrivial closed geodesic – that is, an A-invariant geodesic. Therefore, since M is compact, Fix A is a finite set that has the homotopy type of M_A^I. In particular, the homotopy groups $\pi_q(M_A^I) = 0$ for $q \geq 1$ and the connecting map is an isomorphism for $q \geq 1$. By the description of the connecting map in Lemma 5.22, this proves the assertions (1), (2) and (3).

(4) follows from the fact that, if B is an automorphism of finite order of some $\mathbb{Z}^m$, with m odd, then $\mathrm{id} - B$ is not an isomorphism. In fact, B is represented by a matrix with integer coefficients, which means that the roots of the characteristic polynomial are algebraic integers. On the other hand, since a power of B, say k, is the identity, the eigenvalues are kth roots of unity. Moreover, since the matrix has integer coefficients, if $e^{i\theta_j}$ is

an eigenvalue, then its conjugate $e^{-i\theta_j}$ is too. Since m is odd, at least one eigenvalue is a real number. If 1 is an eigenvalue, then clearly id $-$ B is not an automorphism. So suppose that -1 is an eigenvalue and id $-$ B is an isomorphism. The product $\prod(1 - e^{i\theta_j})(1 - e^{-i\theta_j})$ is an algebraic integer since it is a product of algebraic integers. On the other hand, this product is equal to det(id $-$ B)/2 and, therefore, $\pm 1/2$ must be an algebraic integer. But this is impossible because a rational number that is an algebraic integer must be an integer.

(5) The universal covering is a compact manifold and therefore for some $q \geq 2$ the rank of $\pi_q(M)$ is nonzero. Consider the morphism id $-$ $\pi_q(A)$ on $\pi_q(M)/$torsion. Recall that, over the field F_p with p elements, we have $(x-y)^p = x^p - y^p$. Therefore since $\pi_q(M)$ is abelian and p is a prime number, we have

$$(\mathrm{id} - \pi_q(A))^{p^m} = \mathrm{id} - (\pi_q(A))^{p^m} + pB$$

for some B. Since $(\pi_q(A))^{p^m} = \mathrm{id}$, we see that id $-$ $\pi_q(A)$ is not an isomorphism on $\pi_q(M)/$torsion. Therefore, id $-$ $\pi_q(A)$ is not an isomorphism at all and we now use (2).

(6) Since M is simply connected, the space M_A^I is connected and, therefore, cannot have the homotopy type of Fix A. □

Grove and Halperin used this result to obtain an existence theorem for A-invariant geodesics on a compact rationally elliptic manifold.

Theorem 5.24 ([122]) *Let M be a compact simply connected rationally elliptic Riemannian manifold. If M has no nontrivial A-invariant geodesic, then*

1. *The dimension n is even, and, if $\chi(M) \neq 0$, n is congruent to 0 modulo 4.*
2. *The Euler–Poincaré characteristic is the square of another integer.*

Proof (1) Let $r_i = \mathrm{rank}\, \pi_i(M)$. Recall that the dimension n of the manifold is a formal consequence of the ranks of the homotopy groups (see Theorem 2.75):

$$n = \sum_q (2q-1) \cdot (r_{2q-1} - r_{2q}).$$

Since, by (4) of Theorem 5.23, all the ranks are even, we see that n is also even. Moreover when $\chi(M) \neq 0$, then $\chi_\pi(M) = 0$, so $\sum_q r_{2q-1}(M) = \sum_q r_{2q}(M)$ and therefore n is congruent to 0 modulo 4.

(2) When $\chi(M) = 0$, then $\chi(M)$ is a square. We can thus suppose that $\chi(M) > 0$, so that $\chi_\pi(M) = 0$ (Theorem 2.75). Recall from Theorem 2.79 that the Poincaré series of M and the ranks of its homotopy groups are

related by
$$\sum_q \dim H_q(M;\mathbb{Q})\, t^q = \prod_i (1-t^{2i})^{r_{(2i-1)}-r_{2i}}.$$

Decompose $1-t^{2i}$ in the form $1-t^{2i} = (1+t)(1-t+t^2-\cdots-t^{2i-1})$. Then, since $\sum_i r_{2i-1}-r_{2i}=0$, we have

$$\prod_i(1-t^{2i})^{r_{2i-1}-r_{2i}} = \prod_i(1+t)^{r_{2i-1}-r_{2i}}\prod_i(1-t+\cdots-t^{2i-1})^{r_{2i-1}-r_{2i}}$$
$$= \prod_i(1-t+\cdots-t^{2i-1})^{r_{2i-1}-r_{2i}}.$$

Therefore taking the value at -1, we obtain

$$\chi(M) = \prod_i (2i)^{r_{2i-1}-r_{2i}}.$$

Since the ranks r_i are even, we deduce that $\chi(M)$ is a square. □

Corollary 5.25 *Let A be an isometry of a compact simply connected rationally elliptic Riemannian manifold M. If either the dimension n is odd or the Euler–Poincaré characteristic is not a square, then M admits a nontrivial A-invariant geodesic.*

There are other manifolds that have the property that for any Riemannian metric, every isometry admits a nontrivial invariant geodesic. For instance Papadima has proved that this is the case for the homogeneous space G/K when K is a maximal torus (see [221, Theorem 1.4(i)]), or G is simple (see [222, Corollary 1.4(i)]).

It is quite easy to compute a minimal model for the space M_A^I because M_A^I is the pullback of the following diagram of fibrations

$$\begin{array}{ccc} M_A^I & \longrightarrow & M^I \\ \downarrow & & \downarrow {\scriptstyle (p_0,p_1)} \\ M & \xrightarrow{(1,A)} & M \times M \end{array}$$

Example 5.26 When M is the sphere S^2 and A the antipodal isometry, then A induces multiplication by -1 on $\pi_2(S^2)$, so a model for $(1,A)$ is given by

$$\varphi \colon (\wedge(x,y,x',y'),d) \to (\wedge(x,y),d)$$

with $d(x)=d(x')=0$, $d(y)=x^2$, $d(y')=x'^2$, $\varphi(x)=x$, $\varphi(x')=-x$ and $\varphi(y)=\varphi(y')=y$. A model for M_A^I is thus obtained by taking the pushout in the following diagram (see Theorem 2.70)

$$(\wedge(x,y,x',y',\bar{x},\bar{y}),d)$$
$$\uparrow j$$
$$(\wedge(x,y),d) \xleftarrow{\varphi} (\wedge(x,y,x',y'),d)$$

In this diagram j is the inclusion and $d(\bar{x}) = x-x'$, $d(\bar{y}) = y-y'-x\bar{x}-x'\bar{x}$. It follows that a model (not minimal) for M_A^I is given by

$$(\wedge(x,y,\bar{x},\bar{y}),D), \quad D(\bar{x}) = 2x, \quad D(\bar{y}) = 0.$$

This space has the rational homotopy type of $S^3 \times \Omega S^3$, and is very different from LS^2. For instance, $H^1(LS^2;\mathbb{Q})$ is non-zero, but $H^1((S^2)_A^I;\mathbb{Q}) = 0$.

Since we can take the isometry A to be of finite order, the cyclic group G generated by A acts on M. Therefore, Theorem 3.26 implies that the manifold M admits a G-equivariant minimal model $(\wedge V, d)$. This means that A acts on V and that the action is extended in a multiplicative way to $\wedge V$. In particular, V decomposes as $V^A \oplus J$ where V^A is the subvector space generated by the vectors that are invariant under A, and J is the sum of the eigenspaces corresponding to the eigenvalues different from 1. Since the model is equivariant, we also have $d(J) \subset \wedge V^A \otimes \wedge^+ J$. Let $(\wedge V^A, \bar{d})$ denote the quotient of $(\wedge V, d)$ by the differential ideal generated by J. Therefore the ideal I generated by J and sJ in the model of the free loop space $(\wedge V \otimes \wedge sV, \delta)$ is a differential ideal, and we can form the quotient cdga $(\wedge V^A \otimes \wedge s(V^A), \bar{\delta})$, where $\bar{\delta}(sx) = -\bar{s}d(x)$.

Lemma 5.27 *The cdga $(\wedge V^A \otimes \wedge s(V^A), \bar{\delta})$, where $\bar{\delta}(sx) = -\bar{s}d(x)$ is the minimal model for M_A^I.*

Proof Following the steps in the construction of the minimal model for the free loop space given in Section 5.2, we give a natural G structure to $(\wedge V \otimes \wedge \bar{V} \otimes \wedge \hat{V}, D)$, by $g \cdot \bar{v} = \overline{g \cdot v}$ and $g \cdot \hat{v} = \widehat{g \cdot v}$. The isomorphism of algebras $\psi: \wedge V_1 \otimes \wedge V_2 \otimes \wedge sV \to \wedge V \otimes \wedge \bar{V} \otimes \wedge \hat{V}$ induces a G-structure on $(\wedge V_1 \otimes \wedge V_2 \otimes \wedge sV, D)$. As in the nonequivariant case, a relative model for (p_0, p_1) is given by the inclusion

$$(\wedge V_1 \otimes \wedge V_2, D) \hookrightarrow (\wedge V_1 \otimes \wedge V_2 \otimes \wedge sV, D).$$

A Sullivan model for M_A^I is then given by the tensor product

$$(\wedge V \otimes \wedge sV, \delta_A) = (\wedge V, d) \otimes_{(\wedge V_1 \otimes \wedge V_2, D)} (\wedge V_1 \otimes \wedge V_2 \otimes \wedge sV, D),$$

where the structure of $(\wedge V_1 \otimes \wedge V_2)$-module on $(\wedge V, d)$ comes from the morphism $(1, A): (\wedge V_1, d) \otimes (\wedge V_2, d) \to (\wedge V, d)$.

We extend the coefficient field to $\mathbb{C}$ and we take a basis of J formed by eigenvectors x with eigenvalue $\lambda_x \neq 1$. In this case $\delta_A(sx) - (\lambda_x x - x)$ is a decomposable element, which shows that the ideal I is acyclic. The commutative differential graded algebra $(\wedge V \otimes \wedge sV, \delta_A)$ is therefore quasi-isomorphic to the quotient

$$(\wedge V^A \otimes \wedge s(V^A), \bar{\delta}_A) = (\wedge V^A, \bar{d}) \otimes_{(\wedge V_1^A \otimes \wedge V_2^A)} (\wedge V_1^A \otimes \wedge V_2^A \otimes \wedge s(V^A), \bar{D}).$$

This cdga is the minimal model of the free loop space on a space whose minimal model is $(\wedge V^A, \bar{d})$. This proves the lemma. □

Corollary 5.28 *Let M be a compact simply connected Riemannian manifold and A an isometry. If for some q, $(\pi_q(M) \otimes \mathbb{Q})^A \neq 0$, then M admits a nontrivial A-invariant geodesic.*

Proof The space M_A^I has model $(\wedge V^A \otimes \wedge s(V^A), \bar{\delta})$ by Lemma 5.27. Since $(\pi_q(M) \otimes \mathbb{Q})^A$ is dual to $(V^q)^A$, we have $\pi_q(M_A^I) \neq 0$ and the result follows from Theorem 5.23. □

5.5 Existence of infinitely many A-invariant geodesics

A result of Tanaka extends the Gromoll–Meyer Theorem and defines a bridge between the existence of infinitely many A-invariant closed geodesics and the homology of M_A^I.

Theorem 5.29 ([247]) *Let M be a compact simply connected Riemannian manifold. If the Betti numbers of M_A^I are unbounded, then M admits infinitely many geometrically distinct A-invariant geodesics.*

We now can use the minimal model described in Lemma 5.27 to give a criterion for the Betti numbers of M_A^I to be unbounded.

Theorem 5.30 ([123]) *Let M be a simply connected compact Riemannian manifold, let A be an isometry, and let $(\pi_*(M) \otimes \mathbb{Q})^A$ denote the A-invariant part of rational homotopy. If $\dim (\pi_{\text{odd}}(M) \otimes \mathbb{Q})^A \geq 2$, then the Betti numbers of M_A^I are unbounded, and M admits infinitely many nontrivial geometrically distinct A-invariant geodesics.*

Proof It follows from the description of the minimal model of M_A^I in Lemma 5.27 that M_A^I has the rational homotopy type of the free loop space on a space whose rational homotopy is $(\pi_*(M) \otimes \mathbb{Q})^A$. We then see from the Sullivan–Vigué theorem (Proposition 5.14) that the Betti numbers of M_A^I are unbounded. We then apply Tanaka's Theorem 5.29 to infer the existence of infinitely many geometrically distinct A-invariant geodesics. □

Theorem 5.31 ([122]) *Let M be a simply connected compact Riemannian manifold, and let A be an isometry. If M is rationally hyperbolic, then the Betti numbers of M_A^I are unbounded, and M admits infinitely many geometrically distinct A-invariant geodesics.*

Proof Let $(\wedge V, d)$ be the model of a space of finite rational category (Definition 2.34) and suppose A is an automorphism of $(\wedge V, d)$ of finite order k. Then we will prove that V^A is infinite dimensional and $(\wedge V^A, \bar{d})$ also has finite category.

Since $(\wedge V^A, \bar{d})$ is a quotient of $(\wedge V, d)$, the second assertion follows from Theorem 2.81. By Theorem 2.82 this will imply that $(V^A)^{\text{odd}}$ is also infinite dimensional. This means that $\pi_{\text{odd}}(M) \otimes \mathbb{Q}$ contains infinitely many invariant elements. We then apply Theorem 5.30 to infer the existence of infinitely many geometrically distinct A-invariant geodesics.

We extend the coefficient field to $\mathbb{C}$. We suppose $A^k = 1$ and we first give the proof when k is a prime number p. Then we will only have to write $k = k'p$ where p is a prime number. We deduce that $V^{A^{k'}}$ is infinite dimensional and by induction that V^A is infinite dimensional.

Therefore, consider the case $k = p$. Since the category of $(\wedge V, d)$ is finite, by Theorem 2.82, there is some integer N such that for each $\alpha \in \pi_*(\Omega M) \otimes \mathbb{Q}$ in degree $> N$, there is β with

$$\text{ad}^m(\alpha)(\beta) = [\alpha, [\alpha, \ldots, [\alpha, \beta]\ldots]] \neq 0 \text{ for all } m.$$

We can suppose that α and β are eigenvectors of A having respective eigenvalues λ^q and λ^r with $\lambda = e^{\frac{2\pi i}{p}}$. When $mq + r \equiv 0 \pmod{p}$, then $\text{ad}^m(\alpha)(\beta)$ is an eigenvector with eigenvalue 1. Since we can do this for all α and M is hyperbolic, this proves that V^A is infinite dimensional. $\square$

5.6 Gromov's estimate and the growth of closed geodesics

Once we know that manifolds have many closed geodesics, we can start to study the exact nature of the geodesics. A first thing to ask is whether the growth rate of the number of geodesics is understandable in terms of length, say. With this in mind, let M be a Riemannian manifold and denote by n_T the number of geometrically distinct closed geodesics of length less than or equal to T. The behavior of the sequence n_T is once again related to the Betti numbers of the free loop space, $b_i(LM)$. An important result on n_T was obtained by Gromov for bumpy metrics, so let's review this notion now.

First, recall from Morse theory that, for a smooth function $f\colon W \to \mathbb{R}$ on a manifold W, a submanifold N is called a *nondegenerate critical submanifold* if N is composed of critical points and if the Hessian is nondegenerate at each point in the direction normal to N. We are interested in certain submanifolds of LM; namely, recall that, for a closed geodesic $c\colon S^1 \to M$ and an element of the circle $t \in S^1$, we obtain a new (but not geometrically distinct) closed geodesic $\tilde{c}$ defined by $\tilde{c}(s) = c(s+t)$. The circle S^1 therefore acts on any such c to give a submanifold of LM, $S^1(c)$.

Definition 5.32 *A metric is called bumpy if for each critical point c of the energy function, the manifold $S^1(c)$ is a nondegenerate critical submanifold of LM.*

In fact, this is a generic condition. Indeed, as shown in [12], for each manifold, the set of bumpy metrics is dense in the Frechet space of all the possible metrics on M. The theorem of Gromov, improved by Ballmann and Ziller, can now be stated.

Theorem 5.33 ([18], [119]) *Let M be a simply connected compact manifold. Then for any bumpy metric and any field $\Bbbk$, there exists $a > 0$ and $c > 0$ such that*

$$n_T \geq a \cdot \max_{p \leq cT} \dim H^p(LM; \Bbbk).$$

Proof sketch The idea of Gromov's proof is the following. Denote by $(LM)_T$ the subspace of LM consisting of curves of length less than or equal to T. Let $d(T)$ be the maximal integer d such that, for $i = 0\ldots,d$, the inclusion $(LM)_T \hookrightarrow LM$ induces a surjective map in $\Bbbk$-homology in degree i. Gromov then proved that there are constants c and C depending only on the manifold such that $cT \leq d(T) \leq CT$. Now, when the metric is bumpy, Morse theory shows that $(LM)_T$ has the homotopy type of a CW complex with the number of cells in degree p equal to the number of closed geodesics of length less than or equal to T and with (Morse) index p. Therefore the number of closed geodesics of length $\leq T$, n_T, satisfies the following inequalities.

$$n_T \geq \sum_{i=1}^{d(T)} b_i(Y_T) \geq \sum_{i=1}^{d(T)} b_i(LM) \geq \sum_{i=1}^{cT} b_i(LM).$$

$\square$

Moreover, Gromov conjectured that, for almost all manifolds M, the sequence n_T has exponential growth. One important result in this direction is due to P. Lambrechts.

5.6 Gromov's estimate and the growth of closed geodesics

Theorem 5.34 ([164]) *Let M and N be simply connected compact manifolds of the same dimension n and let P be their connected sum, $P = M\#N$. Suppose that, for some field $\Bbbk$, $H^*(M; \Bbbk)$ is not singly generated as an algebra and $H^*(N; \Bbbk)$ is not the cohomology of a sphere. Then the sequence n_T (for P) has exponential growth.*

Remark 5.35 When $\pi_1(M)$ and $\pi_1(N)$ are not finite, then by Example 5.6, $M\#N$ admits infinitely many geometrically distinct closed geodesics. Theorem 5.34 is a generalization of that result to the case when the manifolds are simply connected.

Although the result of Lambrechts is true for any field, we will restrict ourselves to a proof in the rational setting. First, let's briefly recall the construction of the model of the connected sum $A_{M\#N}$ (also see Example 3.6). Let $(\wedge V, d)$ and $(\wedge W, d)$ be the minimal models for manifolds M and N of the same dimension n. Denote by S a complement of the cocycles in degree n in $\wedge V$ and by I the acyclic differential ideal $I = S \oplus (\wedge V)^{>n}$. We denote by $A_M = (\wedge V)/I$ the quotient of $\wedge V$ by the acyclic ideal. Since I is acyclic, the quotient homomorphism is a quasi-isomorphism $(\wedge V, d) \to A_M$. We do the same for N to obtain a (nonminimal) model A_N for N. Denote by u_M and u_N cocycles in degree n representing the fundamental cohomology classes in A_M and A_N. We deduce from Example 3.6 that a model for the connected sum $M\#N$ is given by

$$A_{M\#N} = ((A_M \oplus_{\mathbb{Q}} A_N) \oplus \mathbb{Q} \cdot z, d) \quad \text{with} \quad d(z) = u_M - u_N.$$

We will compute the cohomology of the free loop space on $M\#N$ by using Hochschild homology. Hochschild homology is defined in Section 5.9, where we recall the fundamental isomorphism

$$H^*(L(M\#N); \mathbb{Q}) \cong HH_*(A_{PL}(M\#N)).$$

Since Hochschild homology transforms quasi-isomorphisms of differential graded algebras into isomorphisms, we deduce the isomorphism

$$H^*(L(M\#N); \mathbb{Q}) \cong HH_*(A_{M\#N}).$$

The proof of Lambrechts's theorem is now a direct corollary of the following proposition.

Proposition 5.36 *The sequence $\dim HH_q(A_{M\#N})$ in Hochschild homology has exponential growth.*

Proof Let x and x' be two indecomposable elements in $A_M^{\leq n}$ that are cocycles whose cohomology classes are linearly independent, and let y be an indecomposable element in $A_N^{\leq n}$ that is a cocycle whose cohomology class is nontrivial. Take $r \geq 1$, $z = (z_1, \ldots, z_r)$ with $z_i = x$ or x', and

$$a_z = [z_1|y|x|y|z_2|y|x|y| \cdots |z_i|y|x|y| \cdots |z_r|y|x|y|x'|y|x'|y] \in \mathbb{B}_q(A_{M\#N}),$$

with $q = 4r+4$. Consider the action of the cyclic group $\mathbb{Z}_q$ on the Hochschild complex $\mathbb{B}_q(A_{M\#N})$ defined by $\tau \bullet a[a_1|\cdots|a_q] = a[a_2|\cdots|a_q|a_1]$.

Lemma 5.37 *For each $z = (z_1, \ldots, z_r)$, there are integers $\varepsilon_1, \ldots, \varepsilon_q$ such that*

$$\langle a_z \rangle = \sum_{i=0}^{q-1} (-1)^{\varepsilon_i} \tau^i \bullet a_z$$

is a cocycle in $\mathbb{B}_*(A_{M\#N})$.

Proof Suppose that $a_1, \ldots, a_q$ are elements of positive degree such that $a_i a_{i+1} = 0$ for $i = 1, \ldots, q-1$ and $a_q a_1 = 0$. Put $a = [a_1|\cdots|a_q]$. Define $\theta_i = |a_i| + (|a_i| + 1)(\sum_{j \neq i} |a_j|)$, $\varepsilon_0 = 1$ and $\varepsilon_i = 1 + \varepsilon_{i-1} + \theta_i$. Then a straightforward computation shows that

$$\sum_{i=0}^{q-1} (-1)^{\varepsilon_i} \tau^i \bullet a$$

is a cocycle. The argument applies in particular to a_z. □

Now, returning to the proof of Proposition 5.36, since $(A_{M\#N})^0 = \mathbb{Q}$, the augmentation ideal I is isomorphic to $(A_{M\#N})^+$. We denote by S a sub-vector space of I containing the decomposable elements which is a complement to the subspace L generated by x, x' and y. Then the space

$$J = I \otimes T(sI) + \sum_{k=0}^{\infty} \sum_{i=1}^{k} A_{M\#N} \otimes ((sI)^{\otimes i} \otimes (sS) \otimes (sI)^{k-i-1})$$

is a sub-complex of $\mathbb{B}_*(A_{M\#N})$ and the quotient complex is $(T(sL), 0)$. Denote by $q: L \to \mathbb{Q}x'$ the projection with kernel the subspace generated by x and y. The map $1 \otimes \ldots \otimes 1 \otimes sq \otimes 1 \otimes sq \otimes 1$ then projects each component $(sL)^{\otimes q}$ to $(sL)^{q-4} \otimes sx' \otimes sL \otimes sx' \otimes sL$.

Now, the 2^r classes $\langle a_z \rangle$ project in $(sL)^{q-4} \otimes sx' \otimes sL \otimes sx' \otimes sL$ to linearly independent classes (because the projections are linearly independent and the differential is zero). Since the degree of $\langle a_z \rangle$ is less than or equal to $(4r+4)n$, where n is the dimension of the manifolds, we have obtained a sequence of classes that grows exponentially. □

5.7 The topological entropy

So far in this chapter, we have used models to elucidate the nature of geodesics on manifolds. Here we want to describe a kind of converse approach due to G. Paternain (see [223] and [224] for all the results of this section). More explicitly, Paternain shows that certain constraints on the geodesic flow of a manifold M entail consequent constraints on the rational homotopy type of M. We then see that the information flow between geometry and topology is two-way.

Let M^n be a closed (i.e. compact without boundary) smooth n-manifold. Define a flow $\phi \colon SM \times \mathbb{R} \to SM$ on the sphere bundle SM associated to the tangent bundle TM by

$$\phi_t(x, v) \stackrel{\text{def}}{=} \phi(x, v, t) = (\gamma_v(t), \dot\gamma_v(t))$$

where v is a unit tangent vector at $x \in M$ and γ_v is the unique unit speed geodesic on M starting at x in the direction v. This is the *geodesic flow* on M. For each $T > 0$, $\hat{x}, \hat{y} \in SM$, define a distance function,

$$d_T(\hat{x}, \hat{y}) = \max_{0 \le t \le T} d(\phi_t(\hat{x}), \phi_t(\hat{y})),$$

where $d(-,-)$ is the distance function on M given by the Riemannian metric. Because M is compact, it makes sense to ask, what is the minimum number of balls of size ϵ with respect to d_T that are needed to cover SM? If this number is denoted by N_ϵ^T, then we can measure the relative growth of this quantity by defining

$$h(\phi, \epsilon) = \limsup_{T \to \infty} \frac{1}{T} \log(N_\epsilon^T).$$

The function $\epsilon \to h(\phi, \epsilon)$ is monotone decreasing so the following definition makes sense.

Definition 5.38 *The topological entropy of the flow ϕ is defined to be*

$$h_{\text{top}}(\phi) = \lim_{\epsilon \to 0} h(\phi, \epsilon).$$

Because the geodesic flow depends only on the Riemannian metric g, the topological entropy $h_{\text{top}}(\phi)$ of the geodesic flow ϕ associated to g is denoted by $h_{\text{top}}(g)$.

Topological entropy, as its name indicates, is a measure of the dynamical complexity of the flow. When $h_{\text{top}}(\phi) > 0$, we see chaos in the orbit structure of ϕ. Of course, the definition of topological entropy applies to any flow, but it is the geodesic flow which seems to hold within it homotopical information. A result of A. Manning (see the discussion in [224, Chapter 5] and Section 6.1 for the definition of sectional curvature $\sec_M$) says that, for $-\kappa \leq \sec_M \leq \kappa$ with respect to a metric g,

$$h_{\text{top}}(g) \leq (\dim(M) - 1)\sqrt{\kappa}.$$

In particular, flat manifolds have $h_{\text{top}}(g) = 0$.

It is perhaps not surprising that the entropy of the geodesic flow is related to the growth in the number of geodesics joining two points when length is increased. Indeed, a result of Mañé characterizes $h_{\text{top}}(g)$ in this way, showing that entropy is truly a manifestation of the underlying Riemannian structure alone. With this in mind, let $n_T(x, y)$ denote the number of geodesic segments from $x \in M$ to $y \in M$ of length $\leq T$ with respect to the metric g. More precisely, for fixed $x \in M$, let $D_T M = \{(x, v) \in T_x M \mid |v| \leq T\}$ and define

$$n_T(x, y) = \#\{\exp^{-1}(y) \cap D_T M\},$$

where $\exp_x : T_x M \to M$ is the exponential map. (Note that the definition of $n_T(x, y)$ just says that you count the number of vectors v of length $\leq T$ such that $\exp_x(v) = y$. Two vectors can be collinear and counted as distinct if they have different lengths.[1]) Then we have (see [224, Corollary 3.22], for instance)

Theorem 5.39 *The topological entropy $h_{\text{top}}(g)$ obeys the following inequality: for fixed $x \in M$,*

$$h_{\text{top}}(g) \geq \limsup_{T \to \infty} \frac{1}{T} \log \left(\int_M n_T(x, y) \, dy \right).$$

This result is proved by using a theorem of Yomdin which shows that the relative growth rate of volume under the geodesic flow is majorized by topological entropy. The inequality in Theorem 5.39 allows us to relate the algebraic topology of M to topological entropy via the following result of M. Gromov (generalizing his result for closed geodesics).

Theorem 5.40 *There exists a constant $C > 0$ such that, for fixed $x, y \in M$,*

$$n_T(x, y) \geq \sum_{i=1}^{[C(T-1)]} b_i(\Omega M),$$

where b_i denotes the ith Betti number and $[-]$ denotes the greatest integer function.

---

[1] Thanks to Gabriel Paternain for making this point clear to us.

5.7 The topological entropy

The ideas behind this result are very similar to what we did leading up to Theorem 5.7. The energy functional E may be defined on the space of paths in M from x to y, $\Omega_M(x, y)$ by

$$E(c(t)) = \frac{1}{2} \int_0^1 \langle c'(t), c'(t) \rangle \, dt. \qquad \text{for } c \in \Omega_M(x, y),$$

and it can be shown that the critical points of E are geodesics joining x and y. Furthermore, these critical points are nondegenerate, so Morse theory can be applied. In fact, as usual, Morse theory is applied to "finite pieces" of loop spaces by defining, for example, $\Omega_M^{T^-}(x, y) = E^{-1}(-\infty, T^2/2)$ and proving that critical points of E correspond to geodesic segments of length less than T. The Morse inequalities say that the sum of the Betti numbers to a given degree are majorized by the number of critical points up to that index. Furthermore, Morse theory tells us that, for a closed Riemannian manifold M and nonconjugate points $x, y \in M$ (see Exercise 5.4), $\Omega_M(x, y)$ has the homotopy type of a countable CW-complex with one cell of dimension k for each geodesic from x to y of index k.

Recall that y is *conjugate* to x along a geodesic α with $\alpha(0) = x$ if the exponential map $\exp_x \colon T_xM \to M$ has $\exp_x(tA) = \alpha(t)$ and y is a critical value for $\exp_x(tA)$. Of course, α is a critical point for the energy E, so we can also consider the Hessian of E (in a suitable sense) at α. The dimension of the nullspace of this Hessian is called the *multiplicity* of the conjugate point y. Now, the *index of a geodesic* from x is the index of the Hessian of the energy (i.e. the maximum dimension of a subspace on which the Hessian is negative definite) and this is the number of conjugate points to x along the geodesic (counted with multiplicities). All of these concepts may be found in [195].

The final thing to notice is that $\Omega_M(x, y)$ has the homotopy type of the ordinary loop space ΩM. To see this, observe that both ΩM and $\Omega_M(x, y)$ are fibers of the "smooth" path space (Serre) fibration $p \colon P(M) \to M$, where

$$P(M) = \{\alpha \colon I \to M \mid \alpha(0) = x, \text{ and } \alpha \text{ is smooth}\}, \qquad p(\alpha) = \alpha(1).$$

Namely, $\Omega M = p^{-1}(x)$ and $\Omega_M(x, y) = p^{-1}(y)$. Therefore, since fibers of a fibration have the same homotopy type, we have $\Omega M \simeq \Omega_M(x, y)$. Alternatively, if γ is a path from x to y, then composition with γ and with the inverse path $\bar\gamma$ gives inverse homotopy equivalences between ΩM and $\Omega_M(x, y)$. In particular, the Betti numbers of $\Omega_M(x, y)$ are the same as those of ΩM. These identifications then give the inequality of Theorem 5.40. Note the following immediate application (due to Serre; see [236]).

5: Geodesics

Theorem 5.41 *If M^n is a closed simply connected Riemannian manifold and $x, y \in M$ are nonconjugate points, then*

$$\lim_{T \to \infty} n_T(x, y) = \infty.$$

Proof Suppose $\lim_{T \to \infty} n_T(x, y)$ is finite. Then, by Theorem 5.40, $\Omega_M(x, y)$ has the homotopy type of a finite complex. Now, since $H^n(M; \mathbb{Q}) \neq 0$, the minimal model $(\wedge V, d)$ is nontrivial and contains a generator of odd degree because, otherwise, the dimension of $H^*(M; \mathbb{Q})$ is infinite. Therefore, $\pi_{k-1}(\Omega M) \otimes \mathbb{Q} = \pi_k(M) \otimes \mathbb{Q} \neq 0$ for some odd integer k. Then the minimal model of ΩM has an even degree generator u. Since the differential of the model of ΩM is zero, we see that u^s is a nontrivial cocycle for all $s \geq 0$, and this contradicts the finiteness of $\Omega_M(x, y) \simeq \Omega M$. □

Now, let's return to entropy and combine the estimates of Theorems 5.39 and 5.40 to see how properties of the geodesic flow on a manifold serve to constrain its homotopy type.

Theorem 5.42 ([223, Theorem 3.2]) *If (M, g) is a simply connected closed Riemannian manifold with $h_{\text{top}}(g) = 0$, then M is rationally elliptic.*

Proof By Theorem 5.40, we have $n_T(x, y) \geq \sum_{i=1}^{[C(T-1)]} b_i(\Omega M)$. If we integrate over M, take $\limsup \frac{1}{T} \log$ and apply Theorem 5.39, we obtain

$$h_{\text{top}}(g) \geq \limsup_{T \to \infty} \frac{1}{T} \log \left(\int_M n_T(x, y) \, dy \right)$$

$$\geq \limsup_{T \to \infty} \frac{1}{T} \log \left(\text{vol}(M) \sum_{i=1}^{[C(T-1)]} b_i(\Omega M) \right).$$

By hypothesis, $h_{\text{top}}(g) = 0$, so we must have

$$\limsup_{m \to \infty} \frac{1}{m} \log \left(\sum_{i=1}^{m} b_i(\Omega M) \right) = 0.$$

But this means that the sum of the Betti numbers of the loop space grows sub-exponentially. By Theorem 2.74 and Theorem 2.75, this is equivalent to M being rationally elliptic. □

Example 5.43 In [223], Paternain provided an important example of a situation where the geodesic flow has $h_{\text{top}}(g) = 0$. Let (M^n, g) be a closed manifold with metric g. Although the cotangent bundle $T^*(M)$ has a canonical symplectic structure, the tangent bundle obtains one only through the duality provided by the metric. The geodesic flow is then the Hamiltonian flow (with respect to the symplectic structure associated to g) corresponding to the Hamiltonian provided by the g-norm, $H = \| - \| \colon T(M) \to \mathbb{R}$. The geodesic flow is *completely integrable* as a Hamiltonian system if there are n smooth functions on $T(M)$, $f_1 = H, f_2, f_3, \ldots, f_n$ such that $\{f_i, f_j\} = 0$, where $\{-,-\}$ is the Poisson bracket on functions given by the symplectic structure. Completely integrable Hamiltonian systems have a very particular structure which allows the flow to be analyzed by restricting to invariant tori in $T(M)$. To each f_j, we may associate a vector field X_j by using the nondegeneracy of the symplectic form ω:

$$i_{X_j}\omega = df_j,$$

where i_{X_j} is interior multiplication by X_j. Of course, the vector field X_j has an associated flow ϕ_j^t. If, for each $j = 1, \ldots, n$, the flow ϕ_j^t actually is the orbit flow given by a circle action, then we say that the geodesic flow is *completely integrable with periodic integrals*. Essentially, we are saying that the Hamiltonian system has T^n as a symmetry group, thus reducing the system to solution by integration (i.e. quadrature). We then have the following.

Theorem 5.44 ([223]) *Let (M, g) be a simply connected closed Riemannian manifold whose geodesic flow is completely integrable with periodic integrals. Then $h_{\text{top}}(g) = 0$ and M is rationally elliptic.*

Now let M be a compact simply connected Riemannian manifold and let N be a compact simply connected submanifold. Fix $p \in M$ and $T > 0$. Paternain and Paternain (see [225]) define $n_T(p, N)$ to be the number of geodesics leaving orthogonally from N and ending at p with length $\leq T$. If p is not a focal point of N, $n_T(p, N)$ is finite. We can therefore define

$$I_N(T) = \int_M n_T(p, N) d\mu,$$

where μ is the measure induced by the Riemannian structure. Then extending the method of Gromov, Paternain and Paternain prove

Theorem 5.45 ([225]) *There exists a constant $C > 0$ depending only on the geometry of M and N such that if $\Omega(p, N)$ denotes the Hilbert manifold of paths from N to p, then for $k \geq 1$,*

$$\sum_{i=1}^{k} b_i(\Omega(p, N)) \leq \frac{1}{\operatorname{Vol} M} I_N(Ck).$$

Corollary 5.46 *Suppose that M is rationally hyperbolic and N is rationally elliptic, then the sequence $I_N(T)$ has exponential growth.*

Proof We have a fibration $\Omega M \to \Omega(N,p) \to N \times \{p\}$ from which we deduce the homotopy fibration

$$\Omega N \to \Omega M \to \Omega(N,p).$$

Since the homology of ΩM has exponential growth and the homology of ΩN polynomial growth, the homology of $\Omega(N,p)$ must have exponential growth. □

5.8 Manifolds whose geodesics are closed

The topological entropy of a smooth diffeomorphism is essentially a measure of the volume growth induced by its iterates. In [100], various types of volume growth invariants are defined and related to the rational ellipticity or hyperbolicity of the manifold of definition. Here, in the spirit of this chapter, we want to discuss a fact that is implicit in [27], but seemingly not observed until its appearance in [100], concerning manifolds with the property that all geodesics are periodic. The basic reference for these manifolds is [27].

Definition 5.47 Let M^n be closed Riemannian manifold and $x \in M$.

1. A geodesic (always parametrized by arclength) issuing from x, $\alpha \colon [0, T] \to M$ is *periodic with period* T if $\alpha(0) = x = \alpha(T)$ and $\dot\alpha(0) = \dot\alpha(T)$.
2. M is a P_T^x-manifold if all geodesics issuing from x are periodic with common period T. (Note that some geodesics are allowed to come back to x before length T.)
3. If M is a P_T^x manifold for all $x \in M$, then M is said to be a *P-manifold*. (It is a fact that if all the geodesics of a manifold are periodic, then they admit a common period. See [27, Lemma 7.11].)
4. The *index* of the geodesic $\alpha \colon [0, T] \to M$ is the number of $t \in (0, T)$ such that $\alpha(t)$ is conjugate to $\alpha(0) = x$ along α. (Note that $\alpha(T)$ is not counted here. This is explained by Proposition 5.48.)

Proposition 5.48 ([27, 7.25]) *If M^n is a P_T^x-manifold, then x is conjugate to itself with multiplicity $n - 1$ along any geodesic from x to x of any length. Furthermore, along any geodesic from x to x of length T, there are k conjugate points for some fixed integer k.*

Now here are the facts about P_T^x-manifolds culled from [27, Chapter 7]. First, note that there always exists some point $y \in M$ that is not conjugate to x along any geodesic. This is a simple consequence of Sard's theorem since conjugate points are critical values of the exponential map $\exp_x$ (see

Exercise 5.4). Then, it can be shown that there are only a finite number of geodesics, $\alpha_1, \ldots, \alpha_N$ from x to y of length strictly less than T. To see this, suppose α_i, $i = 1, \ldots, \infty$ is an infinite set of geodesics from x to y with lengths t_i smaller than T and restrict to the compact set $S(T_x M) \times [0, T]$ (where $S(T_x M)$ denotes the unit sphere in $T_x M$) so that there is a convergent subsequence of the sequence $(\alpha_i'(0), t_i)$ which converges to the limit $(\alpha'(0), t)$. Then $\lim \alpha_i(t_i) = \alpha(t) = y$ is conjugate to x along the geodesic α with initial tangent vector $\alpha'(0)$ (see Exercise 5.4). This is a contradiction, so only a finite number of geodesics exist.

So now, let $k_1, \ldots, k_N$ be the indices of the geodesics $\alpha_1, \ldots, \alpha_N$ from x to y of length strictly less than T. Without loss of generality, let $k_1 \leq k_2 \leq \cdots \leq k_N$. There is some minimizing geodesic of length $d(x, y)$, so the energy Hessian is positive semidefinite along it and, thus, $k_0 = 0$. Now, since M is a P_T^x-manifold and the lengths of the α_i are all less than T, we know there is an opposite geodesic $\tilde{\alpha}_i \colon [0, T - d(x, y)] \to M$ given by $\tilde{\alpha}_i(t) = \alpha_i(T - t)$. (Think of the sphere. A minimal geodesic is the shorter part of a great circle, while the longer part of the geodesic is the opposite geodesic.) Then, since the indices are the numbers of conjugate points and all loops have index k, we see that the opposite geodesic is one of the α_j and that its index is $k - k_i$. In particular, the opposite geodesic to the minimizer then has largest index $k_N = k$ (by Proposition 5.48).

The crucial fact is that every geodesic from x to y is the union of p geodesics from x to x of length T (for arbitrary p) and one of the α_i's. Of course, each geodesic from x to y has $k + n - 1$ conjugate points counted with multiplicity. Therefore, for each integer p, there are exactly N geodesics from x to y whose lengths are between pT and $(p+1)T$ and whose indices are $k_i + p(k + n - 1)$. Note that, since k is the largest index, different p produce a completely different set of indices.

If we now consider $\Omega_M(x, y)$, we see that, for each integer m, the number of critical points (i.e. geodesics) with fixed index m is bounded by N. Since each critical point gives a cell in a CW-decomposition of $\Omega_M(x, y)$, the number of cells, and hence the Betti number, in each dimension is also bounded by N. Because $\Omega_M(x, y) \simeq \Omega M$, we have the following.

Theorem 5.49 *If M is a P_T^x-manifold, then* $\dim H^q(\Omega M; \mathbb{Q}) \leq N$ *for some fixed integer N and all q. This holds in particular for P-manifolds.*

Now, an argument similar to that in the proof of Proposition 5.14 shows that the Betti numbers of the loop space of a simply connected finite complex are bounded if and only if the rational cohomology ring of the complex is singly generated (see Exercise 5.5). Therefore, in particular, we have

Corollary 5.50 *If M is a P_T^x-manifold, then $H^*(M;\mathbb{Q})$ is singly generated. Hence, M is rationally elliptic. This holds in particular for P-manifolds.*

Therefore, we see how the nature of the geodesics on a manifold can have powerful consequences for its homotopy structure.

5.9 Bar construction, Hochschild homology and cohomology

The bar construction and Hochschild homology play important roles in geometric applications of algebraic models. One reason for this is that Hochschild homology is inextricably linked to the cohomology of the free loop space. Furthermore, it will be an important tool for the proof of Lambrechts's result on the cohomology of the free loop space of a connected sum of varieties, as well as for the definition of iterated integrals (see Section 9.6) and for the computation of the Chas–Sullivan loop product (see Section 8.4). Indeed, in Subsection 8.4.5, we shall see some implications that the Chas–Sullivan loop product has for closed geodesics.

Let (A, d) be a differential graded algebra over a field $\Bbbk$, $A = \oplus_{k=0}^\infty A^k$, with an augmentation $\varepsilon \colon A \to \Bbbk$, and augmentation ideal $\bar{A} = \text{Ker } \varepsilon$. We denote by $s\bar{A}$ the suspension of $\bar{A}$, $(s\bar{A})^q = (\bar{A})^{q+1}$.

Definition 5.51 *The double bar construction on (A, d) is the complex*

$$(B(A, A, A), D) = (\oplus_n B_n(A, A, A), D)$$

with $B_n(A, A, A) = A \otimes (s\bar{A} \otimes \cdots \otimes s\bar{A}) \otimes A$, (n copies of $s\bar{A}$). The element $a \otimes (sa_1 \otimes \cdots \otimes sa_n) \otimes a'$ is denoted by $a[a_1|\cdots|a_n]a'$. The differential D decomposes into two terms $D = b + d$.

$$b(a[a_1|\ldots|a_n]a') = -\hat{a}a_1[a_2|\ldots|a_n]a'$$
$$- \sum_{i=1}^{n-1}(-1)^i \hat{a}[\hat{a}_1|\ldots|\hat{a}_i a_{i+1}|\ldots|a_n]a'$$
$$+ (-1)^{n-1} \hat{a}[\hat{a}_1|\ldots|\hat{a}_{n-1}]a_n a',$$

$$d(a[a_1|\ldots|a_n]a') = d(a)[a_1|\ldots|a_n]a'$$
$$+ \sum_{i=1}^n (-1)^i \hat{a}[\hat{a}_1|\ldots|\hat{a}_{i-1}|d(a_i)|a_{i+1}|\ldots|a_n]a'$$
$$+ (-1)^n \hat{a}[\hat{a}_1|\ldots|\hat{a}_n]d(a').$$

Here $\hat{a} = (-1)^{|a|}a$.

5.9 Bar construction, Hochschild homology and cohomology

The complex $B(A, A, A)$ is a right and a left differential A-module, quasi-isomorphic to A as an A-bimodule. Therefore for each right A-module M, and each left A-module N, we can form the complex

$$B(M, A, N) = M \otimes_A B(A, A, A) \otimes_A N.$$

When $M = N = \mathbb{Q}$, then $\bar{\mathbb{B}}(A) = B(\mathbb{Q}, A, \mathbb{Q})$ is a coalgebra. The comultiplication is defined by

$$\nabla : T(s\bar{A}) \to T(s\bar{A}) \otimes T(s\bar{A}),$$

$$\nabla[a_1|a_2|\ldots|a_n] = \sum_{i=0}^{n} [a_1|\ldots|a_i] \otimes [a_{i+1}|\ldots|a_n].$$

We have

Theorem 5.52 ([1]) *If $A = C^*(X; \mathbb{Q})$, then $H_*(\bar{\mathbb{B}}(A)) = H^*(B(\mathbb{Q}, A, \mathbb{Q}))$ is isomorphic to $H^*(\Omega X; \mathbb{Q})$ as a coalgebra.*

When $M = A$ and $N = \mathbb{Q}$ the complex $B(A, A, \mathbb{Q})$ is contractible and is therefore a good model for the path space PX. Our last example is related to the free loop space. The multiplication $A \otimes A \to A$ also makes (A, d) into a left and a right A-module.

Definition 5.53 *The Hochschild complex $\mathbb{B}_*(A)$ is the tensor product*

$$\mathbb{B}_*(A) := A \otimes T(s\bar{A}) = (A, d) \otimes_{A \otimes A} B(A, A, A)$$

where $T(V)$ denotes the tensor algebra on V. Its homology is denoted by $HH_*(A)$. The induced differential D on $A \otimes T(s\bar{A})$ is the sum of an internal differential d and an external one b defined by

$$d(a[a_1|\ldots|a_n]) = d(a)[a_1|\ldots|a_n]$$
$$+ \sum_{i=1}^{n} (-1)^i \hat{a}[\hat{a}_1|\ldots|\hat{a}_{i-1}|d(a_i)|a_{i+1}|\ldots|a_n],$$

$$b(a[a_1|\ldots|a_n]) = -\hat{a}a_1[a_2|\ldots|a_n] - \sum_{i=1}^{n-1} (-1)^i \hat{a}[\hat{a}_1|\ldots|\hat{a}_i a_{i+1}|\ldots|a_n]$$
$$- (-1)^{n+|a_n|(|a|+|a_1|+\ldots+|a_{n-1}|+n-1)} a_n \hat{a}[\hat{a}_1|\ldots|\hat{a}_{n-1}].$$

The relation between the homology of the free loop space and the Hochschild homology has been proved by different authors and can be found in [47], [109] and [149].

5 : Geodesics

Theorem 5.54 *If X is a finite simply connected space with finite Betti numbers, then*
$$H^*(LX; \Bbbk) \cong HH_*(C^*(X; \Bbbk)),$$
where $C^*(X; \Bbbk)$ denotes the usual singular cochain algebra of X with coefficients in the field $\Bbbk$.

It is a classical result that the Hochschild homology is invariant along quasi-isomorphisms of differential graded algebras (see [109]), so we have

Corollary 5.55 *If X is a simply connected space with minimal model $(\wedge V, d)$, then*
$$H^*(LX; \mathbb{Q}) \cong HH_*(A_{PL}(X)) \cong HH_*(\wedge V, d).$$

When M is a manifold, we then have

Corollary 5.56 *If M is a simply connected compact manifold with de Rham algebra of forms $A_{DR}(M)$ and rational minimal model $(\wedge V, d)$, then*
$$H^*(LM; \mathbb{R}) \cong HH_*(A_{DR}(M)), \quad H^*(LM; \mathbb{Q}) \cong HH_*(\wedge V, d).$$

In the rational case, the isomorphism between the Hochschild homology of the minimal model $(\wedge V, d)$ of M and the cohomology of the free loop space is quite easy to understand. Indeed, the bar construction $B((\wedge V, d), (\wedge V, d), (\wedge V, d))$ and the differential algebra $(\wedge V \otimes \wedge V \otimes \wedge sV, D)$ constructed in the proof of Theorem 5.11 are quasi-isomorphic semifree $(\wedge V \otimes \wedge V)$-modules. If we take the tensor product with $(\wedge V, d)$ over $(\wedge V \otimes \wedge V)$, we also obtain quasi-isomorphic complexes (see Exercise 2.7). The first tensor product gives the Hochschild homology and the second one is the model of the free loop space (Theorem 5.11).

As we saw, when $A = C^*(M; \Bbbk)$, the double bar construction leads to complexes that give the homology of the various usual path spaces. Table 5.1 is a summary of the connections.

In Section 8.4 we will also need the Hochschild cohomology algebra of a differential graded algebra. We recall the definition here.

Table 5.1 Bar constructions and spaces.

Bar constructions	Spaces
$B(A, A, A)$	$M^{[0,1]}$
$B(A, A, \Bbbk)$	$P(M)$
$\bar{\mathbb{B}}(A) = B(\Bbbk, A, \Bbbk)$	$\Omega(M)$
$\mathbb{B}(A)$	LM

Definition 5.57 *The Hochschild cohomology of a graded differential algebra (A, d), $HH^*(A, A)$, is the cohomology of the complex $\mathrm{Hom}_A(\mathbb{B}(A), A)$. This cohomology is a graded algebra. The product $\star$ of two elements $f, g \in \mathrm{Hom}(T(s\bar{A}), A) = \mathrm{Hom}_A(\mathbb{B}(A), A)$ is defined by*

$$f \star g \colon T(s\bar{A}) \xrightarrow{\nabla} T(s\bar{A}) \otimes T(s\bar{A}) \xrightarrow{f \otimes g} A \otimes A \xrightarrow{m} A.$$

Exercises for Chapter 5

Exercise 5.1 Construct the model for the free loop space on the manifolds $S^3 \times S^3$ and $\mathbb{C}P(n)$ and deduce the cohomology algebra of the corresponding free loop spaces.

Exercise 5.2 Let M be the usual torus and A be an isometry inducing in homotopy the morphism

$$\begin{pmatrix} 0 & -1 \\ 1 & 0 \end{pmatrix}$$

Show that M admits at least one nontrivial A-invariant geodesic.

Exercise 5.3 Let G/K be a simply connected homogeneous space. Suppose that rank $G-$ rank K is odd. Show that for every isometry A, there exists a nontrivial A-invariant geodesic.

Exercise 5.4 For a smooth manifold M, the exponential map at a point $p \in M$, $\exp_p \colon T_p M \to M$ is given by $\exp_p(tv) = \gamma(t)$, where $\gamma \colon [0, L] \to M$ is the unique unit-speed geodesic with $\gamma(0) = p$ and $\gamma'(0) = v$. A point $q = \gamma(s)$ is *conjugate* to p if $d\exp_p \colon T_{sv} T_p M = T_p M \to T_q M$ is singular (i.e. has a kernel) at $sv = s\gamma'(0)$. The significance of conjugate points is mainly due to the fact that geodesics cannot minimize arclength beyond the first conjugate point. Classically, a conjugate point was said to be the intersection of infinitesimally close geodesics. The point of this exercise is to explain what this means in the context of the discussion following Proposition 5.48.

Suppose $\alpha_i \colon [0, t_i] \to M$ is a sequence of unit-speed geodesics with length less than T, $\alpha_i(0) = p$ and $\alpha_i(t_i) = q$. Because $S(T_p M) \times [0, T]$ is compact (where $S(T_p M)$ is the unit sphere in $T_p M$), the sequence $(\alpha_i'(0), t_i)$ has a convergent subsequence converging to $(\alpha'(0), t)$. This initial data gives a geodesic α with $\alpha(0) = p$ and $\alpha(t) = q$. The claim now is that q must be conjugate to p along α. Show this using (and verifying) the following hints.

(1) Suppose that q is not a conjugate point. Therefore, there is an open set $U \subset T_p M$ with $t\alpha'(0) \in U$ such that $\exp_p|_U$ is a diffeomorphism onto an open neighborhood of q.

(2) If K is large enough, then, for all $i > K$, $t_i\alpha_i'(0) \in U$. This is what is meant by infinitesimally close.

(3) Show from the definition of the exponential map above that $\exp_p(t_i\alpha_i'(0)) = \exp_p(t\alpha'(0))$.

(4) Why is this a contradiction?

Exercise 5.5 Here is a warm-up for Proposition 5.14. Show that the Betti numbers of the loop space of a simply connected finite complex X are bounded if and only if its rational cohomology ring is singly generated. Hint: suppose $H^*(X;\mathbb{Q})$ is not singly generated and note that (as in Proposition 5.14) there are always at least two odd degree generators in the minimal model for X. Then, since the differential in the model of the loop space is zero, there are two even generators u and v that are cocycles. If the respective degrees are $2k$ and $2s$ with $L = \text{lcm}(2k, 2s)$, $L = 2kp$, $L = 2sq$, then show that $\dim H^{mL}(\Omega X;\mathbb{Q}) \geq m+1$ since $u^{pi}v^{qj} \in H^{mL}(\Omega X;\mathbb{Q})$ for $i,j \geq 0$ with $i+j = m$.

6 Curvature

Recently algebraic models have found a place in differential geometry, especially in questions revolving around types of curvature restrictions. In particular, F. Fang and X. Rong [84] and B. Totaro [256] have approached certain questions of Karsten Grove in this spirit and the work of Belegradek and Kapovitch [23] on the existence of non-negative curvature metrics on vector bundles has also relied on models. Of course, when we speak of algebraic models here, we are speaking of models ranging from simple cohomology all the way to minimal models and relative models. In this chapter, we want to examine these applications of models to curvature questions as well as other relevant connections between Riemannian geometry and algebraic models. A good general reference for Riemannian geometry and, especially, various finiteness theorems for diffeomorphism or homotopy type and topics surrounding Gromov–Hausdorff convergence, is [226].

6.1 Introduction: Recollections on curvature

A smooth n-manifold M^n is *Riemannian* if each tangent space $T_p(M)$ has a Euclidean metric $\langle -, - \rangle$ and these vary smoothly on M. Associated to the Riemannian metric is a covariant derivative ∇ called the *Riemannian connection*.

Proposition 6.1 *Let V, Z and W be tangent vector fields on M and $f: M \to \mathbb{R}$ be a function on M. The following are properties of the covariant derivative of M:*

(i) $\quad \nabla_V(Z + W) = \nabla_V Z + \nabla_V W;$
(ii) $\quad \nabla_{fV} Z = f \nabla_V Z;$
(iii) $\quad \nabla_V fZ = V[f] Z + f \nabla_V Z;$

(iv) $\nabla_V Z - \nabla_Z V = [V, Z]$, *where $[-, -]$ is the usual bracket of vector fields;*

(v) $V(\langle Z, W \rangle) = \langle \nabla_V Z, W \rangle + \langle Z, \nabla_V W \rangle$, *where the vector field V acts on the function $\langle Z, W \rangle$ in the standard "directional derivative" way and $\langle -, - \rangle$ denotes the metric on M.*

In fact, property (v) may be used to define the covariant derivative and we can also then see that the Riemannian connection is unique. We can use the connection to make the following.

Definition 6.2 *Let X, Y and Z be vector fields on the manifold M. The Riemann curvature is defined to be*

$$R(X, Y)Z = \nabla_{[X,Y]}Z + \nabla_Y \nabla_X Z - \nabla_X \nabla_Y Z.$$

A more intuitive version of curvature that still holds within it all of the information provided by R is the *sectional curvature*,

$$\sec(X, Y) = \frac{R(X, Y, X, Y)}{\langle X, X \rangle \langle Y, Y \rangle - \langle X, Y \rangle^2},$$

where we use the notation

$$R(X, Y, Z, W) \stackrel{\text{def}}{=} \langle R(X, Y)Z, W \rangle.$$

The sectional curvature may be interpreted as follows. Suppose X and Y span a tangent plane inside the tangent space $T_p(M)$. The exponential map carries the plane to a surface inside the manifold, so the ordinary Gauss curvature (see, for instance, [216, page 115, Chapter 3]) of the surface may be computed. This Gauss curvature is then the sectional curvature $\sec(X, Y)$. A more technical approach to sectional curvature is to consider it as a function on the Grassmannian bundle associated to the tangent bundle. More specifically, at $x \in M^n$, sectional curvature is a function $\sec \colon G_2(T_x(M)) \to \mathbb{R}$, where $G_2(T_x(M))$ denotes the Grassmann manifold of 2-planes in $T_x(M)$. We can construct a bundle with fiber $G_{2,n} = G_2(\mathbb{R}^n)$ by noting that $G_{2,n} = O(n)/(O(2) \times O(n-2))$ and then using the natural $O(n)$ action on $G_{2,n}$ to form an associated bundle, $G_{2,n}(M) = F(M) \times_{O(n)} G_{2,n}$ to the principal (frame) bundle $O(n) \to F(M) \to M$ underlying the tangent bundle of M. Then sectional curvature becomes a function $\sec \colon G_{2,n}(M) \to \mathbb{R}$. Furthermore (see, for instance, [28, pages 166–167]), the function $\sec$ is smooth and, therefore, continuous. If M is a closed manifold, then $G_{2,n}(M)$ is compact, so its image under $\sec$ is also compact and we obtain the following.

6.1 Introduction: Recollections on curvature

Proposition 6.3 *If M is a compact manifold, then there exist $\lambda, \kappa \in \mathbb{R}$ such that for any plane $P \subset T_x(M)$ and for any $x \in M$, $\kappa \leq \sec_P \leq \lambda$.*

Now, if we scale the metric by multiplying by a constant (i.e. $\langle -, - \rangle \mapsto c\langle -, - \rangle, c > 0$), then sectional curvature is divided by that constant. In this way, for compact manifolds, we can always consider $\operatorname{Im}(\sec) \subset (-1, 1)$ if we desire. The price we pay for this scaling is that the diameter of the manifold increases. So, there must always be a sort of tension between sectional curvature and diameter. In fact, we shall see this reflected in the hypotheses of the main results below.

Once an invariant such as sec is defined, it is standard procedure to restrict the invariant in some way and try to classify the manifolds obeying the restriction. This classification is often a topological one. For example, Hadamard's famous theorem implies that a manifold M with $\sec(X, Y) \leq 0$ for all pairs X, Y must be an Eilenberg–Mac Lane space, $M = K(\pi, 1)$. Similarly, the famous pinching theorem says that bounding sectional curvature as $1/4 < \sec \leq 1$ implies that the manifold is a sphere – a very strong constraint indeed!

Example 6.4 A manifold M^n with constant sectional curvature equal to zero is said to be *flat*; it is also called a Bieberbach manifold since it was L. Bieberbach who characterized such a manifold as one covered by Euclidean space $\mathbb{R}^n$ such that the integral translations $\mathbb{Z}^n$ are a normal subgroup of finite index in $\pi_1(M)$. In other words, there is a short exact sequence of groups,

$$\mathbb{Z}^n \to \pi_1(M) \to F$$

where F is finite. Such a group is called a Bieberbach group. It is a fact that flat manifolds are classified up to diffeomorphism by their fundamental groups, so in this case group theory determines topology.

Example 6.5 If G is a compact Lie group (see Chapter 1), then G has a bi-invariant metric $\langle -, - \rangle$ so that geodesics with respect to this metric are the one-parameter subgroups arising from integration of left invariant vector fields. Therefore, for any $Y \in \mathfrak{g}$ (where $\mathfrak{g}$ is the Lie algebra of G), $\nabla_Y Y = 0$, where ∇ is the associated Riemannian connection. Now, by expanding $\nabla_{X-Y}(X-Y) = 0$, we obtain $\nabla_X Y = -\nabla_Y X$. Then, since ∇ is the Riemannian connection, we have $[X, Y] = \nabla_X Y - \nabla_Y X = 2\nabla_X Y$. Putting this in Definition 6.2, we easily calculate (with the help of the Jacobi identity) $R(X, Y)Z = \frac{1}{4}[[X, Y], Z]$. Substituting this result into the definition of sec gives

$$\sec(X, Y) = \frac{1}{4}|[X, Y]|^2,$$

where $|-|$ is the norm associated to the metric and X, Y are orthonormal. See Exercise 6.2 for details. Thus, compact Lie groups have non-negative sectional curvature. Once we know this, we can create more manifolds with non-negative sectional curvature by using the following result.

Proposition 6.6 (O'Neill's formula: [213]) *Let $\pi: M \to N$ be a Riemannian submersion and suppose that X and Y are orthonormal horizontal vector fields on M. Then*

$$\sec(\pi_* X, \pi_* Y) = \sec(X, Y) + \frac{3}{4} |[X, Y]^v|^2$$

where $[-, -]^v$ denotes the vertical projection. (If X and Y are not orthonormal, then the second term on the right must be divided by $|X \wedge Y| \stackrel{\text{def}}{=} \langle X, X \rangle \langle Y, Y \rangle - \langle X, Y \rangle^2$.)

Therefore, a Riemannian submersion can only increase sectional curvature, never decrease it. Using this, we can see that, for instance, homogeneous spaces G/H and biquotients $G//H$ (where H is closed) also have non-negative sectional curvature metrics.

Various other upper and lower bounds on sectional curvature may be combined with restrictions on, for instance, diameter and volume, to create new classes of manifolds. As an example, let $\mathcal{M}^{\leq D}_{\geq \kappa}(n)$ denote the class of closed n-manifolds with sectional curvature bounded below by κ and diameter bounded above by D. Gromov proved that there are constants $b(n, \kappa, D)$, depending on dimension, sectional curvature and diameter, such that the sum of the Betti numbers of any $M \in \mathcal{M}^{\leq D}_{\geq \kappa}(n)$ is bounded above by $b(n, \kappa, D)$. This puts a strong topological constraint on manifolds in $\mathcal{M}^{\leq D}_{\geq \kappa}(n)$. If we also include a volume restriction, volume $\geq V$ with associated class $\mathcal{M}^{\leq D, \geq V}_{\geq \kappa}(n)$, then, for $n \geq 5$, there are only finitely many diffeomorphism types depending on n, κ, D and V.

Because of this result and various pinching theorems, the class $\mathcal{M}^{\leq D}_{\kappa \leq \sec \leq \lambda}(n)$ of closed n-manifolds with sectional curvature and diameter obeying $\kappa \leq \sec \leq \lambda$ and diam $\leq D$ has proven to be of some interest. In particular, K. Grove asked the following question.

Question 6.7 (Grove's question) *Does the sub-class of $\mathcal{M}^{\leq D}_{\kappa \leq \sec \leq \lambda}(n)$ consisting of simply connected manifolds contain finitely many rational homotopy types?*

Let's now consider this question and see how algebraic models play a key role.

6.2 Grove's question

In this section we consider the work of Fang and Rong [84] and of B. Totaro [256] on Grove's conjecture. The Fang–Rong approach essentially creates the right algebra to contradict the conjecture and then realizes the algebra using minimal models. Totaro, on the other hand, uses biquotients to construct manifolds with the right curvature properties and then shows that a sufficient number of these also have the right algebra to contradict the conjecture. Since the methods are so different, we present both.

Both approaches use a fundamental result due to J.-H. Eschenburg that relates qualities of horizontal projections associated to a certain sequence of submersions to the qualities of the resulting sectional curvatures. More specifically, suppose G, M and M_i ($i = 1, 2, \ldots$) are Riemannian manifolds with submersions $\pi \colon G \to M$ and $\pi_i \colon G \to M_i$. For each $g \in G$, there are projections $h(g) \colon T_g(G) \to H(T_g(G)) \subseteq T_g(G)$ and $h_i(g) \colon T_g(G) \to H_i(T_g(G)) \subseteq T_g(G)$ to the horizontal subspaces determined by the bundle structure of the respective submersions. We may then consider h and h_i as elements of $\Gamma(\mathrm{End}(T(G)))$, the space of sections of the endomorphism bundle associated to the tangent bundle of G. As such, we can consider the convergence properties of the sequence h_i in the C^1-topology on $\Gamma(\mathrm{End}(T(G)))$. Then Eschenburg proved

Theorem 6.8 [81, Proposition 22]) *If $h_i \to h$ in the C^1-topology on $\Gamma(\mathrm{End}(T(G)))$, then for any $g \in G$ and any horizontal 2-plane $P \subset T_g(G)$,*

$$\mathrm{sec}_{M_i}(\pi_{i*}P) \to \mathrm{sec}_M(\pi_* P)$$

and the convergence is uniform in P.

Proof idea Let X and Y be π-horizontal vector fields which span P. Then $h_i(X) \to X$ and $h_i(Y) \to Y$ in the C^1-topology. Then Proposition 6.6 and some simplification of the covariant derivative give

$$\mathrm{sec}_{M_i}(\pi_{i*}(X), \pi_{i*}(Y)) = \mathrm{sec}_G(h_i(X), h_i(Y)) + \frac{3|(I - h_i)(\nabla_{h_i(X)} h_i(Y))|^2}{|h_i(X) \wedge h_i(Y)|^2}.$$

Then it is plain that $\mathrm{sec}_{M_i}(\pi_{i*}(X), \pi_{i*}(Y))$ converges to $\mathrm{sec}_M(X, Y)$ uniformly. □

6.2.1 The Fang–Rong approach

In [84], Fang and Rong answered Grove's question as follows. (Note that the formulation of the result provides an upper bound for diameter in order for the examples of the theorem to have sectional curvature scaled between -1 and 1.)

6: Curvature

Theorem 6.9 *For any $n \geq 22$, there exists $D > 0$ such that the subclass of $\mathcal{M}_{-1 \leq \sec \leq 1}^{\leq D}(n)$ consisting of simply connected manifolds contains infinitely many rational homotopy types.*

In order to prove this result, let's first consider the algebra required. For any integer $\alpha > 0$, let

$$\wedge_\alpha = (\wedge(x_1, x_2, y, z_\alpha), d)$$

be the freely generated commutative differential graded algebra with $\deg(x_1) = \deg(x_2) = 2$, $\deg(y) = 5$, $\deg(z_\alpha) = 7$ and differential defined by

$$d(x_1) = d(x_2) = 0, \quad d(y) = x_1^2 x_2, \quad d(z_\alpha) = x_1^4 + x_2^4 + (\alpha x_1 + x_2)^4.$$

Lemma 6.10 *If $\alpha \neq \beta \in \mathbb{Z}$, then $H^*(\wedge_\alpha) \not\cong H^*(\wedge_\beta)$.*

Proof The argument here is standard. Let's suppose that there is an isomorphism $f: H^*(\wedge_\alpha) \to H^*(\wedge_\beta)$, where $\alpha \neq \beta$. We shall denote cohomology classes by their representatives: x_1 instead of $[x_1]$ etc. Also, the generators of $\wedge_\beta$ will be denoted by $\tilde{x}_1, \tilde{x}_2, \tilde{y}$ and $\tilde{z}_\beta$. Now, degree 2 cohomology is generated by the x's, so we must have $f(x_1) = a\tilde{x}_1 + b\tilde{x}_2$ and $f(x_2) = c\tilde{x}_1 + e\tilde{x}_2$, where a, b, c and e are rational numbers. Note that, because f is an isomorphism, coefficients a and b cannot both be zero and coefficients c and e cannot both be zero.

Now, because of the y-differentials, in cohomology we have $x_1^2 x_2 = 0 = \tilde{x}_1^2 \tilde{x}_2$ and these relations are the only ones in the respective degree 6 cohomologies. Thus, $\tilde{x}_1^3$, $\tilde{x}_2^3$ and $\tilde{x}_1 \tilde{x}_2^2$ are linearly independent in $H^*(\wedge_\beta)$. We then have

$$f(x_1^2 x_2) = 0$$
$$(a\tilde{x}_1 + b\tilde{x}_2)^2 (c\tilde{x}_1 + e\tilde{x}_2) = 0$$
$$a^2 c \tilde{x}_1^3 + a^2 e \tilde{x}_1^2 \tilde{x}_2 + 2abc \tilde{x}_1^2 \tilde{x}_2 + 2abe \tilde{x}_1 \tilde{x}_2^2 + b^2 c \tilde{x}_1 \tilde{x}_2^2 + b^2 e \tilde{x}_2^3 = 0$$
$$a^2 c \tilde{x}_1^3 + 2abe \tilde{x}_1 \tilde{x}_2^2 + b^2 c \tilde{x}_1 \tilde{x}_2^2 + b^2 e \tilde{x}_2^3 = 0,$$

using $\tilde{x}_1^2 \tilde{x}_2 = 0$. By linear independence, we obtain $a^2 c = 0$, $2abe + b^2 c = 0$ and $b^2 e = 0$. We must consider various cases.

If $a = c = 0$, then the third condition says that either $b = 0$ or $e = 0$, contradicting the fact that f is an isomorphism.

If $a = 0$ and $c \neq 0$, then the second condition says that $b = 0$, again contradicting the fact that f is an isomorphism.

Therefore, we must have $a \neq 0$. The first condition then implies that $c = 0$. Hence $e \neq 0$ since f is an isomorphism and so $b = 0$ by the third condition. We then have $f(x_1) = a\tilde{x}_1$ and $f(x_2) = e\tilde{x}_2$.

Now, the z-differentials provide relations in degree 8 cohomology as follows.

$$x_1^4 + x_2^4 + (\alpha x_1 + x_2)^4 = 0$$
$$(\alpha^4 + 1)x_1^4 + 4\alpha x_1 x_2^3 + 2x_2^4 = 0$$

using $x_1^2 x_2 = 0$. (Of course, the relation $(\beta^4 + 1)\tilde{x}_1^4 + 4\beta \tilde{x}_1 \tilde{x}_2^3 + 2\tilde{x}_2^4 = 0$ also holds in $H^*(\wedge_\beta)$.) If we apply f to the relation, using $f(x_1) = a\tilde{x}_1$ and $f(x_2) = e\tilde{x}_2$, we get

$$(\alpha^4 + 1)a^4 \tilde{x}_1^4 + 4\alpha a e^3 \tilde{x}_1 \tilde{x}_2^3 + 2e^4 \tilde{x}_2^4 = 0.$$

We want to use the fact that $\tilde{x}_1^4$ and $\tilde{x}_1 \tilde{x}_2^3$ are linearly independent elements with the dependence of $\tilde{x}_2^4$ on them given by the relation. So let's simplify this dependence by writing the relation above as

$$(\alpha^4 + 1)\left(\frac{a}{e}\right)^4 \tilde{x}_1^4 + 4\alpha \left(\frac{a}{e}\right) \tilde{x}_1 \tilde{x}_2^3 + 2\tilde{x}_2^4 = 0.$$

We then have two representations of $2\tilde{x}_2^4$:

$$2\tilde{x}_2^4 = -((\alpha^4 + 1)\left(\frac{a}{e}\right)^4 \tilde{x}_1^4 + 4\alpha \left(\frac{a}{e}\right) \tilde{x}_1 \tilde{x}_2^3)$$
$$2\tilde{x}_2^4 = -((\beta^4 + 1)\tilde{x}_1^4 + 4\beta \tilde{x}_1 \tilde{x}_2^3).$$

By uniqueness of representations, we obtain

$$(\alpha^4 + 1)\left(\frac{a}{e}\right)^4 = (\beta^4 + 1) \quad \text{and} \quad \alpha\left(\frac{a}{e}\right) = \beta.$$

(Note that, since α and β are both positive, we must have $a/e > 0$ as well.) If we replace β in the first equation by its expression in terms of α given in the second equation, we obtain

$$\alpha^4 \left(\frac{a}{e}\right)^4 + \left(\frac{a}{e}\right)^4 = \alpha^4 \left(\frac{a}{e}\right)^4 + 1.$$

Hence, we have $(a/e)^4 = 1$. Since a and e are rational and $a/e > 0$, it must be the case that $a = e$. Putting this into the equation $\alpha(a/e) = \beta$ then gives $\alpha = \beta$, contradicting our initial assumption that $\alpha \neq \beta$. Hence, no isomorphism f can exist. $\square$

So, if we can find simply connected manifolds in $\mathcal{M}^{\leq D}_{-1 \leq \sec \leq 1}(n)$ with the cohomology algebras above, then Theorem 6.9 will be proved. It will come as no surprise then that the algebras $\wedge_\alpha$ that appear above arise as (part of) algebraic models for these manifolds. Let's now see how this comes about.

Let $\mathcal{M}_8 = (\wedge(x_1, x_2, x_3, y, z), d)$ with $\deg(x_i) = 2$, $i = 1, 2, 3$, $\deg(y) = 5$, $\deg(z) = 7$ and differential defined by

$$d(x_i) = 0, \quad d(y) = x_1^2 x_2, \quad d(z) = x_1^4 + x_2^4 + x_3^4.$$

We can explicitly find a $\mathbb{Z}$-form for this algebra; that is, we can find a CW-complex whose homotopy groups are finitely generated (over $\mathbb{Z}$) for which $\mathcal{M}_8$ is an algebraic model. The first step is to take $\mathbb{C}P(\infty) \times \mathbb{C}P(\infty) \times \mathbb{C}P(\infty) = K(\mathbb{Z}, 2) \times K(\mathbb{Z}, 2) \times K(\mathbb{Z}, 2) = K$. This models the x_i generators as *integral* cohomology classes. The second step is to take the principal fibration induced by the classifying map representing the integral cohomology class $x_1^2 x_2 \in H^6(K; \mathbb{Z}) = [K, K(\mathbb{Z}, 6)]$. The pullback of the universal fibration $K(\mathbb{Z}, 5) = \Omega K(\mathbb{Z}, 6) \to PK(\mathbb{Z}, 6) \to K(\mathbb{Z}, 6)$ via the classifying map gives a fibration $K(\mathbb{Z}, 5) \to X \to K$ and X is an integral model for the x_i generators and the y generator of $\mathcal{M}_8$. Finally, use the cohomology class $x_1^4 + x_2^4 + x_3^4 \in H^8(X; \mathbb{Z}) = [X, \mathbb{K}(\mathbb{Z}, 8)]$ to induce a principal fibration $K(\mathbb{Z}, 7) \to E \to X$ with the generator of $\pi_7(K(\mathbb{Z}, 7))$ integrally representing $z \in \mathcal{M}_8$. The space E is then a $\mathbb{Z}$-form for $\mathcal{M}_8$. For the following result, recall from Subsection 2.5.4 the notion of a Postnikov tower for a space and its relation to minimal models.

Lemma 6.11 *For each $n \geq 21$, there is a closed simply connected n-manifold B with $H^2(B; \mathbb{Z}) = \mathbb{Z}^3$ whose minimal model $\mathcal{M}_B$ has degree 8 Postnikov piece $\mathcal{M}_B(8) = \mathcal{M}_8$.*

Proof Let $E[10]$ denote the 10-skeleton of $\mathcal{M}_8$'s $\mathbb{Z}$-form E. Because the 10-skeleton carries E's homotopy type through dimension 9, we also have $\mathcal{M}_{E[10]}(8) = \mathcal{M}_8$. But $E[10]$ is a finite CW-complex of dimension 10, so we can embed $E[10]$ in $\mathbb{R}^{n+1}$ with $n \geq 21$ and take a thickening or regular neighborhood N; that is, a compact $(n+1)$-manifold with boundary of the same homotopy type as $E[10]$: $N \simeq E[10]$. Let $B = \partial N$ be the closed n-manifold boundary of N and note that the 10-skeleton of B is $E[10]$ by properties of thickenings (see [159, Section II]). Thus $\mathcal{M}_B(8) = \mathcal{M}_8$. □

Now that we have the n-manifold B ($n \geq 21$), we construct a closed $(n+1)$-manifold M_α as a principal circle bundle over B classified by the Euler class $e_\alpha = \alpha x_1 + x_2 - x_3$, where $\alpha \in \mathbb{Z}^+$ and the x_i are the integral generators of $H^2(B; \mathbb{Z})$. We then have the following result.

Proposition 6.12 *Let $\mathcal{M}_\alpha$ denote the minimal model of M_α. Then*

$$\mathcal{M}_\alpha(8) = \wedge_\alpha,$$

where $\mathcal{M}_\alpha(8)$ is the degree 8 Postnikov piece of $\mathcal{M}_\alpha$.

6.2 Grove's question

Proof The principal S^1-bundle $S^1 \to M_\alpha \to B$ is modeled in a standard way by a relative model of the form

$$(\mathcal{M}_B \otimes \wedge(t), D), \quad D|_{\mathcal{M}_B} = d_{\mathcal{M}_B}, \quad D(t) = \alpha x_1 + x_2 - x_3.$$

This model is, of course, non-minimal, but we can prove the proposition by giving a map

$$\rho \colon \wedge_\alpha \to (\mathcal{M}_B \otimes \wedge(t), D)$$

that induces an isomorphism on cohomology through degree 8 and an injection on degree 9 cohomology. With this in mind, define ρ by:

$$\rho(x_1) = x_1, \qquad \rho(x_2) = x_2, \qquad \rho(y) = y,$$
$$\rho(z_\alpha) = z + t\,(x_3^3 + (\alpha x_1 + x_2)x_3^2 + (\alpha x_1 + x_2)^2 x_3 + (\alpha x_1 + x_2)^3).$$

Direct calculation then shows that ρ satisfies the required properties. □

Since there were an infinite number of distinct algebras $\wedge_\alpha$, we have now obtained an infinite number of manifolds M_α with distinct rational homotopy types. To complete the proof of Theorem 6.9, we must show that these manifolds have metrics with the right curvature (and diameter) properties. This is where we shall use Theorem 6.8. In order to apply this result, we need to write the M_α as image spaces of submersions. With this in mind, define M to be the total space of a principal T^3-bundle over B classified by $x_1+x_2+x_3$ using the isomorphisms $H^2(B; \mathbb{Z}) \cong [B, K(\mathbb{Z}^3, 2)] \cong [B, BT^3]$. Call this map f and note that f induces an isomorphism $f_\# \colon \pi_2(B) \overset{\cong}{\to} \pi_2(BT^3)$.

Now let $T^3 = \{(z_1, z_2, z_3) \in \mathbb{C} \mid |z_i| = 1\}$ and write T_α^2 for the subgroup generated by all $(1, z, z)$ and $(z, 1, z^\alpha)$. The projection $T^3 \to T^3/T_\alpha^2$ induces $p_\alpha \colon BT^3 \to B(T^3/T_\alpha^2) = BS^1$. We then obtain the circle bundle

$$S^1 = T^3/T_\alpha^2 \to M/T_\alpha^2 \to B$$

with classifying map $f_\alpha = p_\alpha \circ f$.

Proposition 6.13 *As circle bundles, $M_\alpha \cong M/T_\alpha^2$.*

Proof The Euler class of the bundle $M/T_\alpha^2 \to B$ is determined on H^2, so we consider the dual map $(f_\alpha)_* \colon H_2(B; \mathbb{Z}) \to H_2(B(T^3/T_\alpha^2); \mathbb{Z}) = H_2(BS^1; \mathbb{Z})$. Because the elements $(1, z, z)$ and $(z, 1, z^\alpha)$ are killed in T^3, on the homology level, we have

$$(f_\alpha)_*(0, 1, 1) = 0 \quad \text{and} \quad (f_\alpha)_*(1, 0, \alpha) = 0.$$

Then, writing $(f_\alpha)_*(a, b, c) = pa+qb+rc$ (with (a, b, c) a coordinate expression with respect to the basis consisting of the homology duals of x_1, x_2

6 : Curvature

and x_3 respectively), the equations above give $q + r = 0$ and $p + \alpha r = 0$. Therefore,

$$(f_\alpha)_*(a, b, c) = -\alpha r a - rb + rc = -r(\alpha a + b - c).$$

Since $(f_\alpha)_*$ is surjective, we must have $r = \pm 1$. Let $\bar{\iota}$ be the generator of $H_2(BS^1; \mathbb{Z})$ with ι the dual generator of $H^2(BS^1; \mathbb{Z})$. The dual map $(f_\alpha)^*$ gives the Euler class e_α by $e_\alpha = (f_\alpha)^*(\iota) = Ax_1 + Bx_2 + Cx_3$. We can compute as follows.

$$A = (Ax_1 + Bx_2 + Cx_3)(1, 0, 0) = (f_\alpha)^*(\iota)(1, 0, 0) = \iota(f_\alpha)_*(1, 0, 0)$$
$$= \iota(\alpha\bar{\iota}) = \alpha.$$

Similarly, we find that $B = 1$ and $C = -1$. Hence, $e_\alpha = (f_\alpha)^*(\iota) = \alpha x_1 + x_2 - x_3$. Because M_α and M/T_α^2 have the same Euler classes, we have $M_\alpha \cong M/T_\alpha^2$. □

Now note that $M_\alpha = M/T_\alpha^2 = N/S_\alpha^1$, where $N = M/S_0^1$ and $S_0^1 \subset T^3$ is the circle generated by $(1, z, z)$. Let's also use the following notation. Let S^1 denote the subgroup generated by $(1, 1, z)$, $S_\alpha^1 = T_\alpha^2/S_0^1$, $\bar{S}^1 = q(S^1)$ and $\bar{S}_\alpha^1 = q(S_\alpha^1)$, where $q \colon T^3 \to T^2 = T^3/S_0^1$ is the projection. Finally, denote by $s_\alpha \colon N \to M_\alpha$ and $s \colon N \to N/\bar{S}^1$ the obvious projections. Since T^3 acts freely on M, note that T^2 acts freely on N. The final step in the construction of the examples of Fang and Rong is the following.

Proposition 6.14 *The M_α have metrics with a uniform bound on the absolute value of sectional curvature and diameter.*

Proof First note that, given a T^2-invariant metric on N, we can choose metrics on M_α and $N/\bar{S}^1$ so that s_α and s are submersions. Then the diameter condition will be fulfilled since $\text{diam}(M_\alpha) \leq \text{diam}(N)$.

Now, by the definitions above, we can see that a small segment of the $\bar{S}_\alpha^1$-orbit through an $x \in N$ converges locally to a small segment of the $\bar{S}^1$-orbit through x. But then we see that r-balls in the horizontal space of s_α converge (uniformly in $T_x(N)$) to the r-ball in the horizontal space of s. This implies that we have convergence of the respective horizontal projections, $h_\alpha \to h$, as in the hypothesis of Theorem 6.8. Therefore, we obtain uniform convergence

$$\sec_{M_\alpha}((s_\alpha)_*(h_\alpha(P))) \to \sec_{N/\bar{S}^1}(s_*(h(P))).$$

This then provides the desired uniform bound on (absolute value of) sectional curvature (using Proposition 6.3). □

Proof of Theorem 6.9 By Lemma 6.10, Lemma 6.11, Proposition 6.12 and Proposition 6.14, the M_α (for all α large enough) provide an infinite number of mutually nonrationally homotopic manifolds in $\mathcal{M}_{-1\leq \sec \leq 1}^{\leq D}(n)$ once we recognize that the metric on N may be scaled to have $-1 < \sec_N < 1$ while only increasing the diameter to some value D. □

Example 6.15 As we saw in Example 2.72 (also see the proof of Proposition 7.17), a relative extension of the form (where subscripts denote degrees)

$$(\wedge(e_2), d = 0) \to (\wedge(e_2, x_4, y_7, z_9), D) \to (\wedge(x_4, y_7, z_9), d)$$

with differential given by choosing $a \in \mathbb{Z}^+$ (say),

$$D(e_2) = 0, \ D(x_4) = 0, \ D(y_7) = x_4^2 + ae_2^4, \ D(z_9) = e_2^5$$

has a topological realization as a principal circle bundle $M \to N_a \to BS^1$ (where BS^1 is modeled by $(\wedge(e_2), d = 0)$) with M having the rational homotopy type of $S^4 \times S^9$ and N being a smooth 12-manifold (see Theorem 3.2). It can easily be seen from the rational cohomology algebra that, for choices a_1 and a_2, N_{a_1} has the same rational homotopy type as N_{a_2} only if $\dfrac{a_1}{a_2}$ is a rational square (see Example 2.38). Therefore, if we choose a_i, $i = 1, \ldots, \infty$ so that no $\dfrac{a_i}{a_j}$ is a square (e.g. a_i is the ith prime), then the N_{a_i} give an infinite number of rational homotopy types. Note, however, that the real minimal models of all of these N_{a_i} are isomorphic since real square roots always exist for positive numbers. This brings up the following.

Question 6.16 *Can the N_a be given metrics with $\sec_{N_a} \geq 0$?*

Now, $M \simeq_\mathbb{Q} S^4 \times S^9$ and $S^4 \times S^9$ has non-negative sectional curvature, so if it were true that this condition propagates across rational equivalences, then M would have $\sec_M \geq 0$ too.

Question 6.17 *If two closed manifolds have the same (rational) homotopy type and one manifold has a metric of non-negative sectional curvature, does the other?*

On the other hand, we may ask whether Eschenburg's method can produce the N_a, thereby providing non-negative curvature.

6.2.2 Totaro's approach

Now let's consider the approach of B. Totaro [256]. Note that this approach allows the construction of counterexamples in the lowest possible dimension where they can exist, dimension 7 (see [259]). This is achieved at the

6: Curvature

cost of using more complex algebra than that presented in Section 6.2.1. (Indeed, in [256], Totaro uses techniques from algebraic geometry to verify that he has an infinite number of mutually nonrationally equivalent six-dimensional biquotients with non-negative sectional curvature.)

Theorem 6.18 ([256]) *There exist real numbers C and D such that the subclass of $\mathcal{M}^{\leq D}_{-1 \leq \sec \leq C}(7)$ consisting of simply connected manifolds contains infinitely many rational homotopy types.*

Proof First we define a commutative graded algebra by

$$H = \wedge(x_0, x_1, \ldots, x_4)/\mathcal{R}$$

where x_i, $i = 0, \ldots, 4$, are of degree 2 and $\mathcal{R}$ is the ideal generated by $x_0^2 = x_1 x_2$, $x_1^2 = x_2 x_3$, $x_2^2 = x_3 x_4$, $x_3^2 = x_4 x_0$, $x_4^2 = x_0 x_1$ and all the other products $x_i x_j$. One can easily see by direct computation that $H^2 = \mathbb{Q}[x_0, \ldots, x_4]$, $H^4 = \mathbb{Q}[x_0^2, \ldots, x_4^2]$ and $H^6 = \mathbb{Q}[x_0^2 x_1 = x_1^2 x_2 = x_2^2 x_3 = x_3^2 x_4 = x_4^2 x_5]$. Notice that H is of finite total dimension and it satisfies Poincaré duality. Therefore, it can be realized by a manifold M of dimension 6 (see Theorem 3.2).

We now construct a principal bundle $(S^1)^5 \longrightarrow E \longrightarrow M$ determined from the classifying map $M \to (BS^1)^5$ defined by the cohomology classes $x_0, \ldots, x_4$. Note that E is a manifold of dimension 11 endowed with a free action of $(S^1)^5$. We choose a Riemannian metric, invariant by the action of $(S^1)^5$, and we scale the metric to have curvature greater than -1. Denote by D the diameter of E.

We now consider quotients of E by tori $(S^1)^4$; these are simply connected manifolds Y of dimension 7 satisfying the following properties:

- equipped with the induced metric, they are of diameter $\leq D$;
- by Proposition 6.6, we know that a submersion increases the sectional curvature; therefore Y is of curvature ≥ -1;
- from Theorem 6.8, we know that, for the set of tori $(S^1)^4 \subset (S^1)^5$, the curvatures of the corresponding Y are uniformily bounded. This means that there exists a real number C such that the curvatures of all these manifolds Y are $\leq C$.

Let's now study some of the spaces Y. We determine their rational homotopy type from a Sullivan model. Denote by $\mathcal{M}_M$ a minimal model of M. Since $p: Y \to M$ is a S^1-principal bundle with base M, Y has a model of the following type:

$$\mathcal{M}_Y = (\mathcal{M}_M \otimes \wedge x, d),$$

where the differential d on $\mathcal{M}_M$ coincides with the differential of the minimal model $\mathcal{M}_M$ and $dx \in \mathcal{M}_M^2$. We choose $dx = a_0 x_0 - a_1 x_1 + \left(\dfrac{a_1^3}{a_0^2}\right) x_3$
where a_0, a_1 are rational numbers. From direct computations, we can check that the multiplication by dx, from $H^2(M)$ to $H^4(M)$ has a kernel of dimension 1 generated by $z = a_0 x_0 + a_1 x_1 + \left(\dfrac{a_0^2}{a_1}\right) x_2$. From the Gysin exact sequence

$$H^r(M) \longrightarrow H^r(Y) \longrightarrow H^{r-1}(M) \overset{\cup dx}{\longrightarrow} H^{r+1}(M)$$

and the fact that $\dim H^2(M) = \dim H^4(M) = 5$, we see that $\mathrm{Coker}(\cup dx)$ also has dimension 1. Now, $H^3(M) = 0$, so the Gysin sequence then says that $H^4(Y)$ has dimension 1. Clearly, we have that $H^2(Y) \cong H^2(M)/(dx)$ has dimension equal to 4. Now, we have a commutative diagram

$$H^2(M) \otimes H^2(M) \overset{p^* \otimes p^*}{\longrightarrow} H^2(Y) \otimes H^2(Y)$$
$$\varphi \downarrow \qquad \swarrow \psi$$
$$H^6(M) \cong \mathbb{Q}$$

where the top arrow is clearly surjective (since $p^* \colon H^2(M) \to H^2(Y)$ is surjective) and the homomorphism φ is the cup-product with the class z, $\varphi(\alpha \otimes \beta) = \alpha \cup \beta \cup z$. Since $dx \cup z = 0$, φ induces a quadratic form ψ on $H^2(Y)$. From the definition of φ as a cup-product, we deduce its matrix on the five-dimensional space $H^2(M)$,

$$\begin{pmatrix} a_1 & a_0 & 0 & 0 & 0 \\ a_0 & a_2 & a_1 & 0 & 0 \\ 0 & a_1 & 0 & a_2 & 0 \\ 0 & 0 & a_2 & 0 & 0 \\ 0 & 0 & 0 & 0 & a_0 \end{pmatrix}$$

where $a_2 = a_0^2/a_1$. The 4×4 matrix in the right bottom corner is a matrix for ψ; its determinant is $-\dfrac{a_0^7}{a_1^3}$. Note that we are looking for isomorphism types so we have to admit a multiplication of the quadratic form ψ by a scalar. Since $\dim H^2(Y) = 4$, which is even, the determinant in $\mathbb{Q}/(\mathbb{Q}^*)^2$ of the quadratic form is unchanged. Therefore, the determinant is an invariant of isomorphism classes only as an element in $\mathbb{Q}/(\mathbb{Q}^*)^2$. The determinant of

ψ gives us an element

$$-\frac{a_0}{a_1} \in \mathbb{Q}/(\mathbb{Q}^*)^2.$$

Finally, note that all values of $\mathbb{Q}/(\mathbb{Q}^*)^2$ are reached for appropriate choice of $(a_0, a_1) \in \mathbb{Q} \times \mathbb{Q}$. Therefore we get an infinite number of different rational homotopy types corresponding to the spaces Y. □

Remark 6.19 Totaro's examples above are biquotients and this is no accident. In fact, almost all known closed manifolds having non-negative sectional curvature are biquotients. The only such manifolds which are not defined as biquotients are the Cheeger manifolds obtained as the connected sums of two spaces from the list: $\mathbb{R}P(n)$, $\mathbb{C}P(n)$, $\mathbb{H}P(n)$ and CayP (where the last is the Cayley plane) and an example of Grove and Ziller. This is discussed in [255], where Totaro also determines the overlap; that is, he classifies which Cheeger manifolds are diffeomorphic to biquotients and which are not. For instance, $\mathbb{C}P(4)\#\mathbb{H}P(2)$ is not even homotopy equivalent to any biquotient. Furthermore, in contrast to Theorems 6.18 and 6.9, Totaro proves in [255] that there are only finitely many diffeomorphism classes of *2-connected* biquotients of a given dimension.

6.3 Vampiric vector bundles

In [56], the following theorem was proved.

Theorem 6.20 (The soul theorem) *Let M be a complete noncompact Riemannian manifold with non-negative sectional curvature. Then there exists a totally geodesic and totally convex compact submanifold $S \stackrel{i}{\hookrightarrow} M$ such that M is diffeomorphic to the normal bundle of the embedding i.*

Remark 6.21 (Cheeger–Gromoll splitting) In [55], it was shown that a manifold M with non-negative Ricci curvature has a universal cover $\widetilde{M}$ that splits isometrically as $\widetilde{M} = M_0 \times \mathbb{R}^n$, with $\mathbb{R}^n$ flat and M_0 compact. See Exercise 6.1 for the definition and properties of Ricci curvature. We shall only use this result peripherally, so we simply mention it in this remark. We do note, however, that the splitting comes from peeling off "lines" from $\widetilde{M}$. Recall that a line is a geodesic that minimizes distance between each pair of points on it.

The submanifold S is called the *soul* of M. Recall that S is *totally geodesic* in M if every geodesic in M starting from a point of S in a direction tangent to S is in fact a geodesic in S. The submanifold S is *totally convex* if, for any two points of S, any geodesic in M joining them (which exists since M is complete) lies in S.

Remark 6.22 Right away, note some immediate consequences of the soul theorems 6.20.

1. Because S is totally geodesic, S also has non-negative sectional curvature.
2. The inclusion $i\colon S \to M$ is a homotopy equivalence. Therefore, if M is contractible (e.g. $M = \mathbb{R}^n$), then $S = *$.
3. If $M^n = K(\pi, 1)$ is a closed manifold, then $\tilde{M}$ is contractible and also has non-negative sectional curvature. By Cheeger–Gromoll splitting, we have $\tilde{M} = M_0 \times \mathbb{R}^n$. But $\tilde{M} \simeq *$, so $M_0 = *$ as well and, thus, $\tilde{M} = \mathbb{R}^n$ isometrically. Since covering transformations act as isometries, we then see that $M = \mathbb{R}^n/\pi$ is flat (see Example 6.4).
4. Given any closed manifold S with non-negative sectional curvature, it is a soul; namely, S is the soul of $S \times \mathbb{R}^n$. Note that we may consider the product $S \times \mathbb{R}^n$ as the trivial $\mathbb{R}^n$-vector bundle over S.

The last item elicits the following question.

Question 6.23 *Given a vector bundle $\mathbb{R}^n \to E \to M$, where M is a closed manifold with non-negative sectional curvature, does E admit a metric with non-negative sectional curvature?*

Recently, algebraic models have been used to give obstructions to the existence of such a metric on the total space of a vector bundle (see [22] and [23]). This will be the focus of this section, but in order to understand the whole situation better, we will first give the original negative examples for Question 6.23 obtained in [219]. Since these examples don't have *souls*, we refer to them as *vampiric vector bundles*.

6.3.1 The examples of Özaydin and Walschap

The basis of the Özaydin–Walschap approach to constructing the examples mentioned above is the following result.

Theorem 6.24 *Let $\xi\colon \mathbb{R}^k \to E \to M^n$ be a vector bundle over a compact, flat manifold M. The following are equivalent:*

1. *E admits a complete metric with $\sec_E \geq 0$.*
2. *E admits a complete metric with $\sec_E = 0$.*
3. *E is diffeomorphic to the total space of a rank k vector bundle over M that admits a flat Riemannian connection.*

4. $E \cong_{\text{diffeo}} \mathbb{R}^n \times_{\pi_1 M} \mathbb{R}^k$, where $\pi_1(M)$ acts on $\mathbb{R}^n$ by covering transformations (as in Example 6.4) and on $\mathbb{R}^k$ by some orthogonal representation.

We can generalize the notion of flat manifold to bundles as follows.

Definition 6.25 *A principal G-bundle $\xi: G \to P \to X$ is flat if the classifying map $X \to BG$ factors up to homotopy as*

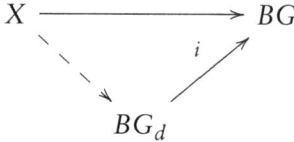

where G_d is the Lie group G taken with the discrete topology and $i: G_d \to G$ is the canonical continuous identification of G_d and G as sets.

See Remark 6.27 for an explanation of the use of the term "flat" and its relation with flat manifolds. The following result gives an equivalent criterion for flatness that leads to better interpretations.

Proposition 6.26 *The principal G-bundle $\xi: G \to P \to X$ is flat if and only if there is a factorization up to homotopy*

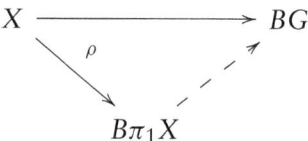

where $B\pi_1 X \to BG$ is induced by a homomorphism $\pi_1 X \to G$ and $\rho: X \to B\pi_1 X = K(\pi_1 X, 1)$ is the "canonical" inclusion obtained by attaching cells of dimension 3 and greater to kill all homotopy groups above degree 1.

Proof Suppose the homomorphism $\pi_1 X \to G$ exists which provides the factorization above. Since $\pi_1 X$ is a discrete group, there is then a factorization

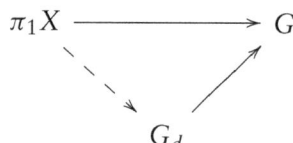

which induces the desired homotopy factorization

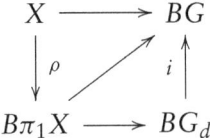

Now assume that $X \to BG$ factors through BG_d. The induced map of fundamental groups is then $\pi_1 X \to G_d$ and this, in turn, induces $B\pi_1 X \to BG_d$. Now, the homotopy class of a map $X \to K(G_d, 1) = BG_d$ is determined by the induced homomorphism on fundamental groups. Thus we obtain a homotopy commutative diagram

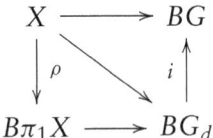

$\square$

Remark 6.27 The principal bundle induced from the canonical map $X \to B\pi_1 X$ is simply the universal covering fibration $\pi_1 X \to \widetilde{X} \to X$. The effect of Proposition 6.26 is to reduce the group of the principal bundle ξ to the fundamental group $\pi_1 X$; hence, $P \cong \widetilde{X} \times_{\pi_1 X} G$, where $\pi_1 X$ acts on $\widetilde{X}$ by covering transformations and on G via the homomorphism of the proposition and multiplication in G.

A vector bundle $\mathbb{R}^k \to E \to X$ is *flat* if the associated principal $O(k)$-bundle $O(k) \to P \to X$ is flat and the consequent homomorphism $\rho \colon \pi_1 X \to O(k)$ gives

$$E = P \times_{O(k)} \mathbb{R}^k = \widetilde{X} \times_{\pi_1 X} \mathbb{R}^k.$$

Of course, the bundle is called flat because it has a flat Riemannian connection (see [194] and [198, Appendix C]). A Riemannian connection ∇ for a vector bundle is the analogue of the covariant derivative on vector fields. The analogue of a vector field is a section of the bundle and the connection "differentiates" the section subject to properties analogous to those in Proposition 6.1. The connection ∇ is *flat* if, in the neighborhood of each point of the base X, there is a basis of sections $s_1, \ldots, s_k$ with $\nabla(s_i) = 0$ for $i = 1, \ldots, k$. This is actually a convenient equivalent condition to the vanishing of the bundle's curvature (see [28, Section 5.3, Theorem 5]). But now consider the change of basis $t_i = \sum_j a_{ji} s_j$ when moving to a new chart with local basis $t_1, \ldots, t_k$. The properties of ∇ (including the hypothesized flatness) give

$$0 = \nabla(t_i) = \sum_j (da_{ji} s_j + a_{ji} \nabla(s_j)) = \sum_j da_{ji} s_j.$$

Hence, $da_{ji} = 0$ for all the a_{ji}. These transition functions are therefore locally constant and factor through the structure group *made discrete*. Since the transition functions are the same for the associated principal bundle, we see that this notion of flatness corresponds to the principal bundle notion. If the flat vector bundle is the tangent bundle $\mathbb{R}^n \to TX \to X^n$, then the flat Riemannian connection is precisely the connection on X associated to the metric. Hence, X is a flat manifold. This is the "connection" between flat manifolds and flat vector bundles. Now, in the situation of (3) and (4) of Theorem 6.24, since X is a flat manifold, the Bieberbach theorem tells us that $\widetilde{X} = \mathbb{R}^n$. Hence, the total space E of the flat vector bundle is

$$E = P \times_{O(k)} \mathbb{R}^k = \widetilde{X} \times_{\pi_1 X} \mathbb{R}^k = \mathbb{R}^n \times_{\pi_1 X} \mathbb{R}^k.$$

Proof of Theorem 6.24 The discussion above shows the equivalence of (3) and (4). Also, (2) clearly implies (1) while (4) implies (2) because $\pi_1 M$ acts on $\mathbb{R}^n \times \mathbb{R}^k = \mathbb{R}^{n+k}$ by isometries (and so the zero curvature of $\mathbb{R}^{n+k}$ is preserved under the quotient operation $E = (\mathbb{R}^n \times \mathbb{R}^k)/\pi_1 M$). Now assume (1) and let S be a soul of E. Then $\sec_S \geq 0$ and $S \simeq E \simeq M = K(\pi, 1)$; therefore, S is flat by Remark 6.22 (3). Now, (see Example 6.4) flat manifolds are classified up to diffeomorphism by their fundamental groups, so S is, in fact, diffeomorphic to M. Now let $p: \widetilde{E} \to E$ denote the universal (Riemannian) covering of E and let $\widetilde{S} = p^{-1}(S) \subset \widetilde{E}$ be the restriction to the soul. The pullback diagram

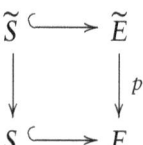

shows that $\widetilde{S} \to S$ is a fibration. Moreover, because $S \hookrightarrow E$ is a homotopy equivalence, it induces an isomorphism of fundamental groups and this means that the map $\widetilde{S} \to S$ is the universal (Riemannian) covering. Since $S = K(\pi, 1)$, $\widetilde{S}$ is contractible; furthermore, by the Cheeger–Gromoll splitting theorem (Remark 6.21), we see that $\widetilde{S} \cong_{\text{isom}} \mathbb{R}^n$. Now observe that $\widetilde{S}$ is totally convex in $\widetilde{E}$. This follows because the geodesic in $\widetilde{E}$ joining two points in $\widetilde{S}$ pushes down via p to a geodesic in E joining two points in S. Then, since S is totally geodesic in E, the whole geodesic must lie in S. Thus, the original geodesic lies in $p^{-1}(S) = \widetilde{S}$. This has the effect that a line (see Remark 6.21) in $\widetilde{S}$ is a line in $\widetilde{E}$. Hence, the Cheeger–Gromoll splitting of $\widetilde{E}$ has the form

$$\widetilde{E} \cong_{\text{isom}} E_0 \times \mathbb{R}^n = E_0 \times \widetilde{S}.$$

Then we see that E splits locally isometrically as a product over S. Therefore, the Riemannian connection on the normal bundle of the inclusion $S \hookrightarrow E$ (induced from the Riemannian connection on E) is flat. Since E is diffeomorphic to the total space of this normal bundle (by the Soul theorem), we have proven (3). □

With these preliminaries out of the way, we can now state the main result of [219].

Theorem 6.28 *Let $\xi \colon \mathbb{R}^2 \to E \to M^n$ be an oriented vector bundle, where M^n is a compact flat manifold. Then E admits a complete metric of non-negative curvature if and only if the rational Euler class $e(\xi)_\mathbb{Q}$ is zero.*

In order to prove the theorem, we need a lemma.

Lemma 6.29 *Let $\xi \colon \mathbb{R}^k \to E \to M$ be an oriented vector bundle over a compact oriented manifold M. If $E \cong_{\text{homeo}} F \times \mathbb{R}$, then $e_\mathbb{Z}(\xi) = 0$, where $e_\mathbb{Z}(\xi)$ is the integral Euler class of ξ.*

Proof Let $h \colon E \xrightarrow{\cong} F \times \mathbb{R}$ be the given homeomorphism. Consider the composition

$$M \xrightarrow{s} E \xrightarrow{h} F \times \mathbb{R} \xrightarrow{\text{pr}} F \xrightarrow{i_t} F \times \{t\} \hookrightarrow F \times \mathbb{R} \xrightarrow{h^{-1}} E,$$

which we denote by ϕ. Here, s is the zero section and t is larger than any t' such that $(f, t') \in \text{Im}(h \circ s)$. This can be done since M is compact. Now, the composition $F \times \mathbb{R} \xrightarrow{\text{pr}} F \xrightarrow{i_t} F \times \mathbb{R}$ is homotopic to $\text{id}_{F \times \mathbb{R}}$ by $H(f, r, u) = (f, r(1-u) + ut)$. Thus, $\phi \simeq h^{-1} \circ \text{id}_{F \times \mathbb{R}} \circ h \circ s \simeq s$.

Now, the (integral) Euler class is defined by $s^*(\overline{\Phi}) = e(\xi)$, where $\Phi \in H^k(E, E/s(M))$ is the Thom class and $\overline{\Phi}$ is its restriction in $H^k(E)$. But, since $\phi \simeq s$, we have $\phi^*(\overline{\Phi}) = e(\xi)$. However, $\phi(M) \cap s(M) = \emptyset$ because of the map i_t. Hence, ϕ factors as

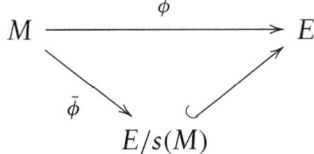

6: Curvature

We then have the commutative diagram

$$H^k(E, E/s(M)) \longrightarrow H^k(E) \longrightarrow H^k(E/s(M))$$

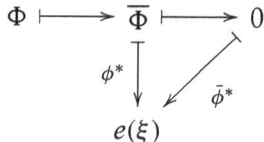

with the Euler class computed two ways: once with the definition and secondly using the factorization of ϕ given above.

$$\Phi \longmapsto \overline{\Phi} \longmapsto 0$$

$$\phi^* \downarrow \quad \swarrow \bar{\phi}^*$$

$$e(\xi)$$

Hence $e(\xi) = 0$ since the composition $H^k(E, E/s(M)) \to H^k(E) \to H^k(E/s(M))$ is zero by exactness. □

Proof of Theorem 6.28 First, let $e(\xi)_\mathbb{Q} = 0$. By Corollary 6.48, the associated principal S^1-bundle is flat. Therefore, the vector bundle is flat and from the equivalence of (3) and (1) in Theorem 6.24, we obtain a non-negatively curved metric on E.

Now suppose $\sec_E \geq 0$. Because M^n is flat, there exists a finite cover $f: T^n \to M^n$. Take the pullback bundle

$$\begin{array}{ccc} E(f^*\xi) & \longrightarrow & E \\ \downarrow & & \downarrow \\ T^n & \xrightarrow{f} & M^n \end{array}$$

and a pullback metric on $E(f^*\xi)$ with $\sec \geq 0$. Now, a finite cover induces an injection in rational cohomology, $f^*: H^*(M; \mathbb{Q}) \to H^*(T^n; \mathbb{Q})$. (Therefore, if $e(\xi)_\mathbb{Q}$ were nonzero, then $e(f^*(\xi))_\mathbb{Q}$ would also be non-zero.) By the equivalence of (3) and (1) in Theorem 6.24, $E(f^*\xi)$ is diffeomorphic to the total space of a flat $\mathbb{R}^2$-bundle η over T^n. But then we know that $e(\eta)_\mathbb{Q} = 0$ and, since $H^2(T^n; \mathbb{Z})$ is torsion-free, in fact $e(\eta)_\mathbb{Z} = 0$ as well. Of course, the associated principal S^1-bundle is classified by $e(\eta)_\mathbb{Z}$, so η is trivial: $E(\eta) = T^n \times \mathbb{R}^2$. But then we have

$$E(f^*\xi) \cong_{\text{diffeo}} E(\eta) = T^n \times \mathbb{R}^2 \cong (T^n \times \mathbb{R}) \times \mathbb{R}.$$

By Lemma 6.29, we have $e(f^*\xi) = 0$ and, therefore, $e(\xi)_\mathbb{Q} = 0$. □

6.3 Vampiric vector bundles

Example 6.30 We can now apply Theorem 6.28 to give explicit examples of vector bundles over compact non-negatively curved manifolds whose total spaces have no metrics of non-negative curvature. Oriented $\mathbb{R}^2$-bundles over T^2 are classified by homotopy classes of maps,

$$[T^2, BSO(2)] = [T^2, BS^1] = [T^2, K(\mathbb{Z}, 2)] = H^2(T^2; \mathbb{Z}) = \mathbb{Z}.$$

Explicitly, each $k \in \mathbb{Z}$ corresponds to an Euler class $e(\xi_k)$. Thus, since the rational Euler class must vanish in order for the total space of the bundle to have non-negative curvature, we have the following.

Theorem 6.31 *Only the total space of the trivial $\mathbb{R}^2$-bundle over T^2 possesses a metric with non-negative curvature.*[2]

6.3.2 The method of Belegradek and Kapovitch

Now that we have seen the particular examples of Özaydin and Walschap, we can explore more general (necessary) conditions for the existence of total spaces of bundles with metrics of non-negative curvature. In the following, we shall mainly follow [23], but also see [22] and [220]. We shall see that a main tool in the understanding of this problem is the most fundamental algebraic model of all, rational cohomology.

First, the following result of [22] generalizes Theorem 6.20 and Remark 6.21 and, since we are attempting to imitate the normal bundle situation, focusses our attention on bundles over products $C \times T^k$.

Theorem 6.32 *Let M be a complete manifold with $\sec_M \geq 0$. There exists a finite cover $\overline{M} \to M$ such that $\overline{M} \cong_{\text{diffeo}} N \times T$ where N is a complete simply connected open manifold with $\sec_N \geq 0$ and T is a torus. Furthermore, if $\overline{S}$ is a soul of $\overline{M}$, then the diffeomorphism carries $\overline{S}$ to $C \times T$ where C is a soul of N.*

Remark 6.33 Note that, since C is a soul of N (i.e. N is diffeomorphic to the normal bundle of the inclusion $C \hookrightarrow N$), the theorem is saying that $\overline{M}$ is, in fact, a bundle η over $C \times T$ of the form $\eta = \xi_C \times T$, where ξ_C is a bundle over C.

With this theorem in mind, let's make the following.

Definition 6.34 *Suppose C is a closed simply connected manifold, T is a torus and $k \in \mathbb{Z}$. The triple (C, T, k) is said to be splitting rigid if, given a rank k vector bundle $\xi : E(\xi) \to C \times T$ with $\sec_{E(\xi)} \geq 0$, then there exists*

---

[2] So the other bundle total spaces do not have non-negative curvature and, hence, have no souls. That is, they are vampires!

a finite covering map $p \colon T \to T$ such that $(\mathrm{id}_C \times p)^*(\xi) \cong \xi_C \times T$, where ξ_C is a rank k bundle over C.

This definition can be used to obtain an obstruction (i.e. a necessary condition) for the existence of a metric of non-negative sectional curvature on the total space of a bundle. Namely, since most bundles over $C \times T$ do not split after passing to a finite cover, the following result says that any such non-splitting bundle over the product of a biquotient and a torus has a total space which does not have sec ≥ 0.

Theorem 6.35 ([23]) *Let $G//H$ be a simply connected biquotient, where G is compact connected and H is semisimple and connected. Then, for any torus T and $k \leq 4$, $(G//H, T, k)$ is splitting rigid.*

Let's begin to understand this result (with the proof following later). Denote by $\mathrm{Char}(C, k)$ the following *characteristic algebra* inside $H^*(C; \mathbb{Q})$.

$$\mathrm{Char}(C, k) = \begin{cases} \bigoplus_{i=1}^{[m/2]} H^{4i}(C; \mathbb{Q}) & \text{if } k = 2m+1 \\ \bigoplus_{i=1}^{[m/2]} H^{4i}(C; \mathbb{Q}) \oplus H^{2m}(C; \mathbb{Q}) & \text{if } k = 2m. \end{cases}$$

Here, $[-]$ denotes the greatest integer function. The important point is that, for a rank k bundle over C, $\mathrm{Char}(C, k)$ contains the Pontryagin classes when k is odd and the Pontryagin classes and Euler class when k is even. (We refer to these classes as the *rational characteristic classes* of a bundle.) This is important because of the following result.

Theorem 6.36 ([22]) *Let ξ and η be vector bundles over $C \times T$ of the same rank. If ξ and η have the same rational characteristic classes and $\xi|_C \cong \eta|_C$, then there exists a finite covering $f \colon C \times T \to C \times T$ such that $f^*\xi \cong f^*\eta$.*

Now, given a bundle ξ over $C \times T$ whose rational characteristic classes lie in $\mathrm{Char}(C, k)$, clearly the rational characteristic classes of ξ and $\xi|_C \times T$ are the same and, by definition $\xi|_C = \xi|_C$. Therefore, Theorem 6.36 implies that there is a finite covering $f \colon C \times T \to C \times T$ with $f^*\xi \cong f^*(\xi|_C \times T)$. But since C is simply connected, the finite covering has the form $\mathrm{id}_C \times \tilde{f}$; thus, $f^*(\xi|_C \times T) = \xi|_C \times T$. This means that we have the following result.

Lemma 6.37 *A bundle ξ over $C \times T$ (as above) splits if and only if the rational characteristic classes lie in $\mathrm{Char}(C, k)$.*

We will now use this result to obtain a criterion for splitting rigidity.

Proposition 6.38 ([23]) *If every self-homotopy equivalence of $C \times T$ preserves the characteristic algebra $\mathrm{Char}(C,k)$ in $H^*(C \times T;\mathbb{Q}) \cong H^*(C;\mathbb{Q}) \otimes H^*(T;\mathbb{Q})$, then (C,T,k) is splitting rigid.*

Proof Suppose ξ is a vector bundle over $C \times T$ with $\sec_{E(\xi)} \geq 0$. By Remark 6.33, there is a finite cover $E' \to E(\xi)$ with $E' \cong N \times T'$, where N is simply connected and $\sec_N \geq 0$. Since N is simply connected, we see in fact that $T' = T$. Also, for S' the soul of E', we have $S' \cong C_N \times T$ where C_N is a soul of N, so $N \cong E(\zeta)$, where ζ is the normal bundle of $C_N \hookrightarrow N$. Therefore, without loss of generality, we take $E(\xi) \cong E(\eta) = E(\zeta) \times T$, where ζ is a vector bundle over a simply connected manifold C'. Thus, ξ and η are two different vector bundle structures on the same manifold $X = E(\xi) = E(\eta)$. The respective zero sections $C \times T \to E(\xi) = X$, $C' \times T \to E(\eta) = X$ identify $C \times T$, $C' \times T$ as submanifolds of X *and we consider the bundles ξ and η as the normal bundles of these respective inclusions*. Note that the composition of the first zero section map with the projection of the bundle η gives a homotopy equivalence $g \colon C \times T \to C' \times T$ which can be taken to have degree one by orienting X properly. Furthermore, the obvious maps $g_C \colon C \to C'$, $g_T \colon T \to T$ are homotopy equivalences, so we may form the self-homotopy equivalence $h = (g_C^{-1} \times g_T^{-1}) \circ g \colon C \times T \to C \times T$. The inclusions and projections give maps

$$h_C \colon C \hookrightarrow C \times T \overset{h}{\to} C \times T \to C$$

$$h_T \colon T \hookrightarrow C \times T \overset{h}{\to} C \times T \to T.$$

These are, in fact, homotopic to the respective identity maps. For instance, if the inclusions and projections for C, C' are denoted by i_C, p_C, $i_{C'}$, $p_{C'}$ respectively, then we have

$$h_C = p_C h i_C = p_C (g_C^{-1} \times g_T^{-1}) g i_C \simeq g_C^{-1} p_{C'} g i_C \simeq g_C^{-1} g_c \simeq \mathrm{id}_C$$

since $p_C \circ (g_C^{-1} \times g_T^{-1}) \simeq g_C^{-1} \circ p_{C'}$ and $p_{C'} \circ g \circ i_C = g_C$. By assumption h^* maps $\mathrm{Char}(C,k)$ to itself, so it is plain that g^* maps $\mathrm{Char}(C',k)$ to $\mathrm{Char}(C,k)$.

Since ξ and η are normal bundles to the respective inclusions of the zero sections, we know by Poincaré duality that the Euler classes obey $e(\xi) = [C \times T]$, $e(\eta) = [C' \times T]$. Therefore, since $\deg(g) = 1$, for any class $y \in H_*(C \times T)$, we obtain

$$\langle g^*(e(\eta)), y \rangle = \langle e(\eta), g_*(y) \rangle = \langle [C' \times T], g_*(y) \rangle$$
$$= \langle g^*([C' \times T]), y \rangle = \langle [C \times T], y \rangle = \langle e(\xi), y \rangle.$$

So we see that $g^*(e(\eta)) = e(\xi)$. Now, $e(\eta) \in \mathrm{Char}(C',k)$ since $\eta \cong \zeta \times T$ and g^* maps $\mathrm{Char}(C',k)$ to $\mathrm{Char}(C,k)$, so $e(\xi) \in \mathrm{Char}(C,k)$.

6: Curvature

Now let's consider Pontryagin classes. Since $g\colon C \times T \to C' \times T$ is defined by

$$C \times T \xrightarrow{\text{zero}} X \xrightarrow{\pi_\eta} C' \times T$$

and π_η is a homotopy inverse to the η-zero section $C' \times T \to X$, we see that g, thought of as a map $g\colon C \times T \to X$ is homotopic to the zero section inclusion $C \times T \to X$. Then we obtain $g^*(TX|_{C' \times T}) = TX|_{C \times T}$. Now, because we view ξ as the normal bundle to the inclusion $C \times T \to X$, and we may apply the Whitney sum formula for the total Pontryagin class p to the bundle $TX|_{C \times T} = T(C \times T) \oplus \xi$ to obtain

$$\begin{aligned}
p(T(C \times T))p(\xi) &= p(T(C \times T) \oplus \xi) \\
&= p(TX|_{C \times T}) \\
&= p(g^*(TX|_{C' \times T})) \\
&= g^*(p(TX|_{C' \times T})) \\
&= g^*(p(T(C' \times T) \oplus \eta)) \\
&= g^*(p(T(C' \times T))g^*(p(\eta)).
\end{aligned}$$

Now, T is parallelizable, so $p(T(C \times T)) = p(TC) \in H^*(C)$ and $p(T(C' \times T)) = p(TC') \in H^*(C')$. Also, $p(T(C \times T))$ is a unit in $H^*(C)$, so we can write

$$p(\xi) = p(T(C \times T))^{-1}g^*(p(T(C' \times T))g^*(p(\eta)).$$

Now $p(\eta) \in H^*(C')$ since $\eta \cong \zeta \times T$ and $g^*(\text{Char}(C',k)) \subseteq \text{Char}(C,k)$, so each factor of $p(\xi)$ lies in $H^*(C)$. Hence, $p(\xi) \in \text{Char}(C,k)$ as well.

Because the rational characteristic classes are in $\text{Char}(C,k)$, by the discussion following Theorem 6.36, we see that (C,T,k) is splitting rigid. $\square$

Remark 6.39 Note that the proof shows that, without loss of generality, we can always assume that the restriction to C of the self-homotopy equivalence is homotopic to id_C.

The link between self-homotopy equivalences of $C \times T$ and algebraic operations on cohomology is provided by the the structure of $H^*(T)$ (where we will understand coefficients to be $\mathbb{Q}$ from this point forward). The torus cohomology $H^*(T)$ is an exterior algebra $\wedge(e_1,\ldots,e_s)$ with vector space basis $t_0 = 1, t_1 = e_1,\ldots,t_s = e_s, t_{s+1} = e_1 e_2$ etc. (in some chosen, perhaps lexicographic, ordering). Now, $H^*(C \times T) \cong H^*(C) \otimes H^*(T)$, so a homotopy equivalence (with $h|_C = \simeq \text{id}_C$ by Remark 6.39) $h\colon C \times T \to C \times T$

induces an automorphism $h^*: H^*(C \times T) \to H^*(C \times T)$ such that, for $c \in H^*(C) \subset H^*(C \times T)$,

$$h^*(c) = c + \sum_i \delta_i(c) t_i$$

where we have not written the $\otimes$ in the free $H^*(C)$-module $H^*(C \times T)$. It can be shown that δ_1 is a derivation and, further, the fact that $t_1^2 = 0$ shows that the mapping $\phi_1 : H^*(C \times T) \to H^*(C \times T)$ defined by $\phi_1(ct) = ct - \delta_1(c)t_1 t$ is a graded algebra isomorphism (with inverse $\phi_1^{-1}(ct) = ct + \delta_1(c)t_1 t$). Then

$$\phi_1(h^*(c)) = \phi_1(c + \sum_{i \geq 1} \delta_i(c) t_i)$$

$$= c - \delta_1(c) t_1 + \delta_1(c) t_1 + \sum_{i \geq 2} \phi_1^i(c) t_i$$

$$= c + \sum_{i \geq 2} \phi_1^i(c) t_i.$$

So we see that $\phi_1(h^*(c))$ resembles the identity with error terms involving only the t_i with $i \geq 2$. The idea is now to continue this process until the error terms are zero. We then shall have written h^* in a very precise way that will permit a connection with derivations of $H^*(C)$.

Now consider the formula for $\phi_1(h^*(c))$. As usual, the first perturbation term ϕ_1^2 is a derivation, so $\phi_2(ct) = ct - \phi_1^2(c) t_2 t$ is an automorphism of $H^*(C \times T)$ with

$$\phi_2(\phi_1(h^*(c))) = c + \sum_{i \geq 3} \phi_2^i(c) t_i.$$

We continue in this fashion to obtain automorphisms ϕ_k for $k = 1, \ldots, N = \binom{s}{k}$ with $\phi_k(c) = c - \phi_{k-1}^k(c) t_k$ and such that

$$(\phi_N \circ \ldots \circ \phi_1 \circ h^*)(c) = c.$$

Then the isomorphism h^* evaluated on $c \in H^*(C)$ may be written as

$$h^*(c) = (\phi_N^{-1} \circ \ldots \circ \phi_1^{-1})(c)$$

where $\phi_k^{-1}(c) = c + \phi_{k-1}^k(c) t_k$.

Theorem 6.40 *If every negative degree derivation of $H^*(C)$ vanishes on the characteristic algebra $\mathrm{Char}(C, k)$, then (C, T, k) is splitting rigid for each T.*

Proof By Proposition 6.38, we must show that every self-homotopy equivalence of $C \times T$ preserves $\mathrm{Char}(C, k)$. As noted in Remark 6.39, we may assume the restriction of the equivalence to C is id_C. With this in mind, the discussion above pertains and the induced isomorphism on cohomology has the form

$$h^*(c) = (\phi_N^{-1} \circ \ldots \circ \phi_1^{-1})(c)$$

where $\phi_k^{-1}(c) = c + \phi_{k-1}^k(c)t_k$. But each $\phi_{k-1}^k(c) = 0$ by the hypothesis of the theorem since $c \in \text{Char}(C, k)$. Thus, $h^*(c) = c$, so h^* preserves the characteristic algebra $\text{Char}(C, k)$. $\square$

Corollary 6.41 (also see [220]) *If $H^*(C)$ has no nonzero negative degree derivations, then (C, T, k) is splitting rigid for each T and k.*

In particular, by Theorem 4.36 and [237], we know that simply connected Kähler manifolds and maximal rank homogeneous spaces have cohomology algebras admitting no negative degree derivations. Therefore, such manifolds are prime candidates for C's. As we have seen, splitting rigidity is an obstruction to the total space of a bundle over $C \times T$ having non-negative sectional curvature; so now we see further that the question of the existence of metrics with non-negative curvature is intimately involved with the algebraic structure of $H^*(C)$. Finally, let's use Theorem 6.40 to prove Theorem 6.35.

Proof of Theorem 6.35 The group H is semisimple, so the fibration $G \to G//H \to BH$ shows that G is also semisimple and, since $\pi_2(G) = 0$ and $\pi_2(BH) = \pi_1(H)$ is finite, $\pi_2(G//H)$ is finite. Therefore, over the rationals, $G//H$ is 2-connected. Therefore, the minimal model $\mathcal{M}_{G//H} = (\wedge V, d)$ of $G//H$ has $V^1 = V^2 = 0$. In Corollary 3.51, we saw that $G//H$ has a pure minimal model, so the generators in V^3 appear in no differential. Hence, by Theorem 2.77, $\mathcal{M}_{G//H} = (\wedge V^3, d = 0) \otimes (\wedge \widehat{V}, d)$, where $\widehat{V}^3 = 0$.

Now, because $H^2(G//H) = 0$, by Theorem 6.40 we need only consider rank 3 and rank 4 bundles over $G//H \times T$. Now, by definition, $\text{Char}(G//H, 3) = \text{Char}(G//H, 4) = H^4(G//H) = H^4(\wedge \widehat{V}, d)$, so we only have to consider derivations of degree -1 and degree -4.

Suppose θ is a derivation of degree -4 and let $c \in H^4(G//H)$. Since $G//H$ is finite dimensional, there is some $m \in \mathbb{Z}^+$, such that $c^m \neq 0$ and $c^{m+1} = 0$. Then

$$0 = \theta(c^{m+1}) = (m+1)\theta(c)c^m$$

and, since $\theta(c) \in \mathbb{Q}$ and $c^m \neq 0$, it must be true that $\theta(c) = 0$. Thus θ vanishes on $H^4(G//H)$.

Now suppose that θ is a derivation of degree -1 and let $c \in H^4(G//H)$. Then $\theta(c) = x \in V^3$. Again find m so that $c^m \neq 0$ and $c^{m+1} = 0$. Then

$$0 = \theta(c^{m+1}) = (m+1)\theta(c)c^m = (m+1)xc^m.$$

Note that, if $\dim(G//H) = n$ and $\dim(V^3) = k$, then $H^j(\wedge \widehat{V}, d) = 0$ for $j > n - 3k$. Since $c^m \neq 0$, we have $4m \leq n - 3k$. Also, $x \in V^3$, so $4m+3 \leq n-3k+3 = n-3(k-1)$ and, since $k \geq 1$, $4m+3 < n$. Therefore xc^m is below the top dimension in cohomology. Because $H^*(\wedge V, d) \cong H^*(\wedge V^3, d = 0) \otimes H^*(\wedge \widehat{V}, d)$, for $x \in H^*(\wedge V^3, d = 0)$ and $c^m \in H^*((\wedge \widehat{V}, d))$, we can only have $xc^m = 0$ if $x = 0$ or $c^m = 0$. Since $c^m \neq 0$, we must have $\theta(c) = x = 0$. Thus, θ vanishes on $H^4(G//H)$.

Therefore, we see that all negative degree derivations vanish on $\mathrm{Char}(G//H, k)$ for $k = 3, 4$ and by Theorem 6.40, this means that, for $k = 3, 4$, $(G//H, T, k)$ is splitting rigid for any T. □

Finally, we mention a result of [23] showing that not all spaces give splitting rigid triples. The proof of the result relies essentially on the properties of the minimal model of $SU(6)/(SU(3) \times SU(3))$ (which is an example of a nonformal homogeneous space).

Theorem 6.42 *Let $C = SU(6)/(SU(3) \times SU(3))$ and take $\dim T \geq 2$. Then there exists a rank 6 vector bundle which does not split, but whose total space has non-negative sectional curvature.*

6.4 Final thoughts

Manifolds with $\sec \geq 0$ enjoy many special properties and are conjectured to have many more. For instance, Synge's famous theorem says that a closed manifold M with $\sec_M > 0$ is orientable if $\dim(M)$ is odd and is simply connected if $\dim(M)$ is even and M is orientable. Of course $\mathbb{R}P(2n+1)$ is a prime example of the first case, while $\mathbb{R}P(2n)$ shows that the orientability hypothesis is necessary for the second case. Many years ago, Hopf conjectured that a closed manifold M with $\dim(M)$ even and $\sec_M \geq 0$ must have $\chi(M) \geq 0$. More recently, Gromov asked whether a closed manifold M with $\sec_M \geq 0$ must have $\sum_i b_i(M) \leq 2^{\dim(M)}$, where $b_i(M)$ is the ith Betti number of M. Recall from Theorem 2.75 that these assertions are indeed true for rationally elliptic manifolds (i.e. those with $\dim(\pi_*(M) \otimes \mathbb{Q}) < \infty$). This then presents a tantalizing possibility which is enunciated in a far-reaching conjecture generally attributed to Raoul Bott.

Conjecture 6.43 *If M is a closed manifold with $\sec_M \geq 0$, then M is rationally elliptic.*

The idea that a geometric constraint imposes conditions on the underlying algebraic topology of a manifold is not a new one, of course. We have already mentioned Hadamard's theorem that a manifold with non-positive sectional curvature must be a $K(\pi, 1)$ and Bieberbach's theorem characterizing flat manifolds. There is also Bochner's famous result that a closed manifold of non-negative Ricci curvature must have first Betti number bounded above by the manifold's dimension. The same hypothesis yields a similar restriction on the growth of the fundamental group (see [196]). In a different direction, we have seen in Section 5.7 that qualities associated with geometric entities such as the geodesic flow can radically limit possibilities for homotopy type as well.

In this chapter, we have seen how algebraic models can be *associated* to geometric objects and how they can be used to *construct* geometric entities. Algebraic models are important ingredients in the solutions of geometric problems, so they have proven to be part of the tradition of obtaining geometric information homotopically. Henri Poincaré had a criterion for significance that he claimed for the mathematics that leads to a mathematical law. The beautiful symbiosis between homotopical algebra and geometry exemplifies this criterion and in Poincaré's words, "Ce sont ceux qui nous révèlent des parentés insoupçonnées entre d'autres faits, connus depuis longtemps, mais qu'on croyait à tort étrangers les uns aux autres." ["They reveal the kinship between other facts, long known, but wrongly believed to be strangers to one another."]

6.5 Appendix

In this appendix, we will show that the rational Euler class for a principal circle bundle vanishes exactly when the bundle is flat (see [217]). Let G_d denote the Lie group G with the discrete topology and similarly for the universal covering group $\widetilde{G}$ and $\widetilde{G}_d$.

Proposition 6.44 *Let G be a connected Lie group and let $\pi: \widetilde{G} \to G$ denote the universal covering. Then $\widetilde{G}_d$ is a pullback of π along the map $i: G_d \to G$.*

Proof Note first that making $\widetilde{G}$ discrete unravels all of the topology linking the fibers of the universal covering of G. Now, for discrete spaces, only cardinality matters, so we have $\widetilde{G}_d = G_d \times \pi_1(G)$.

Let **P** denote the pullback of π along the map $i: G_d \to G$. Now, $\pi: \widetilde{G} \to G$ is a principal $\pi_1(G)$-bundle, so its pullback along i is one also. Because $\pi: \widetilde{G} \to G$ is surjective, there is a set map $s: G \to \widetilde{G}$ with $\pi \circ s = \mathrm{id}_G$. But G_d has the discrete topology, so s considered as a map

$s\colon G_d \to \tilde{G}$ is continuous. A principal bundle with section is trivial, so we obtain $\mathbf{P} = G_d \times \pi_1(G) = \tilde{G}_d$. $\square$

Now let $G = S^1$ with universal cover $\mathbb{R}$ and note that $BS^1 = K(\mathbb{Z}, 2)$, $BS_d^1 = K(S_d^1, 1)$ and $B\mathbb{R}_d = K(\mathbb{R}_d, 1)$. Consider the following pullback diagram.

$$\begin{array}{ccc} \mathbb{R}_d & \xrightarrow{\hat{i}} & \mathbb{R} \\ \downarrow & & \downarrow \pi \\ S_d^1 & \xrightarrow{i} & S^1 \end{array}$$

Of course the homomorphisms of groups shown are the usual ones making the sequence $\mathbb{Z} \to \mathbb{R} \to S^1$ exact. Note that the pullback property guarantees that the fiber of $\mathbb{R}_d \to S_d^1$ is $\mathbb{Z}$ included in $\mathbb{R}_d$ in the standard way and that the restriction of $\hat{i}$ to fibers is the identity. *But now we notice an amazing fact.* Because $\mathbb{R} \to S^1$ is a fibration with $\mathbb{R}$ contractible, $\mathbb{R}_d$ is the homotopy fiber of $S_d^1 \to S^1$. Therefore, $i\colon S_d^1 \to S^1$ classifies $\mathbb{R}_d \to S_d^1$ and $BS_d^1 \to BS^1$ classifies $B\mathbb{R}_d \to BS_d^1$ in the sequence of classifying spaces

$$\begin{array}{ccccccc} B\mathbb{Z} & \longrightarrow & B\mathbb{R}_d & \longrightarrow & BS_d^1 & \longrightarrow & BS^1 \\ \downarrow \simeq & & \downarrow \simeq & & \downarrow \simeq & & \downarrow \simeq \\ K(\mathbb{Z},1) & \longrightarrow & K(\mathbb{R}_d,1) & \longrightarrow & K(S_d^1,1) & \longrightarrow & K(\mathbb{Z},2) \end{array} \quad (6.1)$$

Remark 6.45 Note that the first two maps are determined by the corresponding homomorphisms of groups. The last map also comes from the sequence of discrete groups; it is the class in $\mathrm{Ext}(S_d^1, \mathbb{Z}) \subseteq H^2(S_d^1; \mathbb{Z})$ classifying the central extension $\mathbb{Z} \to \mathbb{R}_d \to S_d^1$ (see [137]).

Proposition 6.46 *There exists a map* $\phi\colon K(\mathbb{Z}, 2) \to K(\mathbb{R}_d, 2)$ *so that the sequence of fibrations (6.1) may be extended to a sequence of fibrations*

$$K(\mathbb{Z},1) \to K(\mathbb{R}_d,1) \to K(S_d^1,1) \to K(\mathbb{Z},2) \xrightarrow{\phi} K(\mathbb{R}_d,2).$$

Proof This is, in fact, a standard result in topology (see [139, Theorem 7.1]).

Let ϕ correspond to the usual inclusion of the integers into the reals, $\mathbb{Z} \to \mathbb{R}_d$. Of course, this is the induced homomorphism at the beginning of the sequence $K(\mathbb{Z}, 1) \to K(\mathbb{R}_d, 1)$ as well. To see that we get a fibration on the right end, let F denote the homotopy fiber of ϕ. We must show that $F \to K(\mathbb{Z},2)$ is precisely the map $K(S^1_d,1) \to K(\mathbb{Z},2)$ in the original fibration sequence corresponding to the group extension $\mathbb{Z} \to \mathbb{R}_d \to S^1_d$ as discussed in Remark 6.45.

The long exact sequence in homotopy associated to the fibration shows immediately that $F \simeq K(S^1_d,1)$ since (because $\mathbb{Z} \to \mathbb{R}_d$ is injective) the only nontrivial homotopy group appears in degree one and is a quotient of the standard inclusion $\mathbb{Z} \to \mathbb{R}_d$ (which defines ϕ). We now extend to a *Puppe sequence* (leaving off the $K(\mathbb{R}_d,2)$ on the right),

$$K(\mathbb{Z},1) \to K(\mathbb{R}_d,1) \to F = K(S^1_d,1) \to K(\mathbb{Z},2),$$

where we can see that the homomorphisms corresponding to the first two maps are the usual ones. The first is just a loop of ϕ, so that is clear. The second map is just a spatial realization of the connecting homomorphism in the homotopy sequence by $K(\mathbb{R}_d,1) = \Omega K(\mathbb{R}_d,2) \to K(S^1_d,1)$. Since the connecting homomorphism was the usual group projection onto the quotient, the second map is the usual one too. The last map $F \to K(\mathbb{Z},2)$ is a classifying map for the fibration $K(\mathbb{Z},1) \to K(\mathbb{R}_d,1) \to F = K(S^1_d,1)$ since the Puppe sequence is constructed by taking consecutive homotopy fibers. But the fibration is determined by the group extension $\mathbb{Z} \to \mathbb{R}_d \to S^1_d$ since the spaces are Eilenberg–Mac Lane spaces in degree one. Indeed, the classifying map corresponds to an element in $\text{Ext}(S^1_d, \mathbb{Z}) \subseteq H^2(S^1_d; \mathbb{Z})$ classifying the group extension. But the group extension is the usual one, so the element, and hence the classifying map $F \to K(\mathbb{Z},2)$ are the usual ones as well. $\square$

A principal S^1-bundle $\xi: S^1 \to P \to X$ is classified by a map $\kappa: X \to BS^1 = K(\mathbb{Z},2)$. This map corresponds to a degree 2 cohomology class by the standard identification of homotopy classes $[X, K(\mathbb{Z},n)]$ with cohomology $H^n(X;\mathbb{Z})$. In fact, $H^2(K(\mathbb{Z},2);\mathbb{Z}) \cong \text{Hom}(\mathbb{Z},\mathbb{Z})$, so there is an element $\iota \in H^2(K(\mathbb{Z},2);\mathbb{Z})$ corresponding to the identity homomorphism $\text{id}: \mathbb{Z} \to \mathbb{Z}$. The *Euler class* $e(\xi) \in H^2(X;\mathbb{Z})$ is then defined to be $\kappa^*(\iota)$. Since $e(\xi)$ characterizes the homotopy class of κ, it serves to classify principal S^1-bundles over X. A real Euler class $e(\xi)_{\mathbb{R}_d}$ is obtained by simply extending coefficients to $H^2(X;\mathbb{R}_d)$. This is equivalent to composing with the map $\phi: K(\mathbb{Z},2) \to K(\mathbb{R}_d,2)$ corresponding to the usual inclusion $\mathbb{Z} \to \mathbb{R}_d$. Again, we have $[X, K(\mathbb{R}_d,2)] \cong H^2(X;\mathbb{R}_d)$ obtained by pulling back a class $\hat{\iota}$ corresponding to $\text{id}: \mathbb{R}_d \to \mathbb{R}_d$. (Note that pulling $\hat{\iota}$ back to $K(\mathbb{Z},2)$ gives the class in $H^2(K(\mathbb{Z},2);\mathbb{R}_d)$ corresponding to the usual inclusion $\mathbb{Z} \to \mathbb{R}_d$.) We can now prove the main result using only the standard fibration and pullback theory presented above.

Theorem 6.47 *The real Euler class $e(\xi)_{\mathbb{R}_d}$ is zero if and only if ξ is flat.*

6.5 Appendix

Proof With the notation above, consider the following fibration diagram, which we shall use for both parts of the proof. (Here we write BS^1 for $K(\mathbb{Z}, 2)$ and BS_d^1 for $K(S_d^1, 1)$ to relate back to the definition of flat bundle.)

$$BS_d^1 \xrightarrow{Bi} BS^1 \xrightarrow{\phi} K(\mathbb{R}_d, 2)$$

with $\widehat{\kappa}$ and κ from X.

Suppose ξ is flat. Then $\widehat{\kappa}$ exists by definition and we have $\phi \kappa \simeq \phi\, Bi\, \widehat{\kappa} \simeq *$ since $\phi\, Bi \simeq *$. But $\phi \kappa$ represents the real Euler class, so we obtain $e(\xi)_{\mathbb{R}_d} = 0$.

On the other hand, suppose $e(\xi)_{\mathbb{R}_d} = 0$. Then $\phi \kappa \simeq *$. But then the homotopy lifting property of a fibration allows us to factor κ through the fiber as shown in the diagram: $Bi\, \widetilde{\kappa} \simeq \kappa$. By definition, this means that the bundle is flat. □

Of course, when X has finite type, for an integral cohomology class e, we can identify the conditions that e vanishes when coefficients are extended to the reals and that e is torsion. This holds, in particular, when M is a compact manifold. Therefore, we have

Corollary 6.48 *If X has finite type, then the Euler class $e(\xi)$ is torsion if and only if ξ is flat. Hence, ξ is flat if and only if $e(\xi)_{\mathbb{Q}} = 0$.*

Exercises for Chapter 6

Exercise 6.1 There are many kinds of curvatures for manifolds that are derived from the Riemann curvature. The two most important of these are as follows.

Definition 6.49 *The* Ricci curvature *is defined to be*

$$\mathrm{Ric}(X, Y) = \sum_{i=1}^{k} \langle R(X, \mathcal{E}_i)Y, \mathcal{E}_i \rangle$$

where X and Y are tangent vector fields on M^k and $\{\mathcal{E}_j\}$ is a frame. Recall that a frame $\{\mathcal{E}_j\}$ consists of an orthonormal basis of vector fields in some neighborhood of a point in M. The scalar curvature *is defined to be*

$$\kappa = \sum_{j=1}^{k} \mathrm{Ric}(\mathcal{E}_j, \mathcal{E}_j) = \sum_{i,j=1}^{k} \langle R(\mathcal{E}_j, \mathcal{E}_i)\mathcal{E}_j, \mathcal{E}_i \rangle.$$

While a manifold M^k is said to have non-negative sectional curvature if $\sec(X, Y) \geq 0$ for all X and Y, the Ricci curvature is said to be non-negative when $\mathrm{Ric}(X, X) \geq 0$

6: Curvature

for all X. Show that non-negative sectional curvature on M implies non-negative Ricci curvature on M and, in turn, this implies non-negative scalar curvature on M. Hint: First prove the formula

$$\text{Ric}(X, X) = \langle X, X \rangle \sum_{j=1}^{k-1} \sec(X, \mathcal{E}_j)$$

for a frame $\{\mathcal{E}_j\}$ spanning $X^\perp$, the space of vector fields orthogonal to X.

Exercise 6.2 Let G be a compact (connected) Lie group with bi-invariant metric. It is known that $\nabla_X X = 0$ for X any left invariant vector field. Thus, curves whose tangent vectors belong to left invariant vector fields are geodesics. Also, the following identity is known to hold: $\langle [X, Y], Z \rangle = -\langle Y, [X, Z] \rangle$ for left invariant vector fields X, Y and Z. Here, $\langle \cdot, \cdot \rangle$ is a the bi-invariant metric and $[-, -]$ is the bracket of vector fields (see Exercise A.1).

Now, for a compact, connected Lie group G with left invariant vector fields X, Y and Z, show that

1. $\nabla_X Y = \frac{1}{2}[X, Y]$. (Consider $\nabla_{X-Y}(X - Y)$ and use Proposition 6.1 (v).)

2. $R(X, Y)Z = \frac{1}{4} [[X, Y], Z]$.

    (Start with the definition of R, then use (1) to express R completely in terms of brackets. Finally, use the Jacobi identity for the bracket (Exercise A.1 (3)).)

3. If X and Y are orthonormal, then $\sec(X, Y) = \frac{1}{4} |[X, Y]|^2$.

    Thus, sectional curvature, and hence, Ricci and scalar curvatures are non-negative for a compact, connected Lie group with its natural bi-invariant metric. (Use the definition of sec, (1), Proposition 6.1 (iv) and the identity $\langle [X, Y], Z \rangle = -\langle Y, [X, Z] \rangle$ with $Z = Y$.)

The *Killing form* of a Lie group G is defined to be $b(X, Y) = \text{trace}(\text{Ad}_X \text{Ad}_Y)$, where $\text{Ad}_Z(W) = [Z, W]$ is a linear transformation of the vector space of left invariant vector fields. It can be shown that $-b$ is symmetric, bilinear and invariant under all automorphisms of G. If $-b$ is also nondegenerate, in the sense that $b(V, W) = 0$ for all W implies $V = 0$, then $-b$ is actually the bi-invariant metric $\langle \cdot, \cdot \rangle$ on G. That is, $\langle X, Y \rangle = -b(X, Y)$. A Lie group with nondegenerate Killing form is said to be *semisimple* and this coincides with Definition 1.41. Using (2) above and the definition of Ad_Z, show that a semisimple Lie group with metric $-b$ has Ricci curvature

$$\text{Ric}(X, Y) = -\frac{1}{4} b(X, Y).$$

Exercise 6.3 Show that (S^{2n+1}, T, k) is splitting rigid for all T and k.

7 G-spaces

The history of group actions is intimately tied up with the development of both algebraic topology and Lie group theory. In this chapter, we show how minimal models can be used to study compact Lie group actions on manifolds. We focus on certain aspects of group actions that are especially amenable to model techniques. Throughout the chapter, G will generally denote a Lie group acting on a compact manifold M.

Certain properties of M will be apparent. For instance, when the action is free (i.e. no point in M is left fixed by any $g \in G$), we have a principal G-bundle $M \to M/G$ with the particular consequence that $\chi(M) = 0$. For a general action, we will see that M fits inside the *Borel fibration*, $M \to M_G \to BG$, and this will be the central tool in our later study of transformation groups via models. The construction of the Borel fibration will be given in Section 7.2.

The efficacy of the general theory is best represented in the case where G is a torus $T^r = (S^1)^r$. For $G = T^r$, we essentially focus on issues related to the following two questions:

- When does M admit a free or an almost free G-action?
- What are the homological and the homotopical properties of the fixed point set M^G of a G-action on M?

For instance, here is a conjecture (generally attributed to S. Halperin [129]) that relates the rank of an almost freely acting torus with the dimension of the manifold's cohomology.

Toral rank conjecture (TRC). If M is a nilpotent compact manifold, then

$$\dim H^*(M; \mathbb{Q}) \geq 2^{\mathrm{rk}_0(M)},$$

where $\mathrm{rk}_0(M)$ is the rational toral rank of M.

The rational toral rank $\mathrm{rk}_0(M)$ is defined as follows. The action of T^r on M is said to be *almost free* if, for any point in the manifold, the subgroup of G fixing that point is a finite group. The largest integer r for which M

admits an almost free T^r-action is called the toral rank of M, and is denoted $\mathrm{rk}(M)$. If M does not admit any almost free torus action, then $\mathrm{rk}(M) = 0$. Unfortunately the invariant $\mathrm{rk}(M)$ is not a homotopy invariant and is quite difficult to compute. To obtain a homotopy invariant, we introduce the rational toral rank, $\mathrm{rk}_0(M)$; that is, the maximum of $\mathrm{rk}(Y)$ among all finite CW complexes Y in the same rational homotopy type as M. *An important advantage of using $\mathrm{rk}_0(M)$ is that it can be explicitly computed from the minimal model of M.* For instance, in the case of a Lie group G, we have $\mathrm{rk}_0(G) = \mathrm{rank}\, G$, and in the case of a homogeneous space G/H, we have $\mathrm{rk}_0(G/H) = \mathrm{rank}\, G - \mathrm{rank}\, H$.

In Section 7.3.3, we will prove the toral rank conjecture for certain families of spaces including homogeneous spaces and compact manifolds whose cohomology satisfies the hard Lefschetz property. The TRC can be reformulated as a conjecture on an upper bound for the toral rank:

$$\mathrm{rk}_0(M) \leq \log_2 \dim H^*(M; \mathbb{Q}).$$

We will derive other upper bounds for $\mathrm{rk}_0(M)$. First, we have

$$\mathrm{rk}_0(M) \leq \frac{1}{2} \dim H^*(M; \mathbb{Q}).$$

Secondly, denoting by $\mathcal{Z}(L)$ the center of a graded Lie algebra L, then, when $\pi_{\mathrm{even}}(M) \otimes \mathbb{Q} = 0$, we have

$$\mathrm{rk}_0(M) \leq \dim \mathcal{Z}\left(\pi_*(\Omega M) \otimes \mathbb{Q}\right).$$

For a nilmanifold, the inequality becomes an equality, and so the rational toral rank of a nilmanifold M with associated Lie algebra L is the dimension of the center of L.

The second part of this chapter begins in Section 7.4 and concerns the topology of the fixed point set M^G when G is a torus T^r. From the Borel fibration $M \to M_G \to BG$, we see that the algebra $H^*(M_G; \mathbb{Q})$ becomes a module over $H^*(BG; \mathbb{Q})$. We have seen in Theorem 1.81 that $H^*(BG; \mathbb{Q})$ is a polynomial algebra $H^*(BG; \mathbb{Q}) = \mathbb{Q}[x_1, \ldots, x_r]$ where, since $G = T^r$, the x_i are in degree 2. Let K denote the corresponding field of fractions; that is, the field of rational fractions in the variables $x_1, \ldots, x_r$, $K = \mathbb{Q}(x_1, \ldots, x_r)$. The Borel localization theorem then says that we have an isomorphism of K-vector spaces

$$K \otimes_{H^*(BG;\mathbb{Q})} H^*(M_G; \mathbb{Q}) \cong K \otimes H^*(M^G; \mathbb{Q}).$$

We use rational homotopy theory in conjunction with the Borel localization theorem to obtain results concerning the cohomology and the homotopy of M^G. Concerning cohomology, we prove the equality $\chi(M) = \chi(M^G)$, and, for each integer n, the inequality

$$\sum_{p \geq n} \dim H^p(M^G; \mathbb{Q}) \leq \sum_{p \geq n} \dim H^p(M; \mathbb{Q}).$$

From the first equality we deduce, in particular, that each G-action has a fixed point when $\chi(M) \neq 0$.

To obtain properties of the homotopy groups of the fixed set, we form the $\mathbb{Z}/2\mathbb{Z}$-graded vector space V (where $(\wedge V, d)$ is a Sullivan model for M), with $V^0 = \pi_{\text{even}}(M) \otimes \mathbb{Q}$ and $V_1 = \pi_{\text{odd}}(M) \otimes \mathbb{Q}$. Then, for each simply connected component F of M^G, we prove that there is a differential D on V, $D: V^0 \to V^1$, and $D: V^1 \to V^0$ such that $H^0(V, D) = \pi_{\text{even}}(F) \otimes \mathbb{Q}$ and $H^1(V, D) = \pi_{\text{odd}}(F) \otimes \mathbb{Q}$. In particular, if M is rationally elliptic, then each simply connected component F of M^G is also rationally elliptic and, moreover, $\chi_\pi(F) = \chi_\pi(M)$.

Beginning in Section 7.6, we consider certain types of actions which are particularly relevant to geometers, the symplectic and Hamiltonian actions on symplectic manifolds. We show how a weakened Lefschetz condition is enough to prove the TRC in the symplectic context. Then we consider the notion of Hamiltonian bundle in both geometric and algebraic formulations and give a proof of a result of Stepien [243] that says Hamiltonian bundles are TNCZ when the fiber is a nilmanifold.

Although we are mainly interested in smooth G-manifolds, torus actions can be defined on more general spaces with important results. In particular, this is the case when we consider toral rank and, in Section 7.3, we discuss toral rank in the more general context of finite CW complexes. However, from Section 7.4 onwards, we only consider smooth actions on manifolds.

This chapter is essentially an expository chapter. The book of Kawakubo [152] is a good reference for basic topological aspects of group actions. The books of Hsiang [144], Bredon [40] and tom Dieck [252] are the classical references for algebraic topological aspects. Finally, the book of Allday and Puppe [10] is the main reference for the use of rational homotopy theory in the study of transformation groups and the present chapter is, in some sense, preparation for studying that work.

7.1 Basic definitions and results

A topological group G acts on a space X if there is a continuous map $\varphi: G \times X \to X$ satisfying $ex = x$ and $g_1(g_2 x) = (g_1 g_2)x$. Here we have adopted the usual notational convention that $gx = \varphi(g, x)$. When G is a Lie group (*which we shall take to be compact unless stated otherwise*) and M is a manifold, the action is called *smooth* if φ is a smooth map. The manifold M is then called a smooth G-*manifold*, or for short, simply a G-manifold.

The subgroup $G_x = \{g \mid gx = x\}$ is called the *isotropy subgroup* at the point x. The *orbit* of x is $G(x) = \{gx \mid g \in G\}$. Note that the isotropy subgroups corresponding to points in the same orbit $G(x)$ are conjugate because $G_{gx} = gG_xg^{-1}$. The canonical map $f: G/G_x \to G(x)$ induced by the action on x is clearly a diffeomorphism. We can consider the set of orbits as a topological space by taking the quotient space defined by the following equivalence relation on M: $x \sim y$ if and only if there exists some $g \in G$ with $gx = y$. This *space of orbits* is denoted by M/G. A point $x \in M$ is a *fixed point of the action* if $G_x = G$; that is, $gx = x$ for all $g \in G$. The set of fixed points under the action of G on M is denoted by M^G. An action is *free* if $G_x = \{e\}$ for all $x \in M$ and is *almost free* if G_x is a finite group for all $x \in M$.

Example 7.1 Let $G = S^1$ act by horizontal rotations on the sphere $M = S^2$. The fixed point set M^G is the union of the North and South poles, the orbit space M/G is a semicircle connecting the poles, the isotropy subgroups are G at the fixed poles and the trivial group $\{e\}$ elsewhere. Note that, in this case, M/G is not a closed manifold, but rather, a manifold with boundary.

Basic properties of actions that will be useful for our purposes are summarized in the following proposition.

Proposition 7.2 ([152]) *If G is a Lie group and M a compact (smooth) G-manifold, then*

1. $G(x)$ *is a submanifold of M, and the action of G induces a diffeomorphism $G/G_x \to G(x)$.*
2. *Up to conjugacy, there are only finitely many isotropy groups. Therefore there are only finitely many diffeomorphism types of orbits.*
3. *The fixed point set M^G is a possibly disconnected compact submanifold of M whose components may have different dimensions.*
4. *If the action is free, then M/G is a manifold and the projection $M \to M/G$ is a principal G-bundle.*

The next lemma enables us to reduce the study of fixed sets of torus actions to the study of fixed sets of circle actions.

Lemma 7.3 *Let $T = (S^1)^r$ be a torus acting smoothly on a compact manifold M. Then there is some circle $S^1 \subseteq T$ such that $M^{S^1} = M^T$.*

Proof A smooth action of a torus T on a compact manifold M has a finite number of orbit types. So let the isotropy groups be $G_1, \ldots, G_s$ not including T itself. In the Lie algebra $\mathfrak{T}$ associated to T, we have the lie algebras of the isotropy groups: $\mathfrak{g}_1, \ldots, \mathfrak{g}_s$. Since we don't include T itself, the $\mathfrak{g}_i$ are all of lower dimension. Therefore there exists a vector in $\mathfrak{T}$ not in the union of the $\mathfrak{g}_i$ and exponentiation gives a circle S in T which has the property that it only intersects each isotropy group in a discrete (so finite by compactness) subgroup. This is the circle that works. Clearly, M^T is in M^S.

Now assume x is in M^S. What is the T-isotropy group T_x at x? It has to be one of the G_i or all of T (in which case it is a fixed point for T). Suppose $T_x = G_i$. Since x is in M^S, we have S contained in G_i, and this contradicts the fact that S intersects G_i in a finite set. Therefore, $T_x = T$ and x is in M^T. □

The same result holds in the context of CW-complexes with finitely many orbit types, see [4, Lemma 4.2.1].

7.2 The Borel fibration

Recall from Section 1.10 that, for every compact Lie group G, there exists a universal principal G-bundle

$$G \to EG \to BG.$$

The space BG, called the *classifying space* of G, is EG/G where EG is a contractible space admitting a free G-action. Note that the contractibility of EG combined with the Puppe sequence show that $\Omega BG \simeq G$.

Example 7.4 When G is the circle S^1, the classifying space BS^1 has been constructed in Example 1.74. Recall that the action of the circle on $S^{2n+1} \subset \mathbb{C}^{n+1}$ is given by the complex multiplication $z(z_1, \ldots, z_{n+1}) = (zz_1, \ldots, zz_{n+1})$. The quotient is the complex projective space $\mathbb{C}P(n)$. Let S^∞ be constructed as the direct limit of the inclusions $S^1 \subset S^2 \subset S^3 \subset \ldots \subset \cup_{i=1}^\infty S = S^\infty$. Now it is known that homotopy groups behave well under direct limits: $\lim_j \pi_k(S^j) = \pi_k(\lim_j(S^j))$. Hence $\pi_k(S^\infty) = 0$ for all k and, indeed, S^∞ is contractible. Clearly S^∞ is also obtained by taking the direct limit $S^3 \subset S^5 \ldots \subset \cup S^{2p+1} = S^\infty$. The free actions of S^1 on the odd spheres are compatible with the inclusions, so they induce a free action of S^1 on S^∞. Since S^∞ is contractible, it is ES^1, and the S^1-bundle $S^\infty \to \mathbb{C}P(\infty)$ is the universal S^1-bundle $ES^1 \to BS^1$. By multiplying together different copies of the universal bundle $S^\infty \to \mathbb{C}P(\infty)$, we obtain the classifying space of an r-torus, $BT^r = (\mathbb{C}P(\infty))^r$.

Definition 7.5 Let G be a Lie group and let M be a G-manifold. The associated **Borel fibration** is the fiber bundle

$$M \to EG \times_G M \xrightarrow{p} BG,$$

where $EG \times_G M$ is the quotient of $EG \times M$ under the action of G defined by $g(x,m) = (xg^{-1}, gm)$. The projection p associates to $[x,m]$ the class of x in BG; $p([x,m]) = [x]$. The total space of the Borel fibration, $EG \times_G M$, is usually denoted by M_G and its rational cohomology is called the rational equivariant cohomology of M; $H_G^*(M;\mathbb{Q}) \stackrel{\text{def}}{=} H^*(M_G;\mathbb{Q})$.

Theorem 7.6

1. When the action of G on M is free, then the projection

$$q \colon M_G = EG \times_G M \to M/G, \quad [x,m] \mapsto [m],$$

is a homotopy equivalence.
2. When the action of G on M is almost free, then the projection $q \colon M_G \to M/G$ is a rational homotopy equivalence. That is, $H^*(q;\mathbb{Q}) \colon H^*(M/G;\mathbb{Q}) \to H^*(M_G;\mathbb{Q})$ is an isomorphism.
3. When the action of G on M is trivial, then $M_G = BG \times M$.
4. When H is a compact subgroup of G, then $(G/H)_G \simeq BH$.

Proof (1) and (3) are clear while a proof of (2) is outlined in Exercise 7.1. The following sequence of homotopy equivalences (the last one coming from Example 1.79) proves (4):

$$(G/H)_G = EG \times_G (G/H) \cong (EG \times_G G)/H \cong EG/H \simeq BH.$$

□

Since (2) above says that, from the viewpoint of rational homotopy theory, almost free actions are equivalent to free actions, it is important to have a criterion to identify them. The following characterization of almost free actions achieves this aim.

Theorem 7.7 (Hsiang's theorem; see [144], [10, Proposition 4.1.7]) *Let G be a compact Lie group and let M be a compact G-manifold. Then the group G acts almost freely on M if and only if the rational equivariant cohomology of M, $H_G^*(M;\mathbb{Q})$, is finite dimensional.*

Note that, if the action is free, then we have $M_G \simeq M/G$, so $H^*(M_G;\mathbb{Q}) \cong H^*(M/G;\mathbb{Q})$ is finite dimensional because M/G is a compact manifold.

7.3 The toral rank

In this section we give conditions for the existence of almost free torus actions on a manifold M. Our first condition concerns the Euler–Poincaré characteristic $\chi(M)$.

7.3 The toral rank

Theorem 7.8 *If a compact manifold M admits an almost free G-action with G a compact connected Lie group, then the Euler–Poincaré characteristic of M is zero.*

Of course, for example, any compact Lie group G admits a free action by its maximal torus via translations and we know that $\chi(G) = 0$ since G is rationally a product of odd spheres. On the other hand, by Theorem 7.8, an even dimensional sphere does not admit any almost free circle action.

Proof Suppose that G acts almost freely on M. Since $\Omega BG \simeq G$, the homotopy fiber of the injection $M \hookrightarrow EG \times_G M$ has the homotopy type of G. We obtain in this way a homotopy fibration

$$G \to M \to EG \times_G M = M_G.$$

The associated Serre spectral sequence satisfies

$$E_2 \cong H^*(M_G; \mathbb{Q}) \otimes H^*(G; \mathbb{Q}) \Rightarrow H^*(M; \mathbb{Q}).$$

By Theorem 7.7, the term E_2 is finite dimensional. Since the Euler characteristic is preserved in a spectral sequence and since $\chi(G) = 0$, we have $\chi(M) = \chi(M_G)\chi(G) = 0$. □

As we explained in the introduction to this chapter, the main "invariant" for the study of almost free actions is the toral rank of a manifold. Let's now define this notion in the context of topological spaces.

Definition 7.9 *The toral rank of a space X, rk(X), is the largest integer r such that a torus T^r acts almost freely on X.*

The toral rank may be bounded for reasons that depend only on topology. For instance, a wedge of spheres has toral rank zero because the basepoint must remain fixed under the action. The latter claim follows simply from the topology of the wedge. The basepoint of the wedge is the only point of the wedge whose removal increases the number of components, so any homeomorphism must fix that point. Unfortunately, the toral rank is not a homotopy invariant.

Example 7.10 (Free S^1-action on $Y \simeq S^2 \vee S^3 \vee S^3$) Let S^1 act freely on S^3 by the Hopf action, and on $S^1 \times S^2$ by translation on the first factor. In each space select a free S^1-orbit and glue together S^3 and $S^1 \times S^2$ along these orbits. The result is a three dimensional CW complex Y admitting a free circle action. Since the orbits in S^3 can be contracted to a point, we see that $Y \simeq S^3 \vee ((S^1 \times S^2)/(S^1 \times *))$. The second summand, which for convenience we denote by Z, is the half-smash product of S^1 and S^2. We

also have (from a homotopy pushout argument)
$$Z = (S^1 \times S^2)/(S^1 \times *) \simeq \Sigma\, (S^1 \times S^1)/(S^1 \times *).$$
But the half-smash product $(S^1 \times S^1)/(S^1 \times *)$ is easy to describe; it is simply a torus with a meridian collapsed and, homotopically, this is a 2-sphere with North and South poles identified or, equivalently, a 2-sphere with an arc attached to the North and South poles. This last identification is clearly of the homotopy type of $S^2 \vee S^1$, so we see that
$$Z = \Sigma\, (S^1 \times S^1)/(S^1 \times *) \simeq \Sigma\, (S^2 \vee S^1) = S^3 \vee S^2.$$
Hence, the CW complex Y has the homotopy type of the wedge $S^2 \vee S^3 \vee S^3$. By the discussion above, $\mathrm{rk}(S^2 \vee S^3 \vee S^3) = 0$, but $\mathrm{rk}(Y) \geq 1$.

Remark 7.11 We can also define the integer $\mathrm{rk}_s(M)$ to be the maximal r such that there is a torus T^r that acts smoothly on M. This invariant obeys $\mathrm{rk}_s(M) \leq \mathrm{rk}(M)$, but it is very difficult to compute and the exact relationship does not seem to be known. The "almost free" condition is not a standard condition in transformation groups, but work of R. Schultz [233] has produced, for instance, an exotic sphere N of dimension 17 that admits no *free* smooth S^1-action. Of course, N *does* admit a *continuous* free circle action since N and S^7 are homeomorphic. For general remarks concerning the difference between smooth and continuous degrees of symmetry, see [144, Chapter VII].

Since we are interested in obtaining a homotopy invariant and since we are only considering the rational homotopy type of spaces, we are led to the following variation of the definition of the toral rank.

Definition 7.12 *The rational toral rank of a space X, $\mathrm{rk}_0(X)$, is the maximum of $\mathrm{rk}(Y)$ for all finite CW complexes Y in the rational homotopy type of X.*

7.3.1 Toral rank for rationally elliptic spaces

The first computations on the rational toral rank are due to Allday and Halperin [6]. Recall that, if M is a rationally elliptic space, the homotopy Euler characteristic is defined by
$$\chi_\pi(M) = \mathrm{rank}\, \pi_{\mathrm{even}}(M) - \mathrm{rank}\, \pi_{\mathrm{odd}}(M).$$
The first toral rank estimate derived from algebraic models was also one of the first verifications (along with formality for Kähler manifolds and the solution of the closed geodesic problem) that rational homotopy theory could be used to obtain interesting geometric information.

Theorem 7.13 ([6]) *If M is a nilpotent rationally elliptic space, then*

$$\mathrm{rk}_0(M) \leq -\chi_\pi(M).$$

Proof Consider the homotopy fibration $T^r \to M \to M_{T^r}$ associated to an almost free action of T^r on M. Since M and T^r are rationally elliptic spaces, the same is also true for M_{T^r}. This follows for homotopy from the long exact homotopy sequence associated to a fibration and, for cohomology, from Theorem 7.7. The long exact homotopy sequence then gives

$$\chi_\pi(M) = \chi_\pi(M_{T^r}) + \chi_\pi(T^r).$$

Recall now that, for a rationally elliptic space X, we have $\chi_\pi(X) \leq 0$ (see Theorem 2.75). From $\chi_\pi(M_{T^r}) \leq 0$ and $\chi_\pi(T^r) = -r$, we deduce $r \leq -\chi_\pi(M)$. $\square$

Corollary 7.14 *If G is a compact connected Lie group, then $\mathrm{rk}_0(G) = \mathrm{rank}\ G$.*

Proof Recall from Example 2.39 that the minimal model of a Lie group G is an exterior algebra $(\wedge(y_1, \ldots, y_r), d = 0)$ with $r = \mathrm{rank}\ G$. It follows that $-\chi_\pi(G) = \mathrm{rank}\ G$. Therefore $\mathrm{rk}_0(G) \leq \mathrm{rank}\ G$. On the other hand, if T^r is a maximal torus in G, then the left multiplication by T^r yields a free action of T^r on G, so that $\mathrm{rk}_0(G) \geq \mathrm{rank}\ G$. $\square$

Corollary 7.15 *If $M = G/K$ is the quotient of a compact connected Lie group G by a compact connected subgroup K, then*

$$\mathrm{rk}_0(G/K) = \mathrm{rank}\ G - \mathrm{rank}\ K.$$

Proof From the fibration $K \to G \to G/K$ (and its long exact homotopy sequence), we deduce $\chi_\pi(G/K) = \chi_\pi(G) - \chi_\pi(K) = -\mathrm{rank}\ G + \mathrm{rank}\ K$. This implies the inequality

$$\mathrm{rk}_0(G/K) \leq -\chi_\pi(G/K) = \mathrm{rank}\ G - \mathrm{rank}\ K.$$

Now denote by T^s a maximal torus in G and by T^r a maximal torus in K. The maximal tori of a compact Lie group are all conjugate, so we can suppose $T^r \subset T^s$. The left multiplication by $T^{s-r} = T^s/T^r$ on G/K is again a free action. Therefore the rational toral rank of G/K is greater than or equal to the difference $s - r$, $\mathrm{rk}_0(G/K) \geq \mathrm{rank}\ G - \mathrm{rank}\ K$. $\square$

Example 7.16 As we have seen, an odd dimensional sphere S^{2n-1} admits a free circle action. Since $\chi_\pi(S^{2n-1}) = -1$, we have $\mathrm{rk}_0(S^{2n-1}) = 1$.

7.3.2 Computation of $\mathrm{rk}_0(M)$ with minimal models

Let $G = T^r$ be an r-torus, M a nilpotent compact G-manifold, and

$$M \to EG \times_G M \xrightarrow{p} BG$$

the associated Borel fibration. Our first aim in this section is the description of the relative minimal model of the Borel fibration.

Let's first recall the rational cohomology of T^r and BT^r (Theorem 1.81). The cohomology of T^r is an exterior algebra on r generators in degree 1,

$$H^*(T^r; \mathbb{Q}) = \wedge(y_1, \ldots, y_r),$$

and the cohomology of the classifying space BT^r is a polynomial algebra on r generators in degree 2,

$$H^*(BT^r; \mathbb{Q}) = \mathbb{Q}[x_1, \ldots, x_r].$$

In particular, the minimal model of T^r is $(\wedge(y_1, \ldots, y_r), 0)$, and the minimal model of BT^r is $(\wedge(x_1, \ldots x_r), 0)$.

Let $(\wedge V, d)$ be the minimal model for M. Then a relative minimal model for the Borel fibration is

$$(\wedge(x_1, \ldots, x_r), 0) \to (\wedge(x_1, \ldots x_r) \otimes \wedge V, D) \to (\wedge V, d).$$

When the action is almost free, $H^*(\wedge(x_1, \ldots, x_r) \otimes \wedge V, D)$ is finite dimensional by Theorem 7.7. This yields a characterization of almost free actions in terms of minimal models.

Proposition 7.17 ([130, Proposition 4.2]) *If M is a nilpotent compact m-dimensional manifold, then $\mathrm{rk}_0(M) \geq r$ if and only if there is a relative minimal model of the form*

$$(\wedge(x_1, \ldots, x_r), 0) \to (\wedge(x_1, \ldots, x_r) \otimes \wedge V, D) \to (\wedge V, d),$$

where $|x_i| = 2$ for $i = 1, \ldots, r$, $(\wedge V, d)$ is the minimal model of M and the cohomology $H^*(\wedge(x_1, \ldots, x_r) \otimes \wedge V, D)$ is finite dimensional.

Moreover, if $\mathrm{rk}_0(M) \geq r$, then T^r acts freely on a finite CW complex X that has the same rational homotopy type as M, and if $m - r \not\equiv 0 \mod 4$, then we can choose X to be a compact manifold.

Proof If we have an almost free T^r action on M, the discussion preceding the statement of the proposition shows that we have a relative minimal model with the required properties.

Conversely, suppose we have such a relative minimal model, and let N be a finite nilpotent CW complex whose minimal model is $(\wedge(x_1, \ldots, x_r) \otimes$

$\wedge V, D)$ (see Section 2.6.1). The classes $[x_i]$ define classes in $H^2(N;\mathbb{Q})$. Multiply by integers k_i to obtain $[k_i x_i] \in H^2(N;\mathbb{Z})$ and use these classes to define a map $\varphi \colon N \to K(\mathbb{Z},2)^r = (\mathbb{C}P(\infty))^r$. The pullback of the universal T^r-bundle with basis $(\mathbb{C}P(\infty))^r$ gives a finite nilpotent CW complex X in the rational homotopy type of M that admits a free T^r-action.

The cohomology of $(\wedge(x_1,\ldots,x_r) \otimes \wedge V, D)$ satisfies Poincaré duality and the fundamental class is in degree $m-r$. Therefore by Theorem 3.2, when $m-r \not\equiv 0 \bmod 4$ we can choose N to be a compact manifold. In this case, the classifying map may be compressed into $\mathbb{C}P(m-r)$ and then may be replaced by a smooth map. The pullback of the Hopf principal bundle over $\mathbb{C}P(m-r)$ then is a principal circle bundle whose total space is a smooth manifold of the rational homotopy type of M. $\square$

Example 7.18 Denote by $S^{11} \to N \to S^{12}$ the sphere bundle associated to the tangent bundle of the sphere S^{12}. We then denote by

$$S^{11} \to E \to (S^3)^4$$

the pullback of this fibration along a map $(S^3)^4 \to S^{12}$ of degree one. Let's show that $\text{rk}_0(E) = 1$. First of all, the minimal model of E is

$$(\wedge(y_1,y_2,y_3,y_4,z),d), \ |y_i|=3, |z|=11, \ dz=y_1 y_2 y_3 y_4.$$

We then consider the relative minimal model (with $|x|=2$),

$$(\wedge x, 0) \to (\wedge x \otimes \wedge(y_1,y_2,y_3,y_4,z), D) \to (\wedge(y_1,y_2,y_3,y_4,z),d)$$

where $D(z) = dz + x^6$. Clearly, the cohomology $H^*(\wedge x \otimes \wedge(y_1,y_2,y_3,y_4,z), D)$ is finite dimensional, which implies by Proposition 7.17 that there exists a manifold in the rational homotopy type of E that admits a free action of S^1. In particular $\text{rk}_0(E) \geq 1$.

Now suppose that $\text{rk}_0(E) = r \geq 1$. By Proposition 7.17, there exists a relative minimal model

$$(\wedge(x_1,\ldots,x_r),0) \to (\wedge(x_1,\ldots,x_r,y_1,y_2,y_3,y_4,z), D)$$
$$\to (\wedge(y_1,y_2,y_3,y_4,z),d),$$

with $\dim H^*(\wedge(x_1,\ldots,x_r,y_1,y_2,y_3,y_4,z),D) < \infty$. Suppose $D(y_1) = a \neq 0$, $a \in \wedge(x_1,\ldots,x_r)$. Now, $D(z) = dz + \tau = y_1 y_2 y_3 y_4 + \tau$ with $\tau \in \text{Ideal}(x_i)$, so

$$0 = D^2(z) = a y_2 y_3 y_4 + \xi,$$

where ξ belongs to the ideal generated by y_1 and $\wedge^{\geq 3}(x_1,\ldots x_r)$ for degree reasons. This is impossible because $a y_2 y_3 y_4$ is not in this ideal. Therefore, by similar arguments, $D(y_1) = D(y_2) = D(y_3) = D(y_4) = 0$.

Now write $D(z) = \alpha + \beta$ with $\alpha \in \wedge(x_1,\ldots,x_r)$ and $\beta \in \wedge(x_i) \otimes \wedge^+(y_j)$. The projection $(\wedge(x_1,\ldots,x_r,y_1,y_2,y_3,y_4,z), D) \to (\wedge(x_1,\ldots,x_r)/(\alpha), 0)$ is then surjective in cohomology. Therefore $\wedge(x_1,\ldots,x_r)/(\alpha)$ is finite dimensional and this can only happen when $r \leq 1$.

Finally, note that, by Theorem 7.13, we have an upper bound for $\mathrm{rk}_0(E)$ given by $5 = -\chi_\pi(E)$. Therefore, we see that the Allday–Halperin estimate is definitely not sharp.

Example 7.19 In [148], an example, due to Halperin, is given of manifolds M and N with $\mathrm{rk}_0(M) = \mathrm{rk}_0(N) = 0$ and $\mathrm{rk}_0(M \times N) \geq 1$. This shows, at least rationally, that taking products can increase the symmetry of manifolds. The minimal models of $M = S^{12}$ and N are, respectively, $(\wedge(x,y), d)$, $|x| = 12$, $|y| = 23$, $dy = x^2$ and $(\wedge(u_1,u_2,u_3,u_4,u_5,u_6,v,w), d)$ with $|u_1| = |u_2| = |u_3| = |u_4| = 3$, $|u_5| = 5$, $|u_6| = 19$, $|w| = 18$, $|v| = 35$, $dv = w^2 + u_1u_2u_3u_4u_5u_6$, with the other differentials being zero.

Since $\mathrm{rk}_0(M) \leq -\chi_\pi(M) = 0$, we have $\mathrm{rk}_0(M) = 0$. We will show that the cohomology of any Sullivan algebra of the form $(\wedge a \otimes \wedge(u_1,u_2,u_3,u_4,u_5,u_6,v,w), D)$ is infinite dimensional when $|a| = 2$ and the image of $D - d$ belongs to the ideal generated by a. This will imply that $\mathrm{rk}_0(N) = 0$. For degree reasons D must satisfy the following properties

$$Du_1 = \alpha_1 a^2, \; Du_2 = \alpha_2 a^2, \; Du_3 = \alpha_3 a^2, \; Du_4 = \alpha_4 a^2, \; Du_5 = \alpha_5 a^3$$
$$Du_6 = \alpha_6 a^{10} + \mu aw + a^3 P$$
$$Dv = w^2 + u_1u_2u_3u_4u_5u_6 + waF + u_6 a^3 R + T$$
$$Dw \in \wedge^+(a) \otimes \wedge^+(u_1,u_2,u_3,u_4,u_5),$$

with $\alpha_1,\alpha_2,\alpha_3,\alpha_4,\alpha_5,\alpha_6,\mu \in \mathbb{Q}$, R and $P \in \wedge a \otimes \wedge^+(u_1,u_2,u_3,u_4,u_5)$, and F and T in $\wedge^+(a) \otimes \wedge(u_1,u_2,u_3,u_4,u_5)$.

Looking at the coefficient of u_6 in D^2v, we find

$$\sum_{i=1}^{5}(-1)^{i+1}\alpha_i a^2 u_1 \cdots \widehat{u_i} \cdots u_6 - u_6 a^3 D(R) = 0,$$

which implies that $\alpha_1 = \cdots = \alpha_5 = 0$. The equation $0 = D^2 u_6$ gives $\mu a Dw = 0$, so that either $\mu = 0$ or else $Dw = 0$.

Now let's consider the equation $D^2 v = 0$. We have

$$0 = 2wDw - u_1u_2u_3u_4u_5(\alpha_6 a^{10} + \mu aw + a^3 P) + D(w)aF$$
$$+ (\alpha_6 a^{10} + \mu aw + a^3 P)a^3 R.$$

If $\mu = 0$, then the coefficient of w becomes Dw, so that $Dw = 0$. If $Dw = 0$, the coefficient of $u_1u_2u_3u_4u_5 aw$ is μ so that $\mu = 0$. In conclusion, in any case we have $\mu = 0$ and $Dw = 0$. Write $R = \sum_{i=0}^{m} a^i R_i$ with $R_i \in$

$\wedge^+(u_1,\cdots,u_5)$, and observe that the coefficient of a^{13} in D^2v is $\alpha_6 R$. It follows that $\alpha_6 = 0$.

Because of the calculations above, the projection

$$q: (\wedge(a, u_1, u_2, u_3, u_4, u_5, u_6, v, w), D) \to (\wedge(a, w)/(w^2), 0)$$

obtained by mapping the u_i and v to 0 is a cdga morphism. If the cohomology $H^*(\wedge(a, u_1, u_2, u_3, u_4, u_5, u_6, v, w), D)$ is finite dimensional, then some power a^k is a coboundary. But this means that, *in cohomology*, $0 = [a]^k \mapsto a^k \neq 0$ which is impossible. Therefore $rk_0(N) = 0$.

We now show that $rk_0(M \times N) = 1$ by giving an explicit relative minimal model that has finite dimensional cohomology:

$$(\wedge(a, x, y, u_1, u_2, u_3, u_4, u_5, u_6, v, w), D).$$

The differential is defined as follows

$$Du_1 = Du_2 = Du_4 = Du_3 = Dw = 0,$$
$$Dx = u_1 u_2 u_3 a^2,$$
$$Dy = x^2 + 2u_1 u_6 a,$$
$$Du_5 = a^3,$$
$$Du_6 = u_2 u_3 x a,$$
$$Dv = w^2 + u_1 u_2 u_3 u_4 u_5 u_6 + x a u_4 u_6 + y a^2 u_2 u_3 u_4.$$

A straightforward computation shows that $D^2 = 0$. To prove that the cohomology is finite dimensional, we put a new gradation $V^{(p)}$ on V by letting $|u_1| = |u_2| = |u_3| = |u_4| = |u_6| = 3$, $|a| = |x| = |w| = 2$, $|y| = |v| = 3$ and $|u_5| = 5$. Since $D(V^{(p)}) \subset (\wedge V)^{\geq p+1}$, this gives a spectral sequence converging to the cohomology of the algebra. The E_2 term is isomorphic to $\wedge(u_1, u_2, u_3, u_4, u_6) \otimes (\wedge x)/(x^2) \otimes (\wedge a)/(a^3) \otimes (\wedge w)/(w^2)$ and is finite dimensional. This proves that the cohomology is finite and that $rk_0(M \times N) \geq 1$.

7.3.3 The toral rank conjecture

The main problem related to the toral rank is the so called toral rank conjecture (TRC), which is usually attributed to S. Halperin.

Conjecture 7.20 (TRC conjecture) *Let M be a nilpotent finite CW complex. Then*

$$\dim H^*(M; \mathbb{Q}) \geq 2^{rk_0(M)}.$$

The intuition behind the TRC is that an almost free T^r-action somehow injects T^r, at least cohomologically, into M. Of course, $\dim H^*(T^r; \mathbb{Q}) = 2^r$, so the conjecture is an expression of this intuitive notion. To set the record straight, however, we note that examples exist showing that $H^*(M;\mathbb{Q})$ does not necessarily contain an exterior algebra on r generators even when an almost free T^r-action exists on M (see [4, Remarks 4.4.2 (5)]).

From Proposition 7.17 we obtain a reformulation of the TRC conjecture in terms of minimal models.

Conjecture 7.21 (Algebraic TRC) *Let $(\wedge V, d)$ be a minimal cdga, and let $(\wedge(x_1,\ldots,x_r), 0) \to (\wedge(x_1,\ldots,x_r) \otimes \wedge V, D) \to (\wedge V, d)$ be a relative minimal model with $|x_i| = 2$, and such that the cohomology algebra $H^*(\wedge(x_1,\ldots,x_r) \otimes \wedge V, D)$ is finite dimensional. Then $\dim H^*(\wedge V, d) \geq 2^r$.*

The TRC conjecture is open in general, but has been proved in some interesting cases that we will now consider.

Proposition 7.22 *The TRC is true for any product of odd-dimensional spheres.*

Proof Suppose an r-torus T^r acts almost freely on a product of odd-dimensional spheres $M = S^{n_1} \times S^{n_2} \times \ldots \times S^{n_p}$. By Theorem 7.13,

$$r \leq -\chi_\pi(M) = p.$$

This implies that

$$2^r \leq 2^p = \dim H^*(M;\mathbb{Q}).$$

$\square$

Proposition 7.23 *The TRC is true for homogeneous spaces.*

Proof Let G be a compact connected Lie group and let $K \subset G$ be a compact connected subgroup. Denote the rational toral rank of G/K by r, and consider the Serre spectral sequence associated to the fibration $K \to G \to G/K$:

$$H^*(G/K) \otimes H^*(K) \Rightarrow H^*(G).$$

It follows that $\dim H^*(G) \leq \dim H^*(G/K) \cdot \dim H^*(K)$. Since $r = \operatorname{rank} G - \operatorname{rank} K$ (see Corollary 7.15), this gives

$$2^r = 2^{\operatorname{rank} G - \operatorname{rank} K} = \frac{2^{\operatorname{rank} G}}{2^{\operatorname{rank} K}} = \frac{\dim H^*(G)}{\dim H^*(K)} \leq \dim H^*(G/K).$$

$\square$

Recall that a connected graded commutative algebra H satisfies the hard Lefschetz property if H behaves like the cohomology of a compact Kähler manifold (see Theorem 4.35). This means that the following property is satisfied:

- There is an element $\omega \in H^2$ such that for every $p < m$, the multiplication by ω^{m-p} induces an isomorphism $H^p \to H^{2m-p}$.

Theorem 7.24 ([8]) *If a torus T^r acts almost freely on a compact connected manifold M with a cohomology satisfying the hard Lefschetz property, then the injection of an orbit, $T^r \cdot x \hookrightarrow M$, induces an injection in homology. In particular, the TRC is true in this case.*

Proof We consider the Borel fibration associated to the action of T^r on M: $M \to M_{T^r} \to BT^r$. The connecting map of the fibration, $\delta \colon T^r \cong \Omega BT^r \to M$ corresponds to the injection of an orbit of the action. We prove that $H_*(\delta; \mathbb{Q})$ is an injective map.

Write $H^*(BT^r; \mathbb{Q}) = \mathbb{Q}[x_1, \ldots, x_r]$ and note that the image of the transgression $H^1(M; \mathbb{Q}) \to H^2(BT^r; \mathbb{Q})$ is a sub-vector space of the vector space generated by the x_i. By making a change of generators, we suppose that the image is the vector space generated by the elements $x_{s+1}, x_{s+2}, \ldots, x_r$. The quotient map $q \colon (\wedge(x_1, \ldots, x_r), 0) \to (\wedge(x_1, \ldots, x_s), 0)$ defined by

$$q(x_j) = \begin{cases} x_j & \text{for } j \leq s \\ 0 & \text{otherwise} \end{cases}$$

can be realized by a map $g \colon BT^s \to BT^r$. We now pull back the Borel fibration along g to obtain a fibration with base BT^s.

$$\begin{array}{ccccc} M & \longrightarrow & M_{T^r} & \longrightarrow & BT^r \\ \| & & \uparrow f & & \uparrow g \\ M & \longrightarrow & E & \longrightarrow & BT^s \end{array}$$

By construction, the transgression of the new fibration, $H^1(M; \mathbb{Q}) \to H^2(BT^s; \mathbb{Q})$, is zero. This implies by Corollary 4.41 that the Serre spectral sequence of the fibration $M \to E \to BT^s$ collapses at the E^2 term: $H^*(E; \mathbb{Q}) \cong H^*(M; \mathbb{Q}) \otimes H^*(BT^s; \mathbb{Q})$. In particular if $s > 0$, then $H^*(E; \mathbb{Q})$ is infinite dimensional. Notice now that the homotopy fiber of the induced map $f \colon E \to M_{T^r}$ is the homotopy fiber of g, T^{r-s}. Since $H^*(T^{r-s}; \mathbb{Q})$ and $H^*(M_{T^r}; \mathbb{Q})$ are finite dimensional (the latter by Theorem 7.7), the same is true for $H^*(E; \mathbb{Q})$ and, thus, $s = 0$. This means that in the original Borel fibration, the transgression $H^1(M; \mathbb{Q}) \to H^2(BT^r; \mathbb{Q})$ is a surjective map.

Now let

$$(\wedge(x_1, \ldots, x_r), 0) \to (\wedge(x_1, \ldots, x_r) \otimes \wedge W, D) \to (\wedge W, d)$$

be a relative minimal model for the Borel fibration. By what we have said above about the transgression, we obtain a decomposition $\wedge W = \wedge(z_1, \ldots, z_r) \otimes \wedge V$, with $D(z_i) = x_i$. We now let $\bar{x}_1, \ldots, \bar{x}_r$ be variables in degree 1, and we form the cochain algebra

$$(C, D) = (\wedge(x_1, \ldots, x_r, \bar{x}_1, \ldots, \bar{x}_r), D)$$

where $D(x_i) = 0$ and $D(\bar{x}_i) = x_i$. This cochain algebra is contractible, which means that $H^*(C, D) = \mathbb{Q}$. The projection

$$\pi : (C, D) \otimes_{(\wedge(x_1, \ldots, x_r))} (\wedge(x_1, \ldots, x_r) \otimes \wedge W, D) \to (\wedge W, d),$$

obtained by mapping the x_i and the $\bar{x}_i$ to zero, is a morphism of differential graded algebras. Since (C, D) is contractible, π is a surjective quasi-isomorphism, and since $(\wedge W, d)$ is minimal, π admits a section σ. The section is not unique, but in any case we always have $\sigma(z_i) = z_i - \bar{x}_i$. We denote by $\psi : (\wedge(z_1, \cdots, z_r), 0) \hookrightarrow (\wedge W, d)$ the canonical injection, and we define a morphism of commutative differential graded algebras

$$\varphi : (C, D) \otimes_{(\wedge(x_1, \ldots, x_r))} (\wedge(x_1, \ldots, x_r) \otimes \wedge W, D) \to (\wedge(\bar{x}_1, \cdots, \bar{x}_r), 0)$$

by sending the x_i, the z_i and V to 0, and $\bar{x}_i$ to $\bar{x}_i$. The composition $\varphi \circ \sigma \circ \psi$ maps z_i to $-\bar{x}_i$, and is therefore an isomorphism. This shows that $H^*(\varphi) : H^*(M; \mathbb{Q}) \to \wedge(\bar{x}_1, \ldots \bar{x}_n)$ is a surjective map. We now observe that the relative minimal model

$$(\wedge(x_i) \otimes \wedge W, D) \to (C, D) \otimes_{(\wedge(x_i))} (\wedge(x_i) \otimes \wedge W, D) \xrightarrow{\varphi} (\wedge(\bar{x}_i), 0)$$

is a model for the homotopy fibration $T^r \cong \Omega BT^r \to M \to M_{T^r}$. This implies that φ is a model for the injection of an orbit, and so the injection of an orbit induces as well an injection in rational homology. □

Corollary 7.25 *A compact simply-connected manifold M whose cohomology satisfies the hard Lefschetz property does not admit any almost free torus action.*

Proof If there were an almost free T^r-action on M, then there would exist an injection $H_*(T^r; \mathbb{Q}) \to H_*(M; \mathbb{Q})$. Since M is simply connected this is impossible. □

Note that, in the case of a symplectic manifold, this follows from Subsection 7.6.2 (in particular Theorem 7.67) as well. A weaker form of the TRC has been proved by Allday and Puppe in [11].

Theorem 7.26 *If a torus T^r acts almost freely on a compact nilpotent manifold M, then*

$$\dim H^*(M; \mathbb{Q}) \geq 2r.$$

Proof Let us consider the relative minimal model of the Borel fibration:

$$(\wedge(x_1,\cdots,x_r),0) \to (\wedge(x_1,\cdots,x_r) \otimes \wedge V, D) \to (\wedge V, d).$$

We decompose $\wedge V$ into the direct sum $\wedge V = B \oplus H \oplus R$, with $d(H \oplus B) = 0$ and $d\colon R \to B$ an isomorphism. The graded vector space $E = (\wedge(x_1,\cdots,x_r) \otimes R) \oplus D(\wedge(x_1,\cdots,x_r) \otimes R)$ is then a free acyclic $\wedge(x_1,\cdots,x_r)$-module. The quotient map is therefore a quasi-isomorphism of $\wedge(x_1,\ldots,x_r)$-modules,

$$(\wedge(x_1,\cdots,x_r) \otimes \wedge V, D) \xrightarrow{\simeq} (\wedge(x_1,\cdots,x_r) \otimes \wedge V, D)/E$$
$$\cong (\wedge(x_1,\cdots,x_r) \otimes H, \overline{D}).$$

For every element h of H we write

$$\overline{D}(h) = \alpha(h) + \beta(h), \quad \alpha(h) \in \wedge(x_1,\cdots,x_r), \ \beta(h) \in \wedge(x_1,\cdots,x_r) \otimes H^+.$$

Because of the finiteness of $H^*(\wedge(x_1,\ldots,x_r) \otimes H, \overline{D})$, there exists at least r linearly independent elements h_i with $\alpha(h_i) \neq 0$. The degrees of these elements h_i are odd. By construction, H is isomorphic to $H^*(\wedge V, d) = H^*(M; \mathbb{Q})$. Since the Euler–Poincaré characteristic of M is zero by Theorem 7.8, the dimension of the rational cohomology of M is at least $2r$. □

7.3.4 Toral rank and center of $\pi_*(\Omega M) \otimes \mathbb{Q}$

There is a strong relation between the rational toral rank of a space and the dimension of the center, $\mathcal{Z}(L_M)$ of the rational homotopy Lie algebra $L_M = \pi_*(\Omega M) \otimes \mathbb{Q}$.

Proposition 7.27 ([7]) *If M is a nilpotent compact manifold with $\pi_{\text{even}}(M) \otimes \mathbb{Q} = 0$, then*

$$\mathrm{rk}_0(M) \leq \dim \mathcal{Z}(L_M).$$

Proof We suppose we have an almost free action of T^r on M and we denote by

$$(\wedge(x_1,\cdots,x_r),0) \to (\wedge(x_1,\cdots,x_r) \otimes \wedge V, D) \to (\wedge V, d)$$

the relative minimal model of the associated Borel fibration $M \to M_{T^r} \to BT^r$. By hypothesis, $V = V^{\text{odd}}$. Therefore $D(V) \subset \wedge(x_1,\cdots,x_r) \otimes \wedge^{\text{even}}(V)$. For $v \in V$, we write $D(v) = \alpha(v) + \beta(v)$, with

$$\alpha(v) \in \wedge(x_1,\cdots,x_r), \quad \beta(v) \in \wedge(x_1,\cdots,x_r) \otimes \wedge^+ V.$$

By sending V to zero, we obtain a map of differential graded algebras

$$(\wedge(x_1,\cdots,x_r) \otimes \wedge V, D) \to (\wedge(x_1,\cdots,x_r)/(\alpha(V)), 0).$$

Since the x_i are cocycles, this map is surjective in cohomology. Now, because the action is almost free, the algebra $H^*(\wedge(x_1,\cdots,x_r) \otimes \wedge V, D)$ is finite dimensional, so dim $(\wedge(x_1,\cdots,x_r)/\alpha(V)) < \infty$. This implies that the ideal $\alpha(V)$ has at least s independent generators $\alpha(v_1),\ldots,\alpha(v_s)$ with $s \geq r$. We denote by W the graded sub-vector space of V consisting of the elements v such that $\alpha(v)$ belongs to the ideal $\sum_i \alpha(v_i) \cdot \wedge(x_1,\ldots,x_r)^+$. Clearly, $V = W \oplus (v_1,\ldots,v_s)$.

We now prove that $d_1(V) \subset \wedge^2(W)$, where d_1 denotes the quadratic part of the differential d. This implies, by Proposition 2.60, that dim $\mathcal{Z}(L_M) \geq s \geq r$.

Let's fix some $i \leq s$. For $v \in V$, $d_1(v)$ can be decomposed into a sum, $d_1(v) = v_i \sigma + \omega$, where σ and ω belong to the algebra generated by $W \oplus (v_1,\ldots,\hat{v}_i,\ldots,v_s)$. Then a standard, but tedious, computation shows that $D^2(v) - \alpha(v_i) \otimes \sigma$ belongs to

$$\left(R \otimes \wedge^{\geq 2} V\right) \oplus (\alpha(v_i) R^+ \otimes \wedge V) \oplus \left(\sum_{i \neq j} \alpha(v_j) R \otimes \wedge V\right),$$

with $R = \wedge(x_1,\ldots,x_r)$. Since $D^2(v) = 0$, we have $\sigma = 0$ and $d_1(V) \subset \wedge^2(W)$. □

For nilmanifolds the relation is stronger. Recall from Theorem 3.18 that a nilmanifold M with associated Lie algebra L has for minimal model the cochain algebra on L, $C^*(L)$, and that L is also the rational homotopy Lie algebra of M. Denote $\mathcal{Z}(L)$ the center of L. Then we have:

Theorem 7.28 *Let M be a nilmanifold with associated Lie algebra L. Then $\mathrm{rk}_0(M) = \dim \mathcal{Z}(L)$.*

Proof By Proposition 7.27, $\mathrm{rk}_0(M) \leq \dim \mathcal{Z}(L_M) = \dim \mathcal{Z}(L)$. We now show that there exists a principal T^s-bundle, $T^s \to M \to N$, with $s = \dim \mathcal{Z}(L)$. This provides a free T^s-action on M and shows that $\mathrm{rk}_0(M) \geq \dim \mathcal{Z}(L)$.

Indeed the short exact sequence $0 \to \mathcal{Z}(L) \to L \to L/\mathcal{Z}(L) \to 0$ induces a relative minimal model

$$C^*(L/\mathcal{Z}(L)) \to C^*(L) \to C^*(\mathcal{Z}(L)).$$

By definition of the cochain algebra on a Lie algebra, $C^*(\mathcal{Z}(L)) = (\wedge(x_1,\ldots x_s), 0)$ and $C^*(L) = (\wedge(y_1,\ldots,y_m,x_1,\ldots,x_s), D)$ with $D(x_i) =$

$P_i \in \wedge^2(y_1, \ldots, y_m)$. Let $u_1, \ldots, u_s$ be elements of degree 2. We then have a natural morphism of differential graded algebras

$$\varphi: (\wedge(u_1, \ldots, u_s), 0) \to (\wedge(y_1, \ldots, y_m), d)$$

given by $\varphi(u_s) = P_s$ (where the P_s are d-cocycles because $D = d$ on $\wedge(y_1, \ldots, y_m)$). The morphism φ is realized by a map $f: N \to BT^s$ where N is the nilmanifold associated to the Lie algebra $L/\mathcal{Z}(L)$. A computation with minimal models shows that the nilmanifold M has the rational homotopy type of the pullback along f of the universal T^s-bundle:

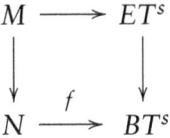

□

7.3.5 The TRC for Lie algebras

By Theorem 7.28, for finite dimensional nilpotent Lie algebras, the TRC conjecture reduces to a conjecture on the center of the Lie algebra.

Conjecture 7.29 (Lie algebra TRC) *If L is finite dimensional nilpotent Lie algebra defined over the rational numbers, then*

$$\dim H^*(L; \mathbb{Q}) \geq 2^{\dim \mathcal{Z}(L)}.$$

While the conjecture is open in general, it has been proved for nilpotent Lie algebras with certain restricted structures. For instance, we have the following result of Deninger and Singhof (see [73]; also [48]).

Theorem 7.30 *The Lie Algebra TRC conjecture is true for 2-step nilpotent Lie algebras.*

Proof A Lie algebra L is called 2-step if there is a short exact sequence $0 \to \mathcal{Z}(L) \to L \to L/\mathcal{Z}(L) \to 0$, where $\mathcal{Z}(L)$ is the center of L and $L/\mathcal{Z}(L)$ is abelian. The cochain algebra on the Lie algebra L therefore has the form

$$C^*(L) = (\wedge U \otimes \wedge V, d)$$

with $d(U) = 0$, and $d(V) \subset \wedge^2 U$. By construction, $(\wedge V, 0)$ is the cochain algebra on $\mathcal{Z}(L)$. We also have $U = U^1$ and $V = V^1$. We equip $C^*(L) =

($\wedge U \otimes \wedge V, d$) with a second (lower) gradation by putting $U = U_0$ and $V = V_1$. Then the differential d satisfies

$$d(C^*(L))_p^q \subset (C^*(L))_{p-1}^{q+1},$$

and the cohomology is bigraded. It follows that $\oplus_{p+q=r} C^*(L)_p^q$ is a subcomplex and, by the usual property of the Euler–Poincaré characteristic of a finite complex, we have the equality of polynomials

$$\sum_{r \geq 0} \left(\sum_{p+q=r} (-1)^p \dim C^*(L)_p^q \right) t^r = \sum_{r \geq 0} \left(\sum_{p+q=r} (-1)^p \dim H_p^q(C^*(L)) \right) t^r.$$

Denote this polynomial by $P_L(t)$. Note that q is the usual degree, so the summation index $r = q + p$ mixes together the degree and lower degree. Since $C^*(L) = \wedge U \otimes \wedge V$, we have

$$P_L(t) = (1+t)^m (1-t^2)^n, \quad m = \dim U, \quad n = \dim V.$$

Now multiply the last equation by $(1-t)^m$ to obtain

$$(1-t)^m P_L(t) = (1-t^2)^{m+n}.$$

By taking the value at the complex number i, we see that $|P_L(i)| \geq 2^n$. Since the modulus of $P_L(t)$ at the point i is less than or equal to the dimension of $H^*(C^*(L))$ (by the triangle inequality), we obtain

$$\dim H^*(C^*(L)) \geq |P_L(i)| \geq 2^n = 2^{\mathcal{Z}(L)}.$$

$\square$

Of course, we immediately obtain the corresponding result for nilmanifolds with appropriately restricted fundamental group.

Corollary 7.31 *The TRC conjecture is true for nilmanifolds M with 2-nilpotent fundamental group.*

The TRC conjecture is also true for low dimensional nilpotent Lie algebras. In [48], it is proved that the TRC conjecture is true for L if one of the following conditions holds:

1. $\dim L \leq 14$;
2. $\dim \mathcal{Z}(L) \leq 5$;
3. $\dim L/\mathcal{Z}(L) \leq 7$.

7.4 The localization theorem

There is a long history in topology and geometry of obtaining global information from local properties. From the Lefschetz theorem to the Poincaré–Hopf theorem to the recent theory of stationary phase, the dynamical philosophy has been that the "interesting points" of a geometric-dynamical system are the stationary points. Compact group actions on manifolds fit into this philosophical framework through their accompanying fundamental vector fields, so we should expect interesting relations between the algebraic topological invariants of the global manifolds and the fixed sets of the actions. The aim of this section is to relate properties of the cohomology of M^G to properties of the cohomology of M when G is a compact Lie group acting on a compact manifold M.

Recall first that if A is a commutative ring without zero divisors, then its field of fractions K is the quotient of $A \times (A \setminus \{0\})$ by the equivalence relation

$$(a, b) \sim (c, d) \text{ if } ad = bc.$$

For instance, the field of fractions of $\mathbb{Z}$ is $\mathbb{Q}$, and the field of fractions of the polynomial algebra $\mathbb{Q}[x_1, \ldots, x_r]$ is the field of rational functions in the variables $x_1, \ldots, x_r$, denoted by

$$K = \mathbb{Q}(x_1, \ldots, x_r).$$

If P is an A-module (or an A-algebra), we can consider the K-vector space (or the K-algebra) $K \otimes_A P$. The kernel of the *localization map*

$$\varphi : P \to K \otimes_A P, \qquad \varphi(p) = 1 \otimes p,$$

is the submodule consisting of elements p such that, for some $a \in A$, we have $ap = 0$.

We denote by $A = \mathbb{Q}[x_1, \ldots, x_r] = H^*(BG; \mathbb{Q})$ the rational cohomology of BG and by K its field of fractions. Note that, since the elements x_i have even degree, the tensor product $K \otimes_A V$ inherits only a $\mathbb{Z}/2\mathbb{Z}$ gradation (i.e. odd or even) for every graded A-module V. In other words, the even elements of V remain in even degree and the odd degree elements remain in odd degree.

7: G-spaces

Let's now come back to our G-manifold M. The injection of $F = M^G$ into M induces a morphism f between Borel fibrations

$$\begin{array}{ccc} F & \longrightarrow & M \\ \downarrow & & \downarrow \\ BG \times F = EG \times_G F & \xrightarrow{f} & EG \times_G M \\ \downarrow & & \downarrow \\ BG & = & BG \end{array}$$

We are now ready to recall *Borel's localization theorem*. Keep in mind that $A = \mathbb{Q}[x_1, \ldots, x_r] = H^*(BG; \mathbb{Q})$ and K denotes the associated field of fractions.

Theorem 7.32 ([144, Proposition 1, page 45]; also [10]) *If G is a torus and M is a compact G-manifold, then the map induced from f by localization,*

$$K \otimes_A H^*(f) \colon K \otimes_A H_G^*(M; \mathbb{Q}) \to K \otimes_A H_G^*(F; \mathbb{Q}) = K \otimes H^*(F; \mathbb{Q}),$$

is an isomorphism of K-algebras.

7.4.1 Relations between G-manifold and fixed set

As a corollary to Theorem 7.32, we deduce the fundamental relation between the Euler–Poincaré characteristics of M and of M^G that was already mentioned in Proposition 3.32.

Theorem 7.33 *Let G be a torus acting on a compact manifold M. Then*

$$\chi(M^G) = \chi(M).$$

Proof Write $A = H^*(BG; \mathbb{Q})$ and K its field of fractions. The Serre spectral sequence associated to the fibration $M \to EG \times_G M \to BG$ satisfies

$$E_2^{p,q} \cong A^p \otimes H^q(M; \mathbb{Q}) \implies H_G^{p+q}(M; \mathbb{Q}).$$

We tensor the terms of the spectral sequence by K over A and obtain a spectral sequence of finite dimensional K-vector spaces. This gives the following sequence of equalities:

$$\chi(M) = \chi_K(K \otimes_A E_2) = \chi_K(K \otimes_A E_\infty) = \chi_K(K \otimes_A H_G^*(M)).$$

By Borel's localization theorem, the last term is equal to $\chi_K(K \otimes H^*(M^G)) = \chi(M^G)$. □

Remark 7.34 Borel's localization theorem is not true for any Lie group G. For instance consider the action of $SO(3)$ on $S^3 \subset \mathbb{R}^4$ given by rotations in the three first coordinates. The fixed point set is composed of the two poles $(0,0,0,1)$ and $(0,0,0,-1)$. Therefore $\chi(S^3) = 0$ and $\chi(S^3)^G = 2$ in contradiction to a possible generalization of Theorem 7.33.

A very useful criterion for the existence of fixed points may be obtained from the localization theorem.

Theorem 7.35 (Borel fixed point criterion) *Suppose M is a compact G-manifold where G is a torus. Then the action has fixed points if and only if the homomorphism $H^*(BG; \mathbb{Q}) \to H^*(M_G; \mathbb{Q})$ is injective.*

Proof Write $A = H^*(BG; \mathbb{Q})$ and let K be its field of fractions. If $H^*(BG; \mathbb{Q}) \to H^*(M_G; \mathbb{Q})$ is injective, then the A-module generated by $1 \in H^0(M_G; \mathbb{Q})$ is free, and the K-module $K \otimes_A H^*(M_G; \mathbb{Q})$ is nonzero. Therefore $K \otimes H^*(M^G; \mathbb{Q})$ is also nonzero, so M^G cannot be empty.

On the other hand, if M^G is not empty, the choice of a point in M^G gives a map $BG \to BG \times M^G$ that is a section for the composition $BG \times M^G = EG \times_G M^G \xrightarrow{f} EG \times_G M \xrightarrow{p} BG$. In particular, $H^*(BG; \mathbb{Q}) \to H^*(M_G; \mathbb{Q})$ admits a retraction and is therefore injective. □

Recall from Definition 4.39 that a fibration $F \xrightarrow{i} E \to B$ is TNCZ (totally noncohomologous to zero) with coefficients in L if $i^*: H^*(E; L) \to H^*(F; L)$ is surjective.

Theorem 7.36 *Suppose M is a compact G-manifold, where G is a torus. Then the Borel fibration, $M \to M_G \to BG$, is TNCZ with coefficients in $\mathbb{Q}$ if and only if $\dim H^*(M; \mathbb{Q}) = \dim H^*(M^G; \mathbb{Q})$.*

Proof Write $A = H^*(BG; \mathbb{Q})$, K its field of fractions and let $(A \otimes H^*(M; \mathbb{Q}), D)$ be a semifree model of the Borel fibration (see Proposition 2.109). Then, for degree reasons, $D = 0$ if and only if the fibration is TNCZ. Therefore, if the fibration is TNCZ, we have by the Borel localization theorem,

$$\dim_{\mathbb{Q}} H^*(M; \mathbb{Q}) = \dim_K(K \otimes H^*(M; \mathbb{Q})) = \dim_K H^*(K \otimes H^*(M; \mathbb{Q}), D)$$
$$= \dim_K(K \otimes H^*(M^G; \mathbb{Q})) = \dim_{\mathbb{Q}} H^*(M^G; \mathbb{Q}).$$

If the fibration is not TNCZ, then we have, in a similar way,

$$\dim_{\mathbb{Q}} H^*(M; \mathbb{Q}) > \dim_K H^*(K \otimes H^*(M; \mathbb{Q}), D) = \dim_{\mathbb{Q}} H^*(M^G; \mathbb{Q}).$$

□

We now come to an interesting relation between the cohomologies of M and the fixed point set M^G when G is a torus.

Theorem 7.37 *For each $n \geq 0$, we have*

$$\sum_{i=0}^{\infty} \dim H^{n+2i}(M^G; \mathbb{Q}) \leq \sum_{i=0}^{\infty} \dim H^{n+2i}(M; \mathbb{Q}).$$

Proof Once again write $A = H^*(BG; \mathbb{Q})$ and let K denote A's field of fractions. The morphism $f: EG \times_G M^G \to EG \times_G M$ is a morphism of fibrations over BG. We construct a morphism of semifree A-modules $\varphi: (A \otimes H^*(M; \mathbb{Q}), D) \to (A \otimes H^*(M^G; \mathbb{Q}), 0)$ inducing $H^*(f; \mathbb{Q})$ in cohomology. As before, let $F = M^G$ and also denote $BG \times F = EG \times_G F = F_G$.

Recall that $C^*(X)$ is the rational singular cochain algebra on a space X. We denote by

$$\varphi_F: C^*(BG) \otimes H^*(F; \mathbb{Q}) \xrightarrow{\simeq} C^*(F_G),$$

and

$$\varphi_M: C^*(BG) \otimes H^*(M; \mathbb{Q}) \xrightarrow{\simeq} C^*(M_G)$$

semifree models of $C^*(F_G)$ and $C^*(M_G)$ (see Proposition 2.109). Then, by Proposition 2.107, there exists a morphism of $C^*(BG)$-modules τ such that the following diagram commutes up to homotopy, and therefore in cohomology:

$$\begin{array}{ccc}
C^*(M_G) & \xrightarrow{C^*(f)} & C^*(F_G) \\
\varphi_M \uparrow \simeq & & \varphi_F \uparrow \simeq \\
(C^*(BG) \otimes H^*(M; \mathbb{Q}), d) & \xrightarrow{\tau} & (C^*(BG) \otimes H^*(F; \mathbb{Q}), d)
\end{array}$$

Since $C^*(BG)$ is related to the minimal model $(A, 0)$ of BG by a sequence of quasi-isomorphisms, there is a morphism of A-semifree modules

$$\varphi: (A \otimes H^*(M; \mathbb{Q}), D) \to (A \otimes H^*(M^G; \mathbb{Q}), 0)$$

inducing $H^*(f; \mathbb{Q})$ in cohomology. Observe that $\varphi(A \otimes H^q(M; \mathbb{Q})) \subset \oplus_{i \geq 0} A \otimes H^{q-2i}(M^G; \mathbb{Q})$. We can therefore consider the restricted morphism of complexes

$$\tilde{\varphi}: (A \otimes H^{\leq n}(M; \mathbb{Q}), D) \to (A \otimes H^{\leq n}(M^G; \mathbb{Q}), 0),$$

and the induced morphism of quotient complexes

$$\bar{\varphi}: (A \otimes H^{\geq n}(M; \mathbb{Q}), \bar{D}) \to (A \otimes H^{\geq n}(M^G; \mathbb{Q}), 0).$$

We tensor by K over A and consider the following diagram of complexes

$$
\begin{array}{ccccc}
(K \otimes H^{\leq n}(M; \mathbb{Q}), D) & \longrightarrow & (K \otimes H^*(M; \mathbb{Q}), D) & \longrightarrow & (K \otimes H^{\geq n}(M; \mathbb{Q}), \bar{D}) \\
{\scriptstyle K \otimes_A \tilde{\varphi}} \downarrow & & {\scriptstyle K \otimes_A \varphi} \downarrow & & {\scriptstyle K \otimes_A \bar{\varphi}} \downarrow \\
(K \otimes H^{\leq n}(M^G; \mathbb{Q}), 0) & \longrightarrow & (K \otimes H^*(M^G; \mathbb{Q}), 0) & \xrightarrow{q} & (K \otimes H^{\geq n}(M^G; \mathbb{Q}), 0)
\end{array}
$$

Since q is surjective in cohomology and $K \otimes_A \varphi$ is a quasi-isomorphism, $K \otimes_A \bar{\varphi}$ is surjective in cohomology.

Note that, for degree reasons, $K \otimes_A \varphi$ maps the vector space $H^n(M; \mathbb{Q})$ into $\oplus_{i \leq 0} K \otimes H^{n+2i}(M^G; \mathbb{Q})$. This implies the surjectivity in cohomology of the map

$$K \otimes_A \bar{\varphi}: (\oplus_{i \geq 0} K \otimes H^{n+2i}(M), D) \longrightarrow (\oplus_{i \geq 0} K \otimes H^{n+2i}(M^G), 0).$$

The result now follows directly. □

Theorem 7.7 and Theorem 7.32 are in fact particular cases of a more general result (see [144, Theorem IV.6], [10, definition 4.1.5]):

Theorem 7.38 Krull-dim $(H^*(M_G; \mathbb{Q})) = \max \{\text{rank } G_x \mid x \in X\}$.

7.4.2 Some examples

Example 7.39 Let M be an even dimensional sphere, $M = S^{2n}$, with $n > 0$, and let $G = S^1$. It follows from Theorems 7.33 and 7.37 that $\chi(M^G) = 2$ and $\dim H^*(M^G; \mathbb{Q}) \leq 2$. Therefore, $\dim H^{\text{even}}(M^G) = 2$ and $\dim H^{\text{odd}}(M^G) = 0$. There are therefore only two possibilities: either M^G has the real homotopy type of the union of two points, or else M^G is a compact connected manifold that has the rational homology of a sphere and therefore the rational homotopy type of an even dimensional sphere.

For instance, fix $0 < p < n$ and write the sphere S^{2n} as

$$\left\{ (z_1, \ldots, z_p, x_{2p+1}, \ldots, x_{2n+1}) \in \mathbb{C}^p \times \mathbb{R}^{2(n-p)+1} \,\middle|\, \sum |z_i|^2 + \sum x_j^2 = 1 \right\}.$$

Then the fixed point set of the action of S^1 on S^{2n} defined by

$$t(z_1, \ldots, z_p, x_{2p+1}, \ldots, x_{2n+1}) = (tz_1, \ldots, tz_p, x_{2p+1}, \ldots, x_{2n+1}), \quad t \in S^1,$$

is the sphere $S^{2(n-p)}$.

7: G-spaces

Example 7.40 Let $M = S^{2n+1}$. If $M^G \neq \emptyset$, then we have $\chi(M^G) = \chi(M) = 0$ by Theorem 7.33, and $\dim H^*(M^G) \leq 2$ by Theorem 7.37. Thus the fixed point set M^G has the rational homotopy type of a sphere S^{2m+1} with $m \leq n$.

Example 7.41 The projective space $\mathbb{C}P(2)$ is the quotient of $\mathbb{C}^3 \setminus \{0\}$ by the equivalence relation identifying the triples (x, y, z) and $(\lambda x, \lambda y, \lambda z)$ for all nonzero $\lambda \in \mathbb{C}$. We consider the action of $G = S^1$ on $M = \mathbb{C}P(2)$ given as follows:

$$t[x, y, z] = [tx, y, z], \quad t \in S^1.$$

The fixed point set M^G is the union of a point $P = [1, 0, 0]$ and of a submanifold S^2 consisting of the elements $[0, y, z]$, $y, z \in \mathbb{C}$. The relative minimal model for the Borel fibration has the form

$$(\wedge x, 0) \to (\wedge x \otimes \wedge(u, v), D) \to (\wedge u, v), d),$$

with $|x| = 2$, $|u| = 2$, $|v| = 5$, $du = 0$, $dv = u^3$. For degree reasons, $D(u) = 0$ and $D(v) = u^3 + \alpha x u^2 + \beta x^2 u + \gamma x^3$, with $\alpha, \beta, \gamma \in \mathbb{R}$. The injection of the fixed point P into $\mathbb{C}P(2)$ induces a commutative diagram of relative Sullivan models

$$\begin{array}{ccccc} (\wedge x, 0) & \longrightarrow & (\wedge x \otimes \wedge(u, v), D) & \longrightarrow & (\wedge u, v), d) \\ \Big\downarrow = & & \Big\downarrow \psi & & \Big\downarrow \\ (\wedge x, 0) & \longrightarrow & (\wedge x, 0) & \longrightarrow & Q \end{array}$$

By making a change of generators, we can suppose that $\psi(u) = 0$. For degree reasons $\psi(v) = 0$. Since ψ is a morphism of differential graded algebras, we deduce that $\gamma = 0$.

The minimal model of the injection j of S^2 into M is given by

$$\rho: (\wedge(u, v), d) \to (\wedge(u, v'), d), \quad d(v') = u^2, \quad \rho(u) = u, \quad \rho(v) = uv'.$$

Since the action of S^1 on S^2 is trivial, j induces a diagram of relative minimal models

$$\begin{array}{ccccc} (\wedge x, 0) & \longrightarrow & (\wedge x \otimes \wedge(u, v), D) & \longrightarrow & (\wedge u, v), d) \\ \Big\downarrow = & & \Big\downarrow \tau & & \Big\downarrow \rho \\ (\wedge x, 0) & \longrightarrow & (\wedge x, 0) \otimes (\wedge(u, v'), d) & \longrightarrow & (\wedge(u, v'), d) \end{array}$$

By making changes of generators, we can suppose that $\tau(u) = u$ and $\tau(v) = \alpha_1 x v' + u v'$. The compatibility with the differentials gives $\beta = 0$.

If $\alpha = 0$, then $H^*(\wedge(x,u,v), D) = \wedge x \otimes \wedge u/u^3$. By the Borel localization theorem, $\mathbb{Q}(x) \otimes_{\wedge x} H(\wedge(x,u,v), D) \cong \mathbb{Q}(x) \otimes H^*(M^G; \mathbb{Q})$, so this would give $H^*(M^G; \mathbb{Q}) = \wedge u/u^3$, which is not correct. Therefore, $\alpha \neq 0$, and by a final change of generators, we can suppose $\alpha = 1$. Therefore, a minimal model for the Borel fibration associated to the action is given by

$$(\wedge x, 0) \to (\wedge x \otimes \wedge(u,v), D) \to (\wedge u, v), d),$$

with $|x| = 2$, $|u| = 2$, $|v| = 5$, $D(u) = 0$ and $D(v) = u^3 + xu^2$. Note that $K \otimes_A H^*(\wedge x \otimes \wedge(v, w), D) \cong K[t]/t^2 \cdot e_1 \oplus K \cdot e_2$ with $e_1 = 1 - \frac{u^2}{x^2}$, $e_2 = \frac{u^2}{x^2}$, $t = u + \frac{u^2}{x}$, $e_1 + e_2 = 1$. This decomposition corresponds to the decomposition $H^*(M^G; \mathbb{Q}) = H^*(\{P\}; \mathbb{Q}) \oplus H^*(S^2; \mathbb{Q})$.

Example 7.42 Let S^1 act on S^2 by horizontal rotations. There are two fixed points, the North pole, NP, and the South pole, SP. Let V denote a small neighborhood of NP which is stable under the action. Now denote by $S_1^2, S_2^2, S_3^2, S_4^2$ four copies of the sphere S^2 with the same S^1-action and let V_1, V_2, V_3, V_4 denote the corresponding neighborhoods of the North poles. We then have a diagonal S^1-action on $S_1^2 \times S_2^2$ and on $S_3^2 \times S_4^2$. We remove the interiors of $V_1 \times V_2$ and of $V_3 \times V_4$ and paste together the boundaries. We obtain in this way a S^1-action on $M = (S_1^2 \times S_2^2) \# (S_3^2 \times S_4^2)$. The manifold M is rationally hyperbolic, and the fixed point set consists of six isolated points. We verify directly that $\chi(M) = \chi(M^G) = 6$ (compare with Theorem 7.33).

Example 7.43 Suppose we are given an action of the circle on $M = (S_1^3 \times S_2^3) \# (S_3^3 \times S_4^3)$. Let's show that the fixed point set is path connected. Since $\chi(M) = -2$, M^G is not empty by Theorem 7.33. The vector space $H^{\text{even}}(M^G; \mathbb{Q})$ therefore has dimension one or two by Theorem 7.37. In the first case, we would have a manifold M^G whose reduced rational cohomology has dimension three and is concentrated in odd degrees. This is not possible by Poincaré duality. We thus have $\dim H^{\text{even}}(M^G) = 2$ and $\dim H^{\text{odd}}(M^G) = 4$. If there are two components, the sum of the Betti numbers in one component is greater than or equal to 3 and the reduced homology is concentrated in odd degrees. Once again this is not possible by Poincaré duality. Therefore M^G is connected and $\dim H^*(M^G) = 6$.

Example 7.44 We construct an action of S^1 on $M = (S_1^3 \times S_2^3) \# (S_3^3 \times S_4^3)$ with fixed point set a torus with two handles. Start with the trivial knot in S^3. The complement is homeomorphic to a product of S^1 with an open disk. The rotation along the meridian circles of this solid torus defines an action of S^1 on S^3 whose fixed point set consists of the original trivial knot. In the next step we choose a closed neighborhood V, stable under the S^1-action, of a closed arc contained in the trivial knot. We take four copies of the sphere S^3 and of the neighborhood V and we construct the connected sum $(S_1^3 \times S_2^3)\#(S_3^3 \times S_4^3)$ by first removing the interiors of the products $V_1 \times V_2$ and $V_3 \times V_4$ and then pasting together their boundaries. The diagonal S^1 action on the products then extends to the connected sum, and the fixed point set is a torus with two handles.

7.5 The rational homotopy of a fixed point set component

In this section, we derive information about the rational homotopy Lie algebra $L_F = \pi_*(\Omega F) \otimes \mathbb{Q}$ of a component F of M^G from knowledge of the rational homotopy Lie algebra of M, $L_M = \pi_*(\Omega M) \otimes \mathbb{Q}$. *Throughout this section, G denotes a torus which acts smoothly on a compact manifold M.* Note that the application of rational homotopy is not easy here because M^G is not necessarily path connected and therefore does not admit a minimal model. The essential part of this section consists of a presentation of theorems due to Allday and Puppe.

We will use $\mathbb{Z}/2\mathbb{Z}$-graded Sullivan models in the proof of the main theorem. For the sake of simplicity, we have concentrated all the algebraic ingredients necessary for the proof in a subsection. The reader who is only interested in topological applications can avoid reading that subsection.

7.5.1 The rational homotopy groups of a component

The fixed point set M^G decomposes as a finite union of submanifolds $M^G = F_1 \coprod F_2 \coprod \ldots \coprod F_n$. The component F_1 is denoted F. The injection of F into M induces a commutative diagram of Borel fibrations and a commutative diagram of corresponding relative minimal models:

$$\begin{array}{ccccc} F & \longrightarrow & F_G & \longrightarrow & BG \\ \downarrow & & \downarrow & & \| \\ M & \longrightarrow & M_G & \longrightarrow & BG \end{array}$$

$$\begin{array}{ccccc} A & \longrightarrow & (A \otimes \wedge X_M, D) & \longrightarrow & (\wedge X_M, d) \\ \| & & \downarrow f & & \downarrow \\ A & \longrightarrow & (A, 0) \otimes (\wedge X_F, d) & \longrightarrow & (\wedge X_F, d) \end{array}$$

Here as usual $A = H^*(BG; \mathbb{Q})$ and K is its field of fractions.

7.5 The rational homotopy of a fixed point set component

Theorem 7.45 ([3]) *With the notation above, there exists a differential $\overline{D}$ on the $\mathbb{Z}/2\mathbb{Z}$-vector space $K \otimes X_M$ such that the cohomology $H^*(K \otimes X_M, \overline{D})$ is isomorphic to $K \otimes X_F$ as a $\mathbb{Z}/2\mathbb{Z}$-vector space.*

Proof The canonical augmentation $\varepsilon \colon (\wedge X_F, d) \to (\mathbb{Q}, 0)$ extends naturally to an A-linear map $\varepsilon \colon (A, 0) \otimes (\wedge X_F, d) \to (A, 0)$. We tensor ε by K over A and denote by I the kernel of the augmentation map

$$\Gamma \colon (K \otimes \wedge X_M, D) \xrightarrow{f} (K, 0) \otimes (\wedge X_F, d) \xrightarrow{\varepsilon} (K, 0).$$

Now let x_i be a basis of X_M indexed by a well ordered set such that $d(x_i) \in A \otimes \wedge(x_j, j < i)$. For each x_i, the element $\tilde{x}_i = x_i - \Gamma(x_i)$ belongs to I, with $\Gamma(x_i) = 0$ when $|x_i|$ is odd. We denote by $\tilde{X}_M$ the vector space generated by the $\tilde{x}_i$. The isomorphism

$$\theta \colon K \otimes \wedge X_M \to K \otimes \wedge \tilde{X}_M$$

defined by $\theta(x_i) = \tilde{x}_i + \Gamma(x_i)$ induces a differential $\tilde{D} = \theta D \theta^{-1}$ on $K \otimes \wedge \tilde{X}_M$. Note that the differential algebra $(K \otimes \wedge \tilde{X}_M, \tilde{D})$ is a $\mathbb{Z}/2\mathbb{Z}$-graded Sullivan algebra, but not necessarily a minimal one. Write $Q(\tilde{D})$ for the differential induced on the indecomposables.

Since $f(I) \subset K \otimes \wedge^+ X_F$, $f(\tilde{X}_M) \subset K \otimes \wedge^+ X_F$, and f induces a map on the homology of indecomposables

$$Q(f) \colon H(K \otimes \tilde{X}_M, Q(\tilde{D})) \longrightarrow K \otimes X_F.$$

We will now prove that this map is an isomorphism and this will imply the result.

By the localization theorem we have an isomorphism

$$\theta \colon K \otimes_A H_G^*(M; \mathbb{Q}) \to K \otimes_A H_G^*(M^G; \mathbb{Q})$$
$$\cong K \otimes H^*(M^G; \mathbb{Q}) \cong \oplus_{i=1}^n (K \otimes H^*(F_i; \mathbb{Q})).$$

If we write $H^0(F_i; \mathbb{Q}) = \mathbb{Q} e_i$ with $e_i^2 = e_i$, then $H^0(M^G; \mathbb{Q})$ is the vector space generated by the orthogonal idempotents e_i:

$$e_i^2 = e_i, \quad e_i \cdot e_j = 0 \text{ if } i \neq j.$$

We choose an element x in $K \otimes_A H_G^*(M; \mathbb{Q})$ such that $\theta(x) = e_1$. By inverting the element $\theta(x)$, we kill the components $H^*(F_i; \mathbb{Q})$ for $i \geq 2$. We get an isomorphism of algebras

$$\bar{\theta} \colon \left(K \otimes_A H_G^*(M; \mathbb{Q})\right)[x^{-1}] \longrightarrow \left(K \otimes H^*(M^G; \mathbb{Q})\right)[\theta(x)^{-1}]$$
$$\cong K \otimes H^*(F; \mathbb{Q}).$$

Now, the isomorphism $H^*(K \otimes \wedge \tilde{X}_M, D) \cong K \otimes_A H_G^*(M; \mathbb{Q})$ shows that there is a cocycle u in $K \otimes \wedge \tilde{X}_M$ such that $[u] = x$. We then consider the $\mathbb{Z}/2\mathbb{Z}$-graded relative minimal model $(K \otimes \wedge \tilde{X}_M \otimes \wedge(v, w), D)$, where $|v| = 0$, $|w| = 1$, $D(v) = 0$, and $D(w) = uv - 1$.

Define a morphism $\psi \colon H^*(K \otimes \wedge \tilde{X}_M, D)[x^{-1}] \to H^*(K \otimes \wedge \tilde{X}_M \otimes \wedge(v, w), D)$ by putting $\psi(x^{-1}) = [v]$. Recall that a localized module $B[x^{-1}]$ is by definition the quotient $B[x^{-1}] = B \otimes \wedge y/(xy - 1)$. In our case, since $uv - 1 = 0$ in $H^*(K \otimes \wedge \tilde{X}_M \otimes \wedge(v, w), D)$, the morphism ψ is well defined and is an isomorphism by Lemma 7.46.

We extend the map $f \colon (K \otimes \wedge X_M, D) \to (K, 0) \otimes (\wedge X_F, d)$ to a map

$$\bar{f} \colon (K \otimes \wedge \tilde{X}_M \otimes \wedge(v, w), D) \to (K, 0) \otimes (\wedge X_F, d),$$

by defining $\bar{f}(v) = 1$ and $\bar{f}(w) = 0$.

The Borel localization isomorphism $H^*(K \otimes \wedge X_M)[x^{-1}] \xrightarrow{\cong} K \otimes H^*(\wedge X_F, d)$ then factors as the composition

$$H^*(K \otimes \wedge \tilde{X}_M)[x^{-1}] \xrightarrow[\cong]{\psi} H^*(K \otimes \wedge \tilde{X}_M \otimes \wedge(v, w), D)$$

$$\xrightarrow{-H^*(\bar{f})} K \otimes H^*(\wedge X_F, d).$$

Since ψ is an isomorphism, $H^*(\bar{f})$ is also an isomorphism.

We introduce the elements $\tilde{u} = u - 1$ and $\tilde{v} = v - 1$ that belong to the kernel of $\varepsilon \circ \bar{f}$, and we have $D(w) = \tilde{u} + \tilde{v} + \tilde{u}\tilde{v}$.

In the same way as in the classical theory of $\mathbb{N}$-graded Sullivan models, an isomorphism between $\mathbb{Z}/2\mathbb{Z}$-graded Sullivan models induces an isomorphism on the homology of indecomposable elements (see Lemma 7.56). In our case this implies that $H(K \otimes \tilde{X}_M, Q(D)) \cong H(K \otimes (\tilde{X}_M \oplus (\tilde{v}, w)), Q(D)) \cong K \otimes X_F$. Since $K \otimes X_M$ is isomorphic to $K \otimes \tilde{X}_M$ as a graded vector space, we obtain a differential $\overline{D}$ on $K \otimes X_M$ such that $H(K \otimes X_M, \overline{D}) \cong H(K \otimes \tilde{X}_M, Q(D))$. □

Lemma 7.46 *Let $(B \otimes \wedge(v, w), D)$ be a differential $\mathbb{Z}/2\mathbb{Z}$-graded commutative algebra with $|v| = 0$, $|w| = 1$, $D(v) = 0$ and $D(w) = uv - 1$, where u is a cocycle of degree 0 in B. Then the morphism $\psi \colon H^*(B)[u^{-1}] \to H^*(B \otimes \wedge(v, w), D)$ defined by $\psi(u^{-1}) = [v]$ is an isomorphism.*

Proof We filter the cochain algebra $(B \otimes \wedge(v, w), D)$ by the degree in $B \otimes \wedge v$. We get a spectral sequence whose E_2-term satisfies

$$E_2 = E_2^{0,*} \oplus E_2^{1,*}, \quad E_2^{0,*} = \bigl(H^*(B, d) \otimes \wedge v\bigr)/(uv - 1) = H^*(B, d)[u^{-1}].$$

A cocycle α in $E_2^{1,*}$ decomposes into a sum $\alpha = \sum_{i=r}^{n} b_i v^i w$, with $b_i \in H^*(B, d)$. The condition $d_1(\alpha) = 0$ implies that $b_r = 0$. Therefore, by induction, we have $E_2^{1,*} = 0$. The spectral sequence therefore collapses at the E_2-level and this shows that ψ is an isomorphism. □

Here is a main consequence of Theorem 7.45. It gives a powerful connection between a G-manifold and its fixed point set when the extra condition of rational ellipticity is added.

Corollary 7.47 ([3]) *Let G be a torus, M a compact nilpotent rationally elliptic G-manifold and F a nilpotent component of M^G. Then*

1. *F is a rationally elliptic manifold;*
2. *$\chi_\pi(F) = \chi_\pi(M)$;*
3. *$\dim \pi_*(F) \otimes \mathbb{Q} \leq \dim \pi_*(M) \otimes \mathbb{Q}$.*

Proof With the above notation, by Theorem 7.45, $\dim X_F = \dim {}_K H^*(K \otimes X_M, \overline{D})$, so clearly we have $\dim X_F \leq \dim X_M$. If we let $\chi(X_M)$, $\chi(\tilde{X}_M)$, $\chi(X_F)$ denote the Euler–Poincaré characteristics of the respective complexes $K \otimes X_M$, $K \otimes \tilde{X}_M$, $K \otimes X_F$, the standard result that the Euler–Poincaré characteristic is invariant under taking homology gives

$$\chi_\pi(M) = \chi(X_M) = \chi(\tilde{X}_M) = \chi(X_F) = \chi_\pi(F).$$

□

As a corollary we recover a special case of the well-known theorem that the inclusion of an orbit for an effective torus action on an aspherical manifold induces an injection at the fundamental group level (see [63]).

Theorem 7.48 *Let G be a torus acting on a nilmanifold $M = K(\pi, 1)$.*

1. *If the action is nontrivial, then $M^G = \emptyset$.*
2. *If the action is effective, then the injection of an orbit $G \cdot x \hookrightarrow M$ induces an injective map $\pi_1(G) \to \pi$.*

Proof The minimal model of a nilmanifold M has the form $(\wedge X_M, d)$, where X_M is a finite dimensional vector space concentrated in degree one. Moreover, the dimension of X_M is equal to the dimension of the manifold M (see Remark 3.21). Suppose $M^G \neq \emptyset$ and let F be a path component of M^G, and $(\wedge X_F, d)$ its minimal model. By Theorem 7.45, $H^*(K \otimes X_M, \overline{D}) \cong K \otimes X_F$. But X_M is concentrated in degree 1, so $K \otimes X_M$ is in odd degrees only. Thus, we have $\overline{D} = 0$ and, therefore, $K \otimes X_M \cong K \otimes X_F$ as a $\mathbb{Z}/2\mathbb{Z}$-graded vector space. Thus, the dimension of X_F is equal to the dimension of X_M. Now, X_F is concentrated in odd degrees, but if there is an element in an odd degree larger than 1, then the dimension of F would exceed the dimension of M by Theorem 2.75 (6). Thus, X_F is concentrated in degree one. This implies that

F is a closed submanifold of the same dimension as M, so $M^G = F = M$, which is not possible. Therefore $M^G = \emptyset$.

Suppose now that the action is effective. We shall show that the action is, in fact, almost free. If not, then there exists some S^1 in some isotropy group at a point $x \in M$, say. Since we assume the action is effective, we know that this S^1 does not act trivially on the whole manifold M. By (1), we also know that $M^{S^1} = \emptyset$, but this contradicts the fact that S^1 is in the isotropy group of x. Hence, no isotropy group contains an S^1 and the action is almost free. By Hsiang's theorem, Theorem 7.7, we have that $H_G^*(M; \mathbb{Q})$ is finite dimensional. Now let

$$(\wedge(x_1, \ldots, x_r), 0) \to (\wedge(x_1, \ldots, x_r, y_1, \ldots, y_n), D) \to (\wedge(y_1, \ldots, y_n), \bar{D})$$

be a relative minimal model for the Borel fibration $M \to M_G \to BG$, with $G = T^r$. We have, as usual, $|x_i| = 2$ and, because M is a nilmanifold, $|y_i| = 1$. Because $(\wedge(x_1, \ldots, x_r), d = 0)$ is a polynomial algebra and all the other generators of $(\wedge(x_1, \ldots, x_r, y_1, \ldots, y_n), D)$ are in degree 1, the only way to achieve finite dimensionality for cohomology is to kill the entire vector space in degree 2 having basis $\{x_1, \ldots, x_r\}$. Therefore, there must exist r elements $y_{i_1}, \ldots, y_{i_r}$ such that the elements $D(y_{i_j})$ form a basis for the vector space generated by the x_k. In particular, this says that the $D(y_{i_j})$ all have nonzero linear parts. Recall from Proposition 2.65 that the linear part of a differential in a relative model represents the connecting map in the long exact homotopy sequence of the associated fibration. Here this means that the connecting map $\pi_2(BG) \otimes \mathbb{Q} \to \pi_1(M) \otimes \mathbb{Q}$ is injective (since the entire vector space generated by the x_k's is hit by the differential D). Since $\pi_2(BG)$ is torsion-free, the usual connecting map $\pi_2(BG) \to \pi_1(M)$ is also injective. Recalling that the composition $\pi_1(G) = \pi_1(\Omega BG) \cong \pi_2(BG) \to \pi_1(M)$ may be identified with the map induced by the injection of an orbit, we obtain the result. $\square$

Corollary 7.49 *Suppose that G is an r-torus acting on a compact nilpotent rationally elliptic manifold M with $\chi_\pi(M) \neq 0$. Then M^G has no contractible component.*

Proof For each component F of M^G, $\chi_F \neq 0$ and so, by Corollary 7.47, the rational cohomology of F is not trivial. $\square$

Example 7.50 Let G be a torus acting on the projective n-space $M = \mathbb{C}P(n)$. Since the rational cohomology of M is zero in odd degrees, by Theorem 7.37, the cohomology of M^G is concentrated in even degrees and $\dim H^*(M^G; \mathbb{Q}) = n + 1$. Denote by F a simply connected component of M^G. Since $\chi_\pi(F) = \chi_\pi(M) = 0$, and $\dim \pi_*(F) \otimes \mathbb{Q} \leq \dim \pi_*(M) \otimes \mathbb{Q} = 2$, F

has the rational homotopy type of a point or of a complex projective space $\mathbb{C}P(m)$ with $m \leq n$.

Example 7.51 Let G be a torus acting on a nilpotent rationally hyperbolic manifold M, and suppose that $\pi_{even}(M) \otimes \mathbb{Q}$ is finite dimensional. Let F be a component of M^G. Then with the notation of Theorem 7.45, $H^*(K \otimes X_M, \overline{D})$ is infinite dimensional, and so, by Theorem 7.45, $K \otimes X_F \cong H^*(K \otimes X_M, \overline{D})$ is also infinite dimensional. In particular, each nilpotent component of the fixed point set is a rationally hyperbolic manifold.

Example 7.52 Let M be the connected sum of two copies of the product $S^3 \times S^3 \times S^3 \times S^3$ and let G be a torus acting on M. Denote by $z \in \pi_*(M) \otimes \mathbb{Q}$ the class of the image of the boundary sphere in $(S^3)^4 \setminus \{*\}$, and by $y \in \pi_*(\Omega M) \otimes \mathbb{Q}$ the image of z by adjunction. The construction of the differential in the minimal model of the connected sum (see Example 3.6) shows that the element y is a generator in the Lie algebra $\pi_*(\Omega M) \otimes \mathbb{Q}$. Therefore, by Corollary 3.4, the rational homotopy Lie algebra $\pi_*(\Omega M) \otimes \mathbb{Q}$ is a graded Lie algebra generated in degrees 2 and 10. It follows from Example 7.51 that $\pi_*(\Omega F) \otimes \mathbb{Q}$ is also infinite dimensional.

7.5.2 Presentation of the Lie algebra $L_F = \pi_*(\Omega F) \otimes \mathbb{Q}$

When $\pi_{even}(M) \otimes \mathbb{Q} = 0$, or more generally, when the differential D defined in Theorem 7.45 on the $\mathbb{Z}/2\mathbb{Z}$-graded vector space $K \otimes X_M$ vanishes, the Borel localization theorem gives information about the Lie algebra $\pi_*(\Omega F) \otimes \mathbb{Q}$.

Theorem 7.53 ([9]) *Suppose M is a compact smooth T^r-manifold satisfying $\pi_{even}(M) \otimes \mathbb{Q} = 0$, and let F be a nilpotent component of M^G. Denote by*

$$\mathbb{L}(W_M) \to \mathbb{L}(G_M) \to L_M$$

and

$$\mathbb{L}(W_F) \to \mathbb{L}(G_F) \to L_F$$

minimal presentations, where $\mathbb{L}(-)$ denotes the graded free Lie algebra functor. Then,

1. $\dim G_F \leq \dim G_M$;
2. $\dim W_F \leq \dim W_M$.

Proof Since the action is smooth, for every generic circle $S^1 \subset T^r$, we have $X^{S^1} = X^{T^r}$ (see Lemma 7.3). Thus we may assume without loss of generality that $r = 1$. In this case, $A = \mathbb{Q}[t]$ and $K = \mathbb{Q}(t)$. For $\alpha \in \mathbb{Q}$, let $\mathbb{Q}_\alpha$ indicate that $\mathbb{Q}$ has been made into an A module via the map $A \to \mathbb{Q}$ given by $t \mapsto \alpha$.

We use the notation contained in the proof of Theorem 7.45. Since X_M is concentrated in odd degrees, $\tilde{X}_M = X_M$ and the differential $Q(D)$ on the indecomposable elements is zero. Therefore f induces an isomorphism of $\mathbb{Z}/2\mathbb{Z}$-graded differential algebras

$$K \otimes_A (A \otimes \wedge X_M, D_1) \to (K, 0) \otimes (\wedge X_F, d_1),$$

where $D_1(X_M) \subset A \otimes \wedge^2 X_M$ and d_1 is the quadratic part of the differential d.

On the other hand, by the definition of the relative minimal model of M_G, we have $\mathbb{Q}_0 \otimes_A (A \otimes \wedge X_M, D_1) = (\wedge X_M, d_1)$. Denote by C^* the complex $(A \otimes \wedge^* X_M, D_1)$. The universal coefficient theorem gives a short exact sequence

$$0 \to \mathbb{Q}_0 \otimes_A H^i(C^*) \to H^i(\mathbb{Q}_0 \otimes_A C^*) \to \mathrm{Tor}_1^A(H^{i+1}(C^*), \mathbb{Q}_0) \to 0.$$

Note, in particular, that the indices refer to the gradation $A \otimes \wedge^* X_M$.

Now observe that, for a morphism $\varphi \colon P_1 \to P_2$ of A-modules, $K \otimes_A \varphi$ is an isomorphism if and only if $\mathbb{Q}_1 \otimes_A \varphi$ is an isomorphism. Then, using Lemma 7.54 below, we have the following sequence of equalities.

$$\begin{aligned}
\dim {}_\mathbb{Q} H^q(\wedge X_F, d_1) &= \dim {}_\mathbb{Q} \mathbb{Q}_1 \otimes_A H^q(A \otimes \wedge X_F, d_1) \\
&= \dim {}_\mathbb{Q} \mathbb{Q}_1 \otimes_A H^q(A \otimes \wedge X_M, D_1) \\
&= \dim {}_\mathbb{Q} \mathbb{Q}_0 \otimes_A H^q(C^*) - \dim {}_\mathbb{Q} \mathrm{Tor}_1^A(H^q(C^*), \mathbb{Q}_0) \\
&= \dim {}_\mathbb{Q} H^q(\mathbb{Q}_0 \otimes_A C^*) - \dim {}_\mathbb{Q} \mathrm{Tor}_1^A(H^{q+1}(C^*), \mathbb{Q}_0) \\
&\quad - \dim {}_\mathbb{Q} \mathrm{Tor}_1^A(H^q(C^*), \mathbb{Q}_0) \\
&= \dim {}_\mathbb{Q} H^q(\wedge X_M, d_1) - \dim {}_\mathbb{Q} \mathrm{Tor}_1^A(H^{q+1}(C^*), \mathbb{Q}_0) \\
&\quad - \dim {}_\mathbb{Q} \mathrm{Tor}_1^A(H^q(C^*), \mathbb{Q}_0).
\end{aligned}$$

By Subsection 2.6.2, we have $W_M \cong H^2(\wedge X_M, d_1)$, $W_F \cong H^2(\wedge X_F, d_1)$, $G_M = H^1(\wedge X_M, d_1)$ and $G_F \cong H^1(\wedge X_F, d_1)$. The result then follows directly from the equalities above. □

Lemma 7.54 *Let P be a graded A-module and suppose that* $\dim(\mathbb{Q}_0 \otimes_A P) < \infty$. *Then*

$$\dim {}_\mathbb{Q}(\mathbb{Q}_0 \otimes_A P) = \dim {}_\mathbb{Q}(\mathbb{Q}_1 \otimes_A P) + \dim {}_\mathbb{Q} \mathrm{Tor}_1^A(P, \mathbb{Q}_0).$$

Proof Since $\dim(\mathbb{Q}_0 \otimes_A P) < \infty$, P is finitely generated, and we can write P as the direct sum of a free part and submodules of the form $\mathbb{Q}[t]/(t^n)$. Let $0 \to N \to L \to P \to 0$ be a minimal free presentation of P. We directly deduce the following exact sequences:

$$0 \to \mathrm{Tor}_1^A(P, \mathbb{Q}_0) \to \mathbb{Q}_0 \otimes_A N \to \mathbb{Q}_0 \otimes_A L \to \mathbb{Q}_0 \otimes_A P \to 0$$

and
$$0 \to \mathbb{Q}_1 \otimes_A N \to \mathbb{Q}_1 \otimes_A L \to \mathbb{Q}_1 \otimes_A P \to 0.$$

□

7.5.3 $\mathbb{Z}/2\mathbb{Z}$-Sullivan models

The purpose of this section is to extend to $\mathbb{Z}/2\mathbb{Z}$-graded Sullivan algebras the basic properties of usual Sullivan algebras. All the material of this section is due to C. Allday [3]. In what follows all vector spaces will be defined over a field $\mathbb{k}$ of characteristic zero.

A $\mathbb{Z}/2\mathbb{Z}$-graded vector space A is a vector space with a decomposition $A = A^0 \oplus A^1$. It is a $\mathbb{Z}/2\mathbb{Z}$-graded algebra if there is a multiplication satisfying $A^p \cdot A^q \subset A^{p+q}$, where the addition $p+q$ is taken modulo 2. The algebra A is called *commutative* if for homogeneous elements a and b, we have $a \cdot b = (-1)^{|a| \cdot |b|} b \cdot a$. It is called *free* if it is the tensor product of a polynomial algebra on elements of degree 0 with an exterior algebra on elements of degree 1; in this case we write $A = \wedge V$ where V is the linear span of the generators of the algebra.

A $\mathbb{Z}/2\mathbb{Z}$-graded differential algebra (A,d) is a $\mathbb{Z}/2\mathbb{Z}$-graded algebra A equipped with a differential $d \colon A^p \to A^{p+1}$. For instance, each cochain algebra (A,d) defines a $\mathbb{Z}/2\mathbb{Z}$-graded differential algebra with $A^0 = A^{\text{even}}$ and $A^1 = A^{\text{odd}}$.

A $\mathbb{Z}/2\mathbb{Z}$-*graded Sullivan algebra* is a $\mathbb{Z}/2\mathbb{Z}$-graded differential algebra (A,d) such that A is the free commutative graded algebra on a vector space V that admits a basis indexed by a well-ordered set (x_i) with $d(x_i) \in \wedge(x_j, j < i)$. We write $(A,d) = (\wedge V, d)$. It is called a $\mathbb{Z}/2\mathbb{Z}$-graded minimal Sullivan algebra if $d(V) \subset \wedge^{\geq 2} V$. For instance every Sullivan algebra induces a $\mathbb{Z}/2\mathbb{Z}$-graded Sullivan algebra and a minimal Sullivan algebra $(\wedge V, d)$ induces a $\mathbb{Z}/2\mathbb{Z}$-graded minimal Sullivan algebra. A *morphism* of $\mathbb{Z}/2\mathbb{Z}$-graded differential algebras $f \colon (A,d) \to (B,d)$ is a degree zero linear map that is compatible with the multiplications ($f(ab) = f(a)f(b)$) and with the differentials ($df = fd$). As usual, a quasi-isomorphism is a morphism that induces an isomorphism in cohomology.

Let (A,d) be a $\mathbb{Z}/2\mathbb{Z}$-graded differential algebra. An augmentation is a surjective map of differential graded algebras $\varepsilon \colon (A,d) \to (\mathbb{k}, 0)$, where $\mathbb{k}$ is in degree 0. A $\mathbb{Z}/2\mathbb{Z}$-graded Sullivan algebra $(\wedge V, d)$ is equipped with a natural augmentation sending V to 0. A morphism between $\mathbb{Z}/2\mathbb{Z}$-graded Sullivan algebras $f \colon (\wedge V, d) \to (\wedge W, d)$ preserves the augmentations if $f(V) \subset \wedge^+ W$.

When $(\wedge V, d)$ is a $\mathbb{Z}/2\mathbb{Z}$-graded Sullivan algebra, the vector space V is called the vector space of indecomposable elements. The differential d

induces a differential $Q(d)$ on V defined as the composition

$$Q(d)\colon V \xrightarrow{d} \wedge^+ V \xrightarrow{p} \wedge^+ V / \wedge^{\geq 2} V \cong V.$$

Here, p denotes the canonical projection. The homology $H^*(V, Q(d))$ is called the homology of indecomposable elements. An augmentation preserving map $f\colon (\wedge V, d) \to (\wedge W, d)$ induces a map of complexes $Q(f)\colon (V, Q(d)) \to (W, Q(d))$.

To define homotopies, we consider the contractible Sullivan algebra $(\wedge(t, dt), d)$ with $|t| = 0$, $|dt| = 1$ and $d(t) = dt$. We denote by $p_i \colon (\wedge(t, dt), d) \to (\Bbbk, 0)$, $i = 0, 1$, the augmentations defined by $p_i(t) = i$ and $p_i(dt) = 0$.

Definition 7.55 *The morphisms $f, g\colon (A, d) \to (B, d)$ are homotopic if there exists a morphism $H\colon (A, d) \to (B, d) \otimes \wedge(t, dt)$ such that $f = (1 \otimes p_0) \circ H$ and $g = (1 \otimes p_1) \circ H$.*

Lemma 7.56 ([3]) *An augmentation preserving quasi-isomorphism $f\colon (A, d) \to (B, d)$ between $\mathbb{Z}/2\mathbb{Z}$-graded Sullivan algebras induces an isomorphism on the homology of indecomposable elements.*

7.6 Hamiltonian actions and bundles

The symmetries which are important in symplectic geometry all involve the symplectic form. Although actions of more general compact Lie groups on symplectic manifolds may be considered, here we shall concentrate on circle and torus actions to avoid the machinery of moment maps.

7.6.1 Basic definitions and properties

From now on, (M^{2n}, ω) denotes a closed symplectic manifold.

Definition 7.57

- A smooth action of the circle, $A\colon S^1 \times M \to M$ is *symplectic* if $g^*\omega = \omega$ for each $g \in S^1$.
- An *orbit map* $\alpha\colon S^1 \to M$ is defined by choosing $x \in M$ and taking the orbit through x; $g \mapsto gx$. We may consider $\alpha \in \pi_1(M)$ and we denote its Hurewicz image by $h(\alpha) \in H_1(M; \mathbb{R})$. Note that all orbit maps induce the same map on homology (since M is always connected).
- The *fundamental vector field* X associated to the action is defined at each $x \in M$ as $D_x\alpha(\partial/\partial\theta)$, where $\partial/\partial\theta$ is the unit vector field of S^1 and $D_x\alpha$ is the induced map on tangent spaces.

The main point that is elicited by the definition of symplectic action is that the Lie derivative $\mathcal{L}(X)$ (see Section A.2) vanishes on forms that are invariant under the action (compare Section 1.7, especially Proposition 1.33). This property applied to the symplectic form ω gives

$$0 = \mathcal{L}(X)\omega = (di(X) + i(X)d)\omega = di(X)\omega,$$

where $i(X)$ denotes interior multiplication by X (see Section A.2). Here we use the formula in Proposition A.4 and the fact that $d\omega = 0$. Therefore, the form $i(X)\omega$ is a closed form. Of course, the same holds for every invariant form on M. Moreover, by Theorem 1.28, we know that the cohomology of the invariant forms is the same as the ordinary de Rham cohomology, so we can consider $i(X)$ as a derivation on $H^*(M;\mathbb{R})$. To see this, let β and β' be invariant forms that represent the same cohomology class. Hence, there is an invariant form γ with $\beta - \beta' = d\gamma$. But then,

$$i(X)\beta - i(X)\beta' = i(X)d\gamma = d(-i(X)\gamma)$$

since $\mathcal{L}(X)\gamma = 0$. Therefore, in $H^*(M;\mathbb{R})$, $[i(X)\beta] = [i(X)\beta']$.

Definition 7.58 *The λ-derivation on $H^*(M;\mathbb{R})$ is defined by*

$$\lambda(b) = [i(X)\beta],$$

where $b \in H^*(M;\mathbb{R})$ is represented by the invariant form β.

Proposition 7.59 ([179]) *The λ-derivation may be identified directly from the homomorphism induced on cohomology by the action A;*

$$A^*b = 1 \times b + u_{S^1} \times \lambda(b),$$

where u_{S^1} is the generator of $H^1(S^1;\mathbb{R})$ corresponding to the unit 1-form $d\theta$ on S^1. In particular, when $b \in H^1(M;\mathbb{R})$, $\lambda(b) = \langle \alpha^*(b), [S^1] \rangle$.

So now we have two ways to view the λ-derivation. In fact, the λ-derivation has already appeared in the proof of Theorem 4.98. Recall from Definition 4.96 that (M^{2n}, ω) has *Lefschetz type* if multiplication by $[\omega]^{n-1}$ is an isomorphism: $\mathrm{Lef} = \omega^{n-1} \cup : H^1(M) \xrightarrow{\cong} H^{2n-1}(M)$. For the sake of simplicity, we denote the symplectic form and its cohomology class by ω. Now here are a few properties of the λ-invariant, see [180].

Proposition 7.60 *With the notation of Definition 7.57 and the preceding discussion, we have:*

1. *$h(\alpha)$ is Poincaré dual to $n\lambda(\omega)\,\omega^{n-1} = n\,\mathrm{Lef}(\lambda(\omega))$;*
2. *$\lambda(\omega^n) = 0$ if and only if $\lambda(H^1(M)) = 0$ if and only if $h(\alpha) = 0$;*

3. If (M^{2n}, ω) has Lefschetz type, then the three properties of part 2 are equivalent to $\lambda(\omega) = 0$.

Proof (1) Let $b \in H^1(M)$ and note that $b\omega^n = 0$ for degree reasons. Applying λ, we have

$$0 = \lambda(b)\omega^n - b\lambda(\omega^n) = \lambda(b)\omega^n - bn\lambda(\omega)\omega^{n-1}$$

using the fact that λ is a graded derivation. Hence, $\lambda(b)\omega^n = bn\lambda(\omega)\omega^{n-1}$. Taking the Kronecker product with the fundamental class of M gives $\langle b\lambda(\omega^n), [M]\rangle = \langle \lambda(b)\omega^n, [M]\rangle = \lambda(b)$. But $\lambda(b) = \langle \alpha^*(b), [S^1]\rangle = \langle b, \alpha_*[S^1]\rangle = \langle b, h(\alpha)\rangle$, so

$$\langle b, h(\alpha)\rangle = \lambda(b) = \langle b\lambda(\omega^n), [M]\rangle = \langle bn\lambda(\omega)\omega^{n-1}, [M]\rangle$$
$$= \langle b, n\lambda(\omega)\omega^{n-1} \cap [M]\rangle$$

and we see that $h(\alpha)$ is Poincaré dual to $n\lambda(\omega)\omega^{n-1}$.

(2) The first equivalence follows exactly as in the proof of Theorem 4.36. The second equivalence comes from $\lambda(b) = \langle b, h(\alpha)\rangle$ for all $b \in H^1(M)$.

(3) Suppose M has Lefschetz type. If $\lambda(\omega) = 0$, then it is clear that $\lambda(\omega^n) = 0$. If $\lambda(\omega^n) = 0$, then we have $0 = \lambda(\omega^n) = n\lambda(\omega)\omega^{n-1} = n\text{Lef}(\lambda(\omega))$. But the Lefschetz type assumption then implies that $\lambda(\omega) = 0$. □

7.6.2 Hamiltonian and cohomologically free actions

The importance of Proposition 7.60 comes from the following fundamental definitions in symplectic geometry. If $A: T^k \times M \to M$ is a symplectic toral action, then we may define k λ-derivations by $\lambda_j(\omega) = [i(X_j)\omega]$, where X_j is the fundamental vector field of the jth circle factor in T^k.

Definition 7.61 *For a symplectic manifold (M^{2n}, ω):*

1. *a symplectic action $A: T^k \times M \to M$ is Hamiltonian if $\lambda_j(\omega) = 0$ for each $j = 1, \ldots, k$. In particular, if $H^1(M; \mathbb{R}) = 0$, then all symplectic actions are Hamiltonian;*
2. *a symplectic action $A: T^k \times M \to M$ is cohomologically free if the homomorphism on cohomology, $\psi: H^1(T^k; \mathbb{R}) \to H^1(M; \mathbb{R})$, defined by*

$$\psi(u_{S_j^1}) = \lambda_j(\omega)$$

is injective. Here $u_{S_j^1}$ denotes the generator of $H^1(T^k; \mathbb{R})$ corresponding to the j-th circle factor.

7.6 Hamiltonian actions and bundles

Remark 7.62 Even if (M^{2n}, ω) is *only c-symplectic* with c-symplectic cohomology class ω, we can, by Proposition 7.59, still define $\lambda(\omega)$ for any S^1-action and say that the action is *c-Hamiltonian* if $\lambda(\omega) = 0$. Similarly, we can define a notion of *cohomologically free* for c-symplectic manifolds. See [179] for details.

Recall from Theorem 1.102 that, for an S^1-action on a closed manifold M, the Cartan–Weil model is the complex $(\Omega_X[u], d_X)$, where Ω_X is the sub-complex of $A_{DR}(M)$ consisting of S^1-invariant forms, u is a degree 2 generator and the differential is defined by

$$d_X u = 0, \text{ and } d_X \alpha = d\alpha - u i(X)\alpha,$$

for $\alpha \in \Omega_X$. The Cartan–Weil model induces an isomorphism

$$H^*(\Omega_X[u], d_X) \cong H^*(M_{S^1}; \mathbb{R}).$$

Thus, the Cartan–Weil model calculates the equivariant cohomology associated to the action of S^1 on M.

The equation $d_X \alpha = d\alpha - u i(X)\alpha$ says something interesting. Although α is an invariant form, even if it is closed, this does *not* mean that α is equivariantly closed (i.e. d_X-closed). For that, we also require $i(X)\alpha = 0$.

Now suppose the S^1-action is a symplectic action so that the symplectic form ω is an invariant form. Then we have $d_X \omega = d\omega - u i(X)\omega = -u i(X)\omega$ since $d\omega = 0$. Therefore, the interior multiplication $i(X)\omega$ is the obstruction to ω being equivariantly closed: that is, representing a cohomology class in $H^*_{S^1}(M; \mathbb{R})$. Of course, we might be able to change ω within its cohomology class in $(\Omega_X[u], d_X)$ and achieve equivariant closure. So let's ask; how can we augment ω in $(\Omega_X[u], d_X)$ to make it equivariantly closed? The only possibility, for degree reasons, is to take $\tilde\omega = \omega - fu$ for some smooth function $f: M \to \mathbb{R}$. The definition of d_X then gives

$$\begin{aligned}
d_X \tilde\omega &= d_X \omega - d_X(fu) \\
&= d\omega - u i(X)\omega - (d_X f)u - f d_X u \\
&= -u i(X)\omega - (d_X f)u \\
&= -u i(X)\omega - (df - u i(X)f)u \\
&= -u i(X)\omega - (df)u \\
&= -u(df + i(X)\omega).
\end{aligned}$$

Note that $\tilde\omega$ is closed exactly when $df + i(X)\omega = 0$. Since $\lambda(\omega) = [i(X)\omega]$, this is exactly when the action is Hamiltonian. Thus, denoting the Borel fibration of the S^1-action by $M \xrightarrow{i} M_{S^1} \to BS^1$, we have

Theorem 7.63 ([15]) *A symplectic S^1-action on (M^{2n}, ω) is Hamiltonian if and only if $\omega \in H^2(M; \mathbb{R})$ is in the image of $i^*\colon H^2(M_{S^1}; \mathbb{R}) \to H^2(M; \mathbb{R})$.*

Now, the pullback of the Borel fibration $M \to M_{S^1} \to BS^1$ via the inclusion $S^2 \hookrightarrow BS^1$ gives a fibration $M \to E \to S^2$ and this fibration has an associated Wang sequence:

$$\cdots \to H^2(E) \xrightarrow{i^*} H^2(M) \xrightarrow{\lambda} H^1(M) \to \cdots$$

where λ is the Wang derivation and $H^2(E) \cong H^2(M_{S^1})$. It may be shown (see [180] and Proposition 7.59) that this Wang derivation is precisely the derivation defined in Definition 7.58. This then gives another "formless" proof of Theorem 7.63 since exactness of the Wang sequence implies that $\lambda(\omega) = 0$ holds precisely when $\omega \in \text{Im}(i^*)$. This Wang approach indicates how algebraic topological methods can often take the place of geometric ones and, in the process extend the relevant notions.

Now let's return to consideration of the properties of Hamiltonian actions. In particular, we are interested in the fixed point set.

Proposition 7.64 *A Hamiltonian action on a compact symplectic manifold has at least two components in its fixed point set. If the manifold has Lefschetz type, then the existence of a fixed point implies that the action is Hamiltonian.*

Proof We give the proof for S^1 because fixed point questions always reduce to this case (see Lemma 7.3). If $\lambda(\omega) = [i(X)\omega] = 0$, then $i(X)\omega = dH$ for some smooth function $H\colon M \to \mathbb{R}$. Since M is compact, H has a maximum y and minimum z on M, so $0 = dH_y = i(X_y)\omega$ and $0 = dH_z = i(X_z)\omega$. But ω is nondegenerate, so we can only have these equalities if $X_y = 0 = X_z$. Since a zero for the fundamental vector field corresponds to a fixed point for the action, we are done. If M has Lefschetz type (see Definition 4.96), then the orbit map at a fixed point gives $h(\alpha) = 0$ and (3) of Proposition 7.60 then says that the action is Hamiltonian. □

Remark 7.65 Note by [179] that this theorem also holds in the case of c-symplectic manifolds with c-Hamiltonian action.

Proposition 7.66 *Any symplectic torus action $A\colon T^k \times M \to M$ splits as $T^k = T^r \times T^{k-r}$, where the action restricted to T^r is Hamiltonian and the action restricted to T^{k-r} is cohomologically free.*

For a proof of this result, see [162] and [179] for a cohomological version. These types of splittings have long been used in transformation groups. For instance, the action of a torus can be split into one with finite isotropy groups and one with fixed points. Now let's characterize cohomologically free actions in the case of Lefschetz type.

Theorem 7.67 ([179]) *Suppose that (M^{2n}, ω) has Lefschetz type and that $A \colon T^k \times M \to M$ is a symplectic torus action with orbit map $\alpha \colon T^k \to M$. Then*

1. *The action is cohomologically free if and only if the induced homomorphism $\alpha_* \colon H_1(T^k; \mathbb{R}) \to H_1(M; \mathbb{R})$ is injective.*
2. *The action is cohomologically free if and only if it is almost free.*

Proof In the following, we take $\mathbb{R}$-coefficients in cohomology. (1) Suppose the action is cohomologically free and choose a vector space splitting $s \colon H^1(M) \to H^1(T^k)$ for $\psi \colon H^1(T^k) \to H^1(M)$ (since ψ is injective). Let $PD \colon H_1(M) \xrightarrow{\cong} H^{2n-1}(M)$ denote the Poincaré duality isomorphism. Then for each S^1 factor in T^k, we have the following (using Proposition 7.60 (1)). (Note that multiplication by ω^{n-1}, denoted Lef, is an isomorphism by the Lefschetz type assumption; see Definition 4.96.)

$$\begin{aligned}
s(1/n \operatorname{Lef}^{-1} PD\alpha_*([S^1])) &= s(1/n \operatorname{Lef}^{-1} PD(h(\alpha))) \\
&= s(1/n \operatorname{Lef}^{-1}(n\lambda(\omega)\omega^{n-1})) \\
&= s\lambda(\omega) \\
&= s\psi(u_{S^1}) \\
&= u_{S^1}.
\end{aligned}$$

The composition $H_1(T^k) \to H^1(T^k)$ is then the Poincaré duality *isomorphism*, so α_* must be injective.

On the other hand, suppose α_* is injective and choose a vector space splitting $t \colon H_1(M) \to H_1(T^k)$. Then we can compute.

$$\begin{aligned}
tPD^{-1}(n \operatorname{Lef}(\psi(u_{S^1}))) &= tPD^{-1}(n \operatorname{Lef}(\lambda(\omega))) \\
&= tPD^{-1}(n \lambda(\omega)\omega^{n-1}) \\
&= t(h(\alpha)) \\
&= t(\alpha_*([S^1])) \\
&= [S^1].
\end{aligned}$$

Again, since this expresses the Poincaré duality isomorphism, we see that ψ is injective and the action is cohomologically free.

(2) If the action is cohomologically free, then α_* is injective by (1). Suppose some isotropy group is not finite; thus, there is a subtorus T^s which fixes a point x. The orbit map at x then induces a homology factorization $\alpha_* \colon H_1(T^k) \to H_1(T^k/T^s) \to H_1(M)$. But the first homomorphism is

clearly not injective, so α_* is not injective. Hence, all isotropy groups are finite and the action is almost free.

Suppose the action is almost free and write $T^k = T^r \times T^{k-r}$ as in Proposition 7.66. By Proposition 7.64, since the T^r action is Hamiltonian, it has fixed points. Because the action is almost free, we must have $T^r = *$ and T^k acts cohomologically freely. □

7.6.3 The symplectic toral rank theorem

In Subsection 7.3.3 we discussed the far-reaching toral rank conjecture, which says that the size of a space's cohomology limits its ability to admit large almost free toral actions. In our context, we ask how large the cohomology of a symplectic manifold must be in order to admit a symplectic T^k action. We can answer this question in the case of Lefschetz type.

Theorem 7.68 *If (M^{2n}, ω) is a closed symplectic manifold of Lefschetz type and T^k acts symplectically and almost freely on M, then*

$$\dim H^*(M; \mathbb{R}) \geq 2^k.$$

Proof By (2) and then (1) of Theorem 7.67, we have that $\alpha_*\colon H_1(T^k) \to H_1(M)$ is injective. Take the Borel fibration (see Section 7.2) associated to the action and go one step back in the associated Puppe sequence (using $\Omega BT^k \simeq T^k$) to get

$$T^k \xrightarrow{\alpha} M \to M_{T^k}.$$

Because α_* is injective, duality says that $\alpha^*\colon H^1(M) \to H^1(T^k)$ is surjective. Now, since $H^*(T^k)$ is generated by $H^1(T^k)$, we see that $\alpha^*\colon H^*(M) \to H^*(T^k)$ is surjective as well. Therefore, by Definition 4.39, the fibration is totally noncohomologous to zero (i.e. TNCZ) and we have an isomorphism

$$H^*(M) \cong H^*(T^k) \otimes H^*(M_{T^k}).$$

Thus, $\dim H^*(M) \geq \dim H^*(T^k) = 2^k$. □

Remark 7.69 In Theorem 7.24, we used models to show that the inequality holds for any c-symplectic manifold whose cohomology satisfies the hard Lefschetz property (see Theorem 4.35). Here we have given a more traditional approach using only the weaker Lefschetz type property (Definition 4.96).

7.6.4 Some properties of Hamiltonian actions

When a torus action is Hamiltonian, there are interesting (and restrictive) consequences beyond those mentioned above (i.e. existence of fixed points).

One of the most famous is the following (see, for instance, [161, Corollary 4.8]).

Theorem 7.70 *Let T^k act on M and let $M \to M_{T^k} \to BT^k$ be the associated Borel fibration. If the T^k action on (M^{2n}, ω) is Hamiltonian, then, as vector spaces,*

$$H^*(M_{T^k}; \mathbb{R}) \cong H^*(M; \mathbb{R}) \otimes H^*(BT^k; \mathbb{R}).$$

Of course, we see that this is equivalent to the fibration being TNCZ. Here is a situation where the symplectic and c-symplectic worlds diverge.

Example 7.71 ([5]) By Theorem 7.36, we know that the Borel fibration $X \to X_G \to BG$ is TNCZ if and only if

$$\dim H^*(X^G; \mathbb{Q}) \cong \dim H^*(X; \mathbb{Q})$$

where X^G is the fixed point set.

Now let S^1 act freely on $S^3 \times S^3$ by translation on the first factor. Take an equivariant tube $C = S^1 \times D^5$ about an orbit (via the slice theorem) and note that $\partial C = S^1 \times S^4 = \partial(D^2 \times S^4)$. Let $X = S^3 \times S^3 - C$ and sew in $D^2 \times S^4$ to obtain $N = X \cup_{\partial C} (D^2 \times S^4)$. We let S^1 act by rotations on D^2 and we observe that this is compatible with the action by translations of S^1 on itself (i.e. the action in S^3). Of course, the center of D^2 is fixed by this action, so we see that S^1 acts on N semifreely (i.e. isotropy is either zero or all of S^1) with $N^{S^1} = S^4$. Note that N has the following cohomology.

$$H^j(N; \mathbb{Z}) = \begin{cases} 0 & j = 1, 5 \\ \mathbb{Z} & j = 2, 4, 6 \\ \mathbb{Z} \oplus \mathbb{Z} & j = 3. \end{cases}$$

Now let $\theta[z_0, z_1, z_2] = [e^{i\theta} z_0, z_1, z_2]$ define a semifree action of S^1 on $\mathbb{CP}(3)$ with fixed set $p \cup \mathbb{CP}(2)$. Note that $p = [1, 0, 0]$ is fixed by the homogeneity of the coordinates. The fixed $\mathbb{CP}(2)$ consists of all points $[0, z_1, z_2]$. Since $\mathbb{CP}(3)$ and N have fixed points, we can take small equivariant disks about chosen fixed points and form the equivariant connected sum $M = \mathbb{CP}(3) \# N$. The fixed set of the S^1 action on M is given by $M^{S^1} = p \cup (\mathbb{CP}(2) \# S^4) = p \cup \mathbb{CP}(2)$. Now, M is clearly c-symplectic since the generator of $H^2(\mathbb{CP}(3))$ multiplies to a top class. Furthermore, the action is c-Hamiltonian since $H^1(\mathbb{CP}(3); \mathbb{Z}) = 0$ and $H^1(N; \mathbb{Z}) = 0$ imply $H^1(M; \mathbb{Z}) = 0$. Also, M has Lefschetz type trivially since $H^1(M; \mathbb{Z}) = 0$. However, we plainly see that we have

$$\dim {}_\mathbb{Q} H^*(M^{S^1}; \mathbb{Q}) = 4 < 8 = \dim {}_\mathbb{Q} H^*(M; \mathbb{Q}).$$

Therefore, by Theorem 7.36, the Borel fibration is not TNCZ and the c-symplectic analogue of Theorem 7.70 does not hold.

7.6.5 Hamiltonian bundles

If a T^k action on (M^{2n}, ω) is *Hamiltonian*, then the Borel fibration $M \to M_{T^k} \to BT^k$ is a bundle whose structure group (i.e., the torus T^k) lies inside the *Hamiltonian* diffeomorphisms of M. Recall that a *Hamiltonian diffeomorphism* of a symplectic manifold (M^{2n}, ω) is a symplectomorphism $\phi\colon M \to M$ which is the time-1 map of a flow generated by a family of Hamiltonian vector fields. This means that there are vector fields X_t with $i(X_t)\omega = dH_t$ for all t and

$$\frac{d\phi_t}{dt} = X_t \circ \phi_t.$$

The Hamiltonian diffeomorphism is then given by $\phi = \phi_1$. The Hamiltonian diffeomorphisms form a subgroup of the symplectomorphism group of all diffeomorphisms preserving the symplectic form. For more information, see [188]. The situation of a Hamiltonian action can then be generalized as follows.

Definition 7.72 *A symplectic bundle $F \to E \to B$ is* Hamiltonian *if (F, ω_F) is symplectic and the structure group of the bundle acts by Hamiltonian diffeomorphisms.*

The fundamental properties of Hamiltonian bundles may be found in [162]. In particular, the following is shown there.

Lemma 7.73 ([162, Lemma 2.3]) *If $(F^{2n}, \omega_F) \to E \to B$ is a symplectic bundle and B is simply connected, then the bundle is Hamiltonian if and only if there exists some cohomology class $\alpha \in H^2(E; \mathbb{R})$ which restricts to the class $[\omega_F] \in H^2(F; \mathbb{R})$ by $i^*(\alpha) = [\omega_F]$.*

Note the similarity to Thurston's condition in Theorem 4.91. In the lemma, however, we do not require B to be symplectic, but we already know that the bundle is symplectic. Of course, Thurston's condition is just a sufficient condition, so the assumption that the bundle is symplectic does not automatically imply the existence of a class α. Lalonde and McDuff apply Lemma 7.73 to prove analogues of Theorem 7.70 for various types of Hamiltonian bundles. For instance, they prove that every Hamiltonian bundle over a product of complex projective spaces is TNCZ. Indeed, they ask the

Question 7.74 (Lalonde–McDuff question) *Is every Hamiltonian bundle TNCZ?*

7.6 Hamiltonian actions and bundles

Here we want to consider the special situation where the fiber is a nilmanifold. We have already seen how to build models of nilmanifolds in Section 3.2 and, in particular, we know that all generators, $\{x_j\}$, are in degree 1. Hence, when we build the model of a symplectic nilmanifold, the symplectic form is represented by an element $\omega = \sum c_{jk} x_j x_k$ made up of products of degree 1 elements. We can now give a result due to Stepien [243] that answers a special case of the Lalonde–McDuff question using models.

Theorem 7.75 *Let $(M^{2n}, \omega_M) \to E \to B$ be a Hamiltonian bundle with B simply connected and (M^{2n}, ω) a nilmanifold. Then the bundle is rationally trivial. In particular, the bundle is TNCZ.*

Proof Let $(\wedge Y, d_Y) \to (\wedge Y \oplus X, D) \overset{\rho}{\to} (\wedge X, d_X)$ denote a relative Sullivan model for the bundle $M \to E \to B$. As discussed above, there is an element $\omega = \sum c_{jk} x_j x_k$ in $(\wedge X, d_X)$ made up of products of generators $x_m \in X^1$ whose cohomology class is $[\omega_M]$. Note that, since $\omega^n \neq 0$ and the x_j, $j = 1, \ldots, 2n$ are in degree 1, it must be the case that ω uses all of the generators in X^1 and $\omega^n = C x_1 x_2 \cdots x_{2n}$. Because B is simply connected, $Y^1 = 0$, so the differential on the x_j has the form $D x_j = d_X x_j + y_j$, where $y_j \in Y^2$.

Now, since the bundle is Hamiltonian, Lemma 7.73 says that there is some cohomology class mapping to the class of ω in $H^*(\wedge X, d_X)$ under the projection $\rho \colon (\wedge Y \oplus X, D) \to (\wedge X, d_X)$. If the cocycle τ represents this class, then $\rho(\tau) - \omega = d_X \gamma$, so in $(\wedge Y \oplus X, D)$ we have $\tau - b = \omega + d_X \gamma$ with $b \in Y^2$ (since $Y^1 = 0$). If we apply D, then we find $0 = D(\omega + d_X \gamma)$, since every element of Y^2 is a cocycle and τ is a cocycle. Let $\tilde{\omega} = \omega + d_X \gamma$. So the Hamiltonian hypothesis has allowed us to change the representative of the symplectic cohomology class in M while keeping the same type of expression in terms of generators of $(\wedge X, d_X)$; $\tilde{\omega} = \sum r_{jk} x_j x_k$. Now we can use the fact that $\tilde{\omega}$ is a D-cocycle to infer that $\tilde{\omega}^n$ is also a D-cocycle. Hence, computing $D \tilde{\omega}^n = 0$, we have

$$0 = C \sum_j (-1)^{j-1} D(x_j) x_1 x_2 \cdots \hat{x}_j \cdots x_{2n}$$

$$= C \sum_j (-1)^{j-1} (d_X x_j + y_j) x_1 x_2 \cdots \hat{x}_j \cdots x_{2n}$$

$$= C \sum_j (-1)^{j-1} y_j x_1 x_2 \cdots \hat{x}_j \cdots x_{2n}$$

(where $\hat{x}_j$ denotes deletion of x_j) since $d_X x_j$ consists of products of the x_k not including x_j, the element $x_1 x_2 \cdots \hat{x}_j \cdots x_{2n}$ contains all generators except for x_j and elements in degree 1 have zero squares. But now, since

$\{x_1 x_2 \cdots \hat{x}_j \cdots x_{2n}\}$ is a set of linearly independent elements in $\wedge X$, we see that we must have $y_j = 0$ for $j = 1, \ldots, 2n$. This means that $Dx_j = d_X x_j$ and, since $X = X^1$, the relative Sullivan model is a tensor product of cdga's. Hence, the bundle is rationally trivial. □

Exercises for Chapter 7

Exercise 7.1 Let M be a compact manifold with an *almost free* action of the compact Lie group G. We consider the projection $q \colon EG \times_G M \to M/G$, $q[(x,m)] = [m]$. Show that the fiber of q at x is $(G/G_x)_G$ and that $(G/G_x)_G \cong EG/G_x = BG_x$. Deduce now from the Leray spectral sequence, $H^p(M/G; H^q(BG_x; \mathbb{Q})) \Rightarrow H_G^{p+q}(M; \mathbb{Q})$, an isomorphism

$$H^*(M/G; \mathbb{Q}) \cong H_G^*(M; \mathbb{Q}).$$

Hint: see [144, page 37].

Exercise 7.2 Suppose M_i, $i = 1, \ldots, n$, are even dimensional simply connected manifolds satisfying $\pi_{\text{even}}(M_i) \otimes \mathbb{Q} = 0$ and $\dim \pi_{\text{odd}}(M_i) \otimes \mathbb{Q} \geq 2$. Show that every circle action on the connected sum $M_1 \# M_2 \# \cdots \# M_n$ has a fixed point. (Hint: consider the Euler–Poincaré characteristic and the rational homotopy Lie algebra of the connected sum.)

Exercise 7.3 Fix a map $S^3 \times S^3 \to S^6$ of degree 1 and use it to pull back the unit sphere bundle of S^6 to a S^5-bundle over $S^3 \times S^3$. Compute $\text{rk}_0(X)$.

Exercise 7.4 Let M be a simply connected compact manifold whose minimal model is $(\wedge(x, y, z, t), d)$ with $|x| = 3$, $|y| = 9$, $|z| = 6$, $|t| = 11$, $dx = dy = dz = 0$, $dt = xy - z^2$. Compute $\text{rk}_0(M)$.

Exercise 7.5 Write $n + 1 = (n_1 + 1) + (n_2 + 1) + (n_3 + 1)$. Consider the action of S^1 on $\mathbb{CP}(n)$ defined by

$$t[z_1, \ldots, z_{n+1}] = [t^2 z_1, \ldots, t^2 z_{n_1+1}, t^3 z_{n_1+2}, \ldots, t^3 z_{n_1+n_2+2}, z_{n_1+z_2+3}, \ldots, z_{n+1}].$$

Show that the fixed point set is the union $\mathbb{CP}(n_1) \amalg \mathbb{CP}(n_2) \amalg \mathbb{CP}(n_3)$.

Exercise 7.6 Let $G = S^1$. Show that G can act on $\mathbb{CP}(4)$ with fixed point set the disjoint union of $\mathbb{CP}(3)$ and an isolated point, and that G can act on the quaternionic projective plane $\mathbb{HP}(2)$ with fixed point set S^4 and an isolated point. Make the equivariant connected sum $\mathbb{CP}(4) \# \mathbb{HP}(2)$ by removing small invariant discs around the isolated fixed points and pasting together the boundaries. Show that you have an action of S^1 on a space Z whose cohomology satisfies the hard Lefschetz property, and such that at least one component of the fixed point set does not satisfy the hard Lefschetz property (see [7]).

Blow-ups and Intersection Products

As we mentioned in the Preface, models play an important role in understanding various properties of complex and symplectic blow-ups. They also can be used to study features of the Chas–Sullivan loop product, a rather mysterious new topological tool.

In this chapter we will consider two types of questions. First of all, let $f: N \hookrightarrow M$ be a closed submanifold of a compact manifold and denote by C its complement $M \backslash N$. The natural problem is to know if the rational homotopy type of C is completely determined by the rational homotopy type of the embedding, and in that case to describe a model for the injection $C \hookrightarrow M$ from a model of the initial embedding. Now suppose that the normal bundle to the embedding has a complex structure. We then can take the blow-up $\widetilde{M}$ of M along N, and ask if it is possible to determine the rational homotopy type of $\widetilde{M}$ from the rational homotopy type of the embedding.

P. Lambrechts and D. Stanley have given a positive answer to those questions. They prove that when $\dim M \geq 2 \dim N + 2$, then the rational homotopy type of the complement C is completely determined by the minimal model of f. They also prove that when M is simply connected and $\dim M \geq 2 \dim N + 3$, the rational homotopy type of $\widetilde{M}$ depends only on the rational homotopy type of the embedding and the Chern classes of the normal bundle [165]. We describe in Section 8.2 their construction of a model for $\widetilde{M}$.

We have seen in Chapter 4 that the assumption that a compact manifold is Kähler implies its formality. Since Kähler manifolds are the most basic examples of compact symplectic manifolds, a fundamental question is whether such a statement also holds for symplectic manifolds. In Section 8.2, we describe non-formal simply connected symplectic manifolds that are blow-ups. The key point here is that we can, in some explicit sense, actually understand the complete rational homotopy of the blow-up rather than just subsidiary properties (such as existence or nonexistence of Massey products). Such an understanding should prove important in the future.

Our second question concerns the theory of cycles in a compact manifold. M. Chas and D. Sullivan extended the ideas behind standard intersection theory to an intersection theory of cycles in $LN = N^{S^1}$ for any compact oriented manifold N [53]. They define a product, called the *loop product*, on $H_*(LN)$,

$$H_p(LN) \otimes H_q(LN) \to H_{p+q-n}(LN),$$

that combines the intersection product on the chains on N and the Pontryagin composition of loops in ΩN. Here, minimal models appear as a useful tool for making explicit computations of the loop product.

As a corollary, we give a proof of the Cohen–Jones algebra isomorphism between $H_{*-n}(LN;\mathbb{Q})$ and the Hochschild cohomology

$$HH^*(C^*(N;\mathbb{Q}), C^*(N;\mathbb{Q})),$$

where, as usual, $C^*(N;\mathbb{Q})$ denotes the singular cochains on N with coefficients in $\mathbb{Q}$.

8.1 The model of the complement of a submanifold

Let M be a compact orientable nilpotent m-dimensional manifold and suppose that $f\colon N^n \to M^m$ is a smooth embedding of a compact nilpotent n-dimensional submanifold. The image $f(N)$ admits a tubular neighborhood which is a compact submanifold (with boundary) $T \subset M$ of codimension 0 that deformation retracts to N. Let C be the closure of the complement of T in M, $C = \overline{M \backslash T}$. Then C and T are both compact manifolds with the same boundary ∂T. Denote by $i\colon \partial T \to T$, $j\colon C \to M$, $g\colon \partial T \to C$, $k\colon T \to M$ and $h\colon N \to T$ the canonical injections. We then have a commutative diagram

$$\begin{array}{ccc} \partial T & \xrightarrow{i} & T \\ {\scriptstyle g}\downarrow & & \downarrow{\scriptstyle k} \\ C & \xrightarrow{j} & M \end{array}$$

Of course, C has the homotopy type of $M \backslash f(N)$ and the above diagram is a homotopy pushout.

A key step is to understand the embedding of the complement algebraically. With this in mind, a model for $j\colon C \hookrightarrow M$ will be defined in Theorem 8.11 in terms of shriek maps and algebraic mapping cones.

8.1.1 Shriek maps

To an embedding, we can associate a cohomological shriek map defined as the composition

$$f^!: s^{n-m}H^*(N;\mathbb{Q}) \xrightarrow{s^{-m}\mathcal{D}_N} s^{-m}H_*(N;\mathbb{Q}) \xrightarrow{H_*(f)} s^{-m}H_*(M;\mathbb{Q}) \xleftarrow{\mathcal{D}_M} H^*(M;\mathbb{Q}),$$

where s is the suspension functor $(s^k V)^r = V^{r+k}$, and $\mathcal{D}_N$ and $\mathcal{D}_M$ denote the corresponding Poincaré duality isomorphisms. (Recall that we have the convention on homology that $(H_*)^{-p} = H_p$.) In particular, if $[N]$ and $[M]$ denote the homology orientation classes, then $\mathcal{D}_N(\alpha) = \alpha \cap [N]$ and

$$f^!(s^{n-m}\nu) \cap [M] = H_*(f)(\nu \cap [N]).$$

By construction the cohomological shriek map $f^!$ is a morphism of $H^*(M;\mathbb{Q})$-modules and an isomorphism in degree m.

Cohomological shriek maps are based on the duality between cohomology and homology. We now extend this duality to models. Let (A,d) be a cdga whose cohomology $H^*(A,d)$ satisfies Poincaré duality with a fundamental class $\omega \in A^m$. We denote by $\varepsilon \colon (A,d) \to (\mathbb{Q},0)$ a map of differential A-modules of degree $-m$ satisfying $\varepsilon(\omega) = 1$. The map ε yields a duality map

$$\psi_A \colon A \to s^{-m}A^{\vee} = s^{-m}\mathrm{Hom}(A,\mathbb{Q}),$$

$$\psi_A(1) = s^{-m}\varepsilon, \quad s^m\psi_A(a)(b) = (-1)^{|b|}\varepsilon(a \cdot b).$$

(For this, recall that we have $(s^{-m}\mathrm{Hom}(A,\mathbb{Q}))^r = (\mathrm{Hom}(A,\mathbb{Q}))^{r-m} = \mathrm{Hom}(A^{m-r},\mathbb{Q})$.)

Lemma 8.1 ψ_A *is a quasi-isomorphism of differential graded A-modules.*

Proof This is a reformulation of the fact that $H^*(A)$ satisfies Poincaré duality. □

Let's come back now to the smooth embedding $f \colon N \to M$ and let $\varphi \colon (A,d) \to (B,d)$ be a model for f.

Definition 8.2 *A shriek map for f associated to φ is a morphism of differential A-modules $g \colon s^{n-m}(B,d) \to (A,d)$ that induces in cohomology the cohomological shriek map $f^! \colon s^{n-m}H^*(N;\mathbb{Q}) \to H^*(M;\mathbb{Q})$.*

From now on we will abuse notation and denote both the shriek map and the cohomological shriek map by $f^!$.

8 : Blow-ups and Intersection Products

Example 8.3 Let $\varphi \colon (A,d) \to (A \otimes \wedge W, d)$ denote a relative minimal model for f, and consider the diagram

$$\begin{array}{ccc} s^{-m+n}(A \otimes \wedge W, d) & & (A, d) \\ {\scriptstyle \psi_{A \otimes \wedge W}} \downarrow {\scriptstyle \simeq} & & {\scriptstyle \simeq} \downarrow {\scriptstyle \psi_A} \\ s^{-m}(A \otimes \wedge W, d)^{\vee} & \xrightarrow{\varphi^{\vee}} & s^{-m}(A, d)^{\vee} \end{array}$$

where ψ_A and $\psi_{A \otimes \wedge W}$ are the quasi-isomorphisms defined in Lemma 8.1. Since $(A \otimes \wedge W, d)$ is a semifree A-module, the lifting property of semifree models (Proposition 2.107) gives a map $f^! \colon s^{-m+n}(A \otimes \wedge W, d) \to (A, d)$ making the diagram commutative up to homotopy. This map is a shriek map for f associated to φ.

Lemma 8.4 *The embedding $f \colon N \to M$ admits a model $\varphi \colon (A, d) \to (B, d)$ with the following properties:*

1. $A^{\geq m+1} = 0$ and $B^{\geq n+2} = 0$;
2. *there is a shriek map $f^! \colon s^{-m+n}(B, d) \to (A, d)$ for f associated to φ.*

Proof We first define the cdga (A, d) as the quotient of the minimal model $(\wedge V, d)$ of M by the ideal $I = (\wedge V)^{>m} \oplus C_1$ where C_1 is a complement to the cocycles in $(\wedge V)^m$. We then take a relative minimal model of f of the form $(A \otimes \wedge W, d)$ and we obtain a shriek map $g \colon s^{-m+n}(A \otimes \wedge W, d) \to (A, d)$ as in Example 8.3. We define $J = (A \otimes \wedge W)^{\geq n+2} \oplus C_2$, where C_2 is a complement of the cocycles in $(A \otimes \wedge W)^{n+1}$. For degree reasons, $g(J) = 0$. We therefore define $(B, d) = ((A \otimes \wedge W)/J, d)$. The maps f and g factor to give the required model and shriek map. □

Lemma 8.5 *Let (A, d) be a finite type cdga whose cohomology satisfies Poincaré duality with top dimension m and let (M, d) be a finite type semifree A-module. Then the evaluation in cohomology induces an isomorphism*

$$[(M, d), (A, d)] \xrightarrow{\cong} \mathrm{Hom}_{H(A)}(H^m(M, d), H^m(A, d)).$$

Proof First, recall the functor Ext defined in Section 2.8. If (M, d) and (N, d) are differential graded A-modules, and $(P, d) \to (M, d)$ is a semifree model for (M, d), then

$$\mathrm{Ext}^p_A(M, N) = H^p(\mathrm{Hom}_A((P, d), (M, d))), \text{ and } [(M, d), (N, d)]$$
$$= \mathrm{Ext}^0_A(M, N).$$

There is a spectral sequence due to Moore [200] converging to $\mathrm{Ext}_A(M, N)$ whose E_2-term is $\mathrm{Ext}^{p,q}_{H(A)}(H(M), H(N))$, where the indices are

8.1 The model of the complement of a submanifold

defined as follows. Let $L_n \xrightarrow{d} L_{n-1} \to \cdots L_0 \to H(M)$ be a free resolution of $H(M)$ as an $H(A)$-module. Then, since $H(M)$ is a graded vector space, each L_n is also a graded vector space and $d(L_n)^q \subset (L_{n-1})^{q+1}$. By definition, we then have

$$\mathrm{Ext}^{p,q}_{H(A)}(H(M),H(N)) = \mathrm{Ext}^p_{H(A)}(H(M),H(N))^{p+q}$$
$$= (H^p(\mathrm{Hom}_{H(A)}((L_*,d),H(N))))^{p+q}.$$

In this notation p refers to the resolution degree of L_* and $p+q$ is the total degree. In our case $(N,d) = (A,d)$ and we have

$$\mathrm{Ext}^{p,q}_{H(A)}(H(M),H(A)) = (H^p(\mathrm{Hom}_{H(A)}((L_*,d),H(A))))^{p+q}$$
$$\cong (H^p(\mathrm{Hom}_{H(A)}(H(A)^\vee,(L_*,d)^\vee)))^{p+q}$$
$$= (H^p(\mathrm{Hom}_{H(A)}(H(A),(L_*,d)^\vee)))^{p+q}$$
$$= H^{p,q}((L_*,d)^\vee) = \begin{cases} 0 & \text{if } p > 0 \\ H^q(M) & \text{if } p = 0. \end{cases}$$

Here we have used the fact that when C and D are finite dimensional R-modules, then taking the dual map gives an isomorphism $\mathrm{Hom}_R(C,D) \to \mathrm{Hom}_R(D^\vee, C^\vee)$. Since $E_2^{p,q} = 0$ when $p \neq 0$, the spectral sequence collapses, and therefore, the morphism that gives the above induced map in cohomology,

$$\mathrm{Ext}^p_A(M,A) \to \mathrm{Hom}_{H(A)}(H^{m-p}(M), H^m(A))$$

is an isomorphism. In particular

$$[(M,d),(A,d)] \cong \mathrm{Hom}_{H(A)}(H^m(M), H^m(A)).$$

$\square$

Corollary 8.6 *Let $\varphi: (\wedge V, d) \to (\wedge W, d)$ be the minimal model of an embedding $f: N \hookrightarrow M$ and let $\theta: (P,d) \to (\wedge V, d)$ be a morphism of $(\wedge V, d)$-modules which is an isomorphism on H^m and such that (P,d) is a semifree model for $s^{-m+n}(\wedge W, d)^\vee$. Then there is a constant q such that $q \cdot \theta$ is a shriek map for the embedding.*

This means that shriek maps are defined in a unique way by what they do on cohomology in degree m.

8.1.2 Algebraic mapping cones

An important tool in our construction of a model for the complement C is the mapping cone construction associated to a morphism of differential graded A-modules. Let's recall that now.

8: Blow-ups and Intersection Products

If $f: (X, d_X) \to (Y, d_Y)$ is a morphism of differential graded A-modules, then the algebraic *mapping cone* of f is the differential graded A-module

$$C(f) = Y \oplus_f sX := \oplus_{i=1}^{\infty} Y^i \oplus (sX)^i,$$

with differential given by

$$d(y, sx) = (d_Y y + f(x), -sd_X x).$$

It is easy to see that $d^2 = 0$. Also, we can define the morphisms $i: Y \to Y \oplus_f sX$ and $\tau: Y \oplus_f sX \to X$ by

$$i(y) = (y, 0), \qquad \tau(y, sx) = x.$$

The mapping cone construction is invariant with respect to quasi-isomorphisms as the next result demonstrates.

Lemma 8.7 *Suppose we have a commutative diagram of differential graded A-modules,*

$$\begin{array}{ccc} (X, d) & \xrightarrow{f} & (Y, d) \\ \simeq \uparrow h & & \simeq \uparrow k \\ (Z, d) & \xrightarrow{g} & (T, d) \end{array}$$

where h and k are quasi-isomorphisms. Then we have the following commutative diagram in which the horizontal maps are the natural injections and ℓ is also a quasi-isomorphism.

$$\begin{array}{ccc} (Y, d) & \longrightarrow & C(f) \\ \simeq \uparrow k & & \simeq \uparrow \ell \\ (T, d) & \longrightarrow & C(g) \end{array}$$

In particular $C(f)$ and $C(g)$ are quasi-isomorphic.

Now, in Example 8.3 and Lemma 8.4, we gave different constructions of a shriek map. By Lemma 8.7, we see that, in fact, they have quasi-isomorphic mapping cones.

Lemma 8.8 *Let $0 \to (X, d) \xrightarrow{g} (Y, d) \to (Z, d) \to 0$ be a short exact sequence of differential graded A-modules. Then there is a quasi-isomorphism of A-modules $C(g) \to (Z, d)$ making the following diagram* commutative.

$$(Y,d) \longrightarrow (Z,d)$$
$$\searrow \quad \uparrow \simeq$$
$$C(g)$$

Proof Consider the diagram of short exact sequences

$$\begin{array}{ccccccccc} 0 & \longrightarrow & (X,d) & \xrightarrow{g} & (Y,d) & \longrightarrow & (Z,d) & \longrightarrow & 0 \\ & & \Vert & & \uparrow \varphi & & \uparrow \bar\varphi & & \\ 0 & \longrightarrow & (X,d) & \longrightarrow & (X \oplus Y \oplus sX, D') & \longrightarrow & C(g) & \longrightarrow & 0 \\ & & & & \uparrow \theta \simeq & & \uparrow = & & \\ & & & & (Y,d) & \longrightarrow & C(g) & & \end{array}$$

Here $D'(x) = dx$, $D'(y) = dy$, $D'(sx) = g(x) - sdx - x$, $\varphi(x) = g(x)$, $\varphi(y) = y$, $\varphi(sx) = 0$ and $\theta(y) = y$. Since φ is a quasi-isomorphism, the same is true for $\bar\varphi$. □

Interesting examples of short exact sequence of differential modules arise over the real numbers from the inclusion of relative de Rham forms associated to the injection $j: C \to M$:

$$0 \to A_{DR}(M,C) \to A_{DR}(M) \xrightarrow{A_{DR}(j)} A_{DR}(C) \to 0.$$

Over the rational numbers, the situation is similar. We have that the morphism $A_{PL}(j): A_{PL}(M) \to A_{PL}(C)$ is surjective with kernel denoted by $A_{PL}(M,C)$ and this then gives the short exact sequence of $A_{PL}(M)$-modules,

$$0 \to A_{PL}(M,C) \to A_{PL}(M) \xrightarrow{A_{PL}(j)} A_{PL}(C) \to 0.$$

Lemma 8.9 *If $f,g: (X,d) \to (Y,d)$ are two homotopic maps of differential graded A-modules, then the algebraic mapping cones $C(f)$ and $C(g)$ are quasi-isomorphic differential A-modules and we have a commutative diagram*

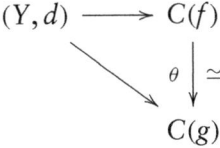

Proof By hypothesis (see Definition 2.106), there is a map of A-modules of degree -1, $H: X \to Y$ such that $f - g = dH + Hd$. We define a map $\theta: C(f) \to C(g)$ by putting $\theta(y) = y$ and $\theta(sx) = sx + Hx$. The morphism θ commutes with the differentials, makes the diagram above commutative and is a quasi-isomorphism of differential A-modules. □

8.1.3 The model for the complement C

In the previous section, we worked with models as differential modules. Here, we add hypotheses to obtain models in the framework of cdga's.

Suppose that we have a smooth embedding $f: N^n \to M^m$ and $\varphi: (A, d) \to (B, d)$ is a model for f with associated shriek map $f^!: s^{n-m}(B, d) \to (A, d)$. Since $H^{\geq m}(C(f^!)) = 0$, $C(f^!)$ contains an acyclic submodule I such that $C(f^!)^{\geq m} \subset I$. When $m \geq 2n+2$, the quotient $C(f^!)/I$ becomes a differential graded algebra with multiplication defined by

$$sx \cdot sy = 0, \quad v \cdot sy = (-1)^{|v|} s(v \cdot y), \quad x, y \in s^{n-m}B, v \in A.$$

Theorem 8.10 ([167]) *Suppose M^m and N^n as above are simply connected. If $m \geq 2n + 1$, then the induced morphism*

$$\varphi: (A, d) \to C(f^!)/I, \quad \varphi(x) = x$$

is a cdga model for the embedding $j: C \to M$.

More generally, we have

Theorem 8.11 ([169]) *Suppose there is an integer $r \geq 0$ such that $H_q(f; \mathbb{Q})$ is an isomorphism for $q \leq r$ and suppose that $m \geq 2n + 2 - r$. Let I be an acyclic submodule in $C(f^!)$ containing $C(f^!)^{\geq m-r}$. Then the complex $C(f^!)/I$ is a differential graded algebra and the induced morphism $(A, d) \to C(f^!)/I$ is a cdga model for the embedding $j: C \hookrightarrow M$.*

As a consequence, under the dimension hypothesis of the theorem, the rational homotopy type of the complement C depends only on the rational homotopy type of f.

Proof Note that, if $H_q(f; \mathbb{Q})$ is an isomorphism for $q \leq r$, then by Poincaré duality, $H^q(f^!)$ is an isomorphism for $q \geq m-r$. In particular the submodule I of $C(f^!)$ defined by $I = S \oplus C(f^!)^{\geq m-r}$, where S is a complement of the cocycles in degree $m - r - 1$, is acyclic.

We use the notation at the beginning of Section 8.1. Denote by $\mu: (\wedge V, d) \to A_{PL}(M)$ a minimal model for M, and by $\varphi: (\wedge V, d) \to$

$(\wedge V \otimes \wedge W, d)$ a relative minimal model for k,

$$\begin{array}{ccccc} A_{PL}(M) & \xrightarrow{A_{PL}(k)} & A_{PL}(T) & \xrightarrow{A_{PL}(h)} & A_{PL}(N) \\ \mu \uparrow \simeq & & \uparrow \simeq & & \\ (\wedge V, d) & \xrightarrow{\varphi} & (\wedge V \otimes \wedge W, d) & & \end{array}$$

By composition with the quasi-isomorphism $A_{PL}(h)$, φ can be seen to be a relative minimal model for f, with associated shriek map $f^!: s^{n-m}(\wedge V \otimes \wedge W, d) \to (\wedge V, d)$.

Observe first that, by the Thom isomorphism, the multiplication by the Thom class induces a quasi-isomorphism of $A_{PL}(M)$-modules

$$s^{n-m} A_{PL}(T) \xrightarrow{\simeq} A_{PL}(T, \partial T).$$

Now, by Mayer–Vietoris, the restriction of forms $A_{PL}(M, C) \to A_{PL}(T, \partial T)$ is also a quasi-isomorphism of $A_{PL}(M)$-modules. By the lifting homotopy property of semifree models (Proposition 2.107 (3)) applied to the diagram

$$\begin{array}{ccccc} s^{n-m} A_{PL}(T) & \xrightarrow{\simeq} & A_{PL}(T, \partial T) & \xleftarrow{\simeq} & A_{PL}(M, C) \\ \simeq \uparrow & & & & \downarrow \\ s^{n-m}(\wedge V \otimes \wedge W, d) & & \xrightarrow{g} & & A_{PL}(M) \end{array}$$

we get a quasi-isomorphism of $(\wedge V, d)$-modules

$$g: s^{n-m}(\wedge V \otimes \wedge W, d) \to A_{PL}(M, C).$$

Denote by $\rho: A_{PL}(M, C) \to A_{PL}(M)$ the canonical injection. Then, if we write $(P, d) = s^{n-m}(\wedge V \otimes \wedge W, d)$, we have two maps of $(\wedge V, d)$-modules from (P, d) to $A_{PL}(M)$: the morphisms $\mu \circ f^!$ and $\rho \circ g$.

$$(P, d) \xrightarrow{f^!} (\wedge V, d) \xrightarrow{\mu} A_{PL}(M)$$

$$(P, d) \xrightarrow{g} A_{PL}(M, C) \xrightarrow{\rho} A_{PL}(M)$$

Both maps induce isomorphisms in cohomology in degree m, so they differ by multiplication by a rational number q. By Lemma 8.5, this implies that $\mu \circ f^!$ is homotopic to $\rho \circ (qg)$.

8: Blow-ups and Intersection Products

Using Lemmas 8.7, 8.8 and 8.9, we have a sequence of quasi-isomorphisms of differential $(\wedge V, d)$-modules.

$$
\begin{array}{ccc}
(\wedge V, d) & \longrightarrow & (\wedge V \oplus_{f^!} sP, D) \\
{\scriptstyle \simeq} \downarrow {\scriptstyle \mu} & & \downarrow {\scriptstyle \simeq} \\
A_{PL}(M) & \longrightarrow & (A_{PL}(M) \oplus_{\mu \circ f^!} sP, D) \\
\Big\Vert & & \downarrow {\scriptstyle \simeq} \\
A_{PL}(M) & \longrightarrow & (A_{PL}(M) \oplus_{\rho \circ (qg)} sP, D) \\
\Big\Vert & & \downarrow {\scriptstyle \simeq} \\
A_{PL}(M) & \longrightarrow & A_{PL}(M) \oplus_\rho sA_{PL}(M, C) \\
\Big\Vert & & \downarrow {\scriptstyle \simeq} \\
A_{PL}(M) & \xrightarrow{A_{PL}(j)} & A_{PL}(C)
\end{array}
$$

Now let $(\wedge V, d) \to (\wedge V \otimes \wedge Z, D)$ be a relative minimal model for the embedding $j \colon C \to M$. Since $(\wedge V \otimes \wedge Z, D)$ is a semifree $(\wedge V, d)$-module, we have a commutative diagram

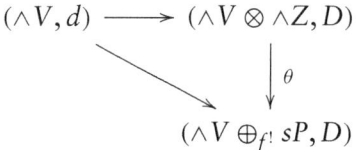

where θ is a quasi-isomorphism of $(\wedge V, d)$-modules. Since $(sP)^{<m-n-1} = 0$, we can suppose that $Z^{<m-n-1} = 0$. Denote by $I \subset C(f^!)$ an acyclic ideal containing $C(f^!)^{\geq m-r}$ and by $\pi \colon C(f^!) \to C(f^!)/I$ the projection. Then π is a quasi-isomorphism of $(\wedge V, d)$-modules. Since $(C(f^!)/I)^{\geq m-r} = 0$, we have $\pi \circ \theta(\wedge^2 Z) \subset (C(f^!)/I)^{\geq 2(m-n-1)} = 0$. So $\pi \circ \theta$ is a morphism of cdga's and this shows that $C(f^!)/I$ is a cdga model for the complement C. □

Proposition 8.12 *With the same notation, a model for the injection $g \colon \partial T \to C$ is given by the map of mapping cones*

$$(\wedge V \oplus_{f^!} P, D) \to (\wedge W \oplus_{\varphi \circ f^!} P, D).$$

Proof The morphism $A_{PL}(g)$ is quasi-isomorphic to the map induced between the mapping cones of α and β in the diagram

$$\begin{array}{ccccccccc}
0 & \longrightarrow & A_{PL}(T, \partial T) & \xrightarrow{\alpha} & A_{PL}(T) & \xrightarrow{A_{PL}(i)} & A_{PL}(\partial T) & \longrightarrow & 0 \\
& & \uparrow & & \uparrow A_{PL}(f) & & \uparrow A_{PL}(g) & & \\
0 & \longrightarrow & A_{PL}(M, C) & \xrightarrow{\beta} & A_{PL}(M) & \xrightarrow{A_{PL}(j)} & A_{PL}(C) & \longrightarrow & 0
\end{array}$$

The result follows directly because $A_{PL}(f) \colon A_{PL}(M, C) \to A_{PL}(T, \partial T)$ is a quasi-isomorphism. $\square$

Example 8.13 Consider the embedding f of $\mathbb{C}P(n)$ into $\mathbb{C}P(m)$ with $n < m$. The complement C is the set of classes $[x_1, \ldots, x_{m+1}]$ with at least one nonzero element in the sequence $x_{n+2}, x_{n+3}, \ldots, x_{m+1}$. The homotopy

$$r_t([x_1, \ldots, x_{n+1}, x_{n+2}, \ldots, x_{m+1}]) = [tx_1, \ldots, tx_{n+1}, x_{n+2}, \ldots, x_{m+1}]$$

defines a deformation retraction of the complement onto $\mathbb{C}P(m-n-1)$.

A model of the embedding f is the projection

$$\varphi \colon (\mathbb{Q}[a]/a^{m+1}, 0) \to (\mathbb{Q}[b]/b^{n+1}, 0), \quad \varphi(a) = b, \quad |a| = |b| = 2.$$

Therefore, a shriek map for f,

$$f^! \colon s^{-2(m-n)} \mathbb{Q}[b]/b^{n+1} \to \mathbb{Q}[a]/a^{m+1},$$

is defined by $f^!(s^{-2(m-n)} b^r) = a^{m-n+r}$. We then have a model for the complement C,

$$(\mathbb{Q}[a]/a^{m+1}) \oplus_{f^!} (ss^{-2(m-n)} \mathbb{Q}[b]/b^{n+1}).$$

Since the ideal generated by a^{m-n} and $ss^{-2(m-n)} \mathbb{Q}[b]$ is acyclic, another model is given by the quotient cdga $(\mathbb{Q}[a]/a^{m-n}, 0)$ — and this is a model for the space $\mathbb{C}P(m-n-1)$.

Example 8.14 With the hypotheses of Theorem 8.10, suppose that the embedding $f \colon N^n \to M^m$ is null homotopic. Let (A, d) and (B, d) be models for M and N, with $A^{>m} = 0$, $A^m = \mathbb{Q} u_M$, $B^{>n} = 0$ and $B^n = \mathbb{Q} u_N$ (where u_M and u_N are the respective fundamental classes). A model for f is given by the trivial map $\varphi \colon (A, d) \to (B, d)$. This means that the A-module structure on B is trivial: $A^+ \cdot B = 0$. A shriek map is then given by

$$f^! \colon s^{n-m}(B, d) \to (A, d), \quad f^!(s^{n-m}(u_N)) = u_M, \quad f^!(s^{n-m} B^{<n}) = 0.$$

The complement C of the embedding therefore has the homotopy type of

$$(A, d) \oplus_{f^!} ss^{n-m}(B, d)/(u_M, ss^{n-m} u_N)$$

Since the product of elements in A with elements in $s^{n-m}B$ is zero in this quotient, the complement C has the rational homotopy type of

$$(M\setminus\{*\}) \vee \Sigma^{m-n-1}(N\setminus\{*\}) \vee S^{m-n-1}.$$

Example 8.15 Let M be the tangent sphere bundle to the product $S^2 \times S^2$ and let $f: S^3 = N \to M$ be the injection of a fiber. By Example 2.69, a model for M is given by

$$(A, d) = (\wedge(a, b, x)/(a^2, b^2), d), \quad |a| = |b| = 2, \quad |x| = 3, \quad d(x) = ab,$$

and a model for f is given by the natural projection $\varphi: (\wedge(a, b, x)/(a^2, b^2), d) \to (\wedge x, 0)$. Therefore, a shriek map is given by

$$f^!: s^{-4}(\wedge x, 0) \to (A, d), \quad f^!(s^{-4}1) = ab, \quad f^!(s^{-4}x) = abx.$$

The mapping cone is quasi-isomorphic to $(\wedge(a, b, x)/(a^2, b^2, ab), 0)$. The complement C of the embedding thus has the rational homotopy type of $(S^2 \vee S^2) \times S^3$.

Note that this space is formal. In this example the embedding f is trivial in cohomology, but is not homotopically trivial. By Example 8.14, the complement of a trivial embedding is $(M\setminus\{*\}) \vee S^2$. This is a very different space. In particular, it is not formal because it has nontrivial Massey products (see Proposition 2.90).

8.1.4 Properties of Poincaré duality models

There are geometric situations for which minimal models alone are not adequate algebraic reflections. Indeed, this chapter shows this quite clearly. With this in mind, for our manifold N choose a Poincaré duality model (A, d) with a fundamental class $\omega \in A^n$ (see Theorem 3.9). By definition of a Poincaré duality model, we have $A^n = \mathbb{Q} \cdot \omega$, $A^{>n} = 0$, and there are graded bases a_i, a_i' such that $(a_i \cdot a_j') = \delta_{ij} \omega$ for all i and j.

Here, the duality map $\psi_A: (A, d) \to s^{-n}(A, d)^\vee$ defined in Lemma 8.1 is an isomorphism of differential graded A-modules (rather than simply being only a quasi-isomorphism).

Definition 8.16 The element $D_A = \sum_i (-1)^{|a_i|} a_i \otimes a_i' \in A \otimes A$ is called the *diagonal cocycle of A*.

The diagonal D_A is a cocycle, satisfies the useful equality

$$(a \otimes 1) \cdot D_A = (1 \otimes a) \cdot D_A,$$

and induces in cohomology the usual diagonal element $D_{H^*(A)}$ defined in the analogous way on $H^*(A, d)$.

8.1 The model of the complement of a submanifold

This diagonal cocycle has a very geometric meaning. Consider the diagonal injection $\Delta\colon N \to N \times N$ of a 2-connected compact manifold N. The complement C is the *configuration space of two points in N*, and we know how to construct a model of this complement from a shriek map associated to Δ.

The multiplication map $(A,d) \otimes (A,d) \to (A,d)$ makes (A,d) into a differential graded $(A \otimes A)$-module. Moreover (see Example 2.48), by the property above, the multiplication by D_A,

$$\mu_{D_A} \colon s^{-n} A \longrightarrow A \otimes A, \qquad s^{-n} a \mapsto D_A \cdot (1 \otimes a)$$

is a morphism of differential graded $(A \otimes A)$-modules. The next result gives the link between the shriek map and the map μ_{D_A}.

Lemma 8.17 *The multiplication by D_A, μ_{D_A} is a shriek map for the diagonal injection $\Delta\colon N \to N \times N$.*

Proof Recall that the multiplication $m\colon A \otimes A \to A$ is a model for the embedding Δ. The lemma follows now from the commutativity of the following diagram in which the vertical maps are isomorphisms.

$$\begin{array}{ccc} s^{-n} A & \xrightarrow{\mu_{D_A}} & A \otimes A \\ {\scriptstyle s^{-n}\psi_A}\downarrow & & \downarrow{\scriptstyle \psi_A \otimes \psi_A} \\ s^{-2n} A^\vee & \xrightarrow{m^\vee} & s^{-n} A^\vee \otimes s^{-n} A^\vee \end{array}$$

$\square$

8.1.5 The configuration space of two points in a manifold

Let N be a 2-connected n-dimensional compact manifold and let $\Delta\colon N \to N \times N$ be the diagonal submanifold. The purpose of this subsection is to give a description of a model for the complement of the diagonal: that is, the configuration space of two points in N: $F(N,2) = \{(x,y) \in N \times N \mid x \neq y\}$.

Theorem 8.18 ([166]) *Let N be a compact 2-connected manifold, and suppose (A,d) is a Poincaré duality model for N with associated diagonal cocycle D_A. Then the quotient map $A \otimes A \to (A \otimes A)/(D_A)$, where (D_A) is the ideal generated by D_A, is a model for the embedding $j\colon F(N,2) \hookrightarrow N \times N$.*

Proof We use the notation of Subsection 8.1.4. Since μ_{D_A} is a shriek map for Δ, a model for $F(N,2)$ as an $A \otimes A$-module is given by the mapping cone

$C(\mu_{D_A})$. Since N is 2-connected, $H_{\leq 2}(j;\mathbb{Q})$ is an isomorphism. Therefore, by Theorem 8.11, there is an acyclic submodule I containing $C(\mu_{D_A})^{\geq m-2}$ such that the quotient map

$$A \otimes A \to C(\mu_{D_A})/I$$

is a model for the injection $F(N,2) \hookrightarrow N \times N$. Note that $D_A = 1 \otimes \omega + \ldots$. This implies that the A-submodule generated by D_A is a free module of rank one. In particular, the multiplication by D_A defines an isomorphism between $s^{-n}A$ and the ideal generated by D_A. Therefore the quotient map $q\colon C(\mu_{D_A})/I \to (A \otimes A)/(D_A)$ is a quasi-isomorphism of differential graded algebras. It follows that the other quotient map $A \otimes A \to (A \otimes A)/(D_A)$ is a model for the injection $j\colon F(N,2) \hookrightarrow N \times N$. □

The homotopy type of the configuration space of k points in N will be described more generally in Section 9.1.

8.2 Symplectic blow-ups

As symplectic topology became an active area in the mid 1980s and 1990s, certain fundamental questions naturally arose. For instance, since a compact Kähler manifold is always symplectic and since the standard symplectic examples are Kähler manifolds (e.g. $\mathbb{C}P(n)$), it was wondered whether all compact symplectic manifolds were Kähler. We have seen that compact Kähler manifolds satisfy the hard Lefschetz property (see Theorem 4.35) and it was shown in Theorem 4.98 that a nilmanifold can only have Lefschetz type if it is a torus. By Proposition 4.94, a nilmanifold is symplectic when it has a degree 2 cohomology class which multiplies up to a top class, so it is therefore fairly easy to find nonsimply connected compact symplectic manifolds that are not Kähler (see [257] for more information on this issue). However, the simply connected case took more work and it wasn't until 1984 that McDuff [188] used the symplectic blow-up construction (modeled after the blow-up in complex geometry) to construct a compact symplectic manifold that did not satisfy the hard Lefschetz Property and, therefore, could not be Kähler (with respect to any metric). So finer questions then arose.

It was shown in [71] (see Theorem 4.43) that compact Kähler manifolds are *formal* spaces. In particular, of course, this entails the vanishing of all Massey products. Even though not every symplectic manifold is Kähler, could every compact symplectic manifold be formal? Again, the non-simply connected case is easy by Proposition 3.20 and the simply connected case is harder. In [257] it was implied that blow-ups were the place to look for counterexamples to formality for symplectic manifolds in the simply connected case and, indeed, Babenko and Taimanov soon thereafter demonstrated that a certain blow-up has Massey products, so could not be formal (see [17]). (This approach was later generalized in [230].) The blow-up construction is fundamental in symplectic (and complex) topology, so if rational homotopy methods are to find real use there, it is essential to know more information about blow-ups than just the existence or non-existence of nontrivial Massey products. Namely, we want to know the rational homotopy type of blow-ups and, for this, the construction of an algebraic model is required. In this section, we shall give background on the blow-up construction and in the next section, we shall describe the model.

8.2.1 Complex blow-ups

The complex *blow-up* of a complex surface (or rather the *blow-down*) originated as a way to remove certain negative intersections in homology obstructing possible embeddings in some $\mathbb{CP}(n)$. See [234] for a brief exposition along these lines. The idea for a complex surface W and a point $p \in W$ is the following. Take a neighborhood T of p given by a holomorphic chart and a disk about $(0,0) \in \mathbb{C}^2$ and, using those identifications, take the bundle over $\mathbb{CP}(1)$,

$$L_T = \{(t,l) \in T \times \mathbb{CP}(1) \mid t \in l\},$$

and note that $T\setminus p$ is bi-holomorphic to $L_T\setminus\{p \times \mathbb{CP}(1)\}$ since $t \neq p$ uniquely determines a complex line through the origin in $\mathbb{C}^2$. This means that the neighborhood T can be replaced by L_T and, effectively, p has been "blown up" to a $\mathbb{CP}(1)$. This $\mathbb{CP}(1)$ is usually denoted by E and is called the *exceptional curve* in the *blow-up of* W at p, $\widetilde{W}$. It has the property that its self-intersection is $E \cdot E = -1$.

Now, again thinking of T as a complex 2-disk centered at p, we see that any complex line l intersects T in a real 2-disk D_l. Hence, T is a union of real 2-disks which intersect only at p. Of course, the intersections of the disks with ∂T are the orbit circles on $\partial T = S^3$ of the Hopf action. The corresponding disks in L_T are simply the D_l, but now taken to be *disjoint*. Indeed, the centers of the disks correspond to the lines l, so the centers form a $\mathbb{CP}(1)$. So we see that L_T is the (total space of the) disk bundle over $\mathbb{CP}(1)$ whose boundary is the associated Hopf sphere bundle $S^3 \to \mathbb{CP}(1)$. To form the blow-up $\widetilde{W}$, cut out T (with $\partial T = S^3$ recall) and glue in L_T (also with boundary S^3). Furthermore, it is well-known that L_T is the normal disk bundle to the embedding $\mathbb{CP}(1) \hookrightarrow \mathbb{CP}(2)$, so we may think of L_T as a tubular neighborhood of $\mathbb{CP}(1)$ in $\mathbb{CP}(2)$. Of course, since $\mathbb{CP}(2) = \mathbb{CP}(1) \cup e^4$, for a 4-cell e^4 attached by the Hopf map, if we delete an open 4-disk D^4 disjoint from $\mathbb{C}P(1)$, we have $\mathbb{C}P(2) - D^4 \simeq L_T$. Then, by attaching L_T to W, we should obtain a connected sum. This heuristic argument can be made precise (see [234, section 7.1], for instance) to obtain

Proposition 8.19
1. *The complex surface $\widetilde{W}$ is diffeomorphic to the connected sum*

$$\widetilde{W} = W \# \overline{\mathbb{C}P}(2).$$

2. *In higher dimensions, the complex manifold $\widetilde{W}^{2n}$ obtained by blowing up a point is diffeomorphic to the connected sum*

$$\widetilde{W} = W \# \overline{\mathbb{C}P}(n).$$

Such a connected sum is often called the *topological blow-up* of W at a point. The bar over $\mathbb{C}P(2)$ (and $\mathbb{C}P(n)$) denotes $\mathbb{C}P(2)$ (and $\mathbb{C}P(n)$) with its orientation reversed. This is a way to obtain an oriented manifold whose orientation is compatible with the orientations of the summands. In symplectic geometry, the role of the blow-up point is taken by a symplectically embedded ball in order to describe a new symplectic form. Since a ball is contractible, the symplectic and complex constructions give the same diffeomorphism type when they both can be defined. See [189, page 239–251] for details.

8.2.2 Blowing up along a submanifold

The blow-up construction can be extended to submanifolds and this is now what we will focus on in the symplectic context. Suppose $f \colon N^n \hookrightarrow M^m$ is a codimension $2k$ submanifold whose normal bundle ν has a complex structure. If $(N, \omega|_N) \subset (M, \omega)$ is a symplectic submanifold, then this hypothesis always holds for the following reasons. First, the nondegeneracy of ω *and of* $\omega|_N$ leads to a splitting $TM = TN \oplus TN^\omega$, where TN^ω denotes the ω-complement of TN in TM. Of course, we have the usual isomorphism $\nu \cong TM/TN$, where ν is the normal bundle of the embedding, so we obtain $\nu \cong TN^\omega$. Now, $\omega|_{TN^\omega}$ is a nondegenerate skew-symmetric bilinear (i.e. symplectic) form on each fiber, thus reducing the structure group of ν to $\text{Sp}(k, \mathbb{R}) \simeq U(k)$ (see Exercise 1.3) and making ν a complex bundle with fiber $\mathbb{C}^k$.

Denote the unit disk bundle of ν by $D\nu$ and the associated sphere bundle by $S\nu$. As above, let T denote a tubular neighborhood of N in M diffeomorphic to $D\nu$ with boundary ∂T diffeomorphic to the sphere bundle $S\nu$. Also, denote by C the closure of the complement of T in M, $C = \overline{M \setminus T}$. The complex structure on ν implies that the circle S^1 (thought of as $S^1 \subset \mathbb{C}^*$) acts on the sphere bundle $S^{2k-1} \to S\nu = \partial T \to N$ to give the projectivized bundle $\mathbb{C}P(k-1) \to P\nu \to N$. Note that there is a bundle map $\partial T \to P\nu$.

Definition 8.20 *The blow-up $\tilde{M}$ of M along N is the smooth manifold*
$$\tilde{M} = C \cup_{\partial T} Pv$$
obtained from the pushout diagram

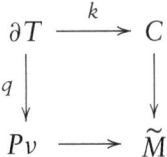

Remark 8.21 The definition above gives the homotopy type of the blow-up because the map $k: \partial T \to C$ is a cofibration and this makes the pushout a homotopy pushout as well. In particular, this means that any of the constituent spaces may be replaced by spaces of the same homotopy type without changing the homotopy type of $\tilde{M}$. To obtain the diffeomorphism type of the blow-up requires a bit more work and a larger space than Pv (which has the same homotopy type). This then allows the definition of the *blow-up projection* $\phi: \tilde{M} \to M$. See [257, Chapter 4] for instance.

The algebraic topology of the blow-up may be understood in terms of the cohomologies of the spaces involved in its construction. In particular, we have the following fundamental result. Recall first that if $c_i \in H^{2i}(N; \mathbb{Z})$ denote the Chern classes of the bundle v, then the cohomology of Pv is given by
$$H^*(Pv; \mathbb{Z}) \cong H^*(V)[a]/\langle a^k + c_1 a^{k-1} + \ldots + c_k \rangle$$
where $a \in H^2(Pv; \mathbb{Z})$. By Van Kampen and Mayer–Vietoris, we have

Theorem 8.22 *With the notation above:*

- $\pi_1(\tilde{M}) \cong \pi_1(M)$.
- *There is a short exact sequence*
$$0 \to H^*(M; \mathbb{Z}) \to H^*(\tilde{M}; \mathbb{Z}) \to A^* \to 0$$
where the quotient A^* is the ideal generated by a in $H^*(Pv; \mathbb{Z})$:
$$A^* = a \cdot H^*(N; \mathbb{Z})[a]/\langle a^k + c_1 a^{k-1} + \ldots + c_k \rangle.$$

Once these algebraic properties of the blow-up have been given, it still remains to show that $\tilde{M}$ has a symplectic structure. This can be done, but we omit details and simply state the result.

Theorem 8.23 ([188]) *If M is compact, then the blow-up $\tilde{M}$ of (M, ω) along $(N, \omega|_N)$ has a symplectic form $\tilde{\omega}$.*

8 : Blow-ups and Intersection Products

8.3 A model for a symplectic blow-up

Throughout this section we assume that M is simply connected and that $\dim M \geq 2\dim N + 3$. This latter condition is only required to guarantee that any model of the embedding $f: N^n \to M^m$ may be used for the construction of the blow-up model. (Indeed, without this condition, homotopic embeddings with isomorphic normal bundles can be defined that give non-rationally equivalent blow-ups [168].)

8.3.1 The basic pullback diagram of PL-forms

The key to obtaining a model for the blow-up is the fundamental pushout diagram

$$\begin{array}{ccc} \partial T & \xrightarrow{k} & C \\ {\scriptstyle q}\downarrow & & \downarrow \\ P\nu & \longrightarrow & \widetilde{M} \end{array}$$

A model for the pushout is given by the pullback of the corresponding homomorphisms of Sullivan PL-forms (see Section 2.4). The pushout diagram induces a commutative diagram

$$\begin{array}{ccc} A_{PL}(\widetilde{M}) & \longrightarrow & A_{PL}(C) \\ \downarrow & & \downarrow {\scriptstyle A_{PL}(k)} \\ A_{PL}(P\nu) & \xrightarrow{A_{PL}(q)} & A_{PL}(\partial T) \end{array}$$

Note that, since $k: \partial T \hookrightarrow C$ is the inclusion of a sub-complex, $A_{PL}(k)$ is surjective. Therefore, the universal property for pullbacks provides a cdga homomorphism

$$\phi: A_{PL}(\widetilde{M}) \to A_{PL}(C) \times_{A_{PL}(\partial T)} A_{PL}(P\nu)$$

and we have the following.

Proposition 8.24 (see [87, Proposition 13.5]) ϕ *is a quasi-isomorphism.*

8.3.2 An illustrative example

Example 8.25 Let $\mathcal{M}_{P\nu} \to \mathcal{M}_{\partial T} \leftarrow \mathcal{M}_C$ denote the induced morphisms of minimal models. It would be very nice if a model for the blow-up could be obtained as a pullback of these morphisms. Unfortunately, the only time we can guarantee this to be the case is when at least one of the morphisms is surjective (see [87, Proposition 13.6]). Of course, there *is* one natural geometric situation where surjectivity holds. Namely, consider blowing up a point p in a complex manifold M^{2n} (where $2n$ is the real dimension). The neighborhood T (as we saw above) is a $2n$-ball with $\partial T = S^{2n-1}$ and trivial normal bundle ν. Therefore, $P\nu \simeq \mathbb{C}P(n-1)$ and the map $\partial T \to P\nu$ is simply the Hopf bundle map $S^{2n-1} \to \mathbb{C}P(n-1)$. The minimal models are, respectively, $(\wedge(z), d = 0)$ and $(\wedge(x,y), dy = x^n)$, where the degrees of generators are $|z| = 2n-1$, $|x| = 2$ and $|y| = 2n-1$. The long exact homotopy sequence of the bundle $S^1 \to S^{2n-1} \to \mathbb{C}P(n-1)$ shows that $\mathbb{Z} = \pi_{2n-1}(S^{2n-1}) \cong \pi_{2n-1}(\mathbb{C}P(n-1))$, so the identification of (the duals of) the rational homotopy groups with the sets of generators (see Theorem 2.50) then implies that y maps to z under the induced model morphism $(\wedge(x,y), dy = x^n) \to (\wedge z, d = 0)$. Since we are over $\mathbb{Q}$ and $(\wedge z, d = 0)$ is an exterior algebra, this means that the morphism is surjective.

By [87, Propositions 13.3, 13.6], we then obtain a model of the blow-up

$$(\wedge(x,y), dy = x^n) \times_{\wedge z} \mathcal{M}_C,$$

where $C = \overline{W - T} = \overline{W - D^4}$ and $\mathcal{M}_C$ is a model for C [87, Proof of Theorem 38.5]. Let (A, d) be a Poincaré duality model for M with fundamental class u_A. A model for the embedding $C \to M$ is thus given by the injection $(A, d) \to (A \oplus \mathbb{Q} \cdot u, d)$ by letting $du = u_A$ and setting $u \cdot A^+ = 0$. Moreover, the model of the injection $S^{2n-1} \to C$ is given by the morphism $(A \oplus \mathbb{Q} \cdot u, d) \to (\wedge(z), d = 0)$ that maps u to z. Together with $y \mapsto z$, we have the pullback of the models,

$$(\wedge(x,y), dy = x^n) \times_{\wedge z} (A \oplus \mathbb{Q} \cdot u, d) = (\wedge(x,y) \oplus_{\mathbb{Q}} A, dy = x^n + u_a).$$

This model is a model for the connected sum $W \# \overline{\mathbb{C}P}(n)$ – just as it should be by Proposition 8.19 (2). We will also see this when we describe the general model for a blow-up. Indeed, this example contains many of the ingredients (at least in spirit) of the general construction of the model for a blow-up.

8.3.3 The model for the blow-up

Let's first fix notation. The morphism $\phi \colon (A, d) \to (B, d)$ is a model for the embedding $f \colon N^n \to M^m$ with $m - n = 2k$. We suppose (without loss of generality by Lemma 8.4) that $A^{\geq m+1} = 0$, $B^{\geq n+1} = 0$ and there is a shriek map $f^! \colon s^{n-m}(B, d) \to (A, d)$. We let $(P, d) = s^{n-m}(B, d)$, and denote by $\gamma_i \in (B, d)$ representatives for the Chern classes $c_i(\nu)$ of the normal bundle ν (with $\gamma_0 = 1$ and $\gamma_k = 0$ since $2k > \dim V$ by assumption).

8: Blow-ups and Intersection Products

Proposition 8.26

1. *A model for $q: \partial T \to Pv$ is given by the projection*
$$\text{proj}: (B \otimes \wedge(x,z), D) \to (B \otimes \wedge(z), \bar{D}), \qquad x \mapsto 0$$
where $|x| = 2$, $|z| = 2k - 1$, $Dx = 0$, $Dz = \sum_{i=0}^{k-1} \gamma_i x^{k-i}$, $\bar{D}z = 0$, and $2k = m - n$.
2. *A model for $k: \partial T \to C$ is given by the homomorphism of algebraic mapping cones*
$$\phi \oplus \text{id}: A \oplus_{f!} sP \to B \oplus_{\phi \circ f!} sP.$$

Proof Since $H^*(Pv; \mathbb{Q}) = H^*(N; \mathbb{Q})[a]/\sum_{i=0}^{k-1} \gamma_i x^{k-i}$, in the relative model for the bundle $Pv \to N$,
$$(B, d) \to (B \otimes \wedge(x,z), D) \to (\wedge(x,z), D),$$
we necessarily have $Dx = 0$ and $Dz = \sum_{i=0}^{k-1} \gamma_i x^{k-i}$. We now have only to recall that $q: \partial T = Sv \to Pv$ is a morphism of bundles over N that maps $\pi_{2k-1}(\mathbb{C}P(k-1))$ isomorphically onto $\pi_{2k-1}(S^{2k-1})$.

By our dimension hypothesis, the second part of the proposition follows directly from Proposition 8.12. □

Note that, by our connectivity hypothesis, the two models for $A_{PL}(\partial T)$ are isomorphic. Since $\phi: A_{PL}(\widetilde{W}) \to A_{PL}(C) \times_{A_{PL}(\partial T)} A_{PL}(Pv)$ is a quasi-isomorphism, we have the following definition and result.

Definition 8.27 (Description of the model for the blow-up)
Assume that $m \geq 2n + 3$. Then, with the notation above, define
$$Bl(A, B) = (A \oplus (B \otimes \wedge^+(x,z)), D)$$
with $|x| = 2$ and $|z| = m - n - 1 = 2k - 1$. The algebra structure on $Bl(A, B)$ is induced by the multiplications on A and $B \otimes \wedge^+(x,z)$ and the formula $a \cdot (b \otimes w) = (\phi(a) \cdot b) \otimes w$.
The differential D on $Bl(A, B)$ is defined by
$$D(a) = d_A a$$
$$D(b \otimes x) = d_B b \otimes x$$
$$D(b \otimes z) = d_B b \otimes z + (-1)^{|b|}\left(\phi'(s^{-2k}b) + \sum_{i=0}^{k-1}(b \cdot \gamma_i) \otimes x^{k-i}\right),$$
where the $\gamma_i \in B$ are representatives for the Chern classes, $c_i(v)$, of the normal bundle v.

Theorem 8.28 ([167]) *With the above notation, when $m \geq 2n + 3$, then $Bl(A, B)$ is a model for $\widetilde{M}$ and the injection $A \hookrightarrow Bl(A, B)$ is a model for the blow-up projection $\widetilde{M} \to M$.*

8.3.4 McDuff's example

A theorem of Tischler (see [251]) says that a closed symplectic manifold (V^{2m}, ω) with integral symplectic form (i.e. $[\omega] \in H^2(V; \mathbb{Z})$) may be embedded symplectically in some $(\mathbb{C}P(n), \alpha)$ for n large enough. So now consider such an embedding for the Kodaira–Thurston manifold, $f: (KT, \omega) \to (\mathbb{C}P(n), \alpha)$ and take the blow-up $\widetilde{\mathbb{C}P}(n)$. In order to satisfy our usual hypothesis $\dim M \geq 2 \dim N + 3$, we take $n \geq 6$, but in fact, simply according to Tischler, we could take $n = 5$. Let's now find the model $Bl(A, B)$ of $\widetilde{\mathbb{C}P}(n)$.

We start by taking the model $(A, d) = (\wedge(a)/(a^{n+1}), d = 0)$ for $\mathbb{C}P(n)$ with $a \mapsto \alpha$. For $V = KT$, we take the model from Example 4.95, $(B, d) = (\wedge(u, v, y, t), du = 0, dy = 0, dv = uy, dt = 0)$. Observe that the nonformality of KT is expressed by the Massey product vy (see Proposition 2.90 and Proposition 3.20). Note that these models satisfy the dimension restrictions imposed by Proposition 8.26. The form of $Bl(A, B)$ is then

$$Bl(A, B) = (\wedge(a)/(a^{n+1}) \oplus \wedge(u, v, y, t) \otimes \wedge^+(x, z), D),$$

with $|z| = (2n - 4) - 1 = 2n - 5$.

We must now define the differential D and for that we need the shriek map and the Chern classes of the normal bundle. By Theorem 8.32, we know that the total Chern class of KT is trivial, $c(KT) = 1$. The total Chern class of $\mathbb{C}P(n)$ is $c(\mathbb{C}P(n)) = (1 + \alpha)^{n+1}$, so the Whitney product formula gives

$$c(\nu) = c(KT)c(\nu) = f^*c(\mathbb{C}P(n)) = f^*((1+\alpha)^{n+1}$$

$$= 1 + (n+1)\omega + \frac{n(n+1)}{2}\omega^2$$

since $f^*(\alpha) = \omega$ and $\dim KT = 4$.

A general fact that we will not prove (see [165, Lemma 8.3]) is that the shriek map for any embedding $f: V^{2m} \to \mathbb{C}P(n)$ is (for $2r = 2n - 2m$),

$$f^!: s^{-2r}B \to A, \qquad f^!(\omega^j) = \ell_V a^{j+r}$$

where $a \mapsto \alpha$ under the model map and ℓ_V is the coefficient relating ω^m and the orientation u_V given by the almost complex structure: $\omega^m = \ell_V u_V$. For $V = KT$, by Proposition 4.94 we have a symplectic class $\omega = uv + yt$ with

8 : Blow-ups and Intersection Products

$\omega^2 = 2uvyt$, so $\ell_{KT} = 2$. Therefore we have $s^{-2r}(1) = 2a^r = 2a^{n-2}$ and $s^{-2r}(\omega) = 2a^{1+r} = 2a^{n-1}$. Since $\omega = uv + yt$, we can define $f'(uv) = a^{n-1}$ and $f'(yt) = a^{n-1}$. Finally, we have $s^{-2r}(\omega^2) = 2a^{2+r} = s^{-2r}(2uvyt) = 2s^{-2r}(uvyt)$ which says $s^{-2r}(u_{KT}) = a^{2+n-2} = a^n = u_{\mathbb{CP}(n)}$ and $f'(s^{-2r}\xi) = 0$ for any other monomial. Here this translates into a differential with

$$D(a) = 0$$
$$D(q \otimes x) = d_B q \otimes x$$
$$D(q \otimes z) = d_B q \otimes z + (-1)^{|q|}\left(f'(s^{-2r}q) + q \otimes x^{n-2}\right.$$
$$\left. + q(n+1)\omega \otimes x^{n-3} + q\frac{n(n+1)}{2}\omega^2 \otimes x^{n-4}\right),$$

using the fact that $\gamma_j = 0$ for $j > 2$ since $\dim KT = 4$. Similarly, the term $q(n(n+1)/2)\omega^2 \otimes x^{n-4} = 0$ unless $q = 1$ for degree reasons. This description of D has the following consequences.

$$D(v \otimes x^2) = D((v \otimes x)(1 \otimes x))$$
$$= D(v \otimes x) \cdot (1 \otimes x) + (-1)^3 (v \otimes x) \cdot D(1 \otimes x)$$
$$= (d_B v \otimes x) \cdot (1 \otimes x)$$
$$= (uy \otimes x) \cdot (1 \otimes x)$$
$$= (u \otimes x) \cdot (y \otimes x).$$

This says that $[u \otimes x] \cdot [y \otimes x] = 0$. Of course, we also have $[y \otimes x] \cdot [y \otimes x] = 0$ since y has odd degree, so the Massey product (see Definition 2.89) $\langle [u \otimes x], [y \otimes x], [y \otimes x]\rangle$ is defined with representative

$$(v \otimes x^2) \cdot (y \otimes x) - (-1)^3 (u \otimes x) \cdot 0 = (v \otimes x^2) \cdot (y \otimes x) = (vy \otimes x^3).$$

Since $v \otimes x^2$ is *not* a cocycle (and from the form of the differential D if $|z| = 7$), we see that $[vy \otimes x^3]$ is not in the ideal generated by $[u \otimes x]$ and $[y \otimes x]$. Hence, $\langle [u \otimes x], [y \otimes x], [y \otimes x]\rangle$ is a nontrivial Massey product. Therefore, by Proposition 2.90, we see that the blow-up of $\mathbb{CP}(n)$ along KT, $\widetilde{\mathbb{CP}}(n)$, is not formal. Thus we have the following result.

Theorem 8.29 ([17]) *There exist closed simply connected symplectic manifolds that are not formal.*

Of course, the description given above using the model $Bl(A, B)$ shows exactly how Massey products can propagate from nonformal submanifolds to blow-ups along them. For a general discussion of this, see [230]. Theorem 8.29 powerfully illustrates the gap between Kähler manifolds and symplectic manifolds, even in the simply connected case.

8.3 A model for a symplectic blow-up

Remark 8.30 In [188], D. McDuff constructed a blow-up of $\mathbb{CP}(5)$ along an embedding of the Kodaira–Thurston manifold $KT \hookrightarrow \mathbb{CP}(5)$. She proved that if the hard Lefschetz property failed for M, then it also failed for the blow-up (see [188, Proposition 2.5] or [257, Chapter 4] for details). G. Cavalcanti [51] has shown that we can't decide whether blow-ups have or don't have the hard Lefschetz property based simply on the ambient space. He also shows that there are compact simply connected symplectic blow-ups which satisfy hard Lefschetz, but which are nonformal. On the other hand, there are compact simply connected symplectic blow-ups (e.g. McDuff's example) which do not satisfy hard Lefschetz, so, by Theorem 4.82, they are simply connected counterexamples to the Brylinski conjecture.

8.3.5 Effect of the symplectic form on the blow-up

How much of an effect does the choice of symplectic form have on the rational homotopy type (and hence the diffeomorphism type) of a blow-up? In [165], embeddings are taken according to the Tischler theorem,

$$\mathbb{CP}(1) \xrightarrow{f_\ell} \mathbb{CP}(5), \quad f_\ell^*(\alpha_5) = \ell\, \alpha_1,$$

where α_i is the standard Kähler form on $\mathbb{CP}(i)$, $i = 1, 5$ and $\ell \in \mathbb{Z}$. Then, denoting the resulting blow-ups by $\widetilde{\mathbb{CP}}_\ell(5)$ and using the model $Bl(A, B)$ (and, in particular, the multiplication derived from the R-dgmodule structure), the following result is shown.

Theorem 8.31 ([165, Section 8.5]) *If $\ell_1 \neq \ell_2$, then the rational homotopy types of $\widetilde{\mathbb{CP}}_{\ell_1}(5)$ and $\widetilde{\mathbb{CP}}_{\ell_2}(5)$ are different. Hence, there are an infinite number of rationally distinct blow-ups of $\mathbb{CP}(5)$ along $\mathbb{CP}(1)$ corresponding to the infinite number of integral symplectic forms on $\mathbb{CP}(1)$, $\ell\, \alpha_1$ for $\ell \in \mathbb{Z}$.*

8.3.6 Vanishing of Chern classes for KT

A key point in the construction of McDuff's example in Subsection 8.3.4 was the triviality of the Kodaira–Thurston manifold's total Chern class. Here we give the details of this vanishing result since it does not seem to be something that is widely known. The proof relies on an old construction that goes back at least to Borel and Hirzebruch's work on the characteristic classes of homogeneous spaces [34]. This construction is known as the tangent bundle along the fibers.

Suppose $F \xrightarrow{i} E \xrightarrow{p} B$ is a fiber bundle of smooth manifolds with associated principal bundle $G \to P \to B$, where G right-acts smoothly on P and G left-acts smoothly on F. Then G left-acts smoothly on the tangent bundle TF of F by derivative maps. We can then form the quotient $\theta = P \times_G TF$

8 : Blow-ups and Intersection Products

and, since $E = P \times_G F$, we obtain the following map of bundles

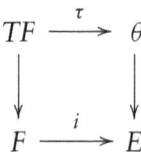

The bundle θ is called the *(tangent) bundle along the fibers*. The essential properties of the bundle along the fibers are delineated in [34]:

1. The tangent bundle TF is a pullback of θ: $i^*(\theta) = TF$.
2. If G preserves some other structure on TF, then θ inherits the same structure and τ is a bundle map preserving that structure. In particular, if F is almost complex and G preserves the almost complex structure, then τ is a bundle map of almost complex structures. By (1), we have $i^*(c(\theta)) = c(TF)$, where $c(-)$ denotes the total Chern class.
3. If $F \to E \xrightarrow{p} B$ is a smooth bundle of manifolds, then the tangent bundle of E is given by $TE \cong \theta \oplus p^*(TB)$ (see [34, Proposition 7.6]).

Our goal is to prove the following result (also see Exercise 8.3).

Theorem 8.32 *The Chern classes of the Kodaira–Thurston manifold vanish. That is, $c_1(KT) = 0$ and $c_2(KT) = 0$.*

Proof First, we note that all nilmanifolds have free circle actions on them, so their Euler characteristics vanish. For KT, this also follows by the product equality for the Euler characteristic of a fibration. Because $c_2(KT)$ is the Euler class, we see that $c_2(KT) = 0$. Hence, we focus on the vanishing of $c_1(KT)$.

We now take the bundle along the fibers of the Borel fibration associated to the free symplectic action of T^2 on KT:

$$\begin{array}{ccc} T(KT) & \xrightarrow{\tau} & \theta = ET^2 \times_{T^2} T(KT) \\ \downarrow & & \downarrow \\ KT & \xrightarrow{i} & ET^2 \times_{T^2} KT & \longrightarrow BT^2 \end{array}$$

By our standing assumption, the action preserves a compatible almost complex structure, so the bundle map gives $i^*(c_1(\theta)) = c_1(KT)$. Now, because the action is free, we have the following commutative diagram with q a homotopy equivalence:

$$
\begin{array}{ccc}
KT & \xrightarrow{i} & ET^2 \times_{T^2} KT \\
& \searrow{\scriptstyle p} & \downarrow{\scriptstyle q} \simeq \\
& & T^2 = KT/T^2
\end{array}
$$

Now, the bundle $T^2 \to KT \xrightarrow{p} T^2$ is classified by the top class in $H^2(T^2;\mathbb{Z})$, so $p^*(H^2(T^2;\mathbb{Z})) = 0$. Since q is a homotopy equivalence, we therefore also have $i^*(H^2(ET^2 \times_{T^2} KT;\mathbb{Z})) = 0$. Because $c_1(\theta) \in H^2(ET^2 \times_{T^2} KT;\mathbb{Z})$, we have $c_1(KT) = i^*(c_1(\theta)) = 0$. □

8.4 The Chas-Sullivan loop product on loop space homology

8.4.1 The classical intersection product

Let N be a compact connected oriented n-dimensional smooth manifold and let $\Delta \colon N \to N \times N$ be the diagonal embedding. Denote by T a tubular neighborhood of $\Delta(N)$ and by ∂T its boundary. The exponential map induces a diffeomorphism between T and the normal disk bundle $D\nu$ to $\Delta(N)$ that restricts to a diffeomorphism between ∂T and the associated sphere bundle $S\nu$. Let $p \colon D\nu \to N$ denote the projection. Since N is oriented, there is an orientation class $O_N \in H^n(D\nu, S\nu)$ such that the cap product with O_N induces an isomorphism $\theta \colon H_*(D\nu, S\nu) \xrightarrow{\cong} H_{*-n}(N)$: $\theta(x) = H_*(p)(O_N \cap x)$.

The homology intersection product on N with coefficients in a field $\mathbb{k}$ is the map $H_k(N) \otimes H_l(N) \to H_{k+l-n}(N)$ obtained as the following composition of maps

$$H_k(N) \otimes H_l(N) \xrightarrow{\cong} H_{k+l}(N \times N) \to H_{k+l}(N \times N, F(N,2))$$

$$\xrightarrow{\mathrm{exc}^{-1}} H_{k+l}(T, \partial T) \xrightarrow{\cong} H_{k+l}(D\nu, S\nu) \xrightarrow{\theta} H_{k+l-n}(N),$$

where exc denotes the excision isomorphism and $F(N,2)$ denotes the complement of the diagonal $\Delta(N)$.

In cohomology, we can take the cup product with O_N to obtain an isomorphism $\theta' \colon H^{*-n}(N) \to H^*(D\nu, S\nu)$: $\theta'(x) = H^*(p)(x) \cup O_N$. With coefficients in a field $\mathbb{k}$, the intersection coproduct is then given by the composition of maps

$$H^{*-n}(N) \xrightarrow{\theta'} H^*(Dv, Sv) \xrightarrow{\cong} H^*(T, \partial T)$$
$$\xrightarrow{\mathrm{exc}^{-1}} H^*(N \times N, F(N,2)) \to H^*(N \times N) \xrightarrow{\cong} H^*(N) \otimes H^*(N).$$

We observe that, since the intersection coproduct is an isomorphism for $* = 2n$, mapping fundamental class to fundamental class, it is a shriek map by Corollary 8.6, and therefore is multiplication by the diagonal cohomology class (see Lemma 8.17).

8.4.2 The Chas–Sullivan loop product

In [53], M. Chas and D. Sullivan defined a product on the desuspension of the homology (with coefficients in a field k) of the free loop space of N:

$$\mathbb{H}_*(LN; k) = H_{*+n}(LN; k).$$

The product, called the *loop product*, is defined at the chain level using both the intersection product on the chains on N and the loop composition. We now give a brief description of the Chas–Sullivan definition without going into details. We present a more homological version afterwards.

The loop product makes the homology of the free loop space a graded commutative and associative algebra [53]. More precisely, let $p: LN \to N$ denote the map that associates to a loop its base point. Let $\sigma: \Delta^p \to LN$ and $\tau: \Delta^q \to LN$ be chains such that their projections in N, $p\sigma$ and $p\tau$, are transverse. We then define a space

$$E = \{(s, t, n) \in \Delta^p \times \Delta^q \times N \mid p\sigma(s) = p\tau(t) = n\}$$

as the pullback of the diagram

$$\begin{array}{ccc} E & \longrightarrow & N \\ \downarrow & & \downarrow \Delta \\ \Delta^p \times \Delta^q & \xrightarrow{(p\sigma, p\tau)} & N \times N \end{array}$$

Note that the composition of the loops $\sigma(s)$ and $\tau(t)$ is defined for $(s, t, n) \in E$. We thus get a map $E \to LN$ by mapping (s, t) to the composition of the loops $\sigma(s)$ and $\tau(t)$. Since N has codimension n in $N \times N$, when σ and τ are transverse, then E has codimension n in $\Delta^p \times \Delta^q$. Chas and Sullivan then used this construction to define their *loop product*,

$$H_p(LN) \otimes H_q(LN) \to H_{p+q-n}(LN).$$

8.4 The Chas-Sullivan loop product on loop space homology

Let's now consider a more homological version of the loop product for a compact connected smooth orientable manifold N. Our presentation looks like the homological presentation given above for the intersection product. For $Z \subset N \times N$, we denote by L_Z the subspace of $L(N \times N) = LN \times LN$ consisting of loops with base point in Z. For instance $L_{\Delta(N)} = LN \times_N LN$, and the composition of loops defines a map that will be very useful later:

$$\mu: LN \times_N LN \to LN.$$

Recall that $(T, \partial T)$ is diffeomorphic to the pair (Dv, Sv) formed by the normal disk bundle and the normal sphere bundle to the embedding Δ. Let p^*Dv and p^*Sv denote the pullbacks over $LN \times_N LN$ of the disk and sphere bundles Dv and Sv along the projection $p: LN \times_N LN \to \Delta(N)$. A point in p^*Dv is a pair (c, v) with $c \in LN \times_N LN$ and $v \in (Dv)_{c(0)}$. Applying the exponential map to v gives a geodesic $u(t) = \exp_{c(0)}(tv)$. Let $\bar{u}(t)$ denote the inverse path, $\bar{u}(t) = u(1-t)$. The correspondence

$$(c, v) \mapsto \bar{u} \cdot c \cdot u$$

gives a homotopy equivalence $p^*Dv \to L_T$, which induces, by restriction, a homotopy equivalence $p^*Sv \xrightarrow{\simeq} L_{\partial T}$.

Denote now by $\bar{O}_N \in H^n(p^*Dv, p^*Sv)$ the pullback of the orientation class $O_N \in H^n(Dv, Sv)$. The cap product with $\bar{O}_N$ defines an isomorphism

$$H_*(L_{\Delta(N)}) \cong H_{*+n}(p^*Dv, p^*Sv).$$

By excision and homotopy equivalence we also have the following isomorphisms:

$$H_*(L(N \times N), L_{F(N,2)}) \xleftarrow{\cong} H_*(L_T, L_{\partial T}) \xrightarrow{\cong} H_*(p^*Dv, p^*Sv).$$

Definition 8.33 *The (Chas–Sullivan) loop product on $H_*(LN)$ is the composition*

$$H_*(LN) \otimes H_*(LN) \to H_*(LN \times LN) \to H_*(LN \times LN, L_{F(N,2)})$$
$$\cong H_*(p^*Dv, p^*Sv) \cong H_{*-n}(L_{\Delta(N)}) \xrightarrow{\mu} H_{*-n}(LN).$$

Definition 8.34 *The dual of the loop product is the composition $\alpha_N \circ H^{*-n}(\mu)$, where*

$$\begin{cases} H^{*-n}(\mu): H^{*-n}(LN) \to H^{*-n}(L_{\Delta(N)}), \\ \alpha_N: H^{*-n}(L_{\Delta(N)}) \cong H^*(LN \times LN, L_{F(N,2)}) \to H^*(LN \times LN). \end{cases}$$

8.4.3 A rational model for the loop product

In this section we will suppose that N is 2-connected. Suppose $(\wedge X, d)$ is a minimal model for N, and let (A, d) be a Poincaré duality model for N. We then have a quasi-isomorphism $\varphi \colon (\wedge X, d) \to (A, d)$. A model for the free loop space fibration $p \colon LN \to N$ has been described in Section 5.2 (where, for the sake of simplicity, we write $\bar{X} = sX$):

$$(\wedge X, d) \to (\wedge X \otimes \wedge \bar{X}, D), \qquad D(\bar{x}) = -sd(x).$$

The pullback

$$(A, d) \to (A \otimes \wedge \bar{X}, D) = A \otimes_{\wedge X} (\wedge X \otimes \wedge \bar{X}, D)$$

is also a model for the projection p.

Since $L_{F(N,2)}$ is the pullback of $LN \times LN$ along the inclusion $i \colon F(N, 2) \hookrightarrow N \times N$, and since the projection $\pi \colon A \otimes A \to A \otimes A/(D_A)$ is a model for i (see Theorem 8.18), a model for the diagram

$$\begin{array}{ccc} L_{F(N,2)} & \longrightarrow & LN \times LN \\ \downarrow & & \downarrow \\ F(N,2) & \xrightarrow{i} & N \times N \end{array}$$

is given by the square

$$\begin{array}{ccc} \big((A \otimes \wedge \bar{X}, D) \otimes (A \otimes \wedge \bar{X}, D)\big)/(D_A) & \xleftarrow{\tilde{\pi}} & (A \otimes \wedge \bar{X}, D) \otimes (A \otimes \wedge \bar{X}, D) \\ \uparrow & & \uparrow \\ (A \otimes A)/(D_A) & \xleftarrow{\pi} & A \otimes A \end{array}$$

where $\tilde{\pi}$ is the map

$$(A, d)^{\otimes 2} \otimes_{A \otimes A} (A \otimes \wedge \bar{X}, D)^{\otimes 2} \to (A, d)^{\otimes 2}/(D_A) \otimes_{A \otimes A} (A \otimes \wedge \bar{X})^{\otimes 2}.$$

This shows that a model for the injection

$$A_{PL}(L(N \times N), L_{F(N,2)}) \to A_{PL}(L(N \times N))$$

is given by the injection of $\mathrm{Ker}\,\tilde{\pi}$ by the map μ_{D_A}, which is the multiplication by D_A (see Subsection 8.1.4):

$$\mu_{D_A} \otimes 1 \colon (s^{-n}A) \otimes_{A \otimes A} (A \otimes \wedge \bar{X}, D) \otimes (A \otimes \wedge \bar{X}, D)$$

$$\to (A \otimes A)_{A \otimes A}(A \otimes \wedge \bar{X}, D)^{\otimes 2}.$$

From the definition of α_N, we deduce the following result.

Proposition 8.35 ([94]) *The map α_N is the map induced in cohomology by $\mu_{D_A} \otimes 1$.*

As we saw in Section 5.9, another convenient model for the free loop space is given by the Hochschild complex $\mathbb{B}(A) = (A \otimes T(s\bar{A}), D)$. This entails the following version of Proposition 8.35.

Proposition 8.36 ([94]) *The map α_N is the map induced in cohomology by the following multiplication by D_A:*

$$\mu_{D_A} \otimes 1 : (s^{-n}A) \otimes_{A \otimes A} (A \otimes T(s\bar{A})) \otimes (A \otimes T(s\bar{A}))$$
$$\to (A \otimes T(s\bar{A})) \otimes (A \otimes T(s\bar{A})).$$

We now give a model for the multiplication $\mu: LN \times_N LN \to LN$. Recall that $T(s\bar{A})$ is a differential coalgebra whose comultiplication is defined by

$$\nabla([a_1|\cdots|a_q]) = \sum_{i=0}^{q} [a_1|\cdots|a_i] \otimes [a_{i+1}|\cdots|a_q].$$

Then we see that

$$1 \otimes \nabla : \mathbb{B}(A) \to (A \otimes T(s\bar{A}) \otimes T(s\bar{A}), D) = A \otimes_{A \otimes A} \mathbb{B}(A) \otimes \mathbb{B}(A)$$

is a morphism of differential A-modules.

Proposition 8.37 *The morphism $H^*(\mu; \mathbb{Q}): H^*(LN; \mathbb{Q}) \to H^*(L_{\Delta(N)}; \mathbb{Q})$ is the map induced in cohomology by the morphism $1 \otimes \nabla$.*

Proof The composition of loops $\mu: L_{\Delta(N)} \to LN$ is the pullback morphism in the diagram

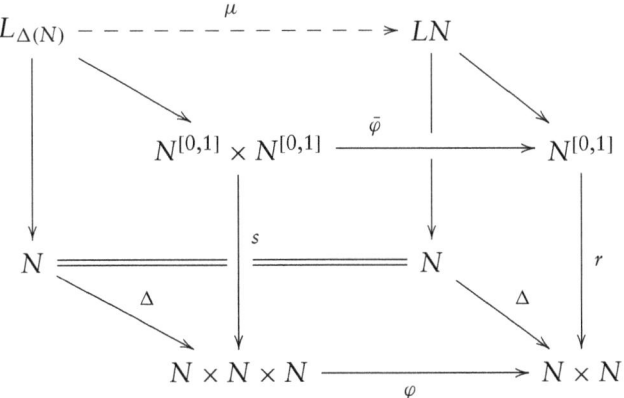

with $\varphi(a,b,c) = (a,c)$, $s(c,c') = (c(0), c(1), c'(1))$ and $r(c) = (c(0), c(1))$. The model for this diagram is

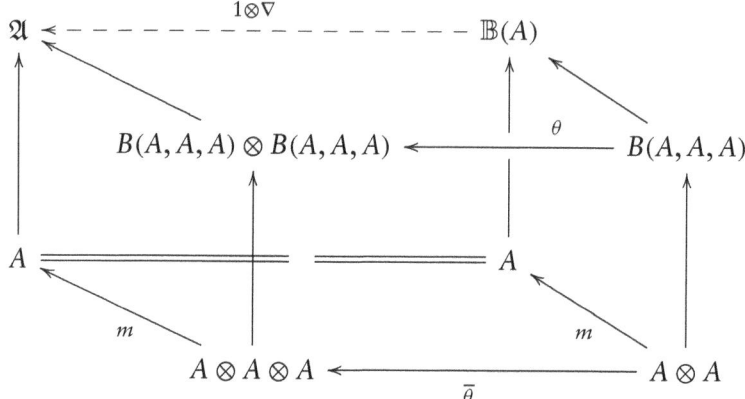

where $\mathfrak{A} = A \otimes_{A \otimes A} (\mathbb{B}(A) \otimes \mathbb{B}(A))$, $B(A, A, A)$ denotes the double bar construction (see Section 5.9), the map m denotes the standard multiplication,

$$\theta(a[a_1|\cdots|a_n]a') = \sum_{i=0}^{n} a[a_1|\cdots|a_i]]1 \otimes 1[[a_{i+1}|\cdots|a_n]a',$$

and $\bar{\theta}(a \otimes a') = a \otimes 1 \otimes a'$. The model for the composition of loops is thus obtained by taking the pushout map in this diagram: that is, the map $1 \otimes \nabla$. □

8.4.4 Hochschild cohomology and Cohen–Jones theorem

Recall that the Hochschild cohomology of a graded differential algebra (A, d), $HH^*(A, A)$, is the homology of the complex $\text{Hom}_A(\mathbb{B}(A), A)$. This homology is a graded algebra whose product $\star$ is defined for $f, g \in \text{Hom}(T(s\bar{A}), A) = \text{Hom}_A(\mathbb{B}(A), A)$ by

$$f \star g : T(s\bar{A}) \xrightarrow{\nabla} T(s\bar{A}) \otimes T(s\bar{A}) \xrightarrow{f \otimes g} A \otimes A \xrightarrow{m} A.$$

Using the isomorphism $\psi_A : A \to s^{-n}A^\vee$, we obtain, by duality, a coproduct on the Hochschild homology of A, $HH_*(A)$. The coproduct has degree n and is the map induced in cohomology by the composition

$$\mathbb{B}(A) = A \otimes_A \mathbb{B}(A) \xrightarrow{\psi_A \otimes 1} s^{-n}A^\vee \otimes_A \mathbb{B}(A)$$

$$= s^{-n}\text{Hom}(T(s\bar{A}), A) \xrightarrow{s^{-n}\star} s^{-n} \left(\text{Hom}(T(s\bar{A}), A) \otimes \text{Hom}(T(s\bar{A}), A)\right)$$

$$= s^{-n}(A^\vee \otimes T(s\bar{A}))^{\otimes 2} \xleftarrow{\psi_A^{\otimes 2}} s^n(A \otimes T(s\bar{A}))^{\otimes 2} = s^n\mathbb{B}(A)^{\otimes 2}.$$

8.4 The Chas-Sullivan loop product on loop space homology

We are now ready to prove the theorem of Cohen and Jones.

Theorem 8.38 ([62], [191], [94]) *The isomorphism*

$$H^*(LN; \mathbb{Q}) \cong HH_*(A)$$

is an isomorphism of coalgebras. By duality, the loop algebra $\mathbb{H}_*(LN; \mathbb{Q})$ is isomorphic as an algebra to the Hochschild cohomology algebra $HH^*(A, A)$.

Proof The theorem is a direct consequence of the commutativity of the following diagrams (the first one coming from Proposition 8.37 and the second one being a combination of Proposition 8.36 and Lemma 8.17).

$$\begin{array}{ccc}
H^*(LN;\mathbb{Q}) & \xrightarrow{\mu} & H^*(L_{\Delta(N)};\mathbb{Q}) \\
\cong \uparrow & & \uparrow \cong \\
H^*(\mathbb{B}(A)) & \xrightarrow{H^*(1\otimes\nabla)} & H^*(A \otimes_{A^{\otimes 2}} \mathbb{B}(A)^{\otimes 2}) \\
H^*(\psi_A\otimes 1) \downarrow \cong & & \cong \downarrow H^*(\psi_A\otimes 1) \\
H^*(s^{-n}A^\vee \otimes_A \mathbb{B}(A)) & \xrightarrow{H^*(1\otimes\nabla)} & H^*(s^{-n}A^\vee \otimes_{A^{\otimes 2}} \mathbb{B}(A)^{\otimes 2})
\end{array}$$

$$\begin{array}{ccc}
s^{-n}H^*(L_{\Delta(N)};\mathbb{Q}) & \xrightarrow{\alpha_N} & H^*(L(N\times N;\mathbb{Q})) \\
\cong \uparrow & & \uparrow \cong \\
H^*(s^{-n}A \otimes_{A^{\otimes 2}} \mathbb{B}(A)^{\otimes 2}) & \xrightarrow{H^*(\mu_{D_A}\otimes 1)} & H^*(A^{\otimes 2} \otimes_{A^{\otimes 2}} \mathbb{B}(A)^{\otimes 2}) \\
H^*(s^{-n}\psi_A\otimes 1) \downarrow \cong & & \cong \downarrow H^*(\psi_A\otimes 1)^{\otimes 2} \\
H^*(s^{-2n}A^\vee \otimes_{A^{\otimes 2}} \mathbb{B}(A)^{\otimes 2}) & \xrightarrow{H^*(m^\vee\otimes 1)} & H^*((s^{-n}A^\vee)^{\otimes 2} \otimes_{A^{\otimes 2}} \mathbb{B}(A)^{\otimes 2})
\end{array}$$

□

Since a quasi-isomorphism of differential graded algebras, $\varphi\colon (A,d) \to (B,d)$, induces an isomorphism of Hochschild algebras $HH^*(A,A) \cong HH^*(B,B)$, we can choose $A = C^*(N;\mathbb{Q})$, $A = A_{PL}(N)$ or $A = (\wedge V, d)$, where $(\wedge V, d)$ is the minimal model of N, in Theorem 8.38. We thus have the following isomorphisms of algebras:

$$H_*(LN; \mathbb{R}) \cong HH^*(A_{DR}(N), A_{DR}(N)),$$
$$H_*(LN; \mathbb{Q}) \cong HH^*((\wedge V, d), (\wedge V, d)).$$

8.4.5 The Chas-Sullivan loop product and closed geodesics

Let $(LN)_a$ denote the subspace of LN consisting of curves of length less than or equal to a. Following [110], we define the critical value $\operatorname{cr}(\alpha)$ of a homology class $\alpha \in H_q(LN; \mathbb{Z})$ to be the number

$$\operatorname{cr}(\alpha) = \inf\{\, a \mid \alpha \in \operatorname{Im}\left(H_q((LN)_a; \mathbb{Z}) \to H_q(LN; \mathbb{Z})\right) \,\}.$$

Denoting the loop product by $\bullet$, Goresky and Hingston prove the

Theorem 8.39 ([110]) *The critical value behaves well with respect to the loop product:*

$$\operatorname{cr}(\alpha \bullet \beta) \leq \operatorname{cr}(\alpha) + \operatorname{cr}(\beta).$$

Moreover, we have the following nilpotency result.

Theorem 8.40 ([110]) *If all closed geodesics are nondegenerate, then every homology class $\alpha \in H_*(LN; \mathbb{Z})$ is level-nilpotent; that is, there is an integer r depending on α such that $\operatorname{cr}(\bullet^r \alpha) < r \operatorname{cr}(\alpha)$.*

This level-nilpotency of the homology classes for the loop product begs the question of whether the classes are nilpotent in the usual sense. In order to investigate this question, we consider the transverse intersection of a homology class in LN with ΩN viewed as a submanifold of LN. This transverse intersection induces a morphism of algebras $I \colon \mathbb{H}_*(LN; \mathbb{Q}) \to H_*(\Omega N; \mathbb{Q})$ [53] that can be computed in the following way. Denote by $(\wedge V, d)$ the minimal model for N, by $(\wedge V \otimes \wedge sV, \delta)$ the minimal model for LN, and by $\varphi \colon (\wedge V, d) \to (A, d)$ a finite dimensional model for M with $A^{>n} = 0$ and $A^n = \mathbb{Q}\omega$. Then we have a quasi-isomorphism

$$\varphi \otimes \operatorname{id} \colon (\wedge V \otimes \wedge sV, \delta) \to (A \otimes \wedge sV, \delta) = (A, d) \otimes_{(\wedge V, d)} (\wedge V \otimes \wedge sV, \delta).$$

Now, the multiplication by ω, $\alpha \mapsto \alpha \cdot \omega$, induces a map

$$I^* \colon H^*(\Omega N; \mathbb{Q}) = \wedge sV \to H^*(A \otimes \wedge sV, \delta) = H^*(LN; \mathbb{Q}).$$

Lemma 8.41 ([93]) *The map I^* is dual to the intersection map I.*

This computation gives a nilpotence result analogous to the result above of Goresky and Hingston, Theorem 8.40.

Theorem 8.42 ([93]) *The image of the intersection map $I \colon \mathbb{H}_*(LN; \mathbb{Q}) \to H_*(\Omega N; \mathbb{Q})$ is a finitely generated algebra, and its kernel is a nilpotent algebra.*

Exercises for Chapter 8

Exercise 8.1 Let G be a connected Lie group. Show that the loop algebra $\mathbb{H}_*(LG;\mathbb{Q})$ is isomorphic to the tensor product $H_*(G;\mathbb{Q}) \otimes H_*(\Omega G;\mathbb{Q})$, where $H_*(G;\mathbb{Q})$ is equipped with the intersection product, and $H_*(\Omega G;\mathbb{Q})$ has the Pontryagin product.

Exercise 8.2 Use the Lambrechts–Stanley model $Bl(A, B)$ (see Definition 8.27) to obtain the rational homotopy type of the blow-up of a point in a manifold M.

Exercise 8.3 A result of F. Peterson says that, in particular, if a real n-dimensional manifold has torsionfree integral cohomology, then any complex $2n$-bundle is complex trivial if and only if its Chern classes vanish. (Note that we are not saying the bundle is complex parallelizable, for this is a restrictive concept requiring analyticity of sections.) Show that we can see the triviality of the tangent bundle of KT directly from the following.

1. Consider the principal bundle $T^2 \to KT \xrightarrow{p} T^2$. Show that $T(T^2)$ is trivial as a complex bundle because $c_1(T^2) = \chi(T^2) = 0$ and the trivialization is accomplished by left translation on T^2. Therefore, the induced action of T^2 on $T(T^2) = T^2 \times \mathbb{C}$ is given by $L_y(x, s) = (yx, s)$.
2. Show that the bundle along the fibers is given by

$$\theta = KT \times_{T^2} (T^2 \times \mathbb{C}) = (KT \times_{T^2} T^2) \times \mathbb{C} = KT \times \mathbb{C}.$$

    Hint: to see that $KT \times_{T^2} T^2 = KT$ is an equivariant homeomorphism, define the inverse maps: $\lambda \colon KT \times_{T^2} T^2 \to KT$, $\lambda([x, g]) = xg$; $\tau \colon KT \to KT \times_{T^2} T^2$, $\tau(x) = [x, e]$. Hence, θ is a trivial complex bundle.
3. Now show that $T(KT) \cong \theta \oplus p^*(T(T^2))$ and therefore it is trivial. Because the tangent bundle is complex trivial, all Chern classes vanish.

A Florilège of geometric applications

This chapter is a survey on the types of models which arise when studying configuration spaces, smooth algebraic varieties, function (or mapping) spaces, and arrangements. We also give a brief introduction to two subjects, Gelfand–Fuchs cohomology and iterated integrals, which provided unexpected connections between models and geometric analysis at the very dawn of rational homotopy theory. Here we explain the relevant models and then refer to the appropriate literature for details or proofs.

Section 9.1 is a quick survey on the rational homotopy type of configuration spaces. When M is a manifold, the configuration space of k points in M is the space $F(M, k) = \{(x_1, \ldots, x_k) \in M^k \,|\, x_i \neq x_j \text{ for } i \neq j\}$. We recall, in particular, how to compute the Betti numbers and the ranks of the homotopy groups of $F(M, k)$. The main problem centers around knowing if the rational homotopy type of $F(M, k)$ depends only on the rational homotopy type of M. By the work of Kriz and Totaro, this is true for complex projective varieties and, by the work of Lambrechts and Stanley, this is true when $k = 2$ and M is a 2-connected compact manifold. Further, we discuss certain chain complexes giving the rational cohomology of unordered configuration spaces.

Next, in Section 9.2, we consider arrangements of hyperplanes and affine subspaces and their accompanying models. The main problem here is to understand the topology of the complement of the arrangement. This is directly related to configuration spaces because the configuration space of k points in $\mathbb{R}^2$ is the complement of the arrangement formed by the complex hyperplanes $z_i = z_j$. Here we describe the atomic model of Yuzvinsky and give concrete examples.

We have seen in Theorem 4.43 that compact Kähler manifolds are formal spaces. We may then wonder if the minimal model of a smooth algebraic manifold also has some special properties. In Section 9.4, with this objective in mind, we first use the existence of a pure Hodge structure on the cohomology of a compact Kähler manifold to give another proof of formality.

In [69], Deligne extended the notion of pure Hodge structure to that of mixed Hodge structure and proved that the cohomology of any complex algebraic manifold carries such a structure. Here, we discuss briefly the result of Morgan (see [201]) which says that the minimal model of a complex algebraic manifold also possesses such a structure. Some homotopy consequences are then given.

Spaces of mappings, and, more generally, spaces of sections of fibrations are very useful in geometry. The free loop space is an example of a space of mappings, and the space of sections of the free loop space fibration $p: LM \to M$ is the loop space on the topological monoid aut(M). By work of D. McDuff [187], the space of finite parts of M with labels in a space X can be identified with the space of sections of a sphere bundle over M. We explain these ideas in Subsection 9.5.1 and show how to compute the rational homotopy type of the space of sections of a given fibration. Finally, the Gelfand–Fuchs cohomology can be interpreted as the cohomology of the space of sections of a certain bundle. We then obtain explicit computations of the Gelfand–Fuchs cohomology in Subsection 9.5.4.

K. Chen's theory of iterated integrals is another geometric way to obtain rational homotopy information for manifolds. Since this theory has been used in many circumstances in geometry, we have included a presentation of the theory in Section 9.6. This gives us the opportunity to make precise the links with the Sullivan approach.

Section 9.7 contains a list of cohomological conjectures that have appeared in different parts of the book and that we present here together with their interrelations.

9.1 Configuration spaces

Let M denote an m-dimensional manifold. The *space of ordered configurations of k points in M* is the space
$$F(M, k) = \{ (x_1, \ldots, x_k) \in M^k \mid x_i \neq x_j \text{ for } i \neq j \}.$$
When two manifolds are homeomorphic, then their configuration spaces are also homeomorphic. The natural question is then: When M and N are homotopy equivalent, is it true that $F(M, k)$ and $F(N, k)$ are also homotopy equivalent?

By Example 9.1 below, $F(\mathbb{R}^n, 2) \simeq S^{n-1}$, so we see that the answer is no in general. Now, if M is compact and nonsimply connected, then the answer to this question is also no as shown by R. Longoni and P. Salvatore [176]. On the other hand, Levitt has proven that the answer is yes for compact 2-connected manifolds when $k = 2$ [172], and M. Aouina and J. Klein have shown that the homotopy type of some suspension of $F(M, k)$ depends only on the homotopy type of M [13]. For general compact simply connected manifolds, the problem remains open.

In this section, we will be concerned with the rational homotopy of configuration spaces. Of course, here we want to know if the rational homotopy type of $F(M,k)$ depends only on the rational homotopy type of M. In Subsection 8.1.5, we show by a theorem of Lambrechts and Stanley, that this is true for $k = 2$ when the manifold is 2-connected. We will also show how to get the rational homotopy groups and the rational Betti numbers of $F(M,k)$ directly from a model of M. In case M is a complex projective manifold, a model for $F(M,k)$ can be derived from a model for M. Now, in certain cases we can obtain homotopy equivalences of $F(M,k)$ with well-known spaces. For instance, we have the following.

Example 9.1 When M admits a multiplication, then the map $(x,y) \mapsto (x, x^{-1}y)$ induces a homeomorphism between $F(M,2)$ and $M \times M\setminus\{e\}$. For instance $F(\mathbb{R}^n, 2) \cong \mathbb{R}^n \times (\mathbb{R}^n\setminus\{0\}) \simeq S^{n-1}$. In the same way, $F(S^1 \times S^1, 2) \simeq S^1 \times S^1 \times (S^1 \vee S^1)$.

Example 9.2 When M is a sphere S^n, $n \geq 2$, then $F(M,2) \simeq S^n$ and $F(M,3)$ has the homotopy type of the tangent sphere bundle to S^n. The map from the sphere bundle to $F(M,3)$ maps (x, v) to $(x, -x, \exp_x(v))$.

9.1.1 The Fadell–Neuwirth fibrations

Let M be a manifold, and let $q_1, \ldots, q_n$ be n distinct fixed points in M. For $i \leq n$, let $Q_i = \{q_1, \ldots, q_i\}$. Observe that the space $M\setminus Q_i$ is an open manifold for $i \geq 1$, and we have a map

$$\pi \colon F(M\setminus Q_{n-1}, k) \to M\setminus Q_{n-1}$$

sending a k-tuple onto its first component: $\pi(x_1, \ldots, x_k) = x_1$. Fadell and Neuwirth prove the following result that is basic in the theory.

Theorem 9.3 ([83]) *The projection π is a locally trivial fiber bundle with fiber $F(M\setminus Q_n, k-1)$. Moreover, when $n \geq 2$, the fiber bundle admits a section.*

Clearly, the fiber over the basepoint q_n is $F(M\setminus Q_n, k-1)$. Also, the section is easy to construct. In M, we first choose an open n-dimensional disk D around q_1, of radius 1 for some metric, and we suppose that D does not contain any of the points $q_2, \ldots, q_{n-1}$. We now choose $k-1$ distinct points $y_1, \ldots, y_{k-1}$ in D on a sphere of radius $1/2$ around q_1. We then define a section σ of π by the following process. When $x \notin D$, we put $\sigma(x) = (x, y_1, \ldots, y_{k-1})$. When $x \in D$, then we put $\sigma(x) = (x, |x|y_1, \ldots, |x|y_{k-1})$, where we denote the distance between q_1 and x by $|x|$.

9.1.2 The rational homotopy of configuration spaces

The homotopy type of the manifold $M\backslash Q_k$ is easy to describe. First, since the group of homeomorphisms of M acts k-transitively (i.e. any two given sets of k distinct points can be mapped one onto another by a single homeomorphism), we can suppose that the k points belong to an open set homeomorphic to an m-dimensional open disk D. The space D minus the points has the homotopy type of a wedge of k spheres S^{m-1}, one of them being the boundary of the disk. Therefore, if $k \geq 2$, the space $M\backslash Q_k$ has the homotopy type of the wedge of $M\backslash Q_1$ with a wedge of $k - 1$ copies of the sphere S^{m-1};

$$M\backslash Q_k \simeq M\backslash Q_1 \vee (\vee_{k-1} S^{m-1}).$$

Proposition 9.4 ([90]) *If the cohomology algebra $H^*(M; \mathbb{Q})$ requires at least two generators, then we have an isomorphism*

$$\pi_* F(M, k) \otimes \mathbb{Q} \cong \oplus_{i=0}^{k-1} \pi_*(M\backslash Q_i) \otimes \mathbb{Q}.$$

Proof We have the following sequence of fibrations

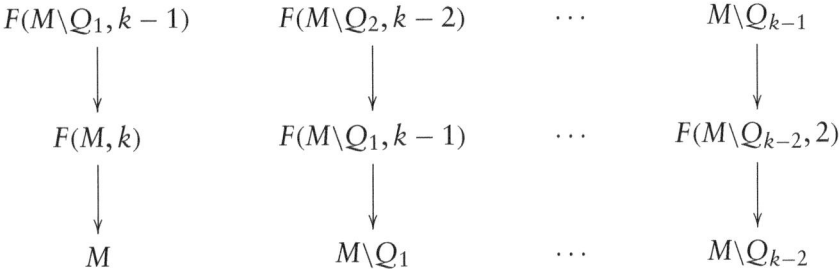

All the fibrations except maybe the first one admit a section and so their long exact homotopy sequences split into short exact sequences.

Recall from Theorem 3.3 that the Lie algebra $\pi_*(\Omega(M\backslash Q_1)) \otimes \mathbb{Q}$ admits a filtration such that the graded associated Lie algebra is the free product of $\pi_*(\Omega M) \otimes \mathbb{Q}$ with a free Lie algebra on one generator. Hence, the center of $\pi_*(\Omega(M\backslash Q_1)) \otimes \mathbb{Q}$ is zero. The same is also true for the center of $\pi_*(\Omega(M\backslash Q_i)) \otimes \mathbb{Q}$ for $i \geq 2$ since, by the above description of $M\backslash Q_i$, we have $\pi_*(\Omega(M\backslash Q_i)) \otimes \mathbb{Q} \cong (\pi_*(\Omega(M\backslash Q_1)) \otimes \mathbb{Q}) \coprod \mathbb{L}(x_2, \ldots, x_{i-1})$. Therefore, by induction on the sequence of fibrations above, we find that the center of $\pi_*(\Omega F(M\backslash Q_1, k-1)) \otimes \mathbb{Q}$ is zero.

Now, we know that the image of the connecting map in the exact rational homotopy sequence for a fibration $\Omega F \to \Omega E \to \Omega B$ is contained in the center of $\pi_*(\Omega F) \otimes \mathbb{Q}$, so the long exact homotopy sequence of the first fibration also splits into short exact sequences. This yields the result. □

9.1.3 The configuration spaces $F(\mathbb{R}^n, k)$

The Betti numbers of $F(\mathbb{R}^n, k)$ can be derived from the Fadell–Neuwirth fibrations. Indeed, we have the

Proposition 9.5 *The integral homology groups of $F(\mathbb{R}^n \setminus Q_m, k)$ are free abelian groups and we have isomorphisms of groups*

$$H_* F(\mathbb{R}^n \setminus Q_m, k) \cong \otimes_{j=0}^{k-1} H_*(\vee_{m+j} S^{n-1}).$$

In particular,

$$H_* F(\mathbb{R}^n, k) \cong H_* F(\mathbb{R}^n \setminus Q_1, k-1) \cong \otimes_{j=1}^{k-1} H_*(\vee_j S^{n-1}).$$

Proof Suppose $n > 2$. We prove the result by induction on k. For $k = 1$, we have

$$H_* F(\mathbb{R}^n \setminus Q_m, 1) \cong H_*(\vee_m S^{n-1}).$$

So suppose the result is true for $k-1$ points, and any m. Then, for degree reasons, the Serre spectral sequence of the Fadell–Neuwirth fibration collapses at the E^2-level and we have isomorphisms

$$H_* F(\mathbb{R}^n \setminus Q_m, k) \cong H_* F(\mathbb{R}^n \setminus Q_{m+1}, k-1) \otimes H_*(\mathbb{R}^n \setminus Q_m)$$

$$\cong \otimes_{j=0}^{k-2} H_*(\vee_{m+1+j} S^{n-1}) \otimes H_*(\vee_m S^{n-1})$$

$$\cong \otimes_{j=1}^{k-1} H_*(\vee_{m+j} S^{n-1}) \otimes H_*(\vee_m S^{n-1}).$$

The case $n = 2$ requires more care and is done in detail in [61]. □

The multiplicative structure of the cohomology of $F(\mathbb{R}^n, k)$ has been described by F. Cohen in [61].

Theorem 9.6 *Let a_{ij} be a sequence of variables of degree $n - 1$, for $i, j = 1, \ldots, k$. Then we have an isomorphism of algebras,*

$$H^*(F(\mathbb{R}^n, k); \mathbb{Z}) \cong \wedge(a_{ij})/I,$$

where I is the ideal generated by the relations $a_{ij} - (-1)^n a_{ji}$, $a_{ij}^2 = 0$ and the Arnold relations

$$a_{ij} a_{jr} + a_{jr} a_{ri} + a_{ri} a_{ij}.$$

This gives the complete rational homotopy type of the configuration spaces of points in $\mathbb{R}^n$ because, by a result of Kontsevich, the spaces $F(\mathbb{R}^n, k)$ are formal.

Theorem 9.7 ([156, Theorem 2]) *The spaces $F(\mathbb{R}^n, k)$ are formal.*

The case $n = 2$ was proved previously by Arnold [42]. Indeed, $F(\mathbb{R}^2, k)$ is the complement in $\mathbb{C}^k$ of the union H of the hyperplanes $H_{ij} = \{(z_1, \ldots, z_n) \mid z_i = z_j\}$. The projection

$$\varphi_{ij} = z_i - z_j \colon \mathbb{C}^k \backslash H_{ij} \to \mathbb{C}\backslash\{0\}$$

is a homotopy equivalence and this implies that $H^1(\mathbb{C}^k\backslash H; \mathbb{R})$ is generated by the classes of the 1-forms

$$e_{ij} = \varphi_{ij}^*\left(\frac{dz}{2\pi i z}\right) = \frac{d\varphi_{ij}}{2\pi i \varphi_{ij}}.$$

A simple computation then shows that the Arnold relations ($\mathcal{A}$) are satisfied:

$$\mathcal{A}\colon e_{ij}e_{jr} + e_{jr}e_{ri} + e_{ri}e_{ij} = 0.$$

This defines, by restriction, a morphism of differential graded algebras

$$(\wedge(e_{ij})/(\mathcal{A}), 0) \to A_{DR}(F(\mathbb{C}, k)),$$

which is a quasi-isomorphism. This then proves the formality of $F(\mathbb{R}^2, k)$.

9.1.4 The configuration spaces of a projective manifold

Independently, and by using different methods, I. Kriz [160] and B. Totaro [254] gave a model for the configuration spaces $F(M, k)$ when M is a complex projective manifold. In order to describe this model, first let $p_i \colon M^k \to M$ denote the projection on the i-th component and let $p_{ij} \colon M^k \to M^2$ denote the projection on components i and j. Since M is a manifold, we can find a graded basis $\{a_i, a'_i\}$ of $H^*(M; \mathbb{Q})$ such that $a_i \cup a'_j = \delta_i^j \omega$ where ω denotes a fundamental class and δ_i^j is the Kronecker delta. The class $D_M = \sum_i (-1)^{|a_i|} a_i \otimes a'_i \in H^*(M \times M; \mathbb{Q})$ is called the *diagonal class* and we set $D_{ij} = p_{ij}^*(D_M) \in H^*(M; \mathbb{Q})^{\otimes k}$. Then we have

Theorem 9.8 *Let M^m be a complex projective manifold. Then a model for $F(M, k)$ is given by the differential graded algebra*

$$(H^*(M; \mathbb{Q})^{\otimes k} \otimes \wedge(x_{ij}, i, j = 1, \ldots k) / I, d),$$

where $|x_{ij}| = m - 1$, $d(x_{ij}) = D_{ij}$ and the ideal I is generated by the relations

$$\begin{cases} x_{ij} = x_{ji}, \\ x_{ij}x_{jr} + x_{jr}x_{ri} + x_{ri}x_{ij}, \\ p_i^*(x) \cdot x_{ij} = p_j^*(x) \cdot x_{ij}. \end{cases}$$

Note that, in this description, we recover the Arnold relations. Also note that the compatibility of the differential with the Arnold relations comes from Poincaré duality. Once again, we have a situation where we obtain a complete description of the rational homotopy type.

Now suppose that M^m is any compact, simply connected manifold and let (A, d) denote a Poincaré duality model for M with diagonal class D_A (see Definition 8.16). Then we can form a new commutative differential graded algebra by replacing $(H^*(M; \mathbb{Q}), 0)$ by (A, d) in the model of Kriz and Totaro. We let $p_i \colon A \to A^{\otimes k}$ denote the injection $p_i(a) = 1 \otimes \cdots \otimes a \otimes \cdots \otimes 1$ with the element a in the ith position. In a similar way $p_{ij} \colon A^{\otimes 2} \to A^{\otimes k}$ denotes the map that sends $a \otimes b$ to the k-tuple $x_1 \otimes \cdots \otimes x_k$ with $x_i = a$, $x_j = b$ and the other x_l equal to 1. We can then form the cochain algebra

$$G(M, k) = (A^{\otimes k} \otimes \wedge(x_{ij}, i, j = 1, \ldots k) / I, D),$$

where $D(a) = d(a)$, $|x_{ij}| = m - 1$, $D(x_{ij}) = p_{ij}(D_A)$ and the ideal I is generated by the relations

$$\begin{cases} x_{ij} = (-1)^m x_{ji}, \; x_{ij}^2 = 0, \\ x_{ij}x_{jr} + x_{jr}x_{ri} + x_{ri}x_{ij}, \\ p_i(x)x_{ij} = p_j(x)x_{ij}. \end{cases}$$

Conjecture 9.9 *$G(M, k)$ is a model for the configuration space $F(M, k)$.*

The conjecture is true for complex projective manifolds. By a result of Lambrechts and Stanley, it is also true for 2-connected manifolds when $k = 2$ (Subsection 8.1.5). Some other results go in the direction of the conjecture; for instance the cohomology of $G(M, k)$ is the right cohomology.

Theorem 9.10 *There is an isomorphism of graded vector spaces*

$$H^*(G(M, k)) \cong H^*(F(M, k); \mathbb{Q}).$$

This result is essentially the work of Bendersky and Gitler [24]. The present presentation comes from a re-writing due independently to B. Berceanu, M. Markl and S. Papadima [26] on one hand, and to Y. Félix and J.-C. Thomas [90] on the other.

Note that the permutation group Σ_k acts freely on $F(M, k)$. The quotient $C_k(M) = F(M, k)/\Sigma_k$ is called the space of *unordered configurations* of k points in M. This construction gives a lot of interesting spaces. For instance, when M is the space $\mathbb{R}^2$, then $\pi_1(C_k(M))$ is the braid group B_k on k generators. Of course, even if M is simply connected, $C_k(M)$ is not simply connected, and so its minimal model gives less information than we might hope. We can, however, compute the rational cohomology of $C_k(M)$ in case the dimension of M is odd.

Theorem 9.11 ([31]) *If* $\dim M = m$ *is odd, then*
$$H^*(C_k(M); \mathbb{Q}) = \wedge^k H^*(M; \mathbb{Q}).$$

For instance, when $M = S^3$, a basis of the vector space of cohomology is given by e_0 in degree 0 and e_3 in degree 3. The only words of length k in $\wedge(e_0, e_3)$ are e_0^k and $e_0^{k-1} e_3$. Therefore for any $k \geq 1$, $H^*(C_k(S^3); \mathbb{Q}) = \wedge(z)$ with z in degree 3. In particular, $C_k(S^3) \simeq_\mathbb{Q} S^3$.

Now let M be a compact simply connected m-dimensional manifold, with m even. Let $V = H^*(M; \mathbb{Q})$, $W = s^{-m+1} H^*(M; \mathbb{Q})$ and let $D_M \in V \otimes V$ be the diagonal class. We consider the differential graded algebra

$$(\wedge(V \oplus W), d),$$

where $d(V) = 0$ and

$$d(s^{-m+1} a) = D_M \cdot (1 \otimes a) \in V \otimes V$$

for $s^{-m+1} a \in W$. In fact we need to write this last expression as an element of $\wedge^2 V$. For that, if $D_M = \sum_i (-1)^{|a_i|} a_i \otimes a_i'$, then we have

$$d(s^{-m+1} a) = \frac{1}{2} \left(\sum_i (-1)^{|a_i|} a_i \wedge (a_i' \cdot a) \right),$$

where the $\cdot$ means the cohomology multiplication. Example 9.13 gives a concrete example of computation.

The algebra $\wedge(V \oplus W)$ can also be equipped with a second gradation

$$(\wedge(V \oplus W))_k = \oplus_{2p+q=k} \wedge^p W \otimes \wedge^q V.$$

Then $d(\wedge(V \oplus W))_k \subset (\wedge(V \oplus W))_k$. Therefore, the cohomology decomposes as a direct sum $H^*(\wedge(V \oplus W), d) \cong \sum_k H_k^*(\wedge(V \oplus W), d)$. We then have

Theorem 9.12 ([91]) *With the above notation, when m is even, there is an isomorphism of graded vector spaces*
$$H_k^*(\wedge(V \oplus W), d) \cong H^*(C_k(M); \mathbb{Q}).$$

Example 9.13 To compute $H^*(C_k(\mathbb{C}P^3); \mathbb{Q})$, we have to form the cdga

$$(\wedge(e_0, e_2, e_4, e_6, x_5, x_7, x_9, x_{11}), d)$$

where the subscripts indicate the degrees, and $d(e_i) = 0$, $d(x_5) = e_0e_6 + e_2e_4$, $d(x_7) = e_2e_6 + (1/2)e_4^2$, $d(x_9) = e_4e_6$, $d(x_{11}) = (1/2)e_6^2$. We define a second (lower) gradation by saying the x_i are in degree 2 and the e_j in degree 1. The differential preserves the gradation, and $H_k^*(\wedge(e_i, x_j), d) = H^*(C_k(\mathbb{CP}^3); \mathbb{Q})$.

The multiplicative structure of $H^*(C_k(M); \mathbb{Q})$ has been explained in [88]. As an illustration, we extract from [88] the rational cohomology algebra of $C_k(\mathbb{CP}^2)$:

$$\begin{cases} k = 1, 2 & \wedge x/x^3 & |x| = 2; \\ k = 3 & \wedge(x, y)/(x^3, yx^2) & |x| = 2, |y| = 3; \\ k \geq 4 & \wedge(x, y)/x^3 & |x| = 2, |y| = 3. \end{cases}$$

9.2 Arrangements

An arrangement $\mathcal{A}$ of linear subspaces in a complex vector space $\mathbb{C}^n$ is simply a finite set of linear subspaces of $\mathbb{C}^n$. The associated intersection lattice $L(\mathcal{A})$ is the poset of intersections among subspaces in $\mathcal{A}$ ordered with reversed inclusions: $x \leq y$ if and only if $y \subset x$. For $x, y \in L(\mathcal{A})$, the *meet* $x \wedge y$ is defined to be $x \wedge y = \cap\{z \in L(\mathcal{A}) \mid x \cup y \subset z\}$, and the *join* is defined to be the intersection $x \vee y = x \cap y$. With these operations, $L(\mathcal{A})$ is a lattice. If σ is a collection of elements in $\mathcal{A}$, we denote by $\vee \sigma$, the join of all the elements in σ. For $x \in L(\mathcal{A})$, the *rank* of x, $\text{rk}(x)$, is the maximal length r of a sequence of the form $\mathbb{C}^n < x_1 < x_2 < \cdots < x_r = x$, with $x_i \in L(\mathcal{A})$.

When $x < y$ and there is no z such that $x < z < y$, we write $x \prec y$. The lattice $L(\mathcal{A})$ is said to be *geometric* if each time $x \prec y$, then, for each z, either $x \vee z \prec y \vee z$ or $x \vee z = y \vee z$. When a lattice is geometric, the following properties hold:

1. If $x < y$, all the maximal sequences $x < x_1 < \cdots < x_n = y$ have the same length.
2. For every $x, y \in L(\mathcal{A})$, we have $\text{rk}(x) + \text{rk}(y) \geq \text{rk}(x \vee y) + \text{rk}(x \wedge y)$.

The arrangement $\mathcal{A}$ is said to be *geometric* if the lattice $L(\mathcal{A})$ is geometric.

The complement, $M(\mathcal{A})$, of an arrangement $\mathcal{A}$ is the complement of the union of the elements of $\mathcal{A}$; that is, $M(\mathcal{A}) = \mathbb{C}^n - \cup_{x \in \mathcal{A}} x$. When $\mathcal{A}$ is an arrangement of codimension one subspaces, then the cohomology of $M(\mathcal{A})$ is well known and the space is formal by a result of Brieskorn [42]. When all the subspaces x have codimension at least 2, then $M(\mathcal{A})$ is simply connected and rational homotopy theory applies.

9.2 Arrangements

Example 9.14 Denote by $\mathcal{A}$ the arrangement of $\mathbb{C}^{2n}$ formed by the subspaces $H_i = \{(z_1, \ldots, z_{2n}) \mid z_{2i-1} = z_{2i} = 0\}$, for $i = 1, \ldots, n$. Then the projection $M(\mathcal{A}) \to \prod_{i=1}^{n} \mathbb{C}^2 \setminus \{0\}$ is a homeomorphism and we obtain a homotopy equivalence $M(\mathcal{A}) \simeq (S^3)^n$.

Example 9.15 Let $\mathcal{A}$ be an arrangement in $\mathbb{C}^n$, and consider the injection j of $\mathbb{C}^n$ into $\mathbb{C}^{n+1}$ as the hyperplane $z_{n+1} = 0$. The image of $\mathcal{A}$ by the map j is an arrangement $\mathcal{A}'$ in $\mathbb{C}^{n+1}$ and we clearly have $M(\mathcal{A}') \simeq \Sigma^2 M(\mathcal{A})$.

Example 9.16 Consider in $\mathbb{C}^6$ the arrangement $\mathcal{A} = \{H_1, H_2, H_3, L\}$ where $H_i = \{(z_1, \ldots, z_6) \mid z_{2i-1} = z_{2i} = 0\}$, and $L = \{(z_1, \ldots, z_6) \mid z_1 = \cdots = z_6\}$. This arrangement is not geometric. The chain $\mathbb{C}^6 < L$ is maximal, but taking intersections with H_1 produces the non-maximal chain $H_1 < H_1 \vee L = 0$ (where, for instance, the subspace $K = \{(z_1, \ldots, z_6) \mid z_1 = z_2 = z_3 = 0\} \in L(\mathcal{A})$ lies between 0 and H_1).

Let $\mathcal{A}$ be an arrangement of linear subspaces in $\mathbb{C}^n$. Since we are interested in the complement of the union of the elements of $\mathcal{A}$, we suppose that $\mathcal{A}$ does not contain any two elements x, y such that $x \subset y$. This implies that *the rank of any element $x \in \mathcal{A}$ is 1*.

The relative atomic complex of $\mathcal{A}$ is the complex (D, d) defined by Yuzvinsky as follows [267]. The complex D is the $\mathbb{Q}$-vector space generated by all the subsets $\sigma = (x_1, x_2, \ldots, x_k)$ of $\mathcal{A}$. We write $|\sigma|$ to denote the number of elements in σ, and we choose an ordering on the elements of $\mathcal{A}$, so that each sequence is always written following this ordering. The differential of $\sigma = (x_1, x_2, \ldots, x_k)$ is defined by

$$d(\sigma) = \sum_{j \in J(\sigma)} (-1)^j \sigma \setminus \{x_j\},$$

where $J(\sigma) = \{j \in \{1, \ldots, k\} \mid \vee \sigma = \vee(\sigma \setminus \{x_j\})\}$. The degree of σ is defined by

$$\deg \sigma = 2 \operatorname{codim}(\vee \sigma) - |\sigma|.$$

Finally we can define a multiplication on D as follows. For two sequences σ, τ, take

$$\sigma \cdot \tau = \begin{cases} (-1)^{\varepsilon(\sigma, \tau)} \sigma \cup \tau & \text{if } \operatorname{codim} \vee \sigma + \operatorname{codim} \vee \tau = \operatorname{codim} \vee (\sigma \cup \tau) \\ 0 & \text{otherwise.} \end{cases}$$

Here "codim" denotes the *complex* codimension and $\varepsilon(\sigma, \tau)$ the sign of the permutation that must be applied to $\sigma \cup \tau$ to put the elements in the chosen linear order.

With those choices, (D, d) is a cochain complex [267]. The following result, due to Yuzvinsky, gives the link between the geometry and the topology of an arrangement [267].

Theorem 9.17 *The relative atomic model of Yuzvinsky is a model of the complement $M(\mathcal{A})$.*

Example 9.18 Let's describe the relative atomic model associated to the arrangement of Example 9.16. By definition we have

$$d(L) = d(H_1) = d(H_2) = d(H_3) = d(H_1, H_2) = d(H_1, H_3)$$
$$= d(H_2, H_3) = 0,$$

$$d(H_1, H_2, H_3) = d(L, H_1) = d(L, H_2) = d(L, H_3) = 0$$

$$d(L, H_1, H_2) = (L, H_2) - (L, H_1), \quad d(L, H_1, H_3) = (L, H_3) - (L, H_1),$$

$$d(L, H_2, H_3) = (L, H_3) - (L, H_2),$$

$$d(L, H_1, H_2, H_3) = -(H_1, H_2, H_3) + (L, H_2, H_3) - (L, H_1, H_3)$$
$$+ (L, H_1, H_2).$$

The elements (H_i) have degree 3 and generate, in cohomology, an exterior algebra $\wedge(H_1, H_2, H_3)$. There are two other cohomology classes: $[(L)]$ in degree 9, and $[(L, H_1)]$ in degree 10. We then consider the cdga

$$(\wedge(x_1, x_2, x_3) \oplus \mathbb{Q}y \oplus \mathbb{Q}z, 0),$$

where $x_i y = x_i z = yz = z^2 = 0$, $|x_i| = 3$, $|y| = 9$ and $|z| = 10$. The morphism $\varphi \colon (\wedge(x_1, x_2, x_3) \oplus \mathbb{Q}y \oplus \mathbb{Q}z, 0) \to (D, d)$ defined by $\varphi(x_i) = (H_i)$, $\varphi(y) = (L)$ and $\varphi(z) = (L, H_1)$, is then a quasi-isomorphism. This shows that $M(\mathcal{A})$ has the rational homotopy type of $(S^3 \times S^3 \times S^3) \vee S^9 \vee S^{10}$.

For geometric arrangements, the following notion is important as we will see in Lemma 9.21.

Definition 9.19 *A sequence $\sigma = (x_1, x_2, \ldots, x_k)$ of elements of $\mathcal{A}$ is said to be independent if, for each k, $\vee \sigma > \vee(\sigma \setminus \{x_k\})$. Note that subsequences of independent sequences are also independent.*

Lemma 9.20 *If the lattice $L(\mathcal{A})$ is geometric, then a sequence $\sigma = (x_1, \ldots, x_k)$ is independent if and only if $\operatorname{rk} \sigma = k$.*

9.2 Arrangements

Proof Suppose σ is independent. Then each subsequence of σ is independent and we have a sequence of inequalities

$$x_1 < x_1 \vee x_2 < x_1 \vee x_2 \vee x_3 < \cdots < \vee \sigma.$$

Therefore $\text{rk}\, \sigma \geq k$. If $\text{rk}\, \sigma > k$, then this sequence is not maximal and there is another element $y \in \mathcal{A}$ such that for some p

$$(x_1 \vee \cdots \vee x_p) < (x_1 \vee \cdots \vee x_p) \vee y < (x_1 \vee \cdots \vee x_p) \vee x_{p+1}.$$

This will imply that

$$\text{rk}\,(x_1 \vee \cdots \vee x_p \vee x_{p+1}) \geq \text{rk}\,(x_1 \vee \cdots \vee x_p) + 2 \tag{9.1}$$
$$> \text{rk}\,(x_1 \vee \cdots \vee x_p) + \text{rk}\,(x_{p+1}), \tag{9.2}$$

in contradiction with the definition of a geometric lattice.

Suppose now that $\text{rk}\, \sigma = k$ and consider the sequence

$$x_{\tau_1} \leq x_{\tau_1} \vee x_{\tau_2} \leq \cdots \leq \vee \sigma,$$

obtained from σ by some permutation τ of the set $\{1, 2, \ldots, k\}$. If all inequalities are strict for any τ, then σ is independent. Otherwise, some inequality is in fact an equality, and, since the rank of σ is k, another part of the sequence extends to the form

$$(x_{\tau_1} \vee \cdots \vee x_{\tau_r}) < (x_{\tau_1} \vee \cdots \vee x_{\tau_r}) \vee y < (x_{\tau_1} \vee \cdots \vee x_{\tau_r}) \vee x_{\tau_{r+1}}.$$

As above, the existence of such a sequence is impossible for a geometric lattice. □

It follows directly from the definition of the differential that independent sequences are cocycles. The key point in the geometric case is the following result.

Lemma 9.21 ([85]) *If the lattice $L(\mathcal{A})$ is geometric, then each cohomology class of (D, d) is represented by a linear combination of independent sequences.*

Of course, now we want to see what the salient properties of arrangements are with respect to rational homotopy theory. The first property is an analogue of Brieskorn's result for geometric arrangements [42].

9.2.1 Formality of the complement of a geometric lattice

Theorem 9.22 ([85]) *Let $\mathcal{A}$ be a subspace arrangement. If $L(\mathcal{A})$ is a geometric lattice, then the space $M(\mathcal{A})$ is formal.*

9: A Florilège of geometric applications

Proof Define a linear map $\psi: D \to H^*(D, d)$ by

$$\sigma \mapsto \begin{cases} [\sigma] & \text{if } \sigma \text{ is independent,} \\ 0 & \text{otherwise.} \end{cases}$$

We first prove that the map ψ is multiplicative. Now, by Lemma 9.20, if σ is not independent, then $\sigma \cup \tau$ is not independent and $\psi(\sigma) = \psi(\sigma \cdot \tau) = 0$. We suppose therefore that σ and τ are independent. If $\sigma \cup \tau$ is not independent, we can suppose there is $x \in \sigma$ such that $\vee(\sigma \cup \tau) = \vee(\sigma \setminus \{x\} \cup \tau)$. In that case

$$\text{codim} \vee (\sigma \cup \tau) = \text{codim} \vee (\sigma \setminus \{x\} \cup \tau)$$

$$\leq \text{codim} \vee (\sigma \setminus \{x\}) + \text{codim} \vee \tau$$

$$< \text{codim} \vee \sigma + \text{codim} \vee \tau.$$

Therefore, by definition of the product and of ψ, we have, $\psi(\sigma \cup \tau) = 0$ and $\psi(\sigma) \cdot \psi(\tau) = [\sigma] \cdot [\tau] = [\sigma \cdot \tau] = 0$. In case σ, τ and $\sigma \cup \tau$ are independent, the result follows from the equality $\psi(\sigma) \cdot \psi(\tau) = [\sigma] \cdot [\tau] = [\sigma \cdot \tau] = \psi(\sigma \cdot \tau)$.

We now prove that ψ is a morphism of complexes: that is, $\psi(d\sigma) = 0$ for every σ. Write $\sigma = (x_1, \ldots, x_k)$ and $\sigma_i = \sigma \setminus \{x_i\}$. When all the σ_i are dependent, $\psi(\sigma_i) = 0$ for all i, and the result is proved. When all the σ_i are such that $\vee \sigma = \vee \sigma_i$ are independent, we have

$$\psi(d\sigma) = \psi\left(\sum_i (-1)^i \sigma_i\right) = \left[\sum_i (-1)^i \sigma_i\right] = [d\sigma] = 0.$$

These two cases are in fact the only possibilities. Suppose indeed that $\vee \sigma = \vee \sigma_i = \vee \sigma_j$ and all the σ_i are independent. Then by Lemma 9.20, $\text{rk}\, \sigma_i = k - 1$. Since $\text{rk}\, \sigma_j = \text{rk}\, \sigma = \text{rk}\, \sigma_i = k - 1$, all the σ_j are also independent.

By Lemma 9.21, ψ induces a surjective map in cohomology. The map $H(\psi)$ is clearly injective and therefore ψ is a quasi-isomorphism. □

9.2.2 Rational hyperbolicity of the space $M(\mathcal{A})$

Let $\mathcal{A} = \{x_1, \ldots, x_q\}$ be a geometric arrangement of subspaces of codimension at least two. When each subspace is transversal to any intersection of the other ones in $\mathcal{A}$, then the complement is a product of odd dimensional spheres. In fact, by a result of G. Debongnie [68], this is the only case where $M(\mathcal{A})$ satisfies Poincaré duality. More precisely, Debongnie's theorem states:

Theorem 9.23 ([68]) *Let $\mathcal{A}$ be a geometric arrangement. Then the following conditions are equivalent:*

1. *The rational cohomology of $M(\mathcal{A})$ satisfies Poincaré duality.*
2. *$M(\mathcal{A})$ is a rationally elliptic space.*
3. *$\operatorname{codim}(\cap_{x \in \mathcal{A}} x) = \sum_{x \in \mathcal{A}} \operatorname{codim}(x)$.*
4. *$M(\mathcal{A})$ is a product of odd dimensional spheres.*

Remark 9.24 The assertion (3) $\Rightarrow$ (4) is easy to understand. Suppose (3) is verified. Let $x_1, \ldots, x_r$ be the subspaces in $\mathcal{A}$. Then the quotient map $p \colon \mathbb{C}^n \to \mathbb{C}^n/(\cap x_i)$ induces a homotopy equivalence $\mathbb{C}^n \setminus (\cup x_i) \to (\mathbb{C}^n/(\cap x_i)) \setminus \cup (x_i/ \cap x_i))$. Hence we can suppose $\cap x_i = \{0\}$. Write $x_i = \operatorname{Ker}(\varphi_i \colon \mathbb{C}^n \to \mathbb{C}^{n_i})$. The map

$$\varphi = (\varphi_1, \varphi_2, \ldots, \varphi_r) \colon \mathbb{C}^n \to \prod_{i=1}^{r} \mathbb{C}^{n_i}$$

is a homeomorphism which induces a homotopy equivalence

$$\mathbb{C}^n \setminus \cup x_i \to \prod_{i=1}^{r} (\mathbb{C}^{n_i} \setminus \{0\}) .$$

Since we also have a homotopy equivalence $\prod_{i=1}^{r}(S^{2n_i-1}) \to \prod_{i=1}^{r}(\mathbb{C}^{n_i} \setminus \{0\})$, the space $M(\mathcal{A})$ has the homotopy type of a product of odd dimensional spheres.

9.3 Toric topology

Recently, ideas and tools from rational homotopy theory have proved to be important in the study of toric spaces. Let's now briefly discuss these notions. Denote $\mathbb{C}^* = \mathbb{C} - \{0\}$. A *toric variety* is a normal algebraic variety M containing the algebraic torus $(\mathbb{C}^*)^n$ as a Zariski open subset in such a way that the natural action of $(\mathbb{C}^*)^n$ on itself extends to an action on M. A fundamental example is given by the so-called moment-angle complex Z_K defined for any simplicial complex K in [45]. When K has m vertices, Z_K is the complex $\cup_\sigma D_\sigma \subset (D^2)^m$, where

$$D_\sigma = \{ (z_1, \ldots, z_m) \in (D^2)^m \mid |z_i| = 1 \text{ if } i \notin \sigma \} .$$

Clearly, each Z_K is a toric variety. In the sequel, we will only consider moment-angle complexes. The *Davis–Januszkiewicz space* associated to K,

$DJ(K)$ is defined by the Borel construction

$$DJ(K) = Z_K \times_{T^m} ET^m.$$

For any coefficient ring R, the cohomology ring $H^*(DJ(K); R)$ is isomorphic to the Stanley–Reisner algebra $R(K)$ [45]. Recall that $R(K) = R[x_1,\ldots,x_m]/I_K$ where $|x_i| = 2$ and the ideal I_K is generated by the square free monomials $x_{i_1}\ldots x_{i_r}$ with $\sigma = \{i_1,\ldots,i_r\} \notin K$. The space $DJ(K)$ can be shown to be homotopy equivalent to the subspace $\cup_{\sigma \in K}(BT)^\sigma$ of $(BT)^m$, with $(BT)^\sigma \cong (BT)^{\dim \sigma}$, see [45]. We then have the following formality result due to Notbohm and Ray.

Theorem 9.25 ([212]) *The space $DJ(K)$ is rationally formal.*

Indeed, this formality result is valid for any ring R. As a corollary, a model for $DJ(K)$ is given by $(\mathbb{Q}(K), 0)$ and a model for the moment-angle complex Z_K is given by $(\mathbb{Q}(K) \otimes \wedge(u_1,\ldots,u_m), d)$ with $d(u_i) = x_i$. However, the spaces Z_K are not always formal as shown by examples of Denham and Suciu [72]. The spaces Z_K are also related to spaces of arrangements. Let K be a simplicial complex on the set $1,\ldots,m$. We associate to it the complex coordinate arrangement $\mathcal{A}_K = \{L_\sigma \mid \sigma \notin K\}$, where, for $\sigma = \{i_1,\ldots,i_r\}$,

$$L_\sigma = \{(z_1,\ldots,z_m) \in \mathbb{C}^m \mid z_{i_1} = z_{i_2} = \cdots = z_{i_r} = 0\}.$$

Then, we have

Theorem 9.26 ([45]) *The spaces $M(\mathcal{A}_K) = \mathbb{C}^m \setminus \cup_\sigma L_\sigma$ and Z_K have the same homotopy type.*

9.4 Complex smooth algebraic varieties

In [201], John Morgan studied models of smooth complex algebraic varieties. A key to his approach is the notion of *pure Hodge structure*.

Definition 9.27 *Let V be a $\mathbb{Q}$-vector space. A pure Hodge structure of weight n on V is a finite bigradation on $V^c = V \otimes_\mathbb{Q} \mathbb{C}$,*

$$V^c = \oplus_{p+q=n} V^{p,q},$$

such that $\overline{V^{p,q}} = V^{q,p}$.

Remark 9.28 The existence of a pure Hodge structure of weight n on V is equivalent to the existence of a decreasing filtration $F^\bullet$ on V^c such that

$$V^c = F^0(V^c) \supset F^1(V^c) \supset \ldots \supset F^n(V^c) \supset F^{n+1}(V^c) = \{0\}$$

9.4 Complex smooth algebraic varieties

and $V^c = F^p(V^c) \oplus \overline{F^{n+1-p}(V^c)}$, for any p. The correspondence goes as follows.

- If $F^\bullet$ is given, one sets $V^{p,q} = F^p \cap \overline{F^q}$.
- If the bigradation is given, we set $F^p(V^c) = \oplus_{k \geq p} V^{k,*}$.

For instance, if M is a connected compact Kähler manifold M, the fact that the different Laplace operators Δ_d, Δ_∂, $\Delta_{\bar\partial}$ coincide implies the existence of a pure Hodge structure of weight k on $H^k(M; \mathbb{Q})$, see [71, Section 5].

Theorem 9.29 *Let M be a compact connected Kähler manifold. Then the complex minimal model $(\wedge V, d)$ of M admits a bigradation such that the following properties hold.*

1. *Let V_k be the subspace of elements of degree k in V. There is a decomposition of V_k into $V_k = C_k \oplus N_k$, with $d_{|C_k} = 0$, d injective on N_k and*

$$C_k = \oplus_{i+j=k} C_k^{i,j} \text{ and } N_k = \oplus_{i+j \geq k+1} N_k^{i,j}.$$

2. *The bigradation is extended multiplicatively to $(\wedge V, d)$ and $d(\wedge V)^{i,j} \subset (\wedge V)^{i,j}$.*
3. *The quasi-isomorphism $\rho: (\wedge V, d) \to (A^c(M), d)$ is compatible with the bigradation $A^c(M) = \oplus_{p,q} A^{p,q}$.*
4. *The bigradation induced on the cohomology of $(\wedge V, d)$ gives the pure Hodge structures of $H(M; \mathbb{C})$.*

The proof is modelled along the steps of construction of a minimal model and uses an argument involving the pure Hodge structure on $H^k(M; \mathbb{C})$. We refer the reader to [71, page 271] for more details. As a consequence, this result gives a short proof of the formality of M.

Corollary 9.30 *A compact connected Kähler manifold M is formal.*

Proof Let $\rho: (\wedge V, d) \to (A^c(M), d)$ be the model of M described above. We define C_i to be the subset of V_i consisting of the elements of degree i and bidegree (r, s) with $r + s = i$. We denote by N_i a complement of C_i in V_i. The elements of N_i are of degree i and of bidegree (r, s) with $r + s > i$. Observe that the elements of C_i are cocycles and that the differential d is injective on N_i.

Denote by I the ideal of $\wedge V$ generated by $\oplus_i N_i$. If z is a cocycle of degree j in $I^{t,u}$, the class $[z]$ is of degree j and bidegree (t, u) with $t + u > j$. Since $H^j(\wedge V, d) \cong H^j(M; \mathbb{C})$ has a pure Hodge structure of weight j, we have $[z] = 0$ and the ideal I is acyclic. The conclusion is now a direct consequence of Exercise 2.3. □

So, we have a bigradation on the minimal model of a compact Kähler manifold which induces the pure Hodge structures on the cohomology. In order to extend this structure to minimal models of smooth algebraic varieties, we have to replace the notion of pure Hodge structure with that of *mixed Hodge structure*. The next remark gives the flavor of what a mixed Hodge structure is.

Remark 9.31 Consider two smooth projective algebraic varieties X_1 and X_2 such that $X_1 \cap X_2$ is also a smooth projective algebraic variety. If we want to study $X = X_1 \cup X_2$, the first elementary tool is the Mayer-Vietoris exact sequence which includes, for instance with $k > 1$,

$$\cdots \longrightarrow H^{k-1}(X_1 \cap X_2) \overset{\delta}{\longrightarrow} H^k(X) \overset{\varphi}{\longrightarrow} H^k(X_1) \oplus H^k(X_2) \longrightarrow \cdots$$

In $H^k(X)$, two types of pure Hodge structures are interfering:

1. a structure of weight $k-1$ coming from $H^{k-1}(X_1 \cap X_2)$;
2. a structure of weight k coming from $H^k(X_1) \oplus H^k(X_2)$.

To separate them, we introduce a filtration $W_\bullet$ on $H^k(X)$ defined by:

1. $W_{k-2} = 0$;
2. $W_{k-1} = \operatorname{Im} \delta$;
3. $W_k = H^k(X)$.

On the associated graded vector spaces, $\operatorname{Gr}_{k-1} H^k(X) = \operatorname{Im} \delta$ and $\operatorname{Gr}_k H^k(X) = H^k(X)/\operatorname{Ker} \varphi$, we get pure Hodge structures of respective weights $k-1$ and k. More generally, the definition of mixed Hodge structure goes as follows.

Definition 9.32 *Let V be a $\mathbb{Q}$-vector space. A mixed Hodge structure on V consists of two filtrations $(W_\bullet, F^\bullet)$ such that:*

1. *$W_\bullet$ is an increasing filtration on V, called the weight filtration;*
2. *$F^\bullet$ is a decreasing filtration on $V^c = V \otimes \mathbb{C}$, called the Hodge filtration;*
3. *the filtration $F^\bullet$ induces a pure Hodge structure of weight k on each of the $\operatorname{Gr}_k^W(V) = W_k/W_{k-1}$.*

In [69], [70], Deligne constructed a mixed Hodge structure on the cohomology of any complex algebraic variety. Morgan adapted the notion of mixed Hodge structure to the context of differential algebras and defined the notion of *mixed Hodge diagrams*, see [201, Definition 3.5]. In particular, any complex smooth algebraic variety gives rise to a mixed Hodge diagram and its model has a bigradation.

Theorem 9.33 *If $\mathcal{M}$ is the minimal model of a simply connected, smooth complex algebraic variety, then*

1. *we can write $\mathcal{M} = \oplus_{0 \leq r,s} \mathcal{M}^{r,s}$ with $\mathcal{M}^{0,0} = \mathbb{C}$, $\mathcal{M}^{r,s} \mathcal{M}^{t,u} \subset \mathcal{M}^{r+t,s+u}$ and $d\mathcal{M}^{r,s} \subset \mathcal{M}^{r,s}$;*
2. *the minimal model $\mathcal{M}$ admits a mixed Hodge structure and the induced mixed Hodge structure on its cohomology coincides with the mixed Hodge structure on the cohomology of the variety.*

From this, Morgan deduces the following properties.

- The homotopy groups $\pi_k(X) \otimes \mathbb{Q}$ of a simply connected, smooth complex algebraic variety are endowed with a mixed Hodge structure coming from the mixed Hodge structure of the variety's minimal model. (For the non-simply connected case, we send the reader to [202, Theorem 9.2].)
- Not all finite, simply connected CW-complexes are homotopy equivalent to smooth complex algebraic varieties since there exist minimal models of finite, simply connected CW-complexes that cannot have bigradations (see [201, page 196]).

Concerning, the existence of mixed Hodge structures on the homotopy groups, a different approach was used by R. Hain. In short, he proved that if A is a connected multiplicative mixed Hodge complex, then the Bar construction on A is a mixed Hodge complex. This structure induces a mixed Hodge structure on the indecomposables that correspond to homotopy groups. The details are in [128].

9.5 Spaces of sections and Gelfand–Fuchs cohomology

9.5.1 The Haefliger model for spaces of sections

Let

$$F \longrightarrow E \underset{\sigma}{\overset{p}{\rightleftarrows}} B$$

be a fibration with section σ where B and F are simply connected and B is finite dimensional. Then the space $\Gamma(p)$ of sections of p *that are homotopic to σ* is a nilpotent space [203], and a model for $\Gamma(p)$ has been constructed by Haefliger [126]. Our goal here is to describe this model. The construction starts with a relative minimal model of p

$$(\wedge V, d) \to (\wedge V \otimes \wedge W, D) \to (\wedge W, \bar{D}). \tag{9.3}$$

We first show that we can always modify this model to obtain a relative minimal model such that a model for the section σ is the map $\rho\colon (\wedge V \otimes \wedge W, D) \to (\wedge V, d)$ that is the identity on V and maps W to 0.

Since p admits a section, there is a map of cdga's $\rho\colon (\wedge V \otimes \wedge W, D) \to (\wedge V, d)$ that is the identity on V. We now change generators by replacing each $w \in W$ by $w - \rho(w)$ and this gives an automorphism of $(\wedge V \otimes \wedge W, D)$ that we use to modify the differential D. Clearly, we now have $\rho(W) = 0$. Also, because we have modified the generators w by elements of $\wedge V$, we see that the differential in the fiber, $\bar{D}$, remains unchanged.

We now choose a connected and *finite* dimensional model (A, d) for $(\wedge V, d)$,

$$(\wedge V, d) \xrightarrow{\simeq} (A, d),$$

and we tensor the models in (9.3) with (A, d) over $(\wedge V, d)$ to obtain a new relative model for the fibration,

$$(A, d) \to (A \otimes \wedge W, D) \to (\wedge W, \bar{D}).$$

Let $A^\vee = \mathrm{Hom}(A, \mathbb{Q})$ be the dual of A. Now denote a graded basis for A by (a_i) and the dual basis of $A^\vee$ by (a'_i). The duality $<,>$ between A and $A^\vee$ means that $\langle a'_i, a_j \rangle = 1$ if $i = j$ and 0 otherwise. We give a'_i the degree $|a'_i| = -|a_i|$.

Since A is finite dimensional, we can consider the morphism of commutative algebras,

$$\theta_W \colon A \otimes \wedge W \to A \otimes \wedge (A^\vee \otimes W),$$

defined by

$$\theta_W(a) = a, \qquad \theta_W(w) = \sum_i a_i \otimes (a'_i \otimes w).$$

Haefliger proves the existence of a unique differential D on $\wedge(A^\vee \otimes W)$ such that $\theta_W \colon (A \otimes \wedge W, D) \to (A, d) \otimes (\wedge(A^\vee \otimes W), D)$ is a morphism of cdga's.

Let $Q(D)$ be the linear part of the differential D in $(\wedge(A^\vee \otimes W), D)$. Denote by I the graded differential ideal of $\wedge(A^\vee \otimes W)$ generated by $(A^\vee \otimes W)^{\leq 0}$ and $Q(D)(A^\vee \otimes W)^0$, and form the quotient cochain algebra $M_\Gamma = (\wedge(A^\vee \otimes W)/I, D)$. The combination of θ_W and the quotient map $(\wedge(A^\vee \otimes W), D) \to M_\Gamma$ gives a map of cdga's

$$\varphi \colon (A \otimes \wedge W, D) \to (A, d) \otimes M_\Gamma.$$

We can now state Haefliger's theorem [126].

9.5 Spaces of sections and Gelfand–Fuchs cohomology

Theorem 9.34 *Let* $\text{ev}: \Gamma(p) \times B \to E$ *denote the evaluation map given by* $(s, b) \mapsto s(b)$. *Then the morphism φ is a model for* ev. *In particular* M_Γ *is a Sullivan model for the space of sections $\Gamma(p)$ homotopic to σ.*

Notice that the space of functions from X into Y may be identified with the space of sections of the trivial bundle $X \times Y \to X$. Therefore, the process above furnishes us with a model for function spaces. We give more details on this construction below.

Example 9.35 *A model for the space of free maps,* $\text{Map}(S^n, X)$, *can be obtained from the Haefliger model.* First of all, a finite dimensional model for S^n is given by the exterior algebra $(\wedge a, 0)$ if n is odd and $((\wedge a)/a^2, 0)$ if n is even. We denote this cdga by $E(a)$. Let $(\wedge W, d)$ be the minimal model for X. A model for the evaluation map $S^n \times \text{Map}(S^n, X) \to S^n \times X$ is then given by

$$\varphi: E(a) \otimes (\wedge W, d) \to E(a) \otimes (\wedge (W \oplus s^n W), D),$$

where we have written $s^n W$ for $a' \otimes W$. By definition, we have $\varphi(x) = x + a \cdot s^n x$ for any $x \in W$. Extend s^n to $\wedge W$ as a derivation of degree $-n$. Then we have for any element $\alpha \in \wedge W$, $\varphi(\alpha) = \alpha + a \cdot s^n(\alpha)$. The compatibility of φ with the differentials then gives, for each element $x \in W$,

$$dx + as^n(dx) = \varphi(dx) = D\varphi(x) = D(x + as^n x) = dx + (-1)^n aD(s^n x).$$

Therefore $D(s^n x) = (-1)^n s^n(dx)$. In the case $n = 1$, we therefore recover the model of the free loop space obtained in Theorem 5.11.

Using the Haefliger model, we can obtain information about the rational homotopy groups and rational cohomology of $\Gamma(p)$ when B is n-dimensional and F is $(n+1)$-connected.

Proposition 9.36 *Let* $F \to E \xrightarrow{p} B$ *be a nilpotent fibration and suppose that B is n-dimensional and F is $(n+1)$-connected. Then the space $\Gamma(p)$ is simply connected and we have an isomorphism of graded Lie algebras,*

$$\pi_q(\Omega \Gamma(p)) \otimes \mathbb{Q} \cong \oplus_{r \geq 0} \text{Hom}(H_r(B; \mathbb{Q}), \pi_{q+r}(\Omega F) \otimes \mathbb{Q}).$$

The Lie bracket on the right-hand side is obtained by taking the diagonal in homology followed by the bracket in $\pi_*(\Omega F) \otimes \mathbb{Q}$:

$$[f, g](x) = \sum_{i,j} [f(x_i), g(x_j)],$$

if the diagonal is given by $\Delta(x) = \sum_{i,j} x_i \otimes x_j$.

In particular, as a graded vector space, we have

$$\pi_q(\Gamma(p)) \otimes \mathbb{Q} \cong \oplus_{r \geq 0} H^r(B; \mathbb{Q}) \otimes \pi_{q+r}(F) \otimes \mathbb{Q}.$$

Denote by $\text{Map}(X, Y, f)$ the component of the mapping space $\text{Map}(X, Y)$ consisting of the maps homotopic to $f: X \to Y$. The rational homotopy groups of $\text{Map}(X, Y, f)$ have been computed independently by G. Lupton and S. Smith [177], U. Buijs and A. Murillo [46] and J. Block and A. Lazarev [30]. Denote by $\varphi: (\wedge V, d) \to (A, d)$ a model of f with $(\wedge V, d)$ a Sullivan model for Y. A linear map $g: \wedge V \to A$ of degree q is a φ-derivation if $g(xy) = g(x)\varphi(y) + (-1)^{q|x|}\varphi(x)g(y)$. Together with the differential D defined by $D(g) = dg - (-1)^{|g|}gd$, the φ-derivations form a complex $\text{Der}((\wedge V, d), (A, d), \varphi)$.

Proposition 9.37 *Suppose X and Y are simply connected. Then there is an isomorphism of graded vector spaces,*

$$H_{-q}\text{Der}((\wedge V, d), (A, d), \varphi) \xrightarrow{\cong} \pi_q \text{Map}(X, Y, f) \otimes \mathbb{Q}.$$

When Y is rationally hyperbolic and X is finite, the space of maps from X into Y has infinite Lusternik–Schnirelmann category and its cohomology is usually very large. For instance M. Vigué-Poirrier proved the following exponential growth law for the Betti numbers of particular mapping spaces.

Theorem 9.38 ([260]) *Let X^M be the space of maps from a compact connected manifold M into a space X that has the rational homotopy type of a wedge of simply connected spheres. Then the Betti numbers of X^M exhibit exponential growth; that is, there is an $A > 1$, such that for k large enough, $\sum_{i=0}^{k} \dim H^i(X^M; \mathbb{Q}) \geq A^k$.*

The free loop space on a manifold M is a very important object in geometry. We have seen its importance for geodesics. It is also important, for instance, for spaces of immersions. Denote by $\text{Imm}(S^1, M)$ the space of immersions of S^1 into a manifold M. There is a map from $\text{Imm}(S^1, M)$ to the free loop space $L(SM)$ of M's tangent sphere bundle SM which associates to an immersion the loop of unit tangent vectors. The Hirsch–Smale theorem (see for instance Smale [239]) says that, for smooth manifolds P and Q, $\text{Imm}(P, Q)$ has the weak homotopy type of the space of bundle maps from TP to TQ that are linear and injective on the fibers. Since the tangent bundle of S^1 is trivial, the space of bundle maps from TS^1 to TM has the homotopy type of the free loop space on M. Hence, after re-parametrizing to have unit speed curves, $\text{Imm}(S^1, M)$ is weakly equivalent to $L(SM)$.

9.5.2 The Bousfield–Peterson–Smith model

The space $\mathrm{Map}(X,Y)$ is in general not path connected. In [39], A. K. Bousfield, C. Peterson and L. Smith constructed a cdga that is a model for $\mathrm{Map}(X,Y)$ in the sense that its geometric realization has the rational homotopy type of $\mathrm{Map}(X,Y)$. This approach has been developed by E. Brown and R. Szczarba in [43].

In fact, if Y is nilpotent of finite type and X is of finite type with finite cohomology, then the component $\mathrm{Map}(X,Y,f)$ of the mapping space $\mathrm{Map}(X,Y)$ containing $f \colon X \to Y$ is a nilpotent space of finite type. Therefore, it has a minimal model, and we describe this model now.

The functor $\mathrm{Map}(X,-)$ is the right adjoint to the product functor $X \times -$:
$$\mathrm{Map}(X \times Z, Y) \cong \mathrm{Map}(Z, \mathrm{Map}(X,Y)).$$

The correspondence associates to a map $g \colon Z \to \mathrm{Map}(X,Y)$ the map $\bar{g} \colon X \times Z \to Y$ obtained as the composition $\mathrm{ev} \circ (1 \times g)$,

$$X \times Z \xrightarrow{1 \times g} X \times \mathrm{Map}(X,Y) \xrightarrow{\mathrm{ev}} Y.$$

Here ev denotes the evaluation map $X \times \mathrm{Map}(X,Y) \to Y$.

Let CDGA be the category of cdga's over $\mathbb{Q}$. If $\mathcal{A} = (A,d)$ is a fixed cdga of finite type, we consider the tensor product functor $\mathcal{A} \otimes - \colon (B,d) \mapsto (A,d) \otimes (B,d)$ of CDGA into itself. Since the category CDGA modelizes topology in a contravariant way, a good model for the function space is given by a left adjoint to the tensor product. Functors of this type were first systematically studied by Jean Lannes. Their utilization for models of mapping spaces is due to A.K. Bousfield, C. Peterson and L. Smith [39] (see also [232] for a coalgebra version).

Proposition 9.39 *Let $\mathcal{A} = (A,d)$ be a fixed cdga of finite type. The tensor product functor $\mathcal{A} \otimes - \colon (B,d) \mapsto (A,d) \otimes (B,d)$ of CDGA into itself admits a left adjoint functor $\tilde{\mathcal{A}}$.*

Proof Let $(\wedge V, d)$ be a minimal algebra. We consider, as in the Haefliger model, the morphism of commutative algebras
$$\theta \colon \wedge V \to A \otimes \wedge (A^{\vee} \otimes V).$$

There is then a differential D on $\wedge(A^{\vee} \otimes V)$ such that $\theta \colon (\wedge V, d) \to (A,d) \otimes (\wedge(A^{\vee} \otimes V), D)$ is a morphism of cdga's. We define
$$\tilde{\mathcal{A}}(\wedge V, d) = (\wedge(A^{\vee} \otimes V), D),$$

and note that this construction extends to any cdga (B,d), making $\tilde{\mathcal{A}}$ into a functor $\tilde{\mathcal{A}} \colon \mathrm{CDGA} \to \mathrm{CDGA}$ (see [39] for details).

The adjunction process works as follows. Let $\varphi \colon (\wedge V, d) \to (A, d) \otimes (B, d)$ be a map of cdga's and write $\varphi(v) = \sum_i a_i \otimes \varphi_i(v)$. We associate to φ the map

$$\overline{\varphi} \colon (\wedge(A^\vee \otimes V), D) \to (B, d)$$

defined by $\overline{\varphi}(a'_i \otimes v) = \varphi_i(v)$. □

Now we describe the process that gives the model of a connected component of function spaces. We first define the maximal connected component of an augmented cdga. Let (A, d) be a cdga and let $\varepsilon \colon (A, d) \to \mathbb{Q}$ be a morphism. We consider the ideal I generated by $A^{<0} \oplus (A^0 \cap \mathrm{Ker}(\varepsilon)) \oplus d(A^0 \cap \mathrm{Ker}(\varepsilon))$. The quotient cdga $(A, d)_\varepsilon = (A/I, d)$ is called the maximal connected component (A, d) with respect to ε. This is the maximal connected quotient through which ε factors

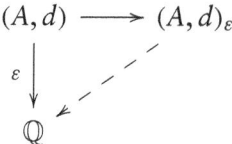

Now consider the component $\mathrm{Map}(X, Y, f)$ of $\mathrm{Map}(X, Y)$ consisting of the maps homotopic to f. Denote by $\varphi \colon (\wedge V, d) \to (A, d)$ a model of f. The injection of the basepoint $\{f\}$ into $\mathrm{Map}(X, Y, f)$ induces the augmentation

$$\varepsilon \colon \tilde{\mathcal{A}}(\wedge V, d) = (\wedge(A^\vee \otimes V), D) \to \mathbb{Q}, \qquad \varepsilon(a'_i \otimes v) = \langle a'_i, \varphi(v) \rangle.$$

Theorem 9.40 ([39], [43]) *With the above notation, $\tilde{\mathcal{A}}(\wedge V, d)_\varepsilon$ is a model of the connected component of f in $\mathrm{Map}(X, Y)$.*

The proof is very short since it is a direct consequence of the next result (implied by the existence of $\tilde{\mathcal{A}}$).

Lemma 9.41 *Let $A_{PL} \colon \mathcal{S} \to \mathrm{CDGA}$ and $\|-\| \colon \mathrm{CDGA} \to \mathcal{S}$ denote the Sullivan functor and its adjoint, spatial realization, (between CDGA and the category of simplicial sets) respectively. If $\mathcal{A} = (A, d)$ and $\mathcal{B} = (B, d)$ are two cdga's with A non-negative of finite type, then $\mathrm{Hom}_{\mathcal{S}}(\|\mathcal{A}\|, \|\mathcal{B}\|)$ has the same homotopy type as $\|\tilde{\mathcal{A}}(\mathcal{B})\|$.*

Proof Let $X \in \mathcal{S}$. The claim follows directly from this sequence of isomorphisms:

$$[X, \mathrm{Hom}_{\mathcal{S}}(\|\mathcal{A}\|, \|\mathcal{B}\|)]_{\mathcal{S}} \cong [\|\mathcal{A}\| \times X, \|\mathcal{B}\|]_{\mathcal{S}}$$
$$\cong [\mathcal{B}, A_{PL}(\|\mathcal{A}\| \times X)]_{\mathrm{CDGA}}$$

9.5 Spaces of sections and Gelfand–Fuchs cohomology

$$\cong [\mathcal{B}, A_{PL}(\|\mathcal{A}\|) \otimes A_{PL}(X)]_{\text{CDGA}}$$
$$\cong [\mathcal{B}, \mathcal{A} \otimes A_{PL}(X)]_{\text{CDGA}}$$
$$\cong \left[\tilde{A}(\mathcal{B}), A_{PL}(X)\right]_{\text{CDGA}}$$
$$\cong \left[X, \|\tilde{A}(\mathcal{B})\|\right]_{S}.$$

□

Now suppose X and Y are pointed spaces and that f preserves the base points. We can then consider the space of pointed maps $\text{Map}_*(X, Y, f)$ and we have

Proposition 9.42 ([43]) *A model for* $\text{Map}_*(X, Y, f)$ *is given by the maximal connected component* $(\wedge((A^+)^\vee \otimes V), D)_\varepsilon$.

Example 9.43 (The models of $\text{Map}(\mathbb{CP}^2, S^6)$ and $\text{Map}_*(\mathbb{CP}^2, S^6)$)
By taking a trivial bundle and using the formula following Proposition 9.36, we see that the mapping space $\text{Map}(\mathbb{CP}^2, S^6)$ is simply connected. Therefore, we do not need to refer to a base point. As a finite dimensional model of $\mathbb{CP}^2$, we take $\mathcal{A} = (A, 0) = (\wedge(x_2)/(x_2^3), 0)$. We follow the procedure described in Example 9.35 and get as model for the mapping space $\text{Map}(\mathbb{CP}^2, S^6)$ the cdga

$$(\wedge(u_6, u_4, u_2, v_{11}, v_9, v_7), D),$$

where indices indicate degrees and the differential is given by $D(u_2) = D(u_4) = D(u_6) = 0$, $D(v_{11}) = u_6^2$, $D(v_9) = 2u_6 u_4$ and $D(v_7) = u_4^2 + 2u_2 u_6$.

Moreover, a model for the *pointed* mapping space is given by the cdga

$$(\wedge(u_4, u_2, v_9, v_7), D), \quad D(u_2) = D(u_4) = 0, D(v_9) = 0, D(v_7) = u_4^2.$$

Therefore, the mapping space $\text{Map}_*(\mathbb{CP}^2, S^6)$ has the rational homotopy type of $S^4 \times S^9 \times \mathbb{CP}^\infty$.

Remark 9.44 Let $\text{Map}_*(X, Y, *)$ be the connected component of the trivial map in the mapping space of based maps. The existence of an H-space structure on this space is studied in [89], [158]. Evidently, such a structure exists if X is a co-H space or Y an H-space. These situations appear as particular cases of a sufficient condition involving the LS-category of X and the differential of the minimal model of Y, see [89, Theorem 2]. The main tool in the proof is Haefliger's model described above.

9.5.3 Configuration spaces and spaces of sections

Let M be a compact m-dimensional manifold and let $(X, *)$ be a pointed space. The space $C(M, X)$ of finite configurations in M with labels in X is the quotient of the disjoint union $\amalg_{k=1}^{\infty} F(M,k) \times_{\Sigma_k} X^k$ by the relation

$$(u_1,\ldots,u_k,x_1,\ldots,x_k) = (u_1,\ldots,u_{k-1},x_1,\ldots,x_{k-1}) \text{ if } x_k = *.$$

Here, Σ_k is the symmetric group acting by permutation on the two factors. A point in $C(M,X)$ can be described in a unique way as a finite set of points in M, without ordering, with a label in $X\setminus\{*\}$ attached to each point. For instance, when $X = S^0$, then $C(M,X) = \amalg_{k=1}^{\infty} C_k(M)$ is the space of finite unordered configurations in M.

Now let $\hat{T}M$ denote the fiberwise compactification of the tangent bundle to M. This gives a sphere bundle

$$S^m \to \hat{T}M \to M,$$

with a natural section obtained by choosing, for each point u in M, the point at infinity in $\hat{T}_u M$. We now take the fiberwise smash product of this bundle with the trivial bundle $M \times X \to M$ and get a new fiber bundle

$$\Sigma^m X \longrightarrow \hat{T}(M,X) \xrightarrow{\gamma_{M,X}} M.$$

The fiber over a point u is the quotient of $(T_u M \times X) \amalg \{\infty\}$ by the relation $(v,x) \sim \infty$ if either $||v|| \geq 1$ or else $x = *$. We denote by $\Gamma(M,X)$ the space of sections of $\gamma_{M,X}$.

To each element $(u,x) = (u_1,\ldots,u_k,x_1,\ldots,x_k)$ in $C(M,X)$ we want to associate a section $\sigma_{(u,x)}$ in $\Gamma(M,X)$. In order to do this, suppose M is equipped with a metric such that each point is the center of a geodesic disk of radius 1. Then let $2r$ denote the maximal distance between two different u_i, and let $d = \min(r,1)$. Let $D_1,\ldots,D_k$ be the geodesic disks of radius d centered at $u_1,\ldots,u_k$. Outside of the disks D_i we put $\sigma_{(u,x)}(y) = \infty$. When $y \in D_r$, we choose the unique unit speed geodesic $g(t)$ such that $g(0) = u_r$ and $g(t_0) = y$ for some $t_0 \leq d$, and we define $\sigma_{(u,x)}(y) = (\frac{t_0}{d}g'(t_0), x)$. This defines a continuous map

$$\sigma : C(M,X) \to \Gamma(M,X).$$

Theorem 9.45 ([187]) *The map σ is a weak homotopy equivalence.*

Remark 9.46 From an explicit model for $\Gamma(M, S^q)$, it is proved in [94] that the rational Betti numbers of the configuration space $C_k(M)$ of k unordered points in an even dimensional orientable closed manifold M depend only on the rational cohomology algebra of M.

9.5.4 Gelfand–Fuchs cohomology

Let M be a compact smooth m-dimensional manifold with L_M the Lie algebra of smooth vector fields on M. Let $C^*(L_M)$ denote the differential graded algebra of *continuous* multilinear forms on L_M with coefficients in $\mathbb{R}$.

Definition 9.47 *The Gelfand–Fuchs cohomology of the manifold M is the cohomology of $C^*(L_M)$.*

Now consider the principal $U(m)$-bundle

$$U(m) \to EU(m)^{(2m)} \to BU(m)^{(2m)}$$

that is the restriction of the universal principal $U(m)$-bundle $U(m) \to EU(m) \to BU(m)$ to the $2m$-skeleton of $BU(m)$. Observe that $EU(m)^{(2m)}$ is not the $2m$-skeleton of $EU(m)$, but $EU(m)^{(2m)}$ is a free $U(m)$-space and the associated fiber bundle (i.e. the Borel construction) over $BU(m)$ is

$$\hat{\gamma}_m : \ EU(m)^{(2m)} \to EU(m) \times_{U(m)} EU(m)^{(2m)} \xrightarrow{f} BU(m).$$

Clearly f has the homotopy type of the inclusion $BU(m)^{(2m)} \hookrightarrow BU(m)$.

Let TM be the tangent bundle of M and denote by T^cM its complexification. The complex vector bundle $T^cM \to M$ is classified by a map $g: M \to BU(m)$. We denote by γ_m the pullback of $\hat{\gamma}_m$ along g,

$$EU(m)^{(2m)} \longrightarrow E \xrightarrow{\gamma_m} M.$$

Solving a conjecture of Bott, Haefliger proved the following result.

Theorem 9.48 ([125]) *The cochain algebra $C^*(L_M)$ is a model for the space of continuous sections of γ_m.*

Thus, the cohomology of the space of sections of γ_m is isomorphic to the Gelfand–Fuchs cohomology. Observe now that the rational homotopy class of the map g is completely determined by what it does in cohomology since $BU(m) \simeq_{\mathbb{Q}} \prod_{i=1}^{m} K(\mathbb{Q}, 2i)$. We have

$$H^*(g; \mathbb{Q}): H^*(BU(m); \mathbb{Q}) = \mathbb{Q}[c_1, \ldots, c_m] \to H^*(M; \mathbb{Q}), \quad |c_i| = 2i.$$

By the properties of complex bundles, $H^*(g; \mathbb{Q})(c_{2i+1}) = 0$ and $H^*(g; \mathbb{Q})(c_{2i}) = p_i$, the ith Pontryagin class. Therefore, if the Pontryagin classes vanish, g is homotopically trivial, $E \cong M \times EU(m)^{(2m)}$, and the space of sections of γ_m is identified with the space of maps from M into $EU(m)^{(2m)}$.

In the case $M = \mathbb{R}^m$, the fiber bundle γ_m is trivial, so $H^*(EU(m)^{(2m)}; \mathbb{Q})$ is the cohomology of the Lie algebra $\mathfrak{A}_m$ of smooth vector fields on $\mathbb{R}^m$.

A model for $EU(m)^{(2m)}$ is easy to describe. Recall that a model for the universal $U(m)$-bundle is given by

$$(\wedge(c_1,\ldots,c_m),0) \to (\wedge(c_1,\ldots,c_m,h_1,\ldots h_m),d) \to (\wedge(h_1,\ldots,h_m),0),$$

with $|c_i| = 2i$ and $d(h_i) = c_i$. Let I denote the ideal of $\wedge(c_1,\ldots,c_m)$ generated by the elements of degree $> 2m$. Then $(\wedge(c_1,\ldots,c_m)/I, 0)$ is a model for $BU(m)^{(2m)}$, and we have the next result.

Proposition 9.49 *The cdga*

$$(\wedge(c_1,\ldots,c_m)/I \otimes \wedge(h_1,\ldots,h_m),d) = (\wedge(c_j)/I,0) \otimes_{\wedge(c_j)} (\wedge(c_j,h_j),d)$$

is a model for $EU(m)^{(2m)}$.

The cohomology of this algebra is well known and a basis for it has been described by Vey [106]. It follows also from [106] that $EU(m)^{(2m)}$ has the rational homotopy type of a wedge of spheres. For instance $EU(1)^{(2)}$ has the rational homotopy type of S^3 and $EU(2)^{(4)}$ has the rational homotopy type of $S^5 \vee S^5 \vee S^7 \vee S^8 \vee S^8$. From Theorem 9.38, we then deduce

Theorem 9.50 ([260]) *Let M be a compact connected nilpotent manifold of dimension $m \geq 2$, whose Pontryagin classes are zero. Then there is $A > 1$ such that for k large enough, $\sum_{i=0}^{k} \dim H^i(C^*(L_M)) \geq A^k$.*

9.6 Iterated integrals

While all of this section can be written in the framework of simplicial sets (see [59], [127] or [248]), for the sake of simplicity, we will stay within the world of manifolds to describe the various ideas introduced by K.T. Chen.

9.6.1 Definition of iterated integrals

Let M be a smooth manifold. Throughout this book, we have been motivated by the same recurrent theme: find homotopical information about a smooth manifold M from its algebra of differential forms. One of the simplest and well-known ways of doing this relies on differential 1-forms. If $\omega \in A_{DR}^1(M)$ is closed, the map $\alpha \in M^{[0,1]} \mapsto \int_\alpha \omega$ respects homotopy classes of smooth loops. In a series of papers, K.T. Chen extended this fact to p-forms and to spaces more general than manifolds, called differential spaces. Since we will be interested only in path spaces on manifolds, we will not use this more general notion. The paper [59] gives a general overview of his theory.

9.6 Iterated integrals

The first objects introduced by Chen that we encounter are the *iterated integrals*. To introduce them, let's begin with a very particular situation. Let ω_1 and ω_2 be two 1-forms on a smooth manifold M. If $\alpha \colon [0,1] \to M$ is a piecewise smooth path on M, for $0 \leq t_1 \leq t_2 \leq 1$, we define

$$\left(\int \omega_1 \omega_2\right)_\alpha = \int_0^1 \left(\int_0^{t_2} \omega_1(\alpha(t_1)) \, d\alpha(t_1)\right) \omega_2(\alpha(t_2)) \, d\alpha(t_2),$$

where $\displaystyle\int_0^{t_2} \omega_1(\alpha(t_1)) \, d\alpha(t_1) = \int_0^{t_2} \omega_1(\alpha(t_1)) \left(\frac{d\alpha(t_1)}{dt_1}\right) dt_1.$

This association is denoted by $\int \omega_1 \omega_2$ and called *an iterated integral*. In the particular case above, this is a function on the space $\mathcal{P}_{\text{free}}(M)$ of (free) piecewise smooth paths on M. Note that $\mathcal{P}_{\text{free}}(M)$ has the homotopy type of the space $M^{[0,1]}$ of free paths, thus the homotopy type of M. We can also see another way to define this function. It is more sophisticated, but it will lead us to the general case. Let

$$\Delta^k = \{(t_1, t_2, \ldots, t_k) \mid 0 \leq t_1 \leq t_2 \leq \ldots t_k \leq 1\}$$

be the euclidian simplex and let $p_i \colon \Delta^k \to [0,1]$ be the canonical projection for $i = 1, \ldots, k$. We denote by $\tilde{\omega}_i$ the 1-form on Δ^2 which is the pullback of ω_i along $\alpha \circ p_i$. We have

$$\left(\int \omega_1 \omega_2\right)_\alpha = \int_{\Delta^2} \tilde{\omega}_1 \wedge \tilde{\omega}_2.$$

This iterated integral is nothing more than an integral on Δ^2 and the qualification *iterated* comes from the manner in which we compute the integral.

In the framework we are going to describe, iterated integrals are not only functions on $\mathcal{P}_{\text{free}}(M)$, but more generally *forms on* $\mathcal{P}_{\text{free}}(M)$. Since $\mathcal{P}_{\text{free}}(M)$ is not a classical smooth manifold, we first have to give a meaning to this notion and that is where Chen's notion of differentiable space comes in. As we mentioned above, our situation is sufficiently simple that we can avoid the introduction of this notion. Recall that a form on a manifold is defined by its trace on any trivializing open set, so the next definitions come naturally.

Definition 9.51 *Let M and N be two smooth manifolds.*

A smooth map $\alpha \colon N \to \mathcal{P}_{\text{free}}(M)$ is a continuous map such that there exists a partition of $[0,1]$, $0 = t_0 < t_1 < \ldots < t_r = 1$, for which the restrictions $\alpha_\sharp \colon N \times [t_i, t_{i+1}] \to M$ of the adjoint map are smooth. The

9 : A Florilège of geometric applications

charts of $\mathcal{P}_{free}(M)$ are the smooth maps α where N is an open subset U of an $\mathbb{R}^n$.

Definition 9.52 Let M be a smooth manifold. A differential p-form on $\mathcal{P}_{free}(M)$ is a correspondence which associates to any chart $\left(\alpha: U \to \mathcal{P}_{free}(M)\right)$ an element $\omega_\alpha \in A^p_{DR}(U)$, such that $\omega_{\alpha \circ \Phi} = \Phi^*(\omega_\alpha)$ for any smooth map $\Phi: U' \to U$. We let $A^p_{DR}(\mathcal{P}_{free}(M))$ denote the space of differential p-forms on M and take

$$A_{DR}(\mathcal{P}_{free}(M)) = \bigoplus_{p \geq 0} A^p_{DR}(\mathcal{P}_{free}(M)).$$

Observe that $A_{DR}(\mathcal{P}_{free}(M))$ has a cdga structure with respect to the obvious differential and the product of forms. Now, we are ready for the definition of iterated integrals as forms on $\mathcal{P}_{free}(M)$.

Definition 9.53 Let $\omega_1, \ldots, \omega_k$ be forms on a smooth manifold M with $\omega_i \in A^{q_i}_{DR}(M)$. The iterated integral $\int \omega_1 \ldots \omega_k$ is a $((q_1 + \cdots + q_k) - k)$-form on $\mathcal{P}_{free}(M)$ defined as follows.

Let $\alpha: U \to \mathcal{P}_{free}(M)$ be a chart with adjoint $\alpha_\sharp: U \times [0, 1] \to M$. Decompose the differential form $\alpha_\sharp^*(\omega_i)$ on $U \times [0, 1]$ into

$$\alpha_\sharp^*(\omega_i)(\mathbf{x}, t) = dt \wedge \omega_i'(\mathbf{x}, t) + \omega_i''(\mathbf{x}, t),$$

where ω_i' and ω_i'' do not contain the factor dt, and set

$$\left(\int \omega_1 \ldots \omega_k\right)_\alpha = \int_{\Delta^k} \omega_1'(\mathbf{x}, t_1) \wedge \cdots \wedge \omega_k'(\mathbf{x}, t_k) \, dt_1 \ldots dt_k.$$

Observe that the form ω_i' is the interior product of the form $\alpha_\sharp^*(\omega_i)$ with the vector field $\frac{\partial}{\partial t}$. Let $\tilde{\omega}_{i,\alpha}$ be the pullback of the form ω_i along the composition $U \times \Delta^k \xrightarrow{\mathrm{id}_U \times p_i} U \times [0, 1] \xrightarrow{\alpha_\sharp} M$. The iterated integral can also be expressed as:

$$\left(\int \omega_1 \ldots \omega_k\right)_\alpha = \oint_{\Delta^k} \tilde{\omega}_{1,\alpha} \wedge \cdots \wedge \tilde{\omega}_{k,\alpha},$$

where $\oint$ is the integral along the fiber in the trivial bundle $U \times \Delta^k \to U$.

9.6.2 The cdga of iterated integrals

Let $\mathfrak{Chen}\,\mathcal{P}_{free}(M)$ be the sub-vector space of forms on $\mathcal{P}_{free}(M)$ generated by the

$$\pi_0^*(\omega_0) \wedge \int \omega_1 \ldots \omega_k \wedge \pi_1^*(\omega_{k+1}),$$

where

- $\omega_i \in A_{DR}(M)$, for $i = 0, \ldots, k+1$;
- $\int \omega_1 \ldots \omega_k$ is the iterated integral of Definition 9.53;
- $\pi_0, \pi_1 \colon \mathcal{P}_{free}(M) \to M$ are the evaluations $\alpha \mapsto \alpha(0)$, $\alpha \mapsto \alpha(1)$ respectively.

Theorem 9.54 *The complex $\mathfrak{Chen}\,\mathcal{P}_{free}(M)$ is a differential graded subalgebra of $A_{DR}(\mathcal{P}_{free}(M))$.*

This comes from Lemma 9.56 and Lemma 9.58 below which show the stability of $\mathfrak{Chen}\,\mathcal{P}_{free}(M)$ under the differential and the product of forms. Before stating these lemmas, we mention an alternative presentation of the elements of $\mathfrak{Chen}\,\mathcal{P}_{free}(M)$.

Remark 9.55 The forms $\pi_0^*(\omega_0) \wedge \int \omega_1 \ldots \omega_k \wedge \pi_1^*(\omega_{k+1})$ on $\mathcal{P}_{free}(M)$ can also be presented as integrals along a fiber, see [105] or [191]. In the diagram,

$$\begin{array}{ccc} \mathcal{P}_{free}(M) \times \Delta^k & \xrightarrow{\widetilde{ev}_k} & M^{k+2} \\ {\scriptstyle p}\downarrow & & \\ \mathcal{P}_{free}(M) & & \end{array}$$

the map $\widetilde{ev}_k$ is the evaluation defined by

$$\widetilde{ev}_k(\tau, (t_1, \ldots, t_k)) = (\tau(0), \tau(t_1), \ldots, \tau(t_k), \tau(1)).$$

Let $(\omega_i)_{0 \leq i \leq k+1}$ be $(k+2)$ forms on M and note that

$$\oint \widetilde{ev}_k^*(\omega_0 \wedge \cdots \wedge \omega_{k+1}) = \pi_0^*(\omega_0) \wedge \int \omega_1 \ldots \omega_k \wedge \pi_1^*(\omega_{k+1}),$$

where $\oint$ is the integral along the fiber in the trivial bundle p.

9 : A Florilège of geometric applications

Lemma 9.56 (Coboundary of an iterated integral; [59])
Let $\omega_i \in A_{DR}^{q_i}(M)$ for $i = 1, \ldots, k$. The following formula holds:

$$d \int \omega_1 \ldots \omega_k = \sum_{i=1}^{k} (-1)^{q_1 + \cdots + q_{i-1} - i} \int \omega_1 \ldots d\omega_i \ldots \omega_k$$

$$- \sum_{i=1}^{k-1} (-1)^{q_1 + \cdots + q_i - i} \int \omega_1 \ldots (\omega_i \wedge \omega_{i+1}) \ldots \omega_k - \pi_0^*(\omega_1) \wedge \int \omega_2 \ldots \omega_k$$

$$+ (-1)^{q_1 + \cdots + q_{k-1} - k + 1} \int \omega_1 \ldots \omega_{k-1} \wedge \pi_1^*(\omega_k).$$

Proof First, let's consider this formula (up to sign) and explain the origin of the terms contained in it. Recall that there is a Stokes theorem for integration along a fiber (see [112, page 311]):

$$d \oint_{\Delta^k} \omega = \oint_{\Delta^k} d\omega \pm \oint_{\partial \Delta^k} \omega.$$

Let's understand the last term in the particular case of Δ^2. The boundary $\partial \Delta^2$ of $\Delta^2 = \{(t_1, t_2) \mid 0 \leq t_1 \leq t_2 \leq 1\}$ has three faces:

- the face $t_1 = 0$ which gives the term $\pi_0^*(\omega_1) \wedge \int \omega_2$;
- the face $t_2 = 1$ which gives the term $\int \omega_1 \wedge \pi_1^*(\omega_2)$;
- the face $t_1 = t_2$ which gives the term $\int \omega_1 \wedge \omega_2$. (Recall that the pull back of two forms on M along the diagonal $M \to M \times M$ is the wedge product of the two forms, [112, page 209].)

Now, let's explain the signs that appear in the expression of the coboundary. Consider the case of the iterated integral of one form $\omega \in A_{DR}(M)$. Let $\alpha \colon U \to \mathcal{P}_{\text{free}}(M)$ be a chart. As in Definition 9.53, we decompose $\alpha_\sharp^*(\omega)$ as $\alpha_\sharp^*(\omega)(\mathbf{x}, t) = dt \wedge \omega'(\mathbf{x}, t) + \omega''(\mathbf{x}, t)$ and observe that

$$\alpha_\sharp^*(d\omega) = -dt \wedge d\omega' + dt \wedge \frac{\partial \omega''}{\partial t} + \sum_j dx_j \wedge \frac{\partial \omega''}{\partial x_j}.$$

This gives:

$$\left(\int d\omega \right)_\alpha = -\int_0^1 d\omega' \, dt + \int_0^1 \frac{\partial \omega''}{\partial t}(\mathbf{x}, t) \, dt$$

$$= -\left(d \int \omega \right)_\alpha + \omega''(\mathbf{x}, 1) - \omega''(\mathbf{x}, 0)$$

$$= -\left(d \int \omega \right)_\alpha + \pi_1^*(\omega)(\alpha) - \pi_0^*(\omega)(\alpha).$$

The general formula follows by induction (see [59, Proposition 1.5.2]). □

It is time to recall certain conventions and definitions concerning permutation of graded objects. First, let A be the free commutative graded algebra generated by elements $(\omega_i)_{1 \leq i \leq n}$ of respective degree q_i. For any permutation σ of the set $\{1, 2, \ldots, n\}$, we denote by $\pm_{\sigma,(q_i)}$ the element of $\{-1, 1\}$ such that

$$\omega_1 \wedge \cdots \wedge \omega_n = \pm_{\sigma,(q_i)} \omega_{\sigma(1)} \wedge \cdots \wedge \omega_{\sigma(n)}.$$

Definition 9.57 *Let k and l be non-negative integers. A shuffle of type (k, l) is a permutation σ of the set $\{1, 2, \ldots, k+l\}$ such that:*

$$\sigma^{-1}(1) < \sigma^{-1}(2) < \cdots < \sigma^{-1}(k)$$

and

$$\sigma^{-1}(k+1) < \sigma^{-1}(k+2) < \cdots < \sigma^{-1}(k+l).$$

We denote by $Sh(k, l)$ the set of shuffles of type (k, l).

Lemma 9.58 (Product of iterated integrals; [59]) *Let $\omega_i \in A_{DR}^{q_i}(M)$ for $i = 1, \ldots, k+l$. The following product formula holds:*

$$\int \omega_1 \ldots \omega_k \wedge \int \omega_{k+1} \ldots \omega_{k+l} = \sum_{\sigma \in Sh(k,l)} \pm_{\sigma,(q_i)} \int \omega_{\sigma(1)} \omega_{\sigma(2)} \cdots \omega_{\sigma(k+l)}.$$

Proof The two terms on the left-hand side correspond to integration over Δ^k and Δ^l respectively. The right side is an integration over Δ^{k+l}. The result comes from the classical triangulation of a product of two simplices:

$$\Delta^k \times \Delta^l = \bigcup_{\sigma \in Sh(k,l)} \{(t_{\sigma(1)}, \ldots, t_{\sigma(k+l)}) | 0 \leq t_1 \leq \cdots \leq t_k \leq 1,$$

$$0 \leq t_{k+1} \leq \cdots \leq t_{k+l} \leq 1\}.$$

□

9.6.3 Iterated integrals and the double bar construction

We now want to compare the double bar construction of Subsection 5.9 with the complex of iterated integrals $\mathfrak{Chen} \, \mathcal{P}_{\text{free}}(M)$. Let M be a connected manifold with basepoint $* \in M$. We denote by (A, d) (or A if there is no ambiguity) the differential algebra $A_{DR}(M)$ and define an augmentation $\varepsilon \colon A \to \mathbb{R}$ by $f \mapsto f(*)$ if $f \in A^0$ and $\varepsilon(\omega) = 0$ if $\omega \in A^{>0}$.

9 : A Florilège of geometric applications

First recall that $B(A, A, A) = (A \otimes (s\overline{A} \otimes \cdots \otimes s\overline{A}) \otimes A, D)$ with $\left(s\overline{A}\right)^q = \overline{A}^{q+1}$ and the differential D as in Definition 5.51. Consider the map:

$$B(A, A, A) \xrightarrow{\psi} A_{DR}(\mathcal{P}_{\text{free}}(M))$$

$$\omega_0 \otimes [\omega_1|\ldots|\omega_k] \otimes \omega_{k+1} \longmapsto \pi_0^*(\omega_0) \wedge \int \omega_1 \ldots \omega_k \wedge \pi_1^*(\omega_{k+1})$$

Lemma 9.56 shows that ψ is compatible with the differentials.

In the case $A = A_{DR}(M)$, the bar construction on the left-hand side can have elements of negative degree. The classical procedure for avoiding this is to replace A by an equivalent augmented complex without elements of degree 0 in the augmentation ideal. Chen uses a normalization of the bar construction that we present now (see [58, page 22] or [105, page 353] for details).

If $f \in A^0$, set $S_i(f)[\omega_0|\ldots|\omega_k] = [\omega_0|\ldots|\omega_{i-1}|f|\omega_i|\ldots|\omega_k]$. Define $\mathcal{D}egen(A)$ as the subspace of $B(A, A, A)$ generated by the $S_i(f)$ and the $[D, S_i(f)]$. Chen proves:

Proposition 9.59 *Let* $A = A_{DR}(M)$. *The space* $\mathcal{D}egen(A)$ *is stable under the differential* D *and is acyclic if* M *is connected. Thus, the quotient* $N(A, A, A) = B(A, A, A)/\mathcal{D}egen(A)$ *has the same cohomology as* $B(A, A, A)$.

Moreover, since $A_{DR}(\mathcal{P}_{\text{free}}(M))$ has no elements of negative degree, the map ψ above induces a map (which we still denote by) $\psi \colon N(A, A, A) \to A_{DR}(\mathcal{P}_{\text{free}}(M))$. Finally observe that $B(A, A, A) = N(A, A, A)$ if $\overline{A}^0 = 0$.

We would also like to consider the other bar constructions presented in Subsection 5.9 and to relate them to certain complexes of iterated integrals. For this, we introduce:

- the subspace $\mathcal{P}_{*,a}(M)$ of $\mathcal{P}_{\text{free}}(M)$ consisting of the paths with origin $* \in M$;
- the subspace $\mathcal{P}_{*,*}(M)$ of $\mathcal{P}_{\text{free}}(M)$ consisting of the based loops on $(M, *)$;
- the cdga's $A_{DR}(\mathcal{P}_{*,a}(M))$ and $A_{DR}(\mathcal{P}_{*,*}(M))$ as in Definition 9.52.

Observe that

- $\mathcal{P}_{\text{free}}(M)$ is of the homotopy type of $M^{[0,1]}$, therefore of M;
- $\mathcal{P}_{*,a}(M)$ is of the homotopy type of $P(M)$, therefore is contractible;
- $\mathcal{P}_{*,*}(M)$ is of the homotopy type of the loop space $\Omega(M)$.

For the proof of the next result, we refer the reader to the paper of Chen [59] (see also [105] with a slightly different sign convention).

Theorem 9.60 (Chen, [59]) *Let M be a simply connected manifold and $A = A_{DR}(M)$. The following maps of complexes are quasi-isomorphisms:*

1. $\psi: N(A, A, A) \to A_{DR}(\mathcal{P}_{free}(M))$,
   $\omega_0 \otimes [\omega_1|\ldots|\omega_k] \otimes \omega_{k+1} \mapsto \pi_0^*(\omega_0) \wedge \int \omega_1 \ldots \omega_k \wedge \pi_1^*(\omega_{k+1})$;

2. $\psi: N(A, A, \mathbb{R}) \to A_{DR}(\mathcal{P}_{*,a}(M))$,
   $\omega_0 \otimes [\omega_1|\ldots|\omega_k] \mapsto \pi_0^*(\omega_0) \wedge \int \omega_1 \ldots \omega_k$;

3. $\psi: N(\mathbb{R}, A, \mathbb{R}) \to A_{DR}(\mathcal{P}_{*,*}(M))$,
   $[\omega_1|\ldots|\omega_k] \mapsto \int \omega_1 \ldots \omega_k$.

As a consequence, the images of these maps give the cohomology of the corresponding spaces. In particular, recalling from Section 5.9 that $B(A, A, A)$ is quasi-isomorphic to A, we have

Corollary 9.61 *If M is a simply connected manifold, then the two cdga's, $A_{DR}(\mathcal{P}_{free}(M))$ and $\mathfrak{Chen}\,\mathcal{P}_{free}(M)$, have the same minimal model as M.*

Let's focus on the last item of Theorem 9.60 which gives a model for the loop space $\Omega(M)$. Let $\mathfrak{Chen}\,\mathcal{P}_{*,*}(M)$ be the subspace of $A_{DR}(\mathcal{P}_{*,*}(M))$ generated by the iterated integrals $\int \omega_1 \ldots \omega_k$, where $\omega_i \in A_{DR}(M)$.

Remark 9.62 This complex can also be expressed using a presentation similar to Remark 9.55. In the diagram

$$\begin{array}{ccc} \mathcal{P}_{*,*}(M) \times \Delta^k & \xrightarrow{\text{ev}_k} & M^k \\ {\scriptstyle p}\downarrow & & \\ \mathcal{P}_{*,*}(M) & & \end{array}$$

the map p is the canonical projection and ev_k is the evaluation defined by $\text{ev}_k(\tau, (t_1, \ldots, t_k)) = (\tau(t_1), \ldots, \tau(t_k))$. Definition 9.53 coincides with:

$$\int \omega_1 \ldots \omega_k = \oint \text{ev}_k^*(\omega_1 \wedge \cdots \wedge \omega_k),$$

where $\oint$ is the integral along the fiber in the trivial bundle p.

Finally, observe that in $\mathcal{P}_{*,*}(M)$ the maps $\pi_0 \circ \alpha$ and $\pi_1 \circ \alpha$ are constant, so that $\pi_0^*(\omega) = \pi_1^*(\omega) = 0$ for any $\omega \in A^{\geq 1}(M)$. This presentation is nothing other than Remark 9.55 restricted to $\mathcal{P}_{*,*}(M)$.

In [59, Theorem 4.1.1], Chen proves that $[\omega_1|\ldots|\omega_k] \mapsto \int \omega_1 \ldots \omega_k$ induces an isomorphism of Hopf algebras between $N(\mathbb{R}, A, \mathbb{R})$ and

$\mathfrak{Chen}\,\mathcal{P}_{*,*}(M)$. The product is the shuffle product of Lemma 9.58 and the diagonal is given by:

$$\Delta\left(\int \omega_1 \ldots \omega_k\right) = \sum_{i=0}^{k} \int \omega_1 \ldots \omega_i \otimes \int \omega_{i+1} \ldots \omega_k.$$

This and Theorem 5.52 directly imply the next statement.

Corollary 9.63 [59]) *Let M be a simply connected manifold with loop space $\Omega(M)$. The map $[\omega_1|\ldots|\omega_r] \mapsto \int \omega_1 \ldots \omega_r$ induces an isomorphism of Hopf algebras between $N(\mathbb{R}, A, \mathbb{R})$ and $\mathfrak{Chen}\,\mathcal{P}_{*,*}(M)$. This gives an isomorphism of Hopf algebras between the cohomology of the loop space $\Omega(M)$ and the cohomology of the complex of iterated integrals:*

$$H^*(\Omega(M); \mathbb{R}) \cong H^*\left(\mathfrak{Chen}\,\mathcal{P}_{*,*}(M)\right).$$

9.6.4 Iterated integrals, the Hochschild complex and the free loop space

We would like now to study the *cohomology of the free loop space with iterated integrals*. Denote by $\mathcal{P}_{\text{free loop}}(M)$ the subspace of $\mathcal{P}_{\text{free}}(M)$ consisting of the free loops on M. Observe that $\mathcal{P}_{\text{free loop}}(M)$ has the homotopy type of the free loop space LM. Let $\mathfrak{Chen}\,\mathcal{P}_{\text{free loop}}(M)$ denote the subspace of forms on $\mathcal{P}_{\text{free loop}}(M)$ generated by the

$$\pi_0^*(\omega_0) \wedge \int \omega_1 \ldots \omega_k.$$

We leave to the reader the descriptions of these forms as integrals along a fiber as in Remarks 9.55 and 9.62. As a complement to Section 5.9, note that the shuffle product on $\mathbb{B}(A_{DR}(M))$ gives a cohomology algebra isomorphism with $H^*(LM; \mathbb{R})$. As in Subsection 9.6.3, we observe that $\mathbb{B}(A_{DR}(M))$ has elements of negative degree. We kill them by quotienting by a differential ideal $\mathcal{D}egen(A)$ which is acyclic when M is connected.

Theorem 9.64 [59, Theorem 4.2.1], [105, Proposition 4.1]) *Let M be a simply connected manifold. The map*

$$\mathbb{B}(A_{DR}(M)) \xrightarrow{\varphi} A_{DR}(\mathcal{P}_{\text{free loop}}(M))$$

$$\omega_0 \otimes [\omega_1|\ldots|\omega_k] \mapsto \pi_0^* \omega_0 \wedge \int \omega_1 \ldots \omega_k$$

induces an isomorphism of algebras between $\mathbb{B}(A_{DR}(M))$ and $\mathfrak{Chen}\,\mathcal{P}_{\text{free loop}}(M)$. Therefore, there is an isomorphism of algebras,

$$H^*\left(\mathfrak{Chen}\,\mathcal{P}_{\text{free loop}}(M)\right) \cong H^*(LM; \mathbb{R}).$$

Table 9.1 Chen's iterated integrals.

Iterated integrals	Bar constructions	Spaces
$\mathfrak{Chen}\,\mathcal{P}_{free}(M)$	$B(A, A, A)$	$M^{[0,1]}$
$\mathfrak{Chen}\,\mathcal{P}_{*,a}(M)$	$B(A, A, \mathbb{R})$	$P(M)$
$\mathfrak{Chen}\,\mathcal{P}_{*,*}(M)$	$B(\mathbb{R}, A, \mathbb{R})$	$\Omega(M)$
$\mathfrak{Chen}\,\mathcal{P}_{free\,loop}(M)$	$\mathbb{B}(A)$	LM

Table 9.1 gives a summary of the correspondences between the various complexes of iterated integrals, bar constructions and particular path spaces.

9.6.5 Formal homology connection and holonomy

In this paragraph, we replace $\mathfrak{Chen}\,\mathcal{P}_{*,*}(M)$ by a differential algebra, free as an algebra, whose homology is isomorphic to the Pontryagin algebra of the loop space, $\Omega(M)$, of a simply connected manifold M with finite Betti numbers. Denote by V the subspace $s^{-1}H_+(M;\mathbb{R})$; that is, $V_n = H_{n+1}(M;\mathbb{R})$ for all $n \geq 1$ and $V_0 = 0$. The free graded Lie algebra on V, denoted by $\mathbb{L}(V)$, is a sub-vector space of the free tensor algebra $T(V)$, and if $T(V)$ is endowed with the Lie algebra structure coming from the commutator bracket, then $\mathbb{L}(V)$ is the sub-Lie algebra generated by V, see [87, Page 289]. A Lie derivation on $\mathbb{L}(V)$ is a map $\partial\colon \mathbb{L}(V)_n \to \mathbb{L}(V)_{n-1}$ such that $\partial[l,l'] = [\partial l, l'] + (-1)^{|l|}[l, \partial l']$ for all homogeneous elements l and l' of $\mathbb{L}(V)$. Finally, we introduce the completion of the tensor product as:

$$A_{DR}(M)\widehat{\otimes}\mathbb{L}(V) = \bigoplus_n \prod_p (A_{DR}^{n-p}(M) \otimes \mathbb{L}(V)_p).$$

This is the set of all formal sums $\sum_i \omega_i \otimes l_i$ such that, for each integer n, there are only a finite number of terms $\omega_i \otimes l_i$ such that $|\omega_i| + |l_i| = n$.

Recall that $A_{DR}(M)\widehat{\otimes}\mathbb{L}(V)$ is a graded Lie algebra with a bracket defined on homogeneous elements by $[a \otimes b, a' \otimes b'] = (-1)^{|a'||b|}aa' \otimes [b,b']$. The map d is extended to the tensor product by setting $d(a \otimes b) = da \otimes b$ on homogeneous elements. Similarly, if ∂ is a Lie derivation on $\mathbb{L}(V)$, we define $\partial(a \otimes b) = (-1)^{|a|}a \otimes \partial b$.

Definition 9.65 *Let M be a compact simply connected manifold. Set $V = s^{-1}H_*(M;\mathbb{R})$ and fix a homogeneous basis $(X_i)_{i\in I}$ of V. A formal homology*

9 : A Florilège of geometric applications

connection *is a pair* (w, ∂) *with*

$$w = \sum_I \omega_i \otimes X_i + \sum_{(i_1,i_2) \in I \times I} \omega_{i_1 i_2} \otimes X_{i_1} X_{i_2} + \cdots$$

$$\in A_{DR}(M) \widehat{\otimes} \mathbb{L}(V) \subset A_{DR}(M) \widehat{\otimes} T(V)$$

and ∂ *a Lie derivation on* $\mathbb{L}(V)$ *such that:*

- *the ω_i are closed elements of $A_{DR}(M)$;*
- *the family $([\omega_i])_{i \in I} \in H^*(M; \mathbb{R})$ is the dual basis of $(sX_i)_{i \in I}$;*
- *w is of degree -1, which means $|\omega_{i_1 \ldots i_r}| = 1 + \sum_{j=1}^{r}(|\omega_{i_j}| - 1)$;*
- *d and ∂ satisfy the Maurer–Cartan equation $dw = \partial w + \frac{1}{2}[w, w]$.*

We give below an equivalent definition in the context of maps instead of tensor products. First, let's consider an easy example.

Example 9.66 Consider the space $\mathbb{CP}(2)$ with its real cohomology denoted by H^* and real homology denoted by H_*. We choose a basis $X_1 \in V_1 = s^{-1} H_2$, $X_2 \in V_3 = s^{-1} H_4$. Let ω be the Kähler form in $A_{DR}^2(\mathbb{CP}(2))$, and let $([\omega], [\omega^2])$ be a dual basis of (X_1, X_2). We set $\partial X_1 = 0$, $\partial X_3 = \frac{1}{2}[X_1, X_1]$ and extend it to $\mathbb{L}(X_1, X_2)$ as a Lie derivation. An easy computation shows that $w = \omega X_1 + \omega^2 X_2$ is a formal homology connection on $\mathbb{CP}(2)$.

Remark 9.67 A formal homology connection can be viewed as a differential form with values in a Lie algebra, the free Lie algebra $\mathbb{L}(V)$. If we consider the classical case of a form $\omega \in A^1(M; \mathfrak{g})$ where $\mathfrak{g}$ is the Lie algebra associated to a Lie group G, then we see that ω can be extended to a morphism of graded algebras $\omega' \colon C^*(\mathfrak{g}) \to A_{DR}(M)$ where $C^*(\mathfrak{g})$ is the classical cochain algebra on $\mathfrak{g}$. This map ω' is compatible with the differentials if and only if $d\omega - \frac{1}{2}[\omega, \omega] = 0$. The analogue of the classical situation exists for the formal homology connection of Chen as shown by the next result.

Theorem 9.68 *Let (ω, ∂) be a formal homology connection on a simply connected compact manifold M. We choose a basis (x_J) of $\mathbb{L}(V)$ and decompose ω as $\omega = \sum_J \omega_J \otimes x_J \in A_{DR}(M) \widehat{\otimes} \mathbb{L}(V)$. Let*

$$\Phi_\omega \colon C^*(\mathbb{L}(V), \partial) \to A_{DR}(M)$$

be defined by $\Phi_\omega(w) = \sum_J \pm \omega_J \langle w, x_J \rangle$, where $C^*(\mathbb{L}(V), \partial) = (\wedge W, d)$ is the cochain algebra introduced in Subsection 2.6.2, $w \in W$ and $\langle , \rangle$ is the duality between $W^{\bullet+1}$ and $\mathbb{L}(V)_\bullet$. Then the following are true:

1. *$\partial^2 = 0$;*
2. *Φ_ω is compatible with the differentials;*
3. *Φ_ω induces an isomorphism in cohomology.*

For the proof, we refer the reader to [59] or [248, Theorem IV. 2.(1)] and observe, as a direct consequence, that a formal homology connection gives a Sullivan model.

Corollary 9.69 *Let (ω, ∂) be a formal homology connection on a simply connected compact manifold M. The map Φ_ω defined in the statement of Theorem 9.68 is a Sullivan model of M.*

We have not introduced differential Lie algebra models here, yet we note that Theorem 9.68 contains within it the fact that $(\mathbb{L}(V), \partial)$ is such a model. In fact, if we let $(T(V), \partial)$ denote the associative tensor algebra obtained as the enveloping algebra of $(\mathbb{L}(V), \partial)$, then we can use a holonomy map $\Theta \colon C_*(\Omega(M)) \to (T(V), \partial)$ constructed by Chen to give an isomorphism of algebras between the Pontryagin algebra $H_*(\Omega(M); \mathbb{R})$ and $H_*(T(V), \partial)$, see [59, Section 3.4] or [248, Proposition IV.3.(1)] for a different approach to the construction of the map Θ.

9.6.6 A topological application

In Subsection 4.6.5, we considered the question of which symplectic manifolds are nilpotent spaces (and thus are amenable to minimal model techniques, see Proposition 4.100). Iterated integrals can be used to approach this question (but not answer it completely yet). We will only state the result and refer the reader to the original paper of Chen [57].

Theorem 9.70 *Let M^n be a connected manifold and suppose $\omega_1, \ldots, \omega_m$ are closed 1-forms on M which satisfy the following conditions:*

- *$\omega_i \wedge \omega_j = 0$ for $i, j = 1, \ldots, m$;*
- *the cohomology classes $[\omega_1], \ldots, [\omega_m]$ are linearly independent.*

Then $\pi_1(M)$ contains a free subgroup G of rank m. Moreover, if ω is a closed q-form on M such that

- *$\omega_i \wedge \omega = 0$ for $i = 1, \ldots, m$;*
- *for some $\alpha \in \pi_q(M)$, the integral $\int_\alpha \omega \neq 0$;*

then $\{\beta \cdot \alpha\}_{\beta \in G}$ is a basis of a free abelian subgroup of $\pi_q(M)$.

This result allows us to see the action of $\pi_1(M)$ on $\pi_q(M)$ in terms of differential forms. Indeed, we can also see Proposition 4.100 from this viewpoint. Can we use Theorem 9.70 or an extension of it to understand nilpotency in the context of symplectic manifolds?

9.7 Cohomological conjectures

Throughout the book we have indicated a series of conjectures related to geometrical problems. Our purpose in this section is to recall them in a more general setting.

9.7.1 The toral rank conjecture

The toral rank conjecture states that if a torus T^r acts almost freely on a compact manifold M, then $\dim H^*(M; \mathbb{Q}) \geq 2^r$. With the help of Borel fibrations, it can be reformulated in terms of fibrations in the following form:

Conjecture 9.71 *If $F \to E \to B$ is a quasi-nilpotent fibration (e.g. if $\pi_1(B) = 0$) where F is a torus T^r and the cohomology of B is finite dimensional, then $\dim H^*(E; \mathbb{Q}) \geq 2^r$.*

This leads to the more general

Problem 9.72 *Let $F \to E \to B$ be a quasi-nilpotent fibration, where F is a rationally elliptic space and the cohomology of B is finite dimensional. Is it true that $\dim H^*(E; \mathbb{Q}) \geq \dim H^*(F; \mathbb{Q})$?*

9.7.2 The Halperin conjecture

The Halperin conjecture is also related to fibrations. Let's recall it and note the strong relation with the conjecture above, see also Section 4.2.

Conjecture 9.73 *If $F \to E \to B$ is a quasi-nilpotent fibration, where F is a rationally elliptic space with $\chi(F) > 0$, then the Serre spectral sequence degenerates at the E_2 term: that is, there is an isomorphism of $H^*(B; \mathbb{Q})$-modules, $H^*(E; \mathbb{Q}) \cong H^*(F; \mathbb{Q}) \otimes H^*(B; \mathbb{Q})$.*

A positive answer to the Halperin conjecture would give a positive answer to Open Problem 9.72 in the case $\chi(F) > 0$.

There are equivalent forms of the Halperin Conjecture: Let X be an elliptic space with $\chi(X) > 0$. Then the following conditions are equivalent (see Proposition 4.40):

1. The Halperin conjecture is true for X.
2. $H^*(X; \mathbb{Q})$ has no non–zero derivation of negative degree.
3. The space of self-homotopy equivalences of X, aut X, has the rational homotopy type of a product of odd dimensional spheres.

The relation between the second and the third characterization follows from the following fact. The commutator of two homotopy self-equivalences gives to $\pi_*(\operatorname{aut} X) \otimes \mathbb{Q}$ the structure of a graded Lie algebra, and there is a morphism of graded Lie algebras $\pi_q(\operatorname{aut} X) \otimes \mathbb{Q} \to \operatorname{Der} H^{-q}(X; \mathbb{Q})$ that is surjective when X is formal.

The derivation viewpoint has a geometric meaning. As we saw in Chapter 6, Belegradek and Kapovitch proved that when M is a compact manifold and $H^*(M; \mathbb{Q})$ has no nonzero negative degree derivation, then, for every torus T and every vector bundle ξ over $C \times T$ with $\sec_{E(\xi)} \geq 0$, there is a finite covering $p \colon T \to T$ such that $(p^* \times \operatorname{id})(\xi) = \xi_C \times T$ where ξ_C is a vector bundle over C. Since bundles do not generally have this property, this then gives an obstruction to positive sectional curvature on total spaces of bundles. *In fact, however, Corollary 6.41 is true even when only the cohomology derivations which descend from derivations of the minimal model vanish.* The question is, how can these be recognized? Let M_C denote the minimal model of C. There is an obvious graded Lie algebra homomorphism

$$\rho \colon H^*(\operatorname{Der}(M_C)) \to \operatorname{Der}(H^*(C)),$$

but knowledge of $\operatorname{Im}(\rho)$ is scant. Of course, if C is formal, then ρ is surjective, but otherwise, we know little.

Problem 9.74 *Determine $\operatorname{Im}(\rho)$ for various classes of spaces C. In particular, determine when $\operatorname{Im}(\rho) = 0$.*

Denote by $\operatorname{Baut}_1(C)$ the classifying space for the monoid of self-equivalences that are homotopic to identity. This space classifies fibrations with simply-connected bases and fibers of the homotopy type of C. The differential graded Lie algebra $(\operatorname{Der}(M_C), d_C)$ is a Lie algebra model for $\operatorname{Baut}_1(C)$. This provides an interesting connection between curvature properties and abstract homotopy theory.

Furthermore, while it was proved in [237] that homogeneous spaces of maximal rank have no negative degree cohomology derivations (and so satisfy the Halperin Conjecture), the following is as yet unknown.

Problem 9.75 *Show that a biquotient $G/\!/H$ has no negative degree cohomology derivations if $\operatorname{rk}(G) = \operatorname{rk}(H)$.*

9.7.3 The Bott conjecture

Recall from Section 6.4 that we have the following conjecture of Bott.

Conjecture 9.76 *If M is a closed manifold with $\sec_M \geq 0$, then M is rationally elliptic.*

Of course, the power of the conjecture (if true) is in the contrapositive statement. Since most manifolds are rationally hyperbolic, this would say that most manifolds cannot carry metrics of non-negative sectional curvature. Also, consider Bott's conjecture in relation to Open Problem 9.72 above.

Problem 9.77 *Let $F \to E \to B$ be a quasi-nilpotent fibration, where F has $\sec_F \geq 0$ and the cohomology of B is finite dimensional. Is it true that $\dim H^*(E; \mathbb{Q}) \geq \dim H^*(F; \mathbb{Q})$?*

Conjecture 9.76 is related to many others as we noted in Section 6.4. At the present time, there seems to be no good approach. Even the results of Chapter 6 lack a certain coherence, at least in terms of their relations to minimal models and the viewpoint of rational homotopy theory. Of course, there is also Paternain's result saying that certain geodesic flows only occur on rationally elliptic manifolds. So we might fairly pose the

Problem 9.78 *Make the connection between geometric constraints on manifolds and algebraic constraints on minimal models systematic.*

The reader may well ask what our precise meaning is here. Evidently, that is part of the problem.

9.7.4 The Gromov conjecture on LM

Let M be a rationally hyperbolic manifold. Gromov conjectured that the cohomology of the free loop space $LM = M^{S^1}$ has exponential growth. On the other hand, Vigué-Poirrier proved that the Betti numbers of the function space from a compact manifold into a wedge of (at least two) spheres has exponential growth. She conjectured [260] that, for any fibration $p: F \to E \to B$, where F has the rational homotopy type of a wedge of spheres, the cohomology of the space of sections of p has exponential growth. Both conjectures merge into a common problem concerning the homology of function spaces. We can state this as

Problem 9.79 *Let $F \to E \to B$ be a quasi-nilpotent fibration where F is rationally hyperbolic and B is a compact manifold. We suppose moreover $\dim B = n$ and F is $(n+1)$-connected. Do the Betti numbers of the space of sections have exponential growth?*

9.7.5 The Lalonde–McDuff question

F. Lalonde and D. McDuff have asked the following

Question 9.80 (Lalonde–McDuff question) *Is every Hamiltonian bundle TNCZ?*

Recall that this says that if $F \to E \to B$ is a symplectic fiber bundle where the structure group of the bundle reduces to the Hamiltonian diffeomorphisms, then the fibration is TNCZ. Of course, we know that all bundles are TNCZ when the fibers are simply connected Kälher manifolds or maximal rank homogeneous spaces. We have also seen in Theorem 7.75 that the Lalonde–McDuff question is answered affirmatively for nilmanifolds.

Problem 9.81 *Use models to examine the Lalonde–McDuff question in other cases of interest.*

Appendix A
De Rham forms

In this appendix we give basic definitions and results about *differential* (also known as *de Rham*) forms. General references for the whole appendix are [107], [112] and [262].

A.1 Differential forms

We shall begin by considering Euclidean space $\mathbb{R}^n$. Of course, $\mathbb{R}^n$ is a smooth manifold with the topology naturally given by the standard vector space dot product structure. We can define vector fields on $\mathbb{R}^n$ by choosing a vector space basis $\{e_1, \ldots, e_n\}$ and defining the vector field E_j by $E_j(x_1, \ldots, x_n) = e_j$ at the point $(x_1, \ldots, x_n) \in \mathbb{R}^n$. In other words, E_j is simply e_j placed at every point of $\mathbb{R}^n$. That such a definition gives a smooth vector field is due to the fact that the coordinates are global. Any vector field on $\mathbb{R}^n$, $V \colon \mathbb{R}^n \to \mathbb{R}^n \times \mathbb{R}^n$, now has the form $V(x_1, \ldots, x_n) = (x_1, \ldots, x_n, \widehat{V})$ with

$$\widehat{V}(x_1, \ldots, x_n) = \sum_{j=1}^n a_j(x_1, \ldots, x_n)\, e_j,$$

where the coefficients a_j are now smooth functions on $\mathbb{R}^n$. It is apparent that we are thinking of $\mathbb{R}^n \times \mathbb{R}^n$ here as pairs (x, v_x), where v_x is an n-vector starting from x. This is the first instance of what is called the *tangent bundle* of a manifold. Here, it is very simple because the topology of $\mathbb{R}^n$ is simple, but it can be more complicated as we shall see. In order to be able to understand these more complicated situations, let's look a little more closely at the $\mathbb{R}^n$ case now.

Let $x \in \mathbb{R}^n$. Consider a smooth curve $\beta \colon (-\epsilon, \epsilon) \to \mathbb{R}^n$ defined on a small interval (which for convenience we take to be $(-\epsilon, \epsilon)$) with $\beta(0) = x$. Of course, we now have n functions $\beta_i \colon (-\epsilon, \epsilon) \to \mathbb{R}$ defined as the coordinate functions of β,

$$\beta(t) = (\beta_1(t), \ldots, \beta_n(t)).$$

The tangent vector of β at 0 is then computed coordinatewise,

$$\beta'(0) = (\beta_1'(0), \ldots, \beta_n'(0)),$$

and we define the collection of all vectors arising as tangent vectors to curves through x to be the *tangent space*, $T_x(\mathbb{R}^n)$, to $\mathbb{R}^n$ at x. It is clear that each vector through x is the tangent vector to a curve, and therefore $T_x(\mathbb{R}^n)$ identifies to the vector space of vectors at x. The *tangent bundle* of $\mathbb{R}^n$ is defined to be

$$T(\mathbb{R}^n) = \{(x, v_x) \mid x \in \mathbb{R}^n \text{ and } v_x \in T_x(\mathbb{R}^n)\}.$$

If $f \colon \mathbb{R}^n \to \mathbb{R}^k$ is a smooth map, then it induces a map $Df \colon T(\mathbb{R}^n) \to T(\mathbb{R}^k)$ of tangent bundles which is a linear map on the tangent space at each point. For $f(x) = y$, the map Df is defined as follows. Let $(x, v_x) \in T(\mathbb{R}^n)$ and suppose β is a curve in $\mathbb{R}^n$ that defines v_x as above. Then the composition $\alpha = f \circ \beta$ is a smooth curve in $\mathbb{R}^k$ with $\alpha(0) = f \circ \beta(0) = f(x) = y$. We then define $Df(x, v_x) = (y, \alpha'(0))$. Note that, in $\mathbb{R}^n$, we can always choose $\beta(t) = x + t v_x$, the line through x in the v_x-direction. If we are only interested in the map Df restricted to a particular tangent space $T_x(\mathbb{R}^n)$, then we denote this linear map by $D_x f$. The vector $\alpha'(0)$ is given by

$$\alpha'(0) = \left.\frac{d(f(\beta(t)))}{dt}\right|_{t=0} = \left.\frac{d(f(x + t v_x))}{dt}\right|_{t=0},$$

where we have taken the straight line through x as β. This is exactly the definition of the usual directional derivative of vector calculus. Therefore, the map Df is just a generalization of that. Indeed, by looking at the coordinate functions α_i of α, we see that Df is just the vector consisting of the directional derivatives of the α_i.

The directional derivative also allows us to transform functions into new functions. Namely, if V is a vector field and $f \colon \mathbb{R}^n \to \mathbb{R}$ is a smooth function, then the smooth function Vf is defined by

$$Vf(x) = \left.\frac{d(f(x + t V_x))}{dt}\right|_{t=0},$$

the usual directional derivative in $\mathbb{R}^n$ in the direction of V_x (the value of V at x). If $V = E_j$, then it is easy to see that $Vf = \partial f / \partial x_j$, the usual partial derivative in the jth coordinate direction.

Now consider the dual space $\mathbb{R}^{n*}$ with dual basis $\{de_1, \ldots, de_n\}$; that is,

$$de_i(e_j) = \begin{cases} 0 & \text{if } i \neq j \\ 1 & \text{if } i = j. \end{cases}$$

A: De Rham forms

We can also take the *exterior algebra* on this basis, $\Lambda(n) = \Lambda(de_1, \ldots, de_n)$. This is the free graded algebra satisfying the anti-commutativity condition $a \wedge b = -b \wedge a$. Note that this means that $de_i \wedge de_i = 0$ for all $i = 1, \ldots, n$. An element ϕ of degree p has the form

$$\phi = \sum a_{i_1 \cdots i_p} de_{i_1} \wedge \cdots \wedge de_{i_p}$$

where the sum is over all $\binom{n}{p}$ choices of p de_j's among the n de_j's available and the $a_{i_1 \cdots i_p}$'s are constant. Note that, by the anti-commutativity property of the exterior algebra, we can always order the i_j so that $i_1 \leq i_2 \leq \ldots \leq i_p$. The algebra $\Lambda(n)$ appears to be the dual of the exterior algebra $\Lambda(\mathbb{R}^n)$. Note that in order to apply an element of the exterior algebra to vectors, we need to have a definition that takes account of the anti-commutativity of the exterior algebra. For instance, if we simply said that $de_1 \wedge de_2(v, w) = de_1(v) \cdot de_2(w)$, then we would not have $de_1 \wedge de_2 = -de_2 \wedge de_1$. Just substitute e_1 for v and e_2 for w and get 1 and 0 for the respective calculations. So, the application of an exterior algebra element $de_{i_1} \wedge \ldots \wedge de_{i_p}$ to vectors $v_1, \ldots, v_p$ is defined by

$$de_{i_1} \wedge \ldots \wedge de_{i_p}(v_1, \ldots, v_p) = \sum_{\sigma \in \Sigma_p} (-1)^{\text{sgn}(\sigma)} de_{i_1}(v_{\sigma(1)}) \cdots de_{i_p}(v_{\sigma(p)}),$$

where the sum is over all permutations σ in the permutation group Σ_p. This definition is compatible with anti-commutativity and will provide a global wedge product below. In fact, we can define 1-forms dE_i dual to the vector fields E_i by requiring

$$dE_i(E_j)(x_1, \ldots, x_n) = \begin{cases} 0 & \text{if } i \neq j \\ 1 & \text{if } i = j. \end{cases}$$

Note that, at any point $(x_1, \ldots, x_n) \in \mathbb{R}^n$, we have $E_i(x_1, \ldots, x_n) = e_i$ for each $i = 1, \ldots, n$. So E_i and 1-forms dE_i are defined globally and evaluate to the corresponding dual vector space basis elements at every point in $\mathbb{R}^n$. Because of the obvious identifications, we may sometimes use the notation de_i for global forms.

Just as for vector fields, we define a *p-differential form* $\alpha \colon \mathbb{R}^n \to \mathbb{R}^n \times \Lambda^p(\mathbb{R}^n)$ by $\alpha(x_1, \ldots, x_n) = (x_1, \ldots, x_n, \widehat{\alpha})$ with

$$\widehat{\alpha}(x_1, \ldots, x_n)$$
$$= \sum a_{i_1 \cdots i_p}(x_1, \ldots, x_n) \, dE_{i_1}(x_1, \ldots, x_n) \wedge \cdots \wedge dE_{i_p}(x_1, \ldots, x_n)$$
$$= \sum a_{i_1 \cdots i_p}(x_1, \ldots, x_n) \, de_{i_1} \wedge \cdots \wedge de_{i_p},$$

where the sum is taken over all multi-indices of length p and the $a_{i_1 \cdots i_p}$ are smooth functions on $\mathbb{R}^n$.

From now on, we will simply denote a form α by its second coordinate, $\hat{\alpha}$. A 1-form $\alpha = \sum a_i \, dE_i$ may be applied to a vector field $V = \sum c_j E_j$ to produce a smooth function on $\mathbb{R}^n$ as follows:

$$\alpha(V)(x_1, \ldots, x_n) = \alpha(x_1, \ldots, x_n)(V(x_1, \ldots, x_n))$$

$$= \left(\sum_{i=1}^n a_i(x_1, \ldots, x_n) \, de_i\right) \left(\sum_{j=1}^n c_j(x_1, \ldots, x_n) \, e_j\right)$$

$$= \sum_{k=1}^n a_k(x_1, \ldots, x_n) \cdot c_k(x_1, \ldots, x_n),$$

using the duality between de_i and e_j.

More generally a p-form $\alpha = \sum a_{i_1 \cdots i_p} \, dE_{i_1} \wedge \cdots \wedge dE_{i_p}$ acts on a p-tuple of vector fields $V_1, \ldots, V_p$ as follows:

$$\alpha(V_1, V_2, \ldots, V_p)(x) = \sum a_{i_1 \cdots i_p}(x) \, (de_{i_1} \wedge \cdots \wedge de_{i_p})((V_1)_x, \ldots, (V_p)_x).$$

Moreover, if $f \colon \mathbb{R}^n \to \mathbb{R}$ is a smooth function, then we may define a 1-form df by $df(V)(x) = Vf(x)$ for all vector fields V. This is our first example of what will be called *exterior differentiation*.

The vector space formed by the forms is equipped with a multiplication, called the *wedge product*. Let $\alpha = \sum a_i \, dE_{i_1} \wedge \cdots \wedge dE_{i_p}$ be a p-form and $\beta = \sum b_j \, dE_{j_1} \wedge \cdots \wedge dE_{j_q}$ be a q-form. The wedge product $\alpha \wedge \beta$ is defined to be the $(p+q)$-form

$$\alpha \wedge \beta = \sum_{i,j} a_i b_j \, dE_{i_1} \wedge \cdots \wedge dE_{i_p} \wedge dE_{j_1} \wedge \cdots \wedge dE_{j_q}.$$

If $Y_1, \ldots, Y_{p+q}$ are vector fields, we have,

$$(\alpha \wedge \beta)(Y_1, \ldots, Y_{p+q})$$
$$= \sum_{\sigma \in \Sigma_{p+q}} (-1)^{\text{sgn}(\sigma)} \alpha(Y_{\sigma(1)}, \ldots, Y_{\sigma(p)}) \cdot \beta(Y_{\sigma(p+1)}, \ldots, Y_{\sigma(p+q)}).$$

Here, the sum is over all permutations σ in the permutation group Σ_{p+q}.

It is also possible to integrate n forms (as the notation de_j suggests). If a form has the expression $\alpha = a \, dE_1 \wedge \ldots \wedge dE_n$, then its integral over some

A: De Rham forms

region $W \subseteq \mathbb{R}^n$ is defined to be

$$\int_W \alpha = \int\int\cdots\int a(x_1,\ldots,x_n)\, dE_1(x_1,\ldots,x_n) \wedge \ldots \wedge dE_n(x_1,\ldots,x_n)$$

$$= \int\int\cdots\int a(x_1,\ldots,x_n)\, de_1 \wedge \ldots \wedge de_n$$

$$= \int\int\cdots\int a(x_1,\ldots,x_n)\, dx_1\ldots dx_n,$$

where we have used the more familiar integral notation using coordinates. Note that, since $dE_i \wedge dE_j = -dE_j \wedge dE_i$, the integral is a *signed* integral depending on the order in which integrals are taken. The $\mathbb{R}^n$-form $\omega = dE_1 \wedge \ldots \wedge dE_n$ is called a *volume form* on $\mathbb{R}^n$ because if we take its integral over any particular region W, we obtain the signed volume of W.

If $f\colon \mathbb{R}^k \to \mathbb{R}^n$ is a smooth function and β is a p-form on $\mathbb{R}^n$, then a new p-form $f^*\beta$ on $\mathbb{R}^k$ is defined by "pulling back" β: namely, if $\beta = \sum \beta_i dE_{i_1}\ldots dE_{i_p}$, then

$$f^*\beta = \sum (\beta_i \circ f)\, df_{i_1} \wedge \ldots \wedge df_{i_p},$$

with $df_{i_r} = \sum \frac{\partial f_{i_r}}{\partial x_k} dx_k$. If $V_1,\ldots,V_p$ are vector fields, then $f^*\beta(V_1,\ldots,V_p)(x) = \beta(Df_x((V_1)_x),\ldots,Df_x((V_p)_x))$, where Df_x is the derivative map of f on tangent vectors. This now leads us to a more general situation.

If M^n is a smooth n-manifold, then M is covered by charts (i.e. homeomorphisms onto open sets) $v_\tau\colon U_\tau \to \mathbb{R}^n$ such that the transition functions $v_{\tau\gamma} = v_\tau v_\gamma^{-1}$ (restricted to the appropriate images) are smooth functions from $\mathbb{R}^n$ to itself. The charts, in effect, parametrize M allowing calculations to be done in euclidean space. For instance, tangent spaces are defined using our definition in $\mathbb{R}^n$. If $\beta\colon (-\epsilon,\epsilon) \to M$ is a curve with $\beta(0) = x \in U_\tau \subset M$, then $\alpha = v_\tau \beta\colon (-\epsilon,\epsilon) \to \mathbb{R}^n$ is a smooth curve with a tangent vector $\alpha'(0)$ at $v(x)$. We then say that $T_x M$ consists of all such vectors. (If M is embedded in some large $\mathbb{R}^N$, then we can visualize the tangent space just as we do for surfaces embedded in $\mathbb{R}^3$.) We then define the *tangent bundle* to M to be

$$TM = \{(x,v_x)\,|\,x \in M \text{ and } v_x \in T_x M\},$$

where $T_x M$ is the tangent space to M at x. Local coordinates provide a basis for $T_x M$, $\{e_1,\ldots,e_n\}$, and we may write $v_x = \sum a_i e_i$ as for $\mathbb{R}^n$. Of course, in the particular chart, the coefficients a_i are functions, changing as x changes. There is a projection $p\colon TM \to M$ given by $p(x,v_x) = x$. A *vector field* V on M is then a smooth mapping $V\colon M \to TM$ such that $pV(x) = x$ for all $x \in M$. It should be mentioned, however, that it is *not* in general the case that the tangent bundle of M is a product $M \times \mathbb{R}^n$ as was true for $\mathbb{R}^n$, because otherwise each manifold would support vector fields that never vanish, which is not the case for instance for S^2. When the vector field is restricted to a chart U_τ, then we can write the tangent bundle over U_τ as such a product.

Charts also tell us whether M has consistent orientations of tangent spaces as we move around the manifold. Each transition function $v_{\tau\gamma}$ is a mapping from an open set in $\mathbb{R}^n$ to another open set in $\mathbb{R}^n$. Therefore, if we consider the linear map on tangent spaces $D_x v_{\tau\gamma}: T_x \mathbb{R}^n \to T_y \mathbb{R}^n$ (where $y = v_{\tau\gamma}(x)$), then we may take $\det(D_x v_{\tau\gamma})$. Since $\det(D_x v_{\tau\gamma})$ is always different from 0, its sign is the same on any connected domain of the transition map, so the question of consistent orientation reduces to looking at all possible overlaps of charts on the manifold. If charts can be chosen covering M so that $\det(D_x v_{\tau\gamma}) > 0$ for all choices of τ and γ, then M is *orientable*. For example, spheres, tori, odd-dimensional real projective spaces and complex projective spaces are orientable. The Klein bottle and even-dimensional real projective spaces are nonorientable. We have the following useful result.

Proposition A.1 *Every simply connected manifold is orientable.*

We may view forms by their images in the different charts. A p-form on M can be seen as a collection of p-forms on the open images of all charts in $\mathbb{R}^n$ obeying a compatibility relation. Namely, for a p-form on M, $\alpha = \{\alpha_\tau\}_\tau$, we must have $v^*_{\tau\gamma}\alpha_\tau = \alpha_\gamma$. With this definition, we now can treat forms locally by fixing a chart (i.e. a coordinate system) and writing the local expression for the form as above for $\mathbb{R}^n$. With this in mind, note that we now can define a wedge product of forms on M simply by taking wedge products in the images of charts in $\mathbb{R}^n$ and then forming the collection: that is, if $\alpha = \{\alpha_\tau\}_\tau$ and $\beta = \{\beta_\tau\}_\tau$, then $\alpha \wedge \beta = \{\alpha_\tau \wedge \beta_\tau\}_\tau$.

A *volume form* on M^n is an n-form ω that is non-zero at every point. It is not true that every manifold has a volume form. For example, for surfaces embedded in $\mathbb{R}^3$, the existence of a volume form is equivalent to the existence of a nonvanishing normal vector field over the whole surface. As the example of the Möbius strip shows, such vector fields sometimes are forced to be zero at points on the surface. The Möbius strip is nonorientable and this is in fact the criterion for nonexistence of a volume form.

Proposition A.2 *A compact manifold M has a volume form if and only if it is orientable.*

In particular, by Proposition A.1, every compact simply connected manifold has a volume form.

Note that a vector field on M also has a local expression in terms of chosen coordinates, so all local computations may be carried out as in $\mathbb{R}^n$.

A: De Rham forms

The *cotangent bundle* and *p*th *exterior algebra bundle* are defined to be

$$T^*M = \{(x, \phi_x) \mid x \in M \text{ and } \phi_x \in T_x^*M\},$$
$$\Lambda^p(T^*M) = \{(x, \phi_x) \mid x \in M \text{ and } \phi_x \in \Lambda^p(T_x^*M)\},$$

where T_x^*M is the dual vector space to T_xM and $\Lambda^*(T_x^*M)$ is the exterior algebra on it. There are obvious projections $q\colon T^*M \to M$ and $\Lambda q\colon \Lambda^p T^*M \to M$ here as well. A *p*-form α on M is then seen to be a smooth mapping $\alpha\colon M \to \Lambda^p(T^*M)$ such that $\Lambda q\, \alpha(x) = x$ for all $x \in M$. In this way, the geometry of M is linearized. We denote the *p*-forms on M by $A^p_{DR}(M)$ and the entire graded algebra of forms by $A_{DR}(M)$.

A.2 Operators on forms

There are several important operations that are performed on forms to create new ones illuminating certain geometric properties.

Let M^n be a manifold with a vector field X and a $(p+1)$-form α defined on it, then we define a *p*-form $i(X)\alpha$ by

$$(i(X)\alpha)(Y_0, \ldots, Y_{p-1}) = \alpha(X, Y_0, \ldots, Y_{p-1}),$$

where $Y_0, \ldots, Y_{p-1}$ are vector fields. The operation on the $(p+1)$-form α that produces the *p*-form $i(X)\alpha$ is called *interior multiplication* by X. Interior multiplication is an anti-derivation (or in more modern terminology, a graded algebra derivation), meaning that

$$i(X)(\alpha \wedge \beta) = (i(X)\alpha) \wedge \beta + (-1)^{|\alpha|}\alpha \wedge (i(X)\beta).$$

If X and Y are vector fields, then their *bracket* $[X, Y]$ is defined by its action on smooth functions, $[X, Y]f = X(Yf) - Y(Xf)$. The bracket is obviously anti-commutative by definition. It also satisfies a Jacobi identity (see Exercise A.1),

$$[[X, Y], Z] + [[Y, Z], X] + [[Z, X], Y] = 0,$$

that will be important when we consider Lie groups. Here the action of a vector field on a function is slightly different from that in $\mathbb{R}^n$. There may be no straight line $x + t\, V_x$ on M in the definition, so we modify the definition as follows. Let X be a vector field and let f be a function on M. At $x \in M$ choose a curve $\beta\colon (-\epsilon, \epsilon) \to M$ such that $\beta'(0) = X_x$. Then

$$Xf(x) = \left.\frac{d(f(\beta(t)))}{dt}\right|_{t=0}.$$

A.2 Operators on forms

If X is a vector field on M, then there is an associated flow $\phi: \mathbb{R} \times M \to M$ given by fixing $x \in M$ and finding the solution of the differential equation

$$\frac{d\phi(t,x)}{dt} = X_{\phi(t,x)},$$

with $\phi(0,x) = x$. The flow is also written $\phi_t(x)$. For fixed x, this is a curve $c(t) = \phi_t(x)$ starting at x and following the vector field X in the sense that its tangent vector at $c(t)$ is $X_{c(t)}$. For fixed t, $\phi_t : M \to M$ is a diffeomorphism of M. Therefore, it can be used to pull back forms. To see the effect of the flow (or equivalently, the vector field X) on forms, we can find the rate of change of the pulled-back form $\phi_t^*\alpha$ minus the original form α at the same point. The definition is

$$\mathcal{L}(X)\alpha(x) = \lim_{t \to 0} \frac{\phi_t^*\alpha_{\phi(t,x)} - \alpha(x)}{t}.$$

The operation $\mathcal{L}(X)$ is called the *Lie derivative* with respect to X. We can also express $\mathcal{L}(X)\alpha$ by its action on vector fields. Let α be a p-form, then

$$\mathcal{L}(X)\alpha(Y_0, \ldots, Y_{p-1})$$
$$= X(\alpha(Y_0, \ldots, Y_{p-1})) - \sum_{j=0}^{p-1} \alpha(Y_0, \ldots, [X, Y_j], \ldots, Y_{p-1}).$$

The Lie derivative $\mathcal{L}(X)$ is a degree 0-derivation on forms:

$$\mathcal{L}(X)(\alpha \wedge \beta) = (\mathcal{L}(X)\alpha) \wedge \beta + \alpha \wedge (\mathcal{L}(X)\beta).$$

In [112, page 157] it is shown that the following formula holds (see Exercise A.2):

$$\phi_t^*\alpha - \alpha = \int_0^t \phi_s^* \mathcal{L}(X)\alpha \, ds.$$

Then we can see that $\phi_t^*\alpha = \alpha$ if and only if $\mathcal{L}(X)\alpha = 0$. In this way the Lie derivative can tell us that α is invariant along trajectories of a vector X. A form α that has $\mathcal{L}(X)\alpha = 0$ for a vector field X is said to be $\mathcal{L}(X)$-*invariant*.

We defined the differential of a function f to be the 1-form df satisfying $df(X) = Xf$ for a vector field X. The operator d is called the *exterior derivative* and is defined on all forms. For a $(p-1)$-form α and vector fields

$Y_0, \ldots, Y_{p-1}$, the p-form $d\alpha$ is defined by,

$$d\alpha(Y_0, \ldots, Y_{p-1}) = \sum_{j=0}^{p-1}(-1)^j Y_j(\alpha(Y_0, \ldots, \widehat{Y_j}, \ldots, Y_{p-1}))$$
$$+ \sum_{i<j}(-1)^{i+j}\alpha([Y_i, Y_j], Y_0, \ldots, \widehat{Y_i}, \ldots, \widehat{Y_j}, \ldots, Y_{p-1}),$$

where "hats" above Y_i's indicate that they are missing. We denote this operation on forms by $d \colon A_{DR}^p(M) \to A_{DR}^{p+1}(M)$. For instance, from the formula, we recover the definition $df(Y) = Yf$ and, for a 1-form α, we find

$$d\alpha(X, Y) = X(\alpha(Y)) - Y(\alpha(X)) - \alpha([X, Y]).$$

The most important property of d is contained in the following theorem.

Theorem A.3 *The vector space $A_{DR}(M)$ equipped with the exterior derivative d and the wedge product is a (commutative graded) cochain algebra. For a p-form α and a q-form β, we have,*

$$d(\alpha \wedge \beta) = d\alpha \wedge \beta + (-1)^p \alpha \wedge d\beta \quad \text{and} \quad d^2 = d \circ d = 0. \quad \text{(A.1)}$$

The following result explains the relation between $i(X)$, d and $\mathcal{L}(X)$.

Proposition A.4 *Let α be a form. Then $\mathcal{L}(X)\alpha = (di(X) + i(X)d)\alpha.$*

Proof Suppose α is a p-form. For vector fields $Y_0, \ldots, Y_{p-1}$, we have

$$di(X)\alpha(Y_0, \ldots, Y_{p-1})$$
$$= \sum_{j=0}^{p-1}(-1)^j Y_j(\alpha(X, Y_0, \ldots, \widehat{Y_j}, \ldots, Y_{p-1}))$$
$$+ \sum_{i<j}(-1)^{i+j}\alpha(X, [Y_i, Y_j], Y_0, \ldots, \widehat{Y_i}, \ldots, \widehat{Y_j}, \ldots, Y_{p-1}),$$

$$i(X)d\alpha(Y_0,\ldots,Y_{p-1}) = d\alpha(X,Y_0,\ldots,Y_{p-1})$$
$$= X(\alpha(Y_0,\ldots,Y_{p-1}))$$
$$- \sum_{j=0}^{p-1}(-1)^j Y_j(\alpha(X,Y_0,\ldots,\widehat{Y_j},\ldots,Y_{p-1}))$$
$$- \sum_{j=0}^{p-1}(-1)^j \alpha([X,Y_j],Y_0,\ldots,\widehat{Y_j},\ldots,Y_{p-1})$$
$$+ \sum_{i<j}(-1)^{i+j}\alpha([Y_i,Y_j],X,Y_0,\ldots,\widehat{Y_i},\ldots,\widehat{Y_j},\ldots,Y_{p-1}).$$

Note that the anti-commutativity of forms forces some terms in the sum $d\,i(X) + i(X)\,d$ to cancel. For instance,

$$\alpha(X,[Y_i,Y_j],Y_0,\ldots,\widehat{Y_i},\ldots,\widehat{Y_j},\ldots,Y_{p-1})$$
$$+ \alpha([Y_i,Y_j],X,Y_0,\ldots,\widehat{Y_i},\ldots,\widehat{Y_j},\ldots,Y_{p-1}) = 0$$

since switching the first two vector field entries changes the sign. We have the following remaining terms:

$$X(\alpha(Y_0,\ldots,Y_{p-1})) - \sum_{j=0}^{p-1}(-1)^j \alpha([X,Y_j],Y_0,\ldots,\widehat{Y_j},\ldots,Y_{p-1})$$
$$= X(\alpha(Y_0,\ldots,Y_{p-1})) - \sum_{j=0}^{p-1}\alpha(Y_0,\ldots,[X,Y_j],\ldots,Y_{p-1})$$

since we make j switches to put the bracket in the jth place and $(-1)^{2j} = 1$. This is then the formula for the Lie derivative $\mathcal{L}(X)$. □

The final operation on forms that will be important for us is *integration*. If M^n is an n-manifold and α is an n-form, then we can integrate α over M. If α is nonzero in one chart U having coordinates $x_1,\ldots,x_n$, then we can write

$$\alpha = a(x_1,\ldots,x_n)\,de_1 \wedge \ldots \wedge de_n.$$

Then integration on M becomes the usual multiple integral in $\mathbb{R}^n$:

$$\int_M \alpha = \int\int \cdots \int a(x_1,\ldots,x_n)\,de_1\ldots de_n.$$

If α is a p-form, $p \neq n$, then $\int_M \alpha = 0$. If α is non-zero in more than one chart, it is necessary to piece together these types of integrals using a partition of unity. We will not go into this here. Integration of forms on M satisfies the following essential property.

Theorem A.5 (Stokes's theorem) *If M^n is an n-manifold with boundary ∂M and α is an $(n-1)$-form, then*

$$\int_M d\alpha = \int_{\partial M} \alpha.$$

A.3 The de Rham theorem

An essential connection between differential geometry and topology is provided by the de Rham Theorem A.6 given below. Recall that the exterior derivative $d: A^p_{DR}(M) \to A^{p+1}_{DR}(M)$ makes $A^*_{DR}(M)$ a cochain complex. Its cohomology is called the *de Rham cohomology* $H^*_{DR}(M)$,

$$H^p_{DR}(M) = \frac{\operatorname{Ker}(d: A^p_{DR}(M) \to A^{p+1}_{DR}(M))}{\operatorname{Im}(d: A^{p-1}_{DR}(M) \to A^p_{DR}(M))}.$$

The forms α with $d\alpha = 0$ are called *closed* forms, while the images under d, $d\beta$, are called *exact* forms. Therefore, de Rham cohomology is the quotient of the closed forms modulo the exact forms. For instance, if $f: M \to \mathbb{R}$ is a smooth function on a connected manifold M, then $df = 0$ only if f is a constant function. Since there are no exact 0-forms, we have $H^0_{DR}(M) = \mathbb{R}$. If M is not connected, then a function could take different constant values on different components. If there are k components, $H^0_{DR}(M) = \mathbb{R}^k$. In some sense, de Rham cohomology is a measure of the solvability of differential equations given by $d\omega$. However, the de Rham theory is not purely analytic, for there is a link to ordinary topology provided by integration and Theorem A.5.

A *singular p-simplex* is a smooth map $\rho: \Delta^p \to M$, where

$$\Delta^p = \{(t_0, \ldots, t_p) \mid t_i \geq 0 \text{ for all } i \text{ and } \sum t_i \leq 1\},$$

is a p-simplex in $\mathbb{R}^{p+1}$. Here by "smooth" we mean that there is some open neighborhood of Δ^p over which ρ can be extended such that the extension is smooth. Usually, singular simplices are continuous maps, but for manifolds we can restrict to smooth ones because the subcomplex consisting of the smooth maps computes the same cohomology. The set of singular p-simplices is the basis for a free abelian group $C_p(M)$ and there is a boundary operator $\partial: C_p(M) \to C_{p-1}(M)$ given by

$$\partial \rho(t_0, \ldots, t_{p-1}) = \sum_i (-1)^i \rho(t_0, \ldots, 0, \ldots, t_{p-1}),$$

where 0 occurs in the ith place. It can be seen that $\partial^2 = 0$, so that

$$\cdots \xrightarrow{\partial} C_{p+1}(M) \xrightarrow{\partial} C_p(M) \xrightarrow{\partial} C_{p-1}(M) \xrightarrow{\partial} \cdots$$

forms a chain complex. An associated cochain complex is formed by taking $C^p(M; \mathbb{R}) = \mathrm{Hom}_{\mathbb{Z}}(C_p(M), \mathbb{R})$ with transposed boundary operator, called the coboundary operator, $\partial^*: C^p(M; \mathbb{R}) \to C^{p+1}(M; \mathbb{R})$, defined by $\partial^*(\phi)(\rho) = \phi(\partial \rho)$, where $\phi \in C^p(M; \mathbb{R})$ and $\rho \in C_{p+1}(M; \mathbb{R})$. Clearly, we also have $(\partial^*)^2 = 0$ and

$$\cdots \xrightarrow{\partial^*} C^{p-1}(M; \mathbb{R}) \xrightarrow{\partial^*} C^p(M; \mathbb{R}) \xrightarrow{\partial^*} C^{p+1}(M; \mathbb{R}) \xrightarrow{\partial^*} \cdots$$

is a cochain complex. In particular, we may define the *singular cohomology* of M with $\mathbb{R}$ coefficients to be

$$H^p(M; \mathbb{R}) = \frac{\mathrm{Ker}(\partial^*: C^p(M; \mathbb{R}) \to C^{p+1}(M; \mathbb{R}))}{\mathrm{Im}(\partial^*: C^{p-1}(M; \mathbb{R}) \to C^p(M; \mathbb{R}))}.$$

Integration defines a homomorphism from p-forms on M^n to p-cochains on M with real coefficients,

$$\int : A^p_{DR}(M) \to C^p(M; \mathbb{R}), \qquad \left(\int \alpha\right)(\rho) = \int_\rho \alpha,$$

where $\rho \in C_p(M)$ is a p-chain in M. The integration makes sense because α is a p-form and ρ is a p-chain. Indeed, each integral is calculated over a simplex by pulling back the form. If α is a p-form and ρ a $(p+1)$-chain, then we have

$$\partial^*\left(\int \alpha\right)(\rho) = \left(\int \alpha\right)(\partial \rho) = \int_{\partial \rho} \alpha = \int_\rho d\alpha = \left(\int d\alpha\right)(\rho),$$

by Stokes's theorem. This shows that integration is a morphism of cochain complexes, and therefore it induces a homomorphism on cohomology, $H^*_{DR}(M) \to H^*(M; \mathbb{R})$. The famous theorem of de Rham is then

Theorem A.6 (de Rham's theorem) *If M is a smooth manifold without boundary, then integration induces an isomorphism of algebras*

$$H^*_{DR}(M) \xrightarrow{\cong} H^*(M; \mathbb{R}).$$

Therefore, in the book, we will typically write $H^*(M;\mathbb{R})$ when we refer to cohomology, even though it may be coming from forms. Note that integration is not an algebra map, however it induces in cohomology an isomorphism of graded algebras where the multiplication in $H^*_{DR}(M)$ is induced by the wedge product of forms, and by the usual cup product in singular cohomology.

A.4 The Hodge decomposition

The Hodge decomposition of forms provides a canonical way to view the vector spaces of cohomology. Let M^n be a compact oriented Riemannian n-manifold. There is a *Hodge star operator*, $*$, that associates to every p-form α an $(n-p)$-form $*\alpha$ so that the following properties hold.

1. The Hodge star applied to the constant function at 1 is the volume form of the manifold: $*1 = \omega$. Note that this immediately implies that the star operator depends on the metric of the manifold.
2. $*$ is linear.
3. $*(f\alpha) = f(*\alpha)$, where f is a function.
4. $**\alpha = (-1)^{np+p}\alpha$, for a p-form α.
5. $\alpha \wedge *\alpha = 0$ if and only if $\alpha = 0$.
6. If β is another p-form, then $\alpha \wedge *\beta = \beta \wedge *\alpha$.

We can indicate the local formula for $*$ by considering p-forms $\alpha = a_I de_I$ and $\beta = b_J de_J$. This notation means that we are just taking monomials here, where $I = (i_1, \ldots, i_p)$ and $J = (j_1, \ldots, j_p)$. Then

$$\alpha \wedge *\beta = \begin{cases} 0 & \text{if } I \neq J \\ \pm a_I b_I \omega & \text{if } I = J, \end{cases}$$

where the ordering of I and its complement determine the sign. In particular, if $\alpha = \sum_I a_I de_I$ with $|I| = p$, then $\alpha \wedge *\alpha = \sum(a_I^2)\omega$.

The Hodge star operator allows us to define an inner product on p-forms by taking

$$(\alpha, \beta) = \int_M \alpha \wedge *\beta.$$

This definition has all of the standard properties of an inner product. For instance, property (6) of $*$ immediately shows that $(-,-)$ is symmetric. The formula for $\alpha \wedge *\alpha$ shows $(\alpha, \alpha) \geq 0$ and property (5) shows $(\alpha, \alpha) = 0$ only when $\alpha = 0$. Finally, note that the star operator is an isometry with respect

A.4 The Hodge decomposition

to the inner product. Here we use property (3), the anti-commutativity of forms and symmetry.

$$(*\alpha, *\beta) = \int_M *\alpha \wedge **\beta = (-1)^{np+p} \int_M *\alpha \wedge \beta$$

$$= (-1)^{np+p} \int_M (-1)^{p(n-p)} \beta \wedge *\alpha = (-1)^{np+p+np-p^2} \int_M \beta \wedge *\alpha$$

$$= \int_M \beta \wedge *\alpha = (\beta, \alpha) = (\alpha, \beta),$$

since $2np - p(p-1)$ is always even.

We can also define a *co-differential* $\delta \colon A_{DR}^p(M) \to A_{DR}^{p-1}(M)$ by setting

$$\delta = (-1)^{np+n+1} *d*.$$

Again note that, as opposed to d, δ depends on the metric (since $*$ does). Also, the presence of $*$ in the definition tells us that δ is not in general a derivation. By property (4) of the star operator and $d^2 = 0$, we see that $\delta^2 = 0$. So we have forms α that are *co-closed*, $\delta \alpha = 0$, and forms β that are *co-exact*, $\beta = \delta \gamma$ for some γ. Let us now prove the key fact relating δ to d.

Lemma A.7 *Let M^n be a compact oriented Riemannian n-manifold without boundary. If α is a $(p-1)$-form and β is a p-form, then $(d\alpha, \beta) = (\alpha, \delta\beta)$.*

Proof Note first that the derivation property of d gives

$$d(\alpha \wedge *\beta) = d\alpha \wedge *\beta + (-1)^{p-1} \alpha \wedge d*\beta.$$

Now we compute using Stokes's theorem,

$$(d\alpha, \beta) = \int_M d\alpha \wedge *\beta = \int_M d(\alpha \wedge *\beta) - (-1)^{p-1} \alpha \wedge d*\beta$$

$$= (-1)^p \int_M \alpha \wedge d*\beta$$

$$= (-1)^p \int_M \alpha \wedge (-1)^{n(n-p+1)+(n-p+1)} **d*\beta$$

$$= (-1)^{p+n(n-p+1)+(n-p+1)} \int_M \alpha \wedge *(*d*)\beta$$

$$= (-1)^{p+n(n-p+1)+(n-p+1)} \int_M \alpha \wedge *(-1)^{np+n+1} \delta\beta$$

$$= (-1)^{np+n+1+p+n(n-p+1)+(n-p+1)} \int_M \alpha \wedge *\delta\beta$$

$$= \int_M \alpha \wedge *\delta\beta = (\alpha, \delta\beta),$$

since the exponent of -1 reduces to $2n + 2 + n(n+1)$ which is always even. $\qquad\square$

The operators d and δ are therefore adjoint to one another. Note for example, that if α is closed, then $(\alpha, \delta\beta) = 0$ for all β. Similarly, if α is co-closed, then $(\alpha, d\beta) = 0$ for all β. So closed forms are orthogonal to Im(δ) and co-closed forms are orthogonal to Im(d). Indeed, the implications may also be reversed here. Suppose α is orthogonal to Im(δ). Then, $(d\alpha, d\alpha) = (\alpha, \delta\, d\alpha) = 0$ implies that $d\alpha = 0$. A similar argument holds for α orthogonal to Im(d). We therefore have the

Proposition A.8
(1) *A form is closed if and only if it is orthogonal to all co-exact forms.*
(2) *A form is co-closed if and only if it is orthogonal to all exact forms.*

Now define an operator $\Delta \colon A^p_{DR}(M) \to A^p_{DR}(M)$ by $\Delta = d\,\delta + \delta\,d$. Note that Δ is self-adjoint; that is, $(\Delta\alpha, \beta) = (\alpha, \Delta\beta)$ by the adjointness of d and δ. We want to know which forms are both closed and co-closed.

Lemma A.9 *Let $\alpha \in A^p_{DR}(M)$. Then $\Delta\alpha = 0$ if and only if $d\alpha = 0$ and $\delta\alpha = 0$.*

Proof We have

$$\begin{aligned}
(\Delta\alpha, \Delta\alpha) &= ((d\,\delta + \delta\,d)\alpha, (d\,\delta + \delta\,d)\alpha) \\
&= (d\,\delta\alpha, d\,\delta\alpha) + (d\,\delta\alpha, \delta\,d\alpha) + (\delta\,d\alpha, d\,\delta\alpha) + (\delta\,d\alpha, \delta\,d\alpha) \\
&= (d\,\delta\alpha, d\,\delta\alpha) + (\delta\,d\alpha, \delta\,d\alpha)
\end{aligned}$$

since $d\,\delta\alpha$ is closed and $\delta\,d\alpha \in \text{Im}(\delta)$. Now, if $\Delta\alpha = 0$, then the positive definiteness of the inner product implies that both $d\,\delta\alpha = 0$ and $\delta\,d\alpha = 0$. But then we have

$$0 = (d\,\delta\alpha, \alpha) = (\delta\alpha, \delta\alpha),$$

which implies $\delta\alpha = 0$ by positive definiteness. Similarly, we have $d\alpha = 0$. On the other hand, if α is closed and co-closed, then by the formula above $(\Delta\alpha, \Delta\alpha) = 0$ and this implies $\Delta\alpha = 0$. $\qquad\square$

A.4 The Hodge decomposition

The forms that are both closed and co-closed are called *harmonic* forms. Obviously, the harmonic p-forms form a sub-vector space of $A^p_{DR}(M)$ and we will denote this subspace by $\mathcal{H}^p$. The Hodge decomposition asserts that all forms can be made up of harmonic forms, exact forms and co-exact forms.

Theorem A.10 (Hodge decomposition theorem) *Let M be a compact oriented Riemannian manifold without boundary. Then*

$$A^p_{DR}(M) = \mathcal{H}^p \oplus \mathrm{Im}(\Delta) = \mathcal{H} \oplus \mathrm{Im}(d) \oplus \mathrm{Im}(\delta).$$

Proof Suppose $\alpha \in \mathcal{H}^p$ and $\beta = \Delta\gamma$ for some γ. Then $(\alpha, \beta) = (\alpha, \Delta\gamma) = (\Delta\alpha, \gamma) = 0$, since Δ is self-adjoint. Hence, $\mathcal{H}$ is orthogonal to $\mathrm{Im}(\Delta)$. Indeed, some hard analysis shows that if a form β is orthogonal to all harmonic forms, then $\beta \in \mathrm{Im}(\Delta)$. Analysis also shows that $\dim(\mathcal{H}^p)$ is finite, so there is a finite basis of harmonic forms $h_1, \ldots, h_k$ which we can take to be orthonormal by scaling appropriately. Now let $\alpha \in A^p_{DR}(M)$ and denote by $\alpha_\mathcal{H}$ the form

$$\alpha_\mathcal{H} = \sum_{i=1}^{k} (\alpha, h_i) h_i.$$

Note that $(\alpha - \alpha_\mathcal{H}, h_j) = (\alpha, h_j) - \sum_i (\alpha, h_i)(h_i, h_j) = (\alpha, h_j) - (\alpha, h_j) = 0$ by orthonormality of the h_i's. By what we said above, this means that $\alpha - \alpha_\mathcal{H} \in \mathrm{Im}(\Delta)$. Hence, $A^p_{DR}(M) = \mathcal{H}^p \oplus \mathrm{Im}(\Delta)$.

We have already seen in Proposition A.8 that $\mathrm{Im}(d)$ and $\mathrm{Im}(\delta)$ are orthogonal. Since forms in $\mathcal{H}^p$ are both closed and co-closed, these three subspaces are mutually orthogonal. Clearly, $\mathrm{Im}(\Delta) \subseteq \mathrm{Im}(d) \oplus \mathrm{Im}(\delta)$ and, since $A^p_{DR}(M) = \mathcal{H}^p \oplus \mathrm{Im}(\Delta)$, we also have $A^p_{DR}(M) = \mathcal{H} \oplus \mathrm{Im}(d) \oplus \mathrm{Im}(\delta)$. □

Corollary A.11 *Let $\alpha \in A^p_{DR}(M)$. Then α has a unique decomposition*

$$\alpha = \mathcal{H}(\alpha) + \alpha_d + \alpha_\delta,$$

where $\mathcal{H}(\alpha) \in \mathcal{H}^p$, $\alpha_d \in \mathrm{Im}(d)$ and $\alpha_\delta \in \mathrm{Im}(\delta)$.

Proposition A.8 says that $\mathcal{H}^p$ and $\mathrm{Im}(d)$ give all closed forms in $A^p_{DR}(M)$. Moreover, we see that α is exact precisely when $\mathcal{H}(\alpha) = 0$, $\alpha_\delta = 0$ and $\alpha = d\beta_\delta$ for a *unique* element $\beta_\delta \in \mathrm{Im}(\delta)$. To see this, suppose $\alpha = dx_\delta = dy_\delta$ with $x_\delta, y_\delta \in \mathrm{Im}(\delta)$. Then we have

$$0 = d(x_\delta - y_\delta) = d(\delta\bar{x} - \delta\bar{y}) = d\delta(\bar{x} - \bar{y}),$$

where $\delta\bar{x} = x_\delta$ and $\delta\bar{y} = y_\delta$. But then

$$0 = (d\delta(\bar{x} - \bar{y}), (\bar{x} - \bar{y})) = (\delta(\bar{x} - \bar{y}), \delta(\bar{x} - \bar{y})),$$

and this implies $x_\delta - y_\delta = \delta(\bar{x} - \bar{y}) = 0$. Hence, $x_\delta = y_\delta$.

The foregoing discussion says that when we calculate $H^p_{DR}(M)$, we can simply ignore the summand $\text{Im}(d)$ and find

Theorem A.12 (de Rham existence) *If M is a compact orientable Riemannian manifold without boundary, then*

$$\mathcal{H}^p = H^p_{DR}(M) = H^p(M; \mathbb{R}).$$

Finally, it is important to note that this theorem only gives an isomorphism of vector spaces in each degree. It is not true in general that the wedge product of harmonic forms is harmonic. Thus, there is no algebra homomorphism in general from harmonic forms to cohomology.

Exercises for Appendix A

Exercise A.1 If V and W are tangent vector fields on M, then we define their *Lie bracket* by

$$[V, W](f) = V(W(f)) - W(V(f)).$$

Apply the definition of bracket to prove the following properties.
(1) $[V, W] = -[W, V]$.
(2) $[aV + bW, Z] = a[V, Z] + b[W, Z]$ where $a, b \in \mathbb{R}$.
(3) $[[V, W], Z] + [[W, Z], V] + [[Z, V], W] = 0$ (Jacobi Identity).
(4) $[fV, gW] = fg[V, W] + fV(g)W - gW(f)V$ for smooth functions f and g.

Exercise A.2 Show that the flow ϕ_t associated to a vector field X satisfies

$$\phi_t^* \alpha - \alpha = \int_0^t \phi_s^* \mathcal{L}(X)\alpha \, ds$$

for a form α and Lie derivative $\mathcal{L}(X)$. Hint: see [112, page 157].

Appendix B
Spectral sequences

Before explaining what is in this appendix, we first mention what is not in it. It does not contain a complete presentation of spectral sequences – nor a historical background of the subject. The readers interested by these points should consult one of the classical books introducing algebraic topology or [186]. But, since in many parts of the book we have been led to natural questions and properties about spectral sequences (see Subsection 4.2.3, Theorem 4.56, Section 4.4, Theorem 7.36, Question 7.74, Section 9.7,...), we must at least give the flavor of what a spectral sequence is and how we can use it (as well as some basic examples for entertainment). This appendix also includes some proofs of theorems used in other chapters. For instance:

- the Zeemann–Moore theorem (Theorem B.15), which is essential for the study of morphisms of fibrations (Corollary B.16);
- the odd spectral sequence (Theorem B.18) that is fundamental for the understanding of the models of homogeneous spaces or biquotients (see Theorem 3.33 or Corollary 3.51 for instance) and more generally for pure models (see Subsection 2.5.3);
- the spectral sequences that appear in the particular case of a double complex, the Frölicher spectral sequence of which is an example (see Definition 4.69).

B.1 What is a spectral sequence?

Definition B.1 *A (homology) spectral sequence $\{E^r, d^r\}$ is a sequence (E^r) of $\mathbb{Z}$-bigraded modules, each of them with a differential $d^r \colon E^r_{p,q} \to E^r_{p-r,q+r-1}$ and such that $E^{r+1} = H(E^r, d^r)$. The module E^r is called the r-level (or r-stage) of the spectral sequence. An element of $E^r_{p,q}$ is said of* total degree $p+q$.

A morphism of spectral sequences $f \colon \{E^r, d^r\} \to \{E'^r, d'^r\}$ is a family of morphisms $f^r \colon E^r \to E'^r$ of bigradation $(0,0)$ and such that $f^{r+1} = H(f^r)$.

B: Spectral sequences

If we start with E^2, we observe that:

- $E^3 = Z^2/B^2$ with $Z^2 = \operatorname{Ker} d^2$ and $B^2 = \operatorname{Im} d^2$;
- $E^4 = Z^3/B^3$ with $Z^3/B^2 = \operatorname{Ker} d^3$ and $B^3/B^2 = \operatorname{Im} d^3$.

If we denote by Z^{r-1} (B^{r-1}) the respective sets of elements of E^2 that survive up to level r (are killed at or before level r), then we get a tower of inclusions

$$0 \subset B^2 \subset B^3 \subset \cdots \subset B^r \subset \cdots \subset Z^r \subset \cdots \subset Z^3 \subset Z^2 \subset E^2,$$

and $E^{r+1} = Z^r/B^r$.

Definition B.2 *Let's denote $Z^\infty = \cap_r Z^r$, $B^\infty = \cup_r B^r$ and $E^\infty_{p,q} = Z^\infty_{p,q}/B^\infty_{p,q}$. The spectral sequence $\{E^r, d^r\}$ converges to a graded module H if there is a filtration $(F_\bullet H)$ of H such that, for any p, $E^\infty_{p,q} = (F_p H/F_{p-1} H)_{p+q}$. A spectral sequence collapses at level r if we have an isomorphism of bigraded modules between E^r and E^∞.*

We represent a spectral sequence by an array or a sequence of arrays. We say that a spectral sequence is a *first quadrant* spectral sequence if $E^r_{p,q} = 0$ for $p < 0$ or $q < 0$. In this case the representation goes like this:

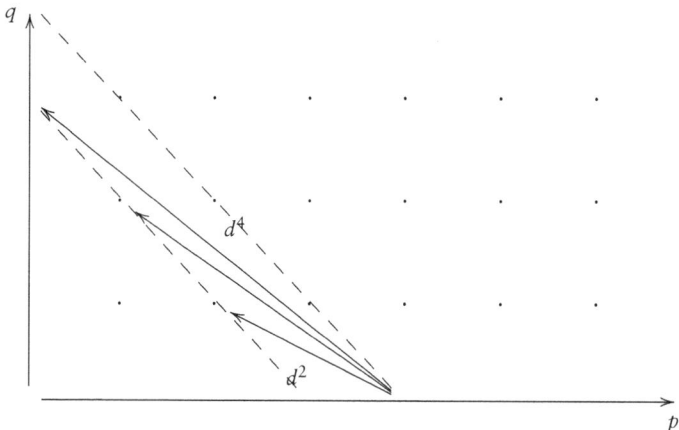

The elements of total degree n are on the line $p + q = n$ and a differential d^i joins a point of the line $p + q = n$ to a point of the line $p + q = n - 1$.

We have represented d^2, d^3 and d^4 above. We also have:
- $E_{p,q}^{r+1} = E_{p,q}^{r}$ if $r > \max(p, q+1)$;
- $E_{p,0}^{r+1}$ is always a submodule of $E_{p,0}^{r}$ and we have a sequence of inclusions

$$E_{p,0}^{\infty} = E_{p,0}^{p+1} \hookrightarrow \cdots \hookrightarrow E_{p,0}^{3} \hookrightarrow E_{p,0}^{2};$$

- $E_{0,q}^{r+1}$ is always a quotient of $E_{0,q}^{r}$ and we have a sequence of surjections

$$E_{0,q}^{2} \twoheadrightarrow E_{0,q}^{3} \twoheadrightarrow \cdots \twoheadrightarrow E_{0,q}^{q+2} = E_{0,q}^{\infty}.$$

Definition B.3 *In a first quadrant spectral sequence, the* transgression *is the relation from $E_{q,0}^{2}$ in $E_{0,q-1}^{2}$ given by the diagram*

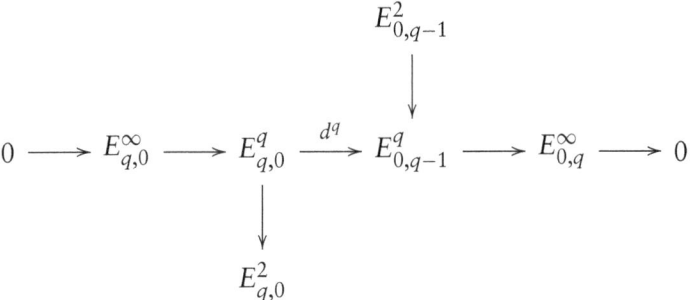

Proposition B.4 *Let $f_r: (E_r^{*,*}, d_r) \to (E_r'^{*,*}, d_r')$ be a morphism between two spectral sequences such that f_n is an isomorphism for some integer n. Then, for any integer m, $n \le m$, f_m is an isomorphism. If we have first quadrant spectral sequences, then f also induces an isomorphism between E^{∞} and E'^{∞}.*

B.2 Spectral sequences in cohomology

Definition B.5 *A spectral sequence in cohomology $\{E_r, d_r\}$ is a sequence, (E_r) of $\mathbb{Z}$-bigraded differential modules such that $d_r: E_r^{p,q} \to E_r^{p+r,q-r+1}$ and $E_{r+1} = H(E_r, d_r)$.*

A morphism of spectral sequences $f: \{E_r, d_r\} \to \{E_r', d_r'\}$ is a family of morphisms $f_r: E_r \to E_r'$ of bigradation $(0, 0)$ and such that $f_{r+1} = H(f_r)$.

In fact, a spectral sequence in cohomology can be viewed as a third quadrant (homology) spectral sequence and all that we said above is easily adapted to this situation. For instance, we say that a spectral sequence in cohomology is a first quadrant spectral sequence if $E_r^{p,q} = 0$ for $p < 0$ or $q < 0$. In this case the representation goes like this:

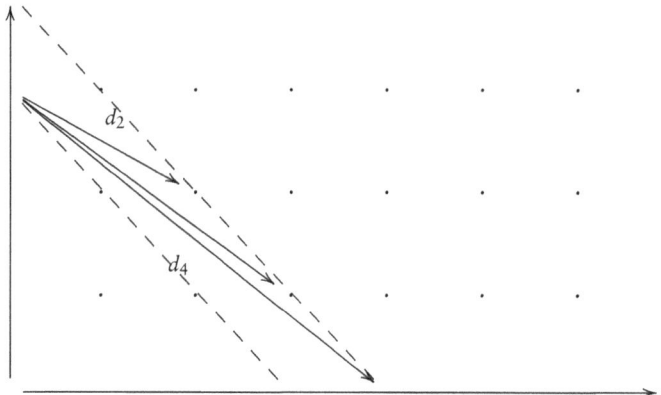

Proposition B.4 is still valid in this context.

B.3 Spectral sequences and filtrations

Most of examples of spectral sequences arise from a filtration on a complex.

Definition B.6 *A filtered differential R-module is a graded R-module M such that* $M = \oplus_{n=0}^{\infty} M^n$, $d(M^n) \subset M^{n+1}$, $F^{p+1}M \subset F^p(M)$ *and* $d(F^p(M)) \subset F^p(M)$.

The filtration is called bounded *if, for any n, there exist two integers $s(n)$ and $t(n)$ such that*

$$0 = F^{s(n)}(M^n) \subset F^{s(n)-1}(M^n) \subset \cdots \subset F^{t(n)}(M^n) = M^n.$$

Definition B.7 *A bigraded differential algebra is a bigraded $\mathbb{Z}$-module $E^{*,*}$ with a multiplication $E^{p,q} \otimes E^{r,s} \to E^{p+r,q+s}$ and a differential d such that*

$$d\left(\bigoplus_{p+q=n} E^{p,q}\right) \subset \bigoplus_{r+s=n+1} E^{r,s}.$$

A filtered differential graded algebra *is a filtered differential module and a differential graded algebra A such that the multiplication sends $F^r A \otimes F^s A$ into $F^{r+s} A$.*

Theorem B.8 *Any filtered differential graded R-module gives rise to a spectral sequence in cohomology $\{E_r^{*,*}, d_r\}$, $r \geq 1$, with $E_1^{p,q} = H^{p+q}(F^p M / F^{p+1} M)$. If the filtration is bounded, the spectral sequence converges to $H(M, d)$.*

In the case of a filtered differential graded algebra, any stage of the spectral sequence is a bigraded differential algebra.

Of course, this is where the use of upper and lower indices shows itself to be powerful. We set:

- $Z_r^{p,q} = F^p M^{p+q} \cap d^{-1}(F^{p+r} M^{p+q+1})$;
- $B_r^{p,q} = F^p M^{p+q} \cap d(F^{p-r} M^{p+q-1})$;
- $E_r^{p,q} = Z_r^{p,q}/(Z_{r-1}^{p+1,q-1} + dZ_{r-1}^{p-r+1,q+r-2})$.

With this notation, we have to prove that we get a spectral sequence. For instance, the differential d sends $Z_r^{p,q}$ to $Z_r^{p+r,q-r+1}$ and induces $d_r \colon E_r^{p,q} \to E_r^{p+r,q-r+1}$. We refer the reader to [186, Chapter 2].

B.4 Serre spectral sequence

Theorem B.9 ([236]) *Let $f \colon E \to B$ be a fibration with F path connected and B simply connected. We consider homology and cohomology with coefficients in any commutative ring.*

1. *There is a filtration of $H_n(E)$,*

$$0 = H_{-1,n+1} \subset H_{0,n} \subset H_{1,n-1} \subset \cdots \subset H_{n-1,1} \subset H_{n,0} = H_n(E),$$

and a first quadrant spectral sequence such that:

$$E_{p,q}^2 = H_p(B; H_q(F)), \quad E_{p,q}^\infty \cong H_{p,q}/H_{p-1,q+1}.$$

2. *There is a filtration on $H^n(E)$*

$$0 = H^{n+1,-1} \subset H^{n,0} \subset \cdots \subset H^{1,n-1} \subset H^{0,n} = H^n(E),$$

and a first quadrant spectral sequence in cohomology such that:

$$E_2^{p,q} = H^p(B; H^q(F)), \quad E_\infty^{p,q} = H^{p,q}/H^{p+1,q-1}.$$

Each (E_r, d_r) is a differential bigraded algebra.

This spectral sequence is called the *Serre spectral sequence*. The key to its construction is a filtration of the chain complex on the total space by a "basic degree". For more general versions using local coefficients, see [236].

Example B.10 In general, a spectral sequence gives little information about the algebra structure of $H(A, d)$ because the convergence uses an associated graded step. We give here an illustration of this fact, see also Remark 3.36.

Consider the Hopf fibration $S^3 \to S^7 \to S^4$. By taking the orbits of the S^1 action on the fiber and the total space, we get a fibration $S^2 \to \mathbb{C}P(3) \to S^4$.

The Serre spectral sequence of this fibration has $E_2^{p,q} = H^p(S^4) \otimes H^q(S^2)$ and thus collapses at level 2. Yet, the cohomology algebra of $\mathbb{CP}(3)$ is not the tensor product of these two algebras.

In this example, we can also observe that the Euler class is zero and that the fibration has no section.

In opposition with the previous example, the next one shows how we can get information about the cohomology algebra of the loop space ΩS^3.

Example B.11 Consider the fibration $\Omega S^3 \to PS^3 \to S^3$. The cohomology ring $H^*(\Omega S^3; \mathbb{Z})$ has a single generator x_{2n} in each even degree with the only relations being $n \cdot x_{2n} = x_2^n$. Thus, all powers of x_2 are nontrivial.

Proof The fibration $\Omega S^3 \to PS^3 \to S^3$, where $PS^3 = \{\gamma : I \to S^3 \mid \gamma(0) = (1,0,0,0)\}$ is the contractible path space on S^3, has the following spectral sequence diagram.

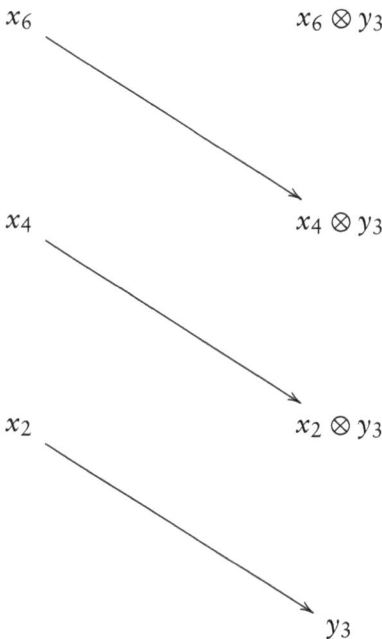

Here we *are not* saying that we know the cohomology of ΩS^3. We are simply indicating the structure that we now determine algebraically for the convenience of the reader.

The diagram has two vertical lines with the cohomology of the fiber along the vertical axis and the cohomology of the base along the horizontal axis. The only possible nonzero differential is the one shown, d_3, which moves over to the right by three and moves down by two. Because the path space is contractible, the class y_3 cannot survive to infinity, for it would then provide a nonzero class in $H^3(PS^3; \mathbb{Z}) = 0$. Thus, there must exist an element x_2 with $d_3(x_2) = y_3$. The differential d_3 is a derivation, so we have $d_3(x_2^2) = 2x_2 \otimes d_3(x_2) = 2x_2 \otimes y_3$. This says that the element $x_2 \otimes y_3$ must still be killed in order to avoid producing a nonzero class in $H^5(PS^3; \mathbb{Z}) = 0$. So there must exist x_4 with $d_3(x_4) = x_2 \otimes y_3$. Hence $d_3(2x_4 - x_2^2) = 0$. The cocycle $2x_4 - x_2^2$ would survive to infinity if it were not zero, so we must have $2x_4 = x_2^2$. We will do one more case to see the general algebraic pattern. We have $d_3(x_2^3) = 3x_2^2 \otimes y_3 = 6x_4 \otimes y_3$, so there must exist x_6 with $d_3(x_6) = x_4 \otimes y_3$. Therefore, $d_3(3 \cdot x_6) = d_3(x_2^3)$ and, by the same argument as before, we must have $3 \cdot x_6 = x_2^3$. Inductively, we can then show that classes x_{2n} exist with $n \cdot x_{2n} = x_2^n$. □

We now state three particular cases of the Serre spectral sequence. Proofs can be found in any classical algebraic topology book or in [186].

Corollary B.12 *With the hypotheses of Theorem B.9, suppose that $H_p(B) = 0$ for $0 < p < p_0$ and $H_q(F) = 0$ for $0 < q < q_0$. Then there is an exact sequence*

$$H_{p_0+q_0-1}(F) \to H_{p_0+q_0-1}(E) \to H_{p_0+q_0-1}(B)$$
$$\to H_{p_0+q_0-2}(F) \cdots \to H_1(E) \to 0.$$

Corollary B.13 (Gysin sequence) *With the hypotheses of Theorem B.9, suppose that the fiber F has the homology of a sphere S^m. Then we have exact sequences,*

$$\cdots \to H_{k+1}(E) \to H_{k+1}(B) \to H_{k-m}(B) \to H_k(E) \to \cdots,$$

and

$$\cdots \to H^k(E) \to H^{k-m}(B) \xrightarrow{\psi} H^{k+1}(B) \to H^{k+1}(E) \to \cdots$$

The morphism ψ satisfies $\psi(v_1 \cup v_2) = v_1 \cup \psi(v_2)$. In fact, ψ is multiplication by the Euler class of the fibration.

Corollary B.14 (Wang sequence) *With the hypotheses of Theorem B.9, suppose that the basis B has the homology of a sphere S^m. Then we have two exact sequences,*

$$\cdots \to H_k(F) \to H_k(E) \to H_{k-m}(F) \to H_{k-1}(F) \to \cdots,$$

and

$$\cdots \to H^{k-1}(F) \xrightarrow{\theta} H^{k-m}(F) \to H^k(E) \to H^k(F) \to \cdots.$$

The morphism θ acts as an algebra derivation.

B.5 Zeeman–Moore theorem

Since we are concerned with rational homotopy type, we *now work with vector spaces over a field of characteristic zero.*

Theorem B.15 (Zeeman–Moore theorem) *If $f_r: (E_r^{*,*}, d_r) \to (E_r'^{*,*}, d_r')$ is a morphism between two first quadrant spectral sequences such that $E_2^{p,q} = E_2^{p,0} \otimes E_2^{0,q}$, $E_2'^{p,q} = E_2'^{p,0} \otimes E_2'^{0,q}$ and $f_2^{p,q} = f_2^{p,0} \otimes f_2^{0,q}$, then any two of the following conditions imply the third:*

(1) $f_2: E_2^{p,0} \to E_2'^{p,0}$ *is an isomorphism for all p;*
(2) $f_2: E_2^{0,q} \to E_2'^{0,q}$ *is an isomorphism for all q;*
(3) $f_\infty: E_\infty^{p,q} \to E_\infty'^{p,q}$ *is an isomorphism for all (p,q).*

This theorem was first proved by J.C. Moore [200, Exposé 22]. Generalizations were given by Zeeman [268], see also [130, Theorem 17.17]. From Theorem B.9 we immediately obtain the following topological interpretation.

Corollary B.16 *Let*

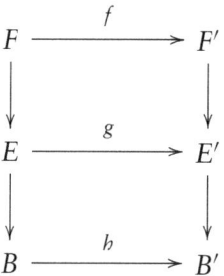

be a morphism between orientable fibrations. Then any two of the following conditions imply the third:

(1) $H^*(h)$ *is an isomorphism;*
(2) $H^*(f)$ *is an isomorphism;*
(3) $H^*(g)$ *is an isomorphism.*

Proof In the statement of Theorem B.15, if properties (1) and (2) are satisfied, then (3) will follow directly from the hypotheses and Proposition B.4.

We suppose now that (1) and (3) are satisfied and prove (2). We follow the proof of [182, Theorem 11.1, Chapter XI] with an appropriate modification in the case of cohomology. We will use the classical Four Lemma of homological algebra that we recall below for the convenience of the reader, see Lemma B.17. We consider the property:

$(2)^j$: the map $f_2: E_2^{0,q} \to E_2'^{0,q}$ is an isomorphism for all q, $0 \leq q \leq j$.

Observe that $E_2^{0,0} = E_\infty^{0,0}$. The hypothesis (3) implies that $(2)_0$ is satisfied. So we now work by induction and assume (1), (3) and $(2)^j$, *with j fixed*.

Step 1. We do a second induction on the integer i, considering the property

$(2)_i^j$: $f_i^{p,q}$ is an isomorphism for all q, $q + i \leq j + 2$, and all p,

and a monomorphism for all q, $q \leq j$, and all p.

Observe that $(2)_2^j$ is true and suppose that $(2)_n^j$ is satisfied for some integer n. We have to prove $(2)_{n+1}^j$ and for that we need to come back to the construction of the $(n+1)$st page from the nth page. First we look at the subspaces $Z_n^{*,*}$. Consider the following morphism between exact sequences:

$$\begin{array}{ccccccc}
0 & \longrightarrow & Z_n^{p,q} & \longrightarrow & E_n^{p,q} & \longrightarrow & E_n^{p+n,q-n+1} \\
& & \downarrow Z(f)_n^{p,q} & & \downarrow f_n^{p,q} & & \downarrow f_n^{p+n,q-n+1} \\
0 & \longrightarrow & Z_n'^{p,q} & \longrightarrow & E_n'^{p,q} & \longrightarrow & E_n'^{p+n,q-n+1}
\end{array}$$

If $q \leq j$, then $(q - n + 1) + n \leq q + 1 \leq j + 2$ and $f_n^{p+n,q-n+1}$ is an isomorphism and $f_n^{p,q}$ a monomorphism, by assumption. Therefore, $Z(f)_n^{p,q}$ is a monomorphism.

If $q \leq j - n + 1$, then $f_n^{p,q}$ is an isomorphism and $f_n^{p+n,q-n+1}$ a monomorphism. Therefore $Z(f)_n^{p,q}$ is an isomorphism.

We look now at the $B_n^{*,*}$:

$$\begin{array}{ccccc}
E_n^{p-n,q+n-1} & \longrightarrow & B_n^{p,q} & \longrightarrow & 0 \\
\downarrow f_n^{p-n,q+n-1} & & \downarrow B(f)_n^{p,q} & & \\
E_n'^{p-n,q+n-1} & \longrightarrow & B_n'^{p,q} & \longrightarrow & 0
\end{array}$$

If $q \leq j$, then $f_n^{p-n,q+n-1}$ is an isomorphism. Therefore $B(f)_n^{p,q}$ is surjective.

Finally, we arrive at the space $E_{n+1}^{*,*}$:

$$\begin{array}{ccccccccc}
0 & \longrightarrow & B_n^{p,q} & \longrightarrow & Z_n^{p,q} & \longrightarrow & E_{n+1}^{p,q} & \longrightarrow & 0 \\
& & \downarrow B(f)_n^{p,q} & & \downarrow Z(f)_n^{p,q} & & \downarrow f_{n+1}^{p,q} & & \\
0 & \longrightarrow & B_n'^{p,q} & \longrightarrow & Z_n'^{p,q} & \longrightarrow & E_{n+1}'^{p,q} & \longrightarrow & 0
\end{array}$$

If $q \leq j$, $Z(f)_n^{p,q}$ is a monomorphism and $B(f)_n^{p,q}$ is surjective. Therefore $f_{n+1}^{p,q}$ is a monomorphism.

If $q \leq j - (n+1) + 2 = j - n + 1$, then $Z(f)_n^{p,q}$ is an isomorphism. Therefore, $f_{n+1}^{p,q}$ is an isomorphism.

We have thus proved $(2)_i^j$ for any i.

Step 2. We establish $(2)^{j+1}$. In order to see this, using a descending induction on i, we prove that $f_i^{0,j+1} : E_i^{0,j+1} \to E_i'^{0,j+1}$ is an isomorphism for any $i \geq 2$. Observe that, in first quadrant spectral sequences, we have:

- $E_i^{0,j+1} \cong E_\infty^{0,j+1}$ for i large enough,
- $Z_i^{0,j+1} = E_{i+1}^{0,j+1}$, for any $i \geq 2$, because the range of $d_i : E_i^{-i,j+i} \to E_i^{0,j+1}$ is zero.

From this first property and the hypothesis (3), we may assume that $f_{n+1}^{0,j+1}$ is an isomorphism and we are reduced to proving that $f_n^{0,j+1}$ is an isomorphism also. We consider the following morphism between two exact sequences:

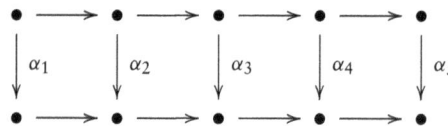

From Step 1, we know that $f_n^{n,j-n+2}$ is an isomorphism. On the other hand, we have $Z_n^{0,j+1} = E_{n+1}^{0,j+1}$, so $Z(f)_n^{0,j+1}$ is an isomorphism by induction. This implies the injectivity of $f_n^{0,j+1}$.

Recall now that $B(f)_n^{n,j-n+2}$ is surjective; this implies the surjectivity of $f_n^{0,j+1}$ by the same argument, replacing $E_n^{n,j-n+2}$ by $B_n^{n,j-n+2}$ in the previous diagram.

We then get that $f_n^{0,j+1}$ is an isomorphism and $(2)^{j+1}$ follows.

The same argument works to prove that (2) and (3) imply (1). □

Lemma B.17 (The Four Lemma [182, Lemma 3.3, Chapter I]) *Consider the following commutative diagram with exact rows.*

$$\begin{array}{ccccccccc}
\bullet & \to & \bullet & \to & \bullet & \to & \bullet & \to & \bullet \\
\downarrow\alpha_1 & & \downarrow\alpha_2 & & \downarrow\alpha_3 & & \downarrow\alpha_4 & & \downarrow\alpha_5 \\
\bullet & \to & \bullet & \to & \bullet & \to & \bullet & \to & \bullet
\end{array}$$

(i) *If α_1 is an epimorphism and α_2 and α_4 monomorphisms, then α_3 is a monomorphism.*

(ii) *If α_5 is a monomorphism and α_2 and α_4 epimorphisms, then α_3 is an epimorphism.*

B.6 An algebraic example: The odd spectral sequence

The following result gives the structure of the cohomology algebra $H^*(G/T; \mathbb{Q})$ where T is the maximal torus of a compact connected Lie group G and, more generally, for the algebra $H^*(G/H; \mathbb{Q})$ if H is a closed subgroup of maximal rank in G, see Theorem 3.33.

Theorem B.18 *Let $(A = \wedge(Q \oplus P), d)$ be a cdga where Q is concentrated in even degrees and P in odd degrees such that $d(Q) = 0$ and $d(P) \subset \wedge Q$. Further, suppose that $H(A, d)$ is of finite dimension. Then the following conditions are equivalent:*

(1) *P and Q have the same dimension;*
(2) *$\chi(H(A, d)) > 0$;*
(3) *$H(A, d)$ is evenly graded.*

If these conditions are satisfied, then $H(A, d) = \wedge Q/I$, where I is the ideal generated by dP.

This statement can be generalized to any minimal model, see [129] or [87, page 444].

Proof We give A a new (lower) gradation defined by $G_s A = \oplus_i (\wedge Q \otimes \wedge^i P)^{s-i}$ and observe from the hypotheses that $dG_s A \subset G_s A$. Therefore, this gradation induces a gradation $G_s H$ on cohomology $H = H(A, d)$ and the corresponding Euler characteristics of $G_s H$ and $G_s A$ are equal. Observe also that an element of $(\wedge Q \otimes \wedge^i P)$ is of odd lower degree if and only if i is odd. As a consequence, we have:

$$\chi(G_s H) = \sum_i (-1)^i \dim H((\wedge Q \otimes \wedge^i P)^{s-i}, d)$$
$$= \sum_i (-1)^i \dim (\wedge Q \otimes \wedge^i P)^{s-i}.$$

Therefore, the global Euler characteristic of $H = \oplus_s G_s H$ is obtained by

$$\chi(H) = \sum_s \chi(G_s H) = \sum_s \sum_i (-1)^i \dim (\wedge Q \otimes \wedge^i P)^{s-i}.$$

To use the finite dimensional hypothesis, we have to take into account a special type of Poincaré polynomial adapted to this situation called the

Koszul–Poincaré polynomial:

$$\mathcal{P}H(t) = \sum_{s=0}^{\infty} \chi(G_s H) t^s = \sum_{s=0}^{\infty} \sum_{i} (-1)^i \dim(\wedge Q \otimes \wedge^i P)^{s-i} t^s.$$

This last term does not depend on the differential and can be computed as:

$$\mathcal{P}H(t) = \frac{\prod_{i=1}^{p}(1 - t^{2u_i})}{\prod_{j=1}^{q}(1 - t^{2v_j})},$$

if Q is of dimension q generated by elements of degrees $(2v_1, \ldots, 2v_q)$ and P of dimension p generated by elements of degrees $(2u_1 - 1, \ldots, 2u_p - 1)$. In order to have H of finite dimension, this expression must be a polynomial and we therefore require $p \geq q$.

If $\chi(H) > 0$ then one has $p = q$ and, in this case, $\chi(H)$ is the quotient $\frac{\prod u_i}{\prod v_j}$. Reciprocally, $p = q$ implies $\chi(H) > 0$. We have thus proved the equivalence between (1) and (2). Obviously (3) implies (2). So we are reduced to proving (1) or (2) implies (3).

For that, we use a classical construction in minimal models in which we kill the vector space Q by adding a copy sQ, shifted by one degree (i.e. $|sq| = |q| - 1$ if $q \in Q$). We construct a cdga $(\wedge Q \otimes \wedge P \otimes \wedge sQ, d)$ with $d(sq) = q$, observe that $(\wedge Q \otimes \wedge sQ, d)$ is acyclic and therefore see that the cohomology of $(\wedge Q \otimes \wedge P \otimes \wedge sQ, d)$ is $\wedge P$.

We now forget all the degrees and consider the degree in word length in P and sQ. In $(\wedge Q \otimes \wedge P \otimes \wedge sQ, d)$, the differential d lowers this word length by exactly one, so we have an induced lower degree on the cohomology. We denote it by H_k.

We now filter $(\wedge Q \otimes \wedge P \otimes \wedge sQ, d)$ by the word length degree in sQ. This gives a spectral sequence with first page $E_1^{i,j} = \oplus_i (\wedge^i sQ) \otimes H_j(\wedge Q \otimes \wedge P)$.

Let k be the greatest integer such that $H_k(\wedge Q \otimes \wedge P) \neq 0$ and let $\alpha \in H_k$. Choose an element a of maximal word length in $\wedge sQ$ that is of degree $\dim Q$. (For this, recall that Q is in even degrees, so sQ is in odd degrees. Therefore, the product of all generators in sQ is one such element a.) The product $a\alpha$ cannot be killed in the spectral sequence (since it is in the corner), so $\dim P \geq k + \dim Q$. From the assumption $\dim P = \dim Q$, we deduce $k = 0$ and, thus, $H = \wedge Q/(dP)$ is evenly graded. □

B.7 A particular case: A double complex

Definition B.19 *A double complex is a bigraded R-module M equipped with two differentials, $d' : M^{p,q} \to M^{p+1,q}$ and $d'' : M^{p,q} \to M^{p,q+1}$, such that*

$d'd'' + d''d' = d'^2 = d''^2 = 0$. The total complex *associated to the double complex* is $(\text{Tot}(M), d)$ with $\text{Tot}(M)^n = \oplus_{p+q=n} M^{p,q}$ and $d = d' + d''$.

Since $d'^2 = d''^2 = 0$, we have two cohomologies associated to a double complex:

$$H_I^{*,*}(M) = H(M^{*,*}, d') \text{ and } H_{II}^{*,*}(M) = H(M^{*,*}, d'').$$

Because $d'd'' + d''d' = 0$, the respective differentials d' and d'' induce differentials $\bar{d}'$ on $H_{II}(M^{*,*})$ and $\bar{d}''$ on $H_I(M^{*,*})$ so that we get two new cohomologies

$$H_I^{*,*} H_{II}(M) = H(H_{II}^{*,*}(M), \bar{d}') \text{ and } H_{II}^{*,*} H_I(M) = H(H_I^{*,*}(M), \bar{d}'').$$

Theorem B.20 *Let* $(M^{*,*}, d', d'')$ *be a double complex. Then we have two spectral sequences*: $\{_I E_r, {}_I d_r\}$ *and* $\{_{II} E_r, {}_{II} d_r\}$ *such that*

$$_I E_2^{*,*} = H_I^{*,*} H_{II}(M) \text{ and } _{II} E_2^{*,*} = H_{II}^{*,*} H_I(M).$$

Moreover, if $M^{p,q} = 0$ *when* $p < 0$ *or* $q < 0$, *then the two spectral sequences converge to* $H^*(\text{Tot}(M), d)$.

The idea is to take the two spectral sequences associated to the following filtrations:

$$F_I^p(\text{Tot}(M))^t = \bigoplus_{r \geq p} M^{r, t-r} \text{ and } F_{II}^p(\text{Tot}(M))^t = \bigoplus_{r \geq p} M^{t-r, r}.$$

A quick (and wrong!) glimpse at this situation might elicit the idea that such spectral sequences always collapse at level 2. A counterexample is given in the case of the Frölicher spectral sequence in Example 4.75.

Exercises for Appendix B

In order to understand spectral sequences, the reader needs to experiment and do concrete examples on his or her own. Here we suggest some "easy" problems for a beginner.

Exercise B.1 Use the Wang exact sequence to prove $H_q(\Omega S^m) = \mathbb{Z}$ if $q = k(m-1)$ and 0 otherwise.

Exercise B.2 Give a generalization of Example B.11 and determine the cohomology algebra of ΩS^m. Hint: see for instance [37, page 204].

Exercise B.3 Let $f: S^3 \to K(\mathbb{Z}, 3)$ be the classifying map of the fundamental class. Let X denote the pullback of the path fibration $PK(\mathbb{Z}, 3) \to K(\mathbb{Z}, 3)$ along f. Compute $H^5(X)$.

Exercise B.4 By using the Serre spectral sequence of the fibration $S^1 \to S^{2n+1} \to \mathbb{C}P(n)$, determine the cohomology *algebra* of $H^*(\mathbb{C}P(n); \mathbb{Z})$. (Remember, the differentials are derivations.)

Exercise B.5 By using the Serre spectral sequence of the path fibration $K(\mathbb{Z}, n) \to PK(\mathbb{Z}, n+1) \to K(\mathbb{Z}, n+1)$, determine the cohomology algebra $H^*(K(\mathbb{Z}, n); \mathbb{Q})$.

Exercise B.6 By using the Serre spectral sequence of the fibration $U(n) \to EU(n) \to BU(n)$, determine the cohomology algebra $H^*(BU(n); \mathbb{Q})$.

Exercise B.7 Compute the cohomology algebras, with coefficients in $\mathbb{Q}$, of the various Stiefel and Grassmann manifolds.

Appendix C
Basic homotopy recollections

In this appendix we collect a few classical notions and results in homotopy theory that are used in various chapters. We denote the set of homotopy classes of maps from the (pointed) space X to the (pointed) space Y by $[X, Y]$, where we take homotopies relative to the basepoint. If $f : X \to Y$, then pre- and post-composition with f induce respective morphisms of homotopy sets $f^\sharp : [Y, Z] \to [X, Z]$ and $f_\sharp : [Z, X] \to [Z, Y]$.

C.1 n-equivalences and homotopy sets

In Chapter 2, we defined a notion of cdga homotopy and considered sets of homotopy classes of cdga morphisms. We saw there that rational homotopy equivalences (i.e. quasi-isomorphisms) induce bijections of homotopy sets. This result is the cdga analogue of the following thread of ideas.

Definition C.1 ([240, page 404]) *A continuous map of path connected spaces $f : X \to Y$ is called an n-equivalence for $n \geq 1$ if for every $x \in X$, the induced map $f_\sharp = \pi_q(f) : \pi_q(X, x) \to \pi_q(Y, f(x))$ is an isomorphism for $0 < q < n$ and an epimorphism for $q = n$. A weak equivalence or ∞-equivalence is an n-equivalence for all $n \geq 1$.*

The effect of an n-equivalence on homotopy sets is described by the

Proposition C.2 ([240, Corollary 7.6.23]) *If $f : X \to Y$ is an n-equivalence, then*
- *$f_\sharp : [P, X] \to [P, Y]$ is bijective if P is a CW-complex of dimension $\leq n - 1$;*
- *$f_\sharp : [P, X] \to [P, Y]$ is surjective if P is a CW-complex of dimension $\leq n$.*

The next result, *Whitehead's theorem*, provides a way to detect n-equivalences of simply connected spaces by looking at homology.

Theorem C.3 ([240, Theorem 7.5.9]) *Let $f \colon (X, x_0) \to (Y, y_0)$ be a map of path connected and pointed spaces. If there is $n \geq 1$ such that $\pi_q(f) \colon \pi_q(X, x_0) \to \pi_q(Y, y_0)$ is an isomorphism for $q < n$ and an epimorphism for $q = n$, then $H_q(f) \colon H_q(X, x_0) \to H_q(Y, y_0)$ is an isomorphism for $q < n$ and an epimorphism for $q = n$. Conversely, if X and Y are simply connected and $H_q(f)$ is an isomorphism for $q < n$ and an epimorphism for $q = n$ then f is an n-equivalence.*

A map $f \colon X \to Y$ is a *homotopy equivalence* if there exists $g \colon Y \to X$ with $fg \simeq \mathrm{id}_Y$ and $gf \simeq \mathrm{id}_X$. The symbol $\simeq$ denotes the homotopy relation between maps and we also write $X \simeq Y$ when there exists a homotopy equivalence $X \to Y$. We then say that X and Y have the same *homotopy type*. Note that we do not need a chain of maps here as in the definition of rational homotopy type (see Definition 2.34) because in this case we require the existence of a map that is a homotopy inverse. The following result provides the link between weak equivalences and homotopy equivalences.

Theorem C.4 ([240, Corollary 7.6.24]) *A map $f \colon X \to Y$ of CW complexes is a weak equivalence if and only if it is a homotopy equivalence.*

C.2 Homotopy pushouts and pullbacks

Throughout the book, we have used the notions of homotopy pushout and homotopy pullback. In this section, we recall these general notions and mention several simple results about them.

Given two maps $f \colon X \to W$ and $g \colon Y \to W$, form the space

$$P(f, g) = \{(x, \theta, y) \in X \times W^I \times Y \mid f(x) = \theta(0), g(y) = \theta(1)\}.$$

This space is the *homotopy pullback* of f and g in the following sense. First, the square

$$\begin{array}{ccc} P(f,g) & \xrightarrow{\bar{g}} & X \\ \bar{f} \downarrow & & \downarrow f \\ Y & \xrightarrow{g} & W \end{array}$$

is homotopy commutative, where $\bar{f}$ and $\bar{g}$ are the canonical projections. Further, given any other homotopy commutative square

$$\begin{array}{ccc} Z & \xrightarrow{h} & X \\ k \downarrow & & \downarrow f \\ Y & \xrightarrow{g} & W \end{array}$$

there exists a map $\omega\colon Z \to P(f,g)$

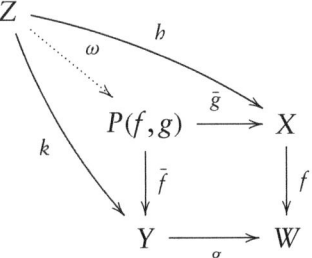

in which $\bar{g}\omega \simeq h\colon Z \to X$ and $\bar{f}\omega \simeq k\colon Z \to Y$. The map ω is not unique, even up to homotopy.

The following are fundamental results for homotopy pullbacks.

Proposition C.5 ([184]) *In the following diagram,*

$$\begin{array}{ccccc} A & \longrightarrow & B & \longrightarrow & C \\ \downarrow & & \downarrow & & \downarrow \\ D & \longrightarrow & E & \longrightarrow & F \end{array}$$

the squares BCFE and ACFD are homotopy pullbacks if and only if the squares BCFE and ABED are homotopy pullbacks.

C: Basic homotopy recollections

Lemma C.6 ([184, Corollary 7, Corollary 9]) *Consider the homotopy commutative cube below:*

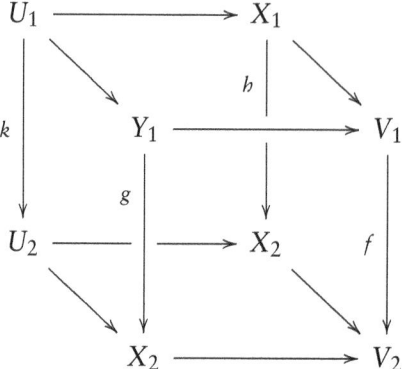

Suppose that the top and bottom squares are homotopy pullbacks. If the maps f, g and h are homotopy equivalences, then so is the map k.

The dual notion is that of a *homotopy pushout*. Given maps $f: W \to X$ and $g: W \to Y$, the *homotopy pushout* of f and g is given by

$$D(f,g) = (X \cup (W \times I) \cup Y)/\sim$$

where $(w, 0) \sim f(w)$ and $(w, 1) \sim g(w)$. This space is also called the the *double mapping cylinder* of f and g and is a homotopy pushout of f and g in the following sense. First, the square

$$\begin{array}{ccc} W & \xrightarrow{f} & X \\ g \downarrow & & \downarrow \bar{g} \\ Y & \xrightarrow{\bar{f}} & D(f,g) \end{array}$$

is homotopy commutative, where $\bar{f}$ and $\bar{g}$ are the canonical inclusions. Further, given any other homotopy commutative square

$$\begin{array}{ccc} W & \xrightarrow{f} & X \\ g \downarrow & & \downarrow h \\ Y & \xrightarrow{k} & Z \end{array}$$

there exists a map $\omega: D(f,g) \to Z$

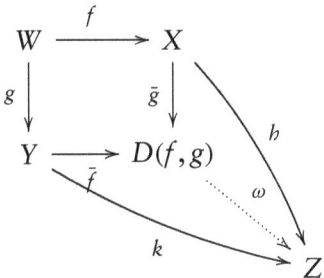

in which $\omega \bar{g} \simeq h: X \to Z$ and $\omega \bar{f} \simeq k: Y \to Z$. In general, the map ω is not unique, even up to homotopy.

Two basic results about homotopy pushouts are the following.

Proposition C.7 *In the following diagram,*

$$
\begin{array}{ccccc}
A & \longrightarrow & B & \longrightarrow & C \\
\downarrow & & \downarrow & & \downarrow \\
D & \longrightarrow & E & \longrightarrow & F
\end{array}
$$

the squares ABED and ACFD are homotopy pushouts if and only if the squares ABED and BCFE are homotopy pushouts.

Lemma C.8 ([184, Corollary 7, Corollary 9]) *Consider the following homotopy commutative cube:*

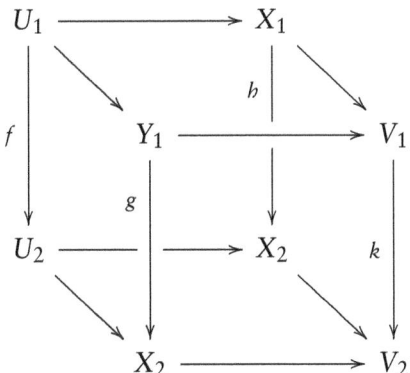

Suppose that the top and bottom squares are homotopy pushouts. If the maps f, g and h are homotopy equivalences, then so is the map k.

As indicated by the two statements above, properties of homotopy pullbacks can often be dualized for homotopy pushouts. Although the next result is a case where duality between homotopy pullbacks and pushouts fails, it nevertheless provides a useful homotopical tool.

Theorem C.9 (Theorem of the cube) *If the bottom of a cube is a homotopy pushout and all sides are homotopy pullbacks, then the top of the cube is a homotopy pushout.*

As a consequence, we have

Corollary C.10 *Let X be a space and ABED be a homotopy pushout. Then the square*

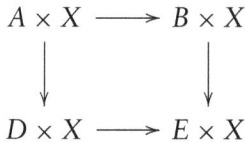

is a homotopy pushout.

We note here that homotopy pullbacks and homotopy pushouts are really only defined up to homotopy equivalence. Replacing any of the spaces or maps by homotopy equivalent ones in the initial diagram produces the same homotopy pushout or pullback *up to homotopy*. In some discussions of homotopy pushouts and pullbacks, such as in [184], the specific homotopies being used are also taken into account.

C.3 Cofibrations and fibrations

Two of the most basic homotopical objects are fibrations and cofibrations. Fibrations are probably more familiar to geometers because they are the homotopical analogues of fiber bundles. Indeed, locally trivial bundles with paracompact base spaces and covering spaces are examples of fibrations. We begin by recalling the notion of cofibration and then see how duality provides the definition of fibration. For this section, we refer to [137] and [265].

C.3 Cofibrations and fibrations

Definition C.11 *A map $j: A \to X$ is a cofibration if for any commutative diagram of solid arrows*

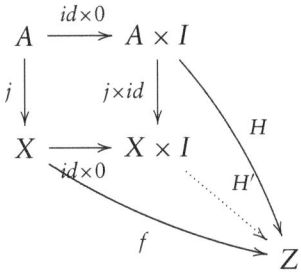

there exists a dotted arrow H' such that the diagram still commutes. The quotient map $q: X \to X/j(A)$ (or the space $X/j(A)$) is called the cofiber of j.

If $j: A \to X$ is a cofibration, the map j is injective and induces a homeomorphism from A to $j(A)$. Therefore, we consider A as a subset of X (closed in X if X is Hausdorff) and denote the cofiber by X/A. For a space A, let CA denote the *cone* on A defined by $CA = (A \times I)/((a,1) \sim *)$. If $f: A \to X$ is any map, recall that the *mapping cone of f* (or the *homotopy cofiber of f*) is the inclusion $q: X \to C_f = X \cup_f CA$ (or the space $C_f = X \cup_f CA$) with $X \cup_f CA = X \cup CA/((f(a) \sim (a,0))$. In the case of a cofibration $j: A \to X$, the cofiber X/A has the homotopy type of the cone C_j. That is, we have a homotopy equivalence of pairs $(X/A, *) \simeq (C_j, CA)$.

A space X has a *non-degenerate basepoint* x_0 if the inclusion $x_0 \hookrightarrow X$ is a cofibration. This implies, at least for compactly generated spaces (see [265] for this definition), that there is an open neighborhood U of x_0 which contracts to x_0 in X relative to x_0. In other words, there is a homotopy $H: U \times I \to X$ such that $H(u, 0) = u$, $H(u, 1) = x_0$ and $H(x_0, t) = x_0$. CW complexes always have nondegenerate basepoints. Indeed, every point of a CW complex is a nondegenerate basepoint. A space is said to be *well-pointed* if it has a nondegenerate basepoint.

If the spaces are well-pointed, $A \to A \vee X$ is a cofibration, called a *trivial cofibration*, with cofiber X. Note also that the inclusion $A \to CA$ is a cofibration with cofiber ΣA. More generally, if $f: A \to X$ is any map, then the *mapping cylinder* $A \to M_f = X \cup (A \wedge I)/(f(a) \sim a \wedge 0)$, $a \mapsto a \wedge 1$, is a cofibration with cofiber C_f and f is the composition of a cofibration and a homotopy equivalence:

$$\begin{array}{ccc} A & \xrightarrow{f} & X \\ & \searrow_{j_f} & \nearrow_{\simeq} \\ & M_f & \end{array}$$

Any sequence $A \xrightarrow{f} X \xrightarrow{q} C$ is called a *cofiber sequence* if $q\colon X \to C$ has the homotopy type of $M_f \xrightarrow{\simeq} X \to C_f$.

For any map $f\colon A \to X$, we have *the Barratt–Puppe sequence*,

$$A \xrightarrow{f} X \xrightarrow{q} C_f \xrightarrow{\delta} \Sigma A \xrightarrow{\Sigma f} \Sigma X \xrightarrow{\Sigma q} \Sigma C_f \longrightarrow \cdots$$

where each three-term sequence is a cofiber sequence. For any space Z, the Barratt–Puppe sequence induces an exact sequence of pointed sets or groups:

$$[A, Z] \xleftarrow{f^\sharp} [X, Z] \xleftarrow{q^\sharp} [C_f, Z] \xleftarrow{\delta^\sharp} [\Sigma A, Z]$$
$$\xleftarrow{(\Sigma f)^\sharp} [\Sigma X, Z] \xleftarrow{(\Sigma q)^\sharp} [\Sigma C_f, Z] \longleftarrow \cdots$$

Finally, we mention that, if either of the maps f or g which define a homotopy pushout is a cofibration, then the homotopy pushout of f and g has the homotopy type of the ordinary topological pushout of f and g.

Now suppose that, instead of considering the problem of extending homotopies, we dualize the definition of cofibration and consider the problem of lifting homotopies. With this in mind, for a space Y, let $Y^I = \{\omega\colon [0, 1] \to Y\}$ be the free path space on Y. Note that any map $f\colon X \to Y$ induces $f^I\colon X^I \to Y^I$, $\omega \mapsto f \circ \omega$.

Definition C.12 *A map $p\colon E \to B$ is a fibration if for any commutative diagram of solid arrows*

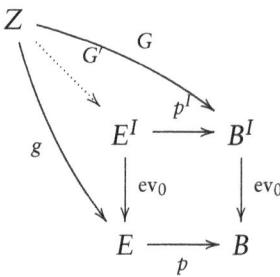

there exists a dotted arrow G' such that the diagram still commutes. Here, for any space Y, the notation ev_0 denotes an evaluation mapping $Y^I \to Y$ given by $\mathrm{ev}_0(\omega) = \omega(0)$. The map $F = p^{-1}(*) \to E$ (or the space F) is called the *fiber of p*.

Note that the diagram above is equivalent to the following one defining the *homotopy lifting property*. The G's in the following diagram have tildes over them to denote that they are the adjoints of the maps in Definition C.12 under the exponential correspondence.

$$\begin{array}{ccc} Z \times 0 & \xrightarrow{g} & E \\ \downarrow & \nearrow_{\widetilde{G}'} & \downarrow p \\ Z \times I & \xrightarrow[\widetilde{G}]{} & B \end{array}$$

Namely, given the commutative diagram of solid arrows, it is the *homotopy lifting property* which ensures the existence of the lift $\widetilde{G}'$.

If $p: E \to B$ is a fibration with B path connected, the map p is onto. If $f: X \to Y$ is any map, the *homotopy fiber* of f is the map $F_f \to X$ (or the space F_f) defined by $F_f = \{(\omega, x) \in Y^I \times X \mid f(x) = \omega(0) \text{ and } \omega(1) = *\}$, $(\omega, x) \mapsto x$. In the case of a fibration p, the fiber F has the homotopy type of the homotopy fiber F_p.

The projection $p_B: E \times B \to B$ is a fibration, called the *trivial fibration*, with fiber E. The path fibration $P(X) = \{\omega \in X^I \mid \omega(0) = *\} \to X$, $\omega \mapsto \omega(1)$, is a fibration with the loop space ΩX as fiber. More generally, if $f: E \to B$ is any map, the *path fibration associated to* f, $p_f: E_f \to B$, is a fibration with fiber F_f, defined by $E_f = \{(\omega, x) \in B^I \times E \mid f(x) = \omega(0)\}$, $p_f((\omega, x)) = \omega(1)$. Any map is the composition of a homotopy equivalence and a fibration as shown below.

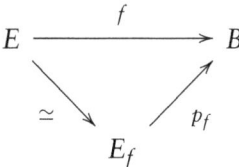

Note that, as we mentioned at the start of Section C.3, locally trivial bundles with paracompact base spaces and covering spaces are fibrations.

Any sequence $F \xrightarrow{\iota} E \xrightarrow{f} B$ is called a *fiber sequence* if $\iota: F \to E$ has the homotopy type of $F_f \to E_f \to E$. For any map $f: E \to B$ we have *the*

Puppe sequence

$$\cdots \longrightarrow \Omega F_f \xrightarrow{\Omega \iota} \Omega E \xrightarrow{\Omega f} \Omega B \xrightarrow{\partial} F_f \xrightarrow{\iota} E \xrightarrow{f} B$$

where each three-term sequence is a fiber sequence. For any space Z, the Puppe sequence induces an exact sequence of pointed sets or groups:

$$[Z, B] \xleftarrow{f_\sharp} [Z, E] \xleftarrow{\iota_\sharp} [Z, F_f] \xleftarrow{\partial_\sharp} [Z, \Omega B] \xleftarrow{(\Omega \iota)_\sharp} [Z, \Omega E] \longleftarrow \cdots$$

Recall that the multiplication in the space of loops ΩB, $\mu \colon \Omega B \times \Omega B \to \Omega B$, is the ordinary composition of loops and the inverse map $\nu \colon \Omega B \to \Omega B$ is given by $\nu(\omega)(t) = \omega(1-t)$, $\omega \in \Omega B$. The group-like properties of ΩB extend to the analogue of a group action of ΩB on the fiber F. This is expressed in the following

Definition C.13 *Let $p\colon E \to B$ be a fibration of fiber F. The holonomy of p is a map $\mathrm{Hol}\colon \Omega B \times F \to F$ obtained by the homotopy lifting property applied to $G(\omega, x, t) = \omega(t)$:*

$$\begin{array}{ccc} \Omega B \times F \times 0 & \longrightarrow & E \\ \downarrow & \nearrow^{G'} & \downarrow p \\ \Omega B \times F \times I & \xrightarrow{G} & B \end{array}$$

with $\mathrm{Hol}(\omega, x) = G'(\omega, x, 1)$.

Finally, we mention that, if either of the two maps f or g defining a homotopy pullback is a fibration, then the homotopy pullback of f and g is of the homotopy type of the topological pullback of f and g. Of course, the topological pullback is both easier to define and more familiar from its role in bundle theory, so this identification is often useful.

References

[1] Adams, J. Frank (1957). On the cobar construction. In *Colloque de topologie algébrique, Louvain, 1956*, pp. 81–87. Georges Thone, Liège.

[2] Adams, J. Frank (1969). *Lectures on Lie Groups*. W. A. Benjamin, Inc., New York.

[3] Allday, Christopher (1978). On the rational homotopy of fixed point sets of torus actions. *Topology*, **17**(1), 95–100.

[4] Allday, Christopher (1994). Circle actions on Kähler spaces. *J. Pure Appl. Algebra*, **91**(1-3), 23–27.

[5] Allday, Christopher (1998). Examples of circle actions on symplectic spaces. In *Homotopy and Geometry (Warsaw, 1997)*, Volume 45 of *Banach Center Publ.*, pp. 87–90. Polish Acad. Sci., Warsaw.

[6] Allday, Christopher and Halperin, Stephen (1978). Lie group actions on spaces of finite rank. *Quart. J. Math. Oxford Ser. (2)*, **29**(113), 63–76.

[7] Allday, Christopher and Puppe, Volker (1985). *On the Localization Theorem at the Cochain Level and Free Torus Actions*, Volume 1172 of *Lecture Notes in Math.*, pp. 1–16. Springer, Berlin.

[8] Allday, Christopher and Puppe, Volker (1986). *Bounds on the Torus Rank*, Volume 1217 of *Lecture Notes in Math.*, pp. 1–10. Springer, Berlin.

[9] Allday, Christopher and Puppe, Volker (1986). On the rational homotopy Lie algebra of a fixed point set of a torus action. *Trans. Amer. Math. Soc.*, **297**(2), 521–528.

[10] Allday, Christopher and Puppe, Volker (1993). *Cohomological Methods in Transformation Groups*, Volume 32 of *Cambridge Studies in Advanced Mathematics*. Cambridge University Press, Cambridge.

[11] Allday, Christopher and Puppe, Volker (2006). The minimal Hirsch–Brown model via classical Hodge theory. *Pacific J. Math.*, **226**(1), 41–51.

[12] Anosov, Dmitri Victorovich (1982). Generic properties of closed geodesics. *Izv. Akad. Nauk SSSR Ser. Mat.*, **46**(4), 675–709, 896.

[13] Aouina, Mokhtar and Klein, John R. (2004). On the homotopy invariance of configuration spaces. *Algebr. Geom. Topol.*, **4**, 813–827.

[14] Atiyah, Michael (1957). Complex analytic connections in fiber bundles. *Trans. Amer. Math. Soc.*, **85**, 181–207.

[15] Atiyah, Michael and Bott, Raoul (1984). The moment map and equivariant cohomology. *Topology*, **23**(1), 1–28.

[16] Audin, Michèle (1991). Exemples de variétés presque complexes. *Enseign. Math. (2)*, **37**(1-2), 175–190.

[17] Babenko, Ivan K. and Taimanov, Iskander A. (2000). On nonformal simply connected symplectic manifolds. *Sibirsk. Mat. Zh.*, **41**(2), 253–269, i.

[18] Ballmann, Werner and Ziller, Wolfgang (1982). On the number of closed geodesics on a compact Riemannian manifold. *Duke Math. J.*, **49**(3), 629–632.

[19] Bangert, Victor (1993). On the existence of closed geodesics on two-spheres. *Internat. J. Math.*, **4**(1), 1–10.

[20] Bangert, Victor and Hingston, Nancy (1984). Closed geodesics on manifolds with infinite abelian fundamental group. *J. Differential Geom.*, **19**(2), 277–282.

[21] Barge, Jean (1976). Structures différentiables sur les types d'homotopie rationnelle simplement connexes. *Ann. Sci. École Norm. Sup. (4)*, **9**(4), 469–501.

[22] Belegradek, Igor and Kapovitch, Vitali (2001). Topological obstructions to nonnegative curvature. *Math. Ann.*, **320**(1), 167–190.

[23] Belegradek, Igor and Kapovitch, Vitali (2003). Obstructions to non-negative curvature and rational homotopy theory. *J. Amer. Math. Soc.*, **16**(2), 259–284.

[24] Bendersky, Martin and Gitler, Sam (1991). The cohomology of certain function spaces. *Trans. Amer. Math. Soc.*, **326**(1), 423–440.

[25] Benson, Chal and Gordon, Carolyn S. (1988). Kähler and symplectic structures on nilmanifolds. *Topology*, **27**(4), 513–518.

[26] Berceanu, Barbu, Markl, Martin, and Papadima, Ştefan. (2005). Multiplicative models for configuration spaces of algebraic varieties. *Topology*, **44**(2), 415–440.

[27] Besse, Arthur L. (1978). *Manifolds all of whose Geodesics are Closed*, Volume 93 of *Ergebnisse der Mathematik und ihrer Grenzgebiete [Results in Mathematics and Related Areas]*. Springer-Verlag, Berlin. With appendices by D. B. A. Epstein, J.-P. Bourguignon, L. Bérard-Bergery, M. Berger and J. L. Kazdan.

[28] Bishop, Richard L. and Crittenden, Richard J. (1964). *Geometry of Manifolds*. Pure and Applied Mathematics, Vol. XV. Academic Press, New York.

[29] Blanchard, André (1956). Sur les variétés analytiques complexes. *Ann. Sci. Ecole Norm. Sup. (3)*, **73**, 157–202.

[30] Block, Jonathan and Lazarev, Andrej (2005). André–Quillen cohomology and rational homotopy of function spaces. *Adv. Math.*, **193**, 18–39.

[31] Bödigheimer, Carl-Friedrich, Cohen, Frederick R., and Taylor, Laurence R. (1989). On the homology of configuration spaces. *Topology*, **28**(1), 111–123.

[32] Borel, Armand (1949-50). Exposés 12-13. Séminaire Cartan, Paris.

[33] Borel, Armand (1960). *Seminar on Transformation Groups*. With contributions by G. Bredon, E. E. Floyd, D. Montgomery, R. Palais. Annals of Mathematics Studies, No. 46. Princeton University Press, Princeton, N.J.

[34] Borel, Armand and Hirzebruch, Friedrich (1958). Characteristic classes and homogeneous spaces. I. *Amer. J. Math.*, **80**, 458–538.

[35] Borel, Armand and Serre, Jean-Pierre (1953). Groupes de Lie et puissances réduites de Steenrod. *Amer. J. Math.*, **75**, 409–448.

[36] Bott, Raoul (1956). An application of the Morse theory to the topology of Lie-groups. *Bull. Soc. Math. France*, **84**, 251–281.

[37] Bott, Raoul and Tu, Loring W. (1982). *Differential Forms in Algebraic Topology*, Volume 82 of *Graduate Texts in Mathematics*. Springer-Verlag, New York.

[38] Bousfield, Aldridge K. and Gugenheim, V. K. A. M. (1976). On PL de Rham theory and rational homotopy type. *Mem. Amer. Math. Soc.*, **8**(179), ix+94.

[39] Bousfield, Aldridge K., Peterson, Chariya, and Smith, Larry (1989). The rational homology of function spaces. *Arch. Math. (Basel)*, **52**(3), 275–283.

[40] Bredon, Glen E. (1972). *Introduction to Compact Transformation Groups*. Academic Press, New York. Pure and Applied Mathematics, Vol. 46.

[41] Bredon, Glen E. (1993). *Topology and Geometry*, Volume 139 of *Graduate Texts in Mathematics*. Springer-Verlag, New York.

[42] Brieskorn, Egbert (1973). Sur les groupes de tresses [d'après V. I. Arnol'd]. In *Séminaire Bourbaki, 24ème année (1971/1972), Exp. No. 401*, pp. 21–44. Lecture Notes in Math., Vol. 317. Springer, Berlin.

[43] Brown, Jr., Edgar H. and Szczarba, Robert H. (1997). On the rational homotopy type of function spaces. *Trans. Amer. Math. Soc.*, **349**(12), 4931–4951.
[44] Brylinski, Jean-Luc (1988). A differential complex for Poisson manifolds. *J. Differential Geom.*, **28**(1), 93–114.
[45] Buchstaber, Victor M. and Panov, Taras E. (2002). *Torus Actions and their Applications in Topology and Combinatorics*, Volume 24 of *University Lecture Series*. American Mathematical Society, Providence, RI.
[46] Buijs, Urtzi and Murillo, Aniceto (2006). Basic constructions in rational homotopy theory of function spaces. *Ann. Inst. Fourier (Grenoble)*, **56**(3), 815–838.
[47] Burghelea, Dan and Fiedorowicz, Zbigniew (1986). Cyclic homology and algebraic K-theory of spaces. II. *Topology*, **25**(3), 303–317.
[48] Cairns, Grant and Jessup, Barry (1997). New bounds on the Betti numbers of nilpotent Lie algebras. *Comm. Algebra*, **25**(2), 415–430.
[49] Calabi, Eugenio (1958). Construction and properties of some 6-dimensional almost complex manifolds. *Trans. Amer. Math. Soc.*, **87**, 407–438.
[50] Cartan, Élie (1936). *La topologie des espaces représentatifs des groupes de Lie*, Volume 358 of *Actualités Scientifiques et Industrielles*. Hermann, Paris.
[51] Cavalcanti, Gil (2007). The Lefschetz property, formality and blowing up in symplectic geometry. *Trans. Amer. Math. Soc.*, **359 no. 1**, 333–348.
[52] Cavalcanti, Gil R. (2004). New aspects of the dd^c-lemma. PhD thesis, University of Oxford, arXiv:math.DG/0501406.
[53] Chas, Moira and Sullivan, Dennis (1999). String topology. arXiv:math. GT/9911159.
[54] Cheeger, Jeff (1973). Some examples of manifolds of nonnegative curvature. *J. Differential Geometry*, **8**, 623–628.
[55] Cheeger, Jeff and Gromoll, Detlef (1971/72). The splitting theorem for manifolds of nonnegative Ricci curvature. *J. Differential Geometry*, **6**, 119–128.
[56] Cheeger, Jeff and Gromoll, Detlef (1972). On the structure of complete manifolds of nonnegative curvature. *Ann. of Math. (2)*, **96**, 413–443.
[57] Chen, Kuo-tsai (1971/72). Differential forms and homotopy groups. *J. Differential Geometry*, **6**, 231–246.
[58] Chen, Kuo Tsai (1976). Reduced bar constructions on de Rham complexes. In *Algebra, Topology, and Category Theory (A Collection of*

Papers in Honor of Samuel Eilenberg), pp. 19–32. Academic Press, New York.

[59] Chen, Kuo Tsai (1977). Iterated path integrals. *Bull. Amer. Math. Soc.*, **83**(5), 831–879.

[60] Chevalley, Claude (1999). *Theory of Lie Groups. I*, Volume 8 of *Princeton Mathematical Series*. Princeton University Press, Princeton, NJ. Fifteenth printing, Princeton Landmarks in Mathematics.

[61] Cohen, Frederick R., Lada, Thomas J., and May, J. Peter (1976). *The Homology of Iterated Loop Spaces*. Springer-Verlag, Berlin. Lecture Notes in Mathematics, Vol. 533.

[62] Cohen, Ralph L. and Jones, John D. S. (2002). A homotopy theoretic realization of string topology. *Math. Ann.*, **324**(4), 773–798.

[63] Conner, P. E. and Raymond, Frank (1970). Actions of compact Lie groups on aspherical manifolds. In *Topology of Manifolds (Proc. Inst., Univ. of Georgia, Athens, Ga., 1969)*, pp. 227–264. Markham, Chicago, Ill.

[64] Cordero, Luis A., Fernández, Marisa, and Gray, Alfred (1986). Symplectic manifolds with no Kähler structure. *Topology*, **25**(3), 375–380.

[65] Cordero, Luis A., Fernández, Marisa, and Gray, Alfred (1987). The Frölicher spectral sequence and complex compact nilmanifolds. *C. R. Acad. Sci. Paris Sér. I Math.*, **305**(17), 753–756.

[66] Cornea, Octav, Lupton, Gregory, Oprea, John, and Tanré, Daniel (2003). *Lusternik–Schnirelmann Category*, Volume 103 of *Mathematical Surveys and Monographs*. American Mathematical Society, Providence, RI.

[67] Datta, Basudeb and Subramanian, Swaminathan (1990). Nonexistence of almost complex structures on products of even-dimensional spheres. *Topology Appl.*, **36**(1), 39–42.

[68] Debongnie, Gery (2007). Rational homotopy type of subspace arrangements. arXiv:0705.1449v1 [math.AT].

[69] Deligne, Pierre (1971). Théorie de Hodge. II. *Inst. Hautes Études Sci. Publ. Math.*, **40**, 5–57.

[70] Deligne, Pierre (1974). Théorie de Hodge. III. *Inst. Hautes Études Sci. Publ. Math.*, **44**, 5–77.

[71] Deligne, Pierre, Griffiths, Phillip, Morgan, John, and Sullivan, Dennis (1975). Real homotopy theory of Kähler manifolds. *Invent. Math.*, **29**(3), 245–274.

[72] Denham, Graham and Suciu, Alexander I. (2007). Moment-angle complexes, monomial ideals, and Massey products. *Pure Appl. Math. Q.*, **3**, 25 – 60.

[73] Deninger, Christopher and Singhof, Wilhelm (1988). On the cohomology of nilpotent Lie algebras. *Bull. Soc. Math. France*, **116**(1), 3–14.

[74] Dold, Albrecht (1963). Partitions of unity in the theory of fibrations. *Ann. of Math. (2)*, **78**, 223–255.

[75] Dwyer, William G. and Wilkerson, Clarence W., Jr. (1998). The elementary geometric structure of compact Lie groups. *Bull. London Math. Soc.*, **30**(4), 337–364.

[76] Eckmann, Beno and Frölicher, Alfred (1951). Sur l'intégrabilité des structures presque complexes. *C. R. Acad. Sci. Paris*, **232**, 2284–2286.

[77] Eells, Jr., James (1966). A setting for global analysis. *Bull. Amer. Math. Soc.*, **72**, 751–807.

[78] Ehresmann, Charles (1952). Sur les variétés presque complexes. In *Proceedings of the International Congress of Mathematicians, Cambridge, Mass., 1950, vol. 2*, Providence, R. I., pp. 412–419. Amer. Math. Soc.

[79] Ehresmann, Charles and Libermann, Paulette (1951). Sur les structures presque hermitiennes isotropes. *C. R. Acad. Sci. Paris*, **232**, 1281–1283.

[80] Ekedahl, Torsten (1986). Two examples of smooth projective varieties with nonzero Massey products. In *Algebra, Algebraic Topology and their Interactions (Stockholm, 1983)*, Volume 1183 of *Lecture Notes in Math.*, pp. 128–132. Springer, Berlin.

[81] Eschenburg, Jost-Hinrich (1982). New examples of manifolds with strictly positive curvature. *Invent. Math.*, **66**(3), 469–480.

[82] Eschenburg, Jost-Hinrich (1992). Cohomology of biquotients. *Manuscripta Math.*, **75**(2), 151–166.

[83] Fadell, Edward and Neuwirth, Lee (1962). Configuration spaces. *Math. Scand.*, **10**, 111–118.

[84] Fang, Fuquan and Rong, Xiaochun (2001). Curvature, diameter, homotopy groups, and cohomology rings. *Duke Math. J.*, **107**(1), 135–158.

[85] Feichtner, Eva Maria and Yuzvinsky, Sergey (2005). Formality of the complements of subspace arrangements with geometric lattices. *Zap. Nauchn. Sem. S.-Peterburg. Otdel. Mat. Inst. Steklov. (POMI)*, **326** (Teor. Predst. Din. Sist. Komb. i Algoritm. Metody. 13), 235–247, 284.

[86] Félix, Yves and Halperin, Stephen (1982). Rational LS category and its applications. *Trans. Amer. Math. Soc.*, **273**(1), 1–38.

[87] Félix, Yves, Halperin, Stephen, and Thomas, Jean-Claude (2001). *Rational Homotopy Theory*, Volume 205 of *Graduate Texts in Mathematics*. Springer-Verlag, New York.

[88] Félix, Yves and Tanré, Daniel (2005). The cohomology algebra of unordered configuration spaces. *J. London Math. Soc. (2)*, **72**(2), 525–544.

[89] Félix, Yves and Tanré, Daniel (2005). H-space structure on pointed mapping spaces. *Algebr. Geom. Topol.*, **5**, 713–724 (electronic).

[90] Félix, Yves and Thomas, Jean-Claude (1994). Homologie des espaces de lacets des espaces de configuration. *Ann. Inst. Fourier (Grenoble)*, **44**(2), 559–568.

[91] Félix, Yves and Thomas, Jean-Claude (2000). Rational Betti numbers of configuration spaces. *Topology Appl.*, **102**(2), 139–149.

[92] Félix, Yves and Thomas, Jean-Claude (2004). Configuration spaces and Massey products. *Int. Math. Res. Not.*, **33**, 1685–1702.

[93] Félix, Yves, Thomas, Jean-Claude, and Vigué-Poirrier, Micheline (2004). The Hochschild cohomology of a closed manifold. *Publ. Math. Inst. Hautes Études Sci.*, **99**, 235–252.

[94] Félix, Yves, Thomas, Jean-Claude, and Vigué-Poirrier, Micheline (2007). Rational string topology. *J. Eur. Math. Soc. (JEMS)*, **9**(1), 123–156.

[95] Fernández, Marisa, de León, Manuel, and Saralegui, Martín (1996). A six-dimensional compact symplectic solvmanifold without Kähler structures. *Osaka J. Math.*, **33**(1), 19–35.

[96] Fernández, Marisa and Gray, Alfred (1990). Compact symplectic solvmanifolds not admitting complex structures. *Geom. Dedicata*, **34**(3), 295–299.

[97] Fernández, Marisa and Muñoz, Vicente (2005). Formality of Donaldson submanifolds. *Math. Z.*, **250**(1), 149–175.

[98] Fine, Benjamin L. and Triantafillou, Georgia (1993). On the equivariant formality of Kähler manifolds with finite group action. *Canad. J. Math.*, **45**(6), 1200–1210.

[99] Franks, John (1992). Geodesics on S^2 and periodic points of annulus homeomorphisms. *Invent. Math.*, **108**(2), 403–418.

[100] Frauenfelder, Urs and Schlenk, Felix (2006). Fiberwise volume growth via Lagrangian intersections. *J. Symplectic Geom.*, **4**(2), 117–148.

[101] Friedlander, John and Halperin, Stephen (1979). Distinct representatives, varieties and rational homotopy. *J. Number Theory*, **11** (3 S. Chowla Anniversary Issue), 321–323.

[102] Frölicher, Alfred (1955). Zur Differentialgeometrie der komplexen Strukturen. *Math. Ann.*, **129**, 50–95.

[103] Fulton, William and Harris, Joe (1991). *Representation Theory*, Volume 129 of *Graduate Texts in Mathematics*. Springer-Verlag, New York.

[104] Gallot, Sylvestre, Hulin, Dominique, and Lafontaine, Jacques (1987). *Riemannian Geometry*. Universitext. Springer-Verlag, Berlin.

[105] Getzler, Ezra, Jones, John D. S., and Petrack, Scott (1991). Differential forms on loop spaces and the cyclic bar complex. *Topology*, **30**(3), 339–371.

[106] Godbillon, Claude (1974). Cohomologies d'algèbres de Lie de champs de vecteurs formels. In *Séminaire Bourbaki, 25ème année (1972/1973), Exp. No. 421*, pp. 69–87. Lecture Notes in Math., Vol. 383. Springer, Berlin.

[107] Goldberg, Samuel I. (1962). *Curvature and Homology*. Pure and Applied Mathematics, Vol. XI. Academic Press, New York.

[108] Gompf, Robert E. (1995). A new construction of symplectic manifolds. *Ann. of Math. (2)*, **142**(3), 527–595.

[109] Goodwillie, Thomas G. (1985). Cyclic homology, derivations, and the free loopspace. *Topology*, **24**(2), 187–215.

[110] Goresky, Mark and Hingston, Nancy (2007). Loop products and closed geodesics. arXiv:0707.3486v [math.AT].

[111] Goyo, John O. (1989). The Sullivan model of the homotopy-fixed-point set. PhD thesis, University of Toronto.

[112] Greub, Werner, Halperin, Stephen, and Vanstone, Ray (1972). *Connections, Curvature, and Cohomology. Vol. I: De Rham Cohomology of Manifolds and Vector Bundles*. Academic Press, New York. Pure and Applied Mathematics, Vol. 47-49.

[113] Greub, Werner, Halperin, Stephen, and Vanstone, Ray (1973). *Connections, Curvature, and Cohomology. Vol. II: Lie Groups, Principal Bundles, and Characteristic Classes*. Academic Press [Harcourt Brace Jovanovich, Publishers], New York. Pure and Applied Mathematics, Vol. 47-II.

[114] Greub, Werner, Halperin, Stephen, and Vanstone, Ray (1976). *Connections, Curvature, and Cohomology. Vol. III: Cohomology of Principal Bundles and Homogeneous Spaces*. Academic Press [Harcourt Brace Jovanovich Publishers], New York. Pure and Applied Mathematics, Vol. 47-III.

[115] Griffiths, Phillip and Harris, Joseph (1978). *Principles of Algebraic Geometry*. Wiley-Interscience [John Wiley & Sons], New York. Pure and Applied Mathematics.

[116] Griffiths, Phillip A. and Morgan, John W. (1981). *Rational Homotopy Theory and Differential Forms*, Volume 16 of *Progress in Mathematics*. Birkhäuser Boston, Mass.

[117] Gromoll, Detlef and Meyer, Wolfgang (1969). Periodic geodesics on compact riemannian manifolds. *J. Differential Geometry*, **3**, 493–510.
[118] Gromoll, Detlef and Meyer, Wolfgang (1974). An exotic sphere with nonnegative sectional curvature. *Ann. of Math. (2)*, **100**, 401–406.
[119] Gromov, Mikhael (1978). Homotopical effects of dilatation. *J. Differential Geom.*, **13**(3), 303–310.
[120] Grove, Karsten (1973). Condition (C) for the energy integral on certain path spaces and applications to the theory of geodesics. *J. Differential Geometry*, **8**, 207–223.
[121] Grove, Karsten (1974). Isometry-invariant geodesics. *Topology*, **13**, 281–292.
[122] Grove, Karsten and Halperin, Stephen (1982). Contributions of rational homotopy theory to global problems in geometry. *Inst. Hautes Études Sci. Publ. Math.*, **56**, 171–177 (1983).
[123] Grove, Karsten, Halperin, Stephen, and Vigué-Poirrier, Micheline (1978). The rational homotopy theory of certain path spaces with applications to geodesics. *Acta Math.*, **140**(3-4), 277–303.
[124] Gualtieri, Marco (2003). Generalized complex geometry. PhD thesis, University of Oxford, arXiv:math.DG/0401221.
[125] Haefliger, André (1976). Sur la cohomologie de l'algèbre de Lie des champs de vecteurs. *Ann. Sci. École Norm. Sup. (4)*, **9**(4), 503–532.
[126] Haefliger, André (1982). Rational homotopy of the space of sections of a nilpotent bundle. *Trans. Amer. Math. Soc.*, **273**(2), 609–620.
[127] Hain, Richard M. (1984). Iterated integrals and homotopy periods. *Mem. Amer. Math. Soc.*, **47**(291), iv+98.
[128] Hain, Richard M. (1987). The de Rham homotopy theory of complex algebraic varieties. I. *K-Theory*, **1**(3), 271–324.
[129] Halperin, Stephen (1977). Finiteness in the minimal models of Sullivan. *Trans. Amer. Math. Soc.*, **230**, 173–199.
[130] Halperin, Stephen (1983). Lectures on minimal models. *Mém. Soc. Math. France (N.S.)*, **9-10**, 261.
[131] Halperin, Stephen (1985). Rational homotopy and torus actions. In *Aspects of Topology*, Volume 93 of *London Math. Soc. Lecture Note Ser.*, Cambridge, pp. 293–306. Cambridge Univ. Press.
[132] Halperin, Stephen and Stasheff, James (1979). Obstructions to homotopy equivalences. *Adv. in Math.*, **32**(3), 233–279.
[133] Halperin, Stephen and Tanré, Daniel (1990). Homotopie filtrée et fibrés C^∞. *Illinois J. Math.*, **34**(2), 284–324.
[134] Hasegawa, Keizo (1989). Minimal models of nilmanifolds. *Proc. Amer. Math. Soc.*, **106**(1), 65–71.

[135] Helgason, Sigurður (1962). *Differential Geometry and Symmetric Spaces*. Pure and Applied Mathematics, Vol. XII. Academic Press, New York.

[136] Helgason, Sigurður (2001). *Differential Geometry, Lie Groups, and Symmetric Spaces*, Volume 34 of *Graduate Studies in Mathematics*. American Mathematical Society, Providence, RI. Corrected reprint of the 1978 original.

[137] Hilton, Peter (1965). *Homotopy Theory and Duality*. Gordon and Breach Science Publishers, New York.

[138] Hilton, Peter, Mislin, Guido, and Roitberg, Joe (1975). *Localization of Nilpotent Groups and Spaces*. North-Holland Publishing Co., Amsterdam. North-Holland Mathematics Studies, No. 15, Notas de Matemática, No. 55. [Notes on Mathematics, No. 55].

[139] Hilton, Peter and Stammbach, Urs (1997). *A Course in Homological Algebra* (Second edn), Volume 4 of *Graduate Texts in Mathematics*. Springer-Verlag, New York.

[140] Hingston, Nancy (1993). On the growth of the number of closed geodesics on the two-sphere. *Internat. Math. Res. Notices*, **9**, 253–262.

[141] Hirzebruch, Friedrich (1966). *Topological Methods in Algebraic Geometry*. Third enlarged edition. New appendix and translation from the second German edition by R. L. E. Schwarzenberger, with an additional section by A. Borel. Die Grundlehren der Mathematischen Wissenschaften, Band 131. Springer-Verlag New York.

[142] Hitchin, Nigel (2003). Generalized Calabi–Yau manifolds. *Quarterly Journal Math.*, **54**(3), 281–308.

[143] Hopf, Heinz (1948). Zur Topologie der komplexen Mannigfaltigkeiten. In *Studies and Essays Presented to R. Courant on his 60th Birthday, January 8, 1948*, pp. 167–185. Interscience Publishers, Inc., New York.

[144] Hsiang, Wu-Yi (1975). *Cohomology Theory of Topological Transformation Groups*. Springer-Verlag, New York. Ergebnisse der Mathematik und ihrer Grenzgebiete, Band 85.

[145] Husemoller, Dale (1994). *Fiber Bundles* (Third edn), Volume 20 of *Graduate Texts in Mathematics*. Springer-Verlag, New York.

[146] Ibáñez, Raoul, Kędra, Jarek, Rudyak, Yuli, and Tralle, Aleksy (2004). On fundamental groups of symplectically aspherical manifolds. *Math. Z.*, **248**(4), 805–826.

[147] Iwasawa, Kenkichi (1949). On some types of topological groups. *Ann. of Math. (2)*, **50**, 507–558.

[148] Jessup, Barry and Lupton, Gregory (2004). Free torus actions and two-stage spaces. *Math. Proc. Cambridge Philos. Soc.*, **137**(1), 191–207.

[149] Jones, John D. S. (1987). Cyclic homology and equivariant homology. *Invent. Math.*, **87**(2), 403–423.

[150] Kapovitch, Vitali (2004). A note on rational homotopy of biquotients. Preprint.

[151] Kapovitch, Vitali and Ziller, Wolfgang (2004). Biquotients with singly generated rational cohomology. *Geom. Dedicata*, **104**, 149–160.

[152] Kawakubo, Katsuo (1991). *The Theory of Transformation Groups* (Japanese edn). The Clarendon Press Oxford University Press, New York.

[153] Kędra, Jarosław (2000). KS-models and symplectic structures on total spaces of bundles. *Bull. Belg. Math. Soc. Simon Stevin*, **7**(3), 377–385.

[154] Klingenberg, Wilhelm (1978). *Lectures on Closed Geodesics*. Springer-Verlag, Berlin. Grundlehren der Mathematischen Wissenschaften, Vol. 230.

[155] Kobayashi, Shoshichi and Nomizu, Katsumi (1969). *Foundations of Differential Geometry. Vol. II*. Interscience Tracts in Pure and Applied Mathematics, No. 15 Vol. II. Interscience Publishers John Wiley & Sons, Inc., New York.

[156] Kontsevich, Maxim (1999). Operads and motives in deformation quantization. *Lett. Math. Phys.*, **48**(1), 35–72. Moshé Flato (1937–1998).

[157] Koszul, Jean-Louis (1985). Crochet de Schouten-Nijenhuis et cohomologie. *Astérisque*, Numero Hors Serie, 257–271. The mathematical heritage of Élie Cartan (Lyon, 1984).

[158] Kotani, Yasusuke (2004). Note on the rational cohomology of the function space of based maps. *Homology Homotopy Appl.*, **6**(1), 341–350 (electronic).

[159] Kreck, M. and Schafer, J. A. (1984). Classification and stable classification of manifolds: some examples. *Comment. Math. Helv.*, **59**(1), 12–38.

[160] Kriz, Igor (1994). On the rational homotopy type of configuration spaces. *Ann. of Math. (2)*, **139**(2), 227–237.

[161] Lalonde, François and McDuff, Dusa (2003). Symplectic structures on fiber bundles. *Topology*, **42**(2), 309–347.

[162] Lalonde, François, McDuff, Dusa, and Polterovich, Leonid (1998). On the flux conjectures. In *Geometry, Topology, and Dynamics*

(Montreal, PQ, 1995), Volume 15 of CRM Proc. Lecture Notes, pp. 69–85. Amer. Math. Soc., Providence, RI.
[163] Lambre, Thierry (1991). Modèle minimal équivariant et formalité. Trans. Amer. Math. Soc., **327**(2), 621–639.
[164] Lambrechts, Pascal (2001). The Betti numbers of the free loop space of a connected sum. J. London Math. Soc. (2), **64**(1), 205–228.
[165] Lambrechts, Pascal and Stanley, Don (2004). The rational homotopy type of a blow-up in the stable case. arXiv:0706.1363v1 [math.AT].
[166] Lambrechts, Pascal and Stanley, Don (2004). The rational homotopy type of configuration spaces of two points. Ann. Inst. Fourier (Grenoble), **54**(4), 1029–1052.
[167] Lambrechts, Pascal and Stanley, Don (2005). Algebraic models of Poincaré embeddings. Algebr. Geom. Topol., **5**, 135–182.
[168] Lambrechts, Pascal and Stanley, Don (2005). Examples of rational homotopy types of blow-ups. Proc. Amer. Math. Soc., **133**(12), 3713–3719.
[169] Lambrechts, Pascal and Stanley, Don (2005). Poincaré duality and commutative differential graded algebras. arXiv:math/0701309v2 [math.AT].
[170] Lehmann, Daniel (1977). *Théorie homotopique des formes différentielles (d'après D. Sullivan)*. Société Mathématique de France, Paris. Astérisque, No. 45.
[171] Lemaire, Jean-Michel (1974). *Algèbres connexes et homologie des espaces de lacets*. Springer-Verlag, Berlin. Lecture Notes in Mathematics, Vol. 422.
[172] Levitt, Norman (1995). Spaces of arcs and configuration spaces of manifolds. Topology, **34**(1), 217–230.
[173] Libermann, Paulette (1954). Sur le problème d'équivalence de certaines structures infinitésimales. Ann. Mat. Pura Appl. (4), **36**, 27–120.
[174] Lichnerowicz, André (1977). Les variétés de Poisson et leurs algèbres de Lie associées. J. Differential Geometry, **12**(2), 253–300.
[175] Lin, Yi and Sjamaar, Reyer (2004). Equivariant symplectic Hodge theory and the $d_G\delta$-lemma. J. Symplectic Geom., **2**(2), 267–278.
[176] Longoni, Riccardo and Salvatore, Paolo (2005). Configuration spaces are not homotopy invariant. Topology, **44**(2), 375–380.
[177] Lupton, Gregory (1990). Intrinsic formality and certain types of algebras. Trans. Amer. Math. Soc., **319**(1), 257–283.
[178] Lupton, Gregory and Oprea, John (1994). Symplectic manifolds and formality. J. Pure Appl. Algebra, **91**(1-3), 193–207.

[179] Lupton, Gregory and Oprea, John (1995). Cohomologically symplectic spaces: toral actions and the Gottlieb group. *Trans. Amer. Math. Soc.*, **347**(1), 261–288.

[180] Lupton, Gregory and Smith, Samuel B. (2007). Rationalized evaluation subgroups of a map i: Sullivan models, derivations and G-sequences. *Journal of Pure and Applied Algebra*, **209**, 159–171.

[181] Lyusternik, Lazar Aronovich and Fet, Abram Il'ich (1951). Variational problems on closed manifolds. *Doklady Akad. Nauk SSSR (N.S.)*, **81**, 17–18.

[182] Mac Lane, Saunders (1995). *Homology*. Classics in Mathematics. Springer-Verlag, Berlin. Reprint of the 1975 edition.

[183] Malcev, Anatoli Ivanovich (1962). On a class of homogeneous spaces. *Amer. Math. Soc. Translations Series 1*, **9**, 276–307.

[184] Mather, Michael (1976). Pull-backs in homotopy theory. *Canad. J. Math.*, **28**(2), 225–263.

[185] Mathieu, Olivier (1995). Harmonic cohomology classes of symplectic manifolds. *Comment. Math. Helv.*, **70**(1), 1–9.

[186] McCleary, John (1985). *User's Guide to Spectral Sequences*, Volume 12 of *Mathematics Lecture Series*. Publish or Perish Inc., Wilmington, DE.

[187] McDuff, Dusa (1975). Configuration spaces of positive and negative particles. *Topology*, **14**, 91–107.

[188] McDuff, Dusa (1984). Examples of simply-connected symplectic non-Kählerian manifolds. *J. Differential Geom.*, **20**(1), 267–277.

[189] McDuff, Dusa and Salamon, Dietmar (1998). *Introduction to Symplectic Topology* (Second edn). Oxford Mathematical Monographs. The Clarendon Press Oxford University Press, New York.

[190] Merkulov, Sergei (1998). Formality of canonical symplectic complexes and Frobenius manifolds. *Internat. Math. Res. Notices*, **14**, 727–733.

[191] Merkulov, Sergei (2004). De Rham model for string topology. *Int. Math. Res. Not.*, **55**, 2955–2981.

[192] Miller, Timothy James (1979). On the formality of $(k-1)$-connected compact manifolds of dimension less than or equal to $4k-2$. *Illinois J. Math.*, **23**(2), 253–258.

[193] Milnor, John (1956). Construction of universal bundles. II. *Ann. of Math. (2)*, **63**, 430–436.

[194] Milnor, John (1958). On the existence of a connection with curvature zero. *Comment. Math. Helv.*, **32**, 215–223.

[195] Milnor, John (1963). *Morse Theory*. Based on lecture notes by M. Spivak and R. Wells. Annals of Mathematics Studies, No. 51. Princeton University Press, Princeton, N.J.

[196] Milnor, John (1968). A note on curvature and fundamental group. *J. Differential Geometry*, **2**, 1–7.

[197] Milnor, John W. and Moore, John C. (1965). On the structure of Hopf algebras. *Ann. of Math. (2)*, **81**, 211–264.

[198] Milnor, John W. and Stasheff, James D. (1974). *Characteristic Classes*. Princeton University Press, Princeton, N. J. Annals of Mathematics Studies, No. 76.

[199] Mimura, Mamoru and Toda, Hirosi (1991). *Topology of Lie Groups. I, II*, Volume 91 of *Translations of Mathematical Monographs*. American Mathematical Society, Providence, RI. Translated from the 1978 Japanese edition by the authors.

[200] Moore, John (1954-55). Exposés 22. Séminaire Cartan, Paris.

[201] Morgan, John W. (1978). The algebraic topology of smooth algebraic varieties. *Inst. Hautes Études Sci. Publ. Math.*, **48**, 137–204.

[202] Morgan, John W. (1986). Correction to: "The algebraic topology of smooth algebraic varieties" [Inst. Hautes Études Sci. Publ. Math. No. 48 (1978), 137–204; MR0516917 (80e:55020)]. *Inst. Hautes Études Sci. Publ. Math.*, **64**, 185.

[203] Møller, Jesper Michael (1987). Nilpotent spaces of sections. *Trans. Amer. Math. Soc.*, **303**(2), 733–741.

[204] Møller, Jesper Michael (2002). Rational equivalences between classifying spaces. *Math. Z.*, **241**(4), 761–799.

[205] Myers, Sumner Byron and Steenrod, Norman E. (1939). The group of isometries of a Riemannian manifold. *Ann. of Math. (2)*, **40**(2), 400–416.

[206] Neisendorfer, Joseph and Miller, Timothy (1978). Formal and coformal spaces. *Illinois J. Math.*, **22**(4), 565–580.

[207] Neisendorfer, Joseph and Taylor, Laurence (1978). Dolbeault homotopy theory. *Trans. Amer. Math. Soc.*, **245**, 183–210.

[208] Newlander, August and Nirenberg, Louis (1957). Complex analytic coordinates in almost complex manifolds. *Ann. of Math. (2)*, **65**, 391–404.

[209] Nomizu, Katsumi (1954). On the cohomology of compact homogeneous spaces of nilpotent Lie groups. *Ann. of Math. (2)*, **59**, 531–538.

[210] Notbohm, Dietrich (1995). Classifying spaces of compact Lie groups and finite loop spaces. In *Handbook of Algebraic Topology*, pp. 1049–1094. North-Holland, Amsterdam.

[211] Notbohm, Dietrich (1995). On the "classifying space" functor for compact Lie groups. *J. London Math. Soc. (2)*, **52**(1), 185–198.

[212] Notbohm, Dietrich and Ray, Nigel (2005). On Davis–Januszkiewicz homotopy types. I. Formality and rationalisation. *Algebr. Geom. Topol.*, **5**, 31–51.

[213] O'Neill, Barrett (1966). The fundamental equations of a submersion. *Michigan Math. J.*, **13**, 459–469.

[214] Onishchik, Arkady L. and Vinberg, Èrnest B. (1993). Foundations of Lie theory. In *Lie Groups and Lie Algebras, I*, Volume 20 of *Encyclopaedia Math. Sci.*, pp. 1–94, 231–235. Springer, Berlin.

[215] Oprea, John (1984). Lifting homotopy actions in rational homotopy theory. *J. Pure Appl. Algebra*, **32**(2), 177–190.

[216] Oprea, John (2004). *Differential Geometry and Its Applications, 2nd edn*. Prentice Hall, England U.K.

[217] Oprea, John and Tanré, Daniel (2005). Flat circle bundles, pullbacks, and the circle made discrete. *Int. J. Math. Math. Sci.*, **21**, 3487–3495.

[218] Osse, Akimou (1992). Sur le classifiant d'un groupe de Lie compact connexe. *C. R. Acad. Sci. Paris Sér. I Math.*, **315**(7), 833–838.

[219] Özaydin, Murad and Walschap, Gerard (1994). Vector bundles with no soul. *Proc. Amer. Math. Soc.*, **120**(2), 565–567.

[220] Pan, Jian Zhong and Wu, Shao Bing (2006). Rational homotopy theory and nonnegative curvature. *Acta Math. Sin. (Engl. Ser.)*, **22**(1), 23–26.

[221] Papadima, Ştefan (1986). Rigidity properties of compact Lie groups modulo maximal tori. *Math. Ann.*, **275**(4), 637–652.

[222] Papadima, Ştefan (1988). Rational homotopy equivalences of Lie type. *Math. Proc. Cambridge Philos. Soc.*, **104**(1), 65–80.

[223] Paternain, Gabriel P. (1992). On the topology of manifolds with completely integrable geodesic flows. *Ergodic Theory Dynam. Systems*, **12**(1), 109–121.

[224] Paternain, Gabriel P. (1999). *Geodesic Flows*, Volume 180 of *Progress in Mathematics*. Birkhäuser Boston Inc., Boston, MA.

[225] Paternain, Gabriel P. and Paternain, Miguel (1994). Topological entropy versus geodesic entropy. *Internat. J. Math.*, **5**(2), 213–218.

[226] Petersen, Peter (1998). *Riemannian Geometry*, Volume 171 of *Graduate Texts in Mathematics*. Springer-Verlag, New York.

[227] Pittie, Harsh V. (1988). The Dolbeault-cohomology ring of a compact, even- dimensional Lie group. *Proc. Indian Acad. Sci. Math. Sci.*, **98**(2-3), 117–152.

[228] Pittie, Harsh V. (1989). The nondegeneration of the Hodge-de Rham spectral sequence. *Bull. Amer. Math. Soc. (N.S.)*, **20**(1), 19–22.

[229] Postnikov, Mikhail Mikhailovich (1990). *Leçons de géométrie.* Traduit du Russe: Mathématiques. [Translations of Russian Works: Mathematics]. Éditions Mir, Moscow. Géométrie différentielle. [Differential geometry], Translated from the Russian by Irina Petrova.
[230] Rudyak, Yuli and Tralle, Aleksy (2000). On Thom spaces, Massey products, and nonformal symplectic manifolds. *Internat. Math. Res. Notices*, **10**, 495–513.
[231] Scheerer, Hans (1968). Homotopieäquivalente kompakte Liesche Gruppen. *Topology*, **7**, 227–232.
[232] Scheerer, Hans and Tanré, Daniel (1992). Homotopie modérée et tempérée avec les coalgèbres. Applications aux espaces fonctionnels. *Arch. Math. (Basel)*, **59**(2), 130–145.
[233] Schultz, Reinhard (1971). The nonexistence of free S^1 actions on some homotopy spheres. *Proc. Amer. Math. Soc.*, **27**, 595–597.
[234] Scorpan, Alexandru (2005). *The Wild World of 4-Manifolds.* American Mathematical Society, Providence, RI.
[235] Scull, Laura (2004). Equivariant formality for actions of torus groups. *Canad. J. Math.*, **56**(6), 1290–1307.
[236] Serre, Jean-Pierre (1951). Homologie singulière des espaces fibrés. Applications. *Ann. of Math. (2)*, **54**, 425–505.
[237] Shiga, Hiroo and Tezuka, Michishige (1987). Rational fibrations, homogeneous spaces with positive Euler characteristics and Jacobians. *Ann. Inst. Fourier (Grenoble)*, **37**(1), 81–106.
[238] Singhof, Wilhelm (1993). On the topology of double coset manifolds. *Math. Ann.*, **297**(1), 133–146.
[239] Smale, Stephen (1959). The classification of immersions of spheres in Euclidean spaces. *Ann. of Math. (2)*, **69**, 327–344.
[240] Spanier, Edwin H. (1966). *Algebraic Topology.* McGraw-Hill, New York.
[241] Stasheff, James (1983). Rational Poincaré duality spaces. *Illinois J. Math.*, **27**(1), 104–109.
[242] Steenrod, Norman (1999). *The topology of fiber bundles.* Princeton Landmarks in Mathematics. Princeton University Press, Princeton, NJ. Reprint of the 1957 edition, Princeton Paperbacks.
[243] Stępień, Zofia (2006). The Lalonde–McDuff conjecture for nilmanifolds. arXiv:math/0608214v1 [math.SG].
[244] Stong, Robert E. (1965). Relations among characteristic numbers. I. *Topology*, **4**, 267–281.
[245] Sullivan, Dennis (1975). Differential forms and the topology of manifolds. In *Manifolds—Tokyo 1973 (Proc. Internat. Conf., Tokyo, 1973)*, pp. 37–49. Univ. Tokyo Press, Tokyo.

[246] Sullivan, Dennis (1977). Infinitesimal computations in topology. *Inst. Hautes Études Sci. Publ. Math.*, **47**, 269–331 (1978).

[247] Tanaka, Minoru (1982). On the existence of infinitely many isometry-invariant geodesics. *J. Differential Geom.*, **17**(2), 171–184.

[248] Tanré, Daniel (1983). *Homotopie rationnelle: modèles de Chen, Quillen, Sullivan*, Volume 1025 of *Lecture Notes in Mathematics*. Springer-Verlag, Berlin.

[249] Tanré, Daniel (1994). Modèle de Dolbeault et fibré holomorphe. *J. Pure Appl. Algebra*, **91**(1-3), 333–345.

[250] Thomas, Jean-Claude (1981). Rational homotopy of Serre fibrations. *Ann. Inst. Fourier (Grenoble)*, **31**(3), v, 71–90.

[251] Tischler, David (1977). Closed 2-forms and an embedding theorem for symplectic manifolds. *J. Differential Geometry*, **12**(2), 229–235.

[252] tom Dieck, Tammo (1987). *Transformation Groups*, Volume 8 of *de Gruyter Studies in Mathematics*. Walter de Gruyter & Co., Berlin.

[253] Toomer, Graham Hilton (1975). Topological localization, category and cocategory. *Canad. J. Math.*, **27**, 319–322.

[254] Totaro, Burt (1996). Configuration spaces of algebraic varieties. *Topology*, **35**(4), 1057–1067.

[255] Totaro, Burt (2002). Cheeger manifolds and the classification of biquotients. *J. Differential Geom.*, **61**(3), 397–451.

[256] Totaro, Burt (2003). Curvature, diameter, and quotient manifolds. *Math. Res. Lett.*, **10**(2-3), 191–203.

[257] Tralle, Aleksy and Oprea, John (1997). *Symplectic Manifolds with no Kähler Structure*, Volume 1661 of *Lecture Notes in Mathematics*. Springer-Verlag, Berlin.

[258] Triantafillou, Georgia (1978). *Äquivariante rationale Homotopietheorie*. Bonner Mathematische Schriften [Bonn Mathematical Publications], 110. Universität Bonn Mathematisches Institut, Bonn. Dissertation, Rheinische Friedrich-Wilhelms-Universität Bonn, Bonn, 1977.

[259] Tuschmann, Wilderich (2002). Geometric diffeomorphism finiteness in low dimensions and homotopy group finiteness. *Math. Ann.*, **322**(2), 413–420.

[260] Vigué-Poirrier, Micheline (1986). Cohomologie de l'espace des sections d'un fibré et cohomologie de Gelfand-Fuchs d'une variété. In *Algebra, Algebraic Topology and their Interactions (Stockholm, 1983)*, Volume 1183 of *Lecture Notes in Math.*, pp. 371–396. Springer, Berlin.

[261] Vigué-Poirrier, Micheline and Sullivan, Dennis (1976). The homology theory of the closed geodesic problem. *J. Differential Geometry*, **11**(4), 633–644.
[262] Warner, Frank W. (1971). *Foundations of Differentiable Manifolds and Lie Groups*. Scott, Foresman and Co., Glenview, Ill.-London.
[263] Weingram, Stephen (1971). On the incompressibility of certain maps. *Ann. of Math. (2)*, **93**, 476–485.
[264] Wells, Raymond O., Jr. (1980). *Differential Analysis on Complex Manifolds* (Second edn), Volume 65 of *Graduate Texts in Mathematics*. Springer-Verlag, New York.
[265] Whitehead, George W. (1978). *Elements of Homotopy Theory*, Volume 61 of *Graduate Texts in Mathematics*. Springer-Verlag, New York.
[266] Yan, Dong (1996). Hodge structure on symplectic manifolds. *Adv. Math.*, **120**(1), 143–154.
[267] Yuzvinsky, Sergey (1999). Rational model of subspace complement on atomic complex. *Publ. Inst. Math. (Beograd) (N.S.)*, **66**(80), 157–164. Geometric combinatorics (Kotor, 1998).
[268] Zeeman, Eric Christopher (1957). A proof of the comparison theorem for spectral sequences. *Proc. Cambridge Philos. Soc.*, **53**, 57–62.

Index

$(B(A, A, A), D)$, 234
A_N, 90
$A_{DR}(\mathcal{P}_{*,*}(M))$, 383
$A_{DR}(\mathcal{P}_{*,a}(M))$, 383
$A_{DR}(\mathcal{P}_{\text{free}}(M))$, 378
A_{PL}, 67
   construction of, 70
D_A, 328, 355
$F(\mathbb{R}^n, k)$, 354
   are formal, 354
$G(x)$, 274
G_x, 274
M^G, 274
$Sh(k, l)$, 381
Top_N, 90
Δ, 406
$\mathcal{L}$, 399
$\mathcal{M}^{\leq D}_{\kappa \leq \sec \leq \lambda}(n)$, 242
$\mathbb{Q}[V]$, 59
$\mathcal{P}_{*,*}(M)$, 383
$\mathcal{P}_{*,a}(M)$, 383
$\mathcal{P}_{\text{free}}(M)$, 377
$\mathfrak{Chen}\,\mathcal{P}_{\text{free}}(M)$, 379
$\mathfrak{Chen}\,\mathcal{P}_{*,*}(M)$, 384
$\mathfrak{Chen}\,\mathcal{P}_{\text{free loop}}(M)$, 385
δ, 405
λ-derivation, 307
$\wedge V$, 58, 59
$\mathcal{CM}$, 141
rk_0, 272
d_1, 77
d_k, 77
n_T, 223
$\mathcal{B}l(A, B)$, 336

abelian
   Lie algebra, 8
   Lie group, 8
action, 273
   almost free, 271, 274, 316
   cohomologically free, 308
   conjugation, 7
   effective, 11
   fixed point of, 274
   free, 11, 274
   Hamiltonian, 308
   of π_1 on π_n, 31
   of Lie group, 11
   orbit of, 274
   smooth, 274
   space of orbits of, 274
   symplectic, 306
adjoint representation, 5
algebra
   commutative differential graded, 58
   commutative graded, 58
   differential graded, 58
algebraic mapping cone, 322
almost complex
   manifold, 150
      integrable, 154
   map, 151
   structure, 150
      obstructions to, 156
      on S^2 and S^6, 151, 156
arrangement, 358
   complement of, 358
   formality of, 361
   geometric, 358, 363
atomic complex
   relative, 359
augmented cdga, 60

bar construction, 236, 385
   normalization, 382
   double, 234, 381
Barratt–Puppe sequence, 430
basepoint
   nondegenerate, 429
basic complex, 48
bi-invariant
   form, 12
   volume form, 13
bigradation, 92, 95, 169, 198
   on $\mathbb{C}$-forms, 153
bigraded
   differential algebra, 412

model, 95
  and Stasheff's theorem, 113
biquotient, 134, 136
  as homotopy pullback, 135
  minimal model of, 138
  model of, 137
  non-negative curvature of, 242, 252
  splitting rigid, 260
  symplectic, 189
blow-up
  along a submanifold, 333
  complex, 331
  example of model, 334
  McDuff's example, 337
  model for, 336, 337
  properties of, 333
  topological, 332
  and hard Lefschetz, 339
Borel
  fibration, 47, 272, 275
    relative model of, 280
  fixed point criterion, 293
  localization theorem, 272, 292
  spectral sequence, 173
bracket, 398
  of vector fields, 398, 408
Brylinski conjecture, 184, 339
  counterexample, 185
bumpy metric, 224
bundle
  along the fibers, 340
  holomorphic, 150
  holomorphic vector, 150
  principal holomorphic, 150

c-split fibration, 165, 313
Calabi–Eckmann manifold, 159, 160
canonical minimal model, 140, 141
  of an isometry, 142
Cartan decomposition, 26
Cartan–Weil model, 51
Cartan–Weil model, 309
cdga, 58
  augmented, 60
  cohomologically symplectic, 188
  filtered, 171
  formal, 92
  homotopy, 61
  minimal, 59
  pure, 419
  r-bigraded, 170
  Sullivan, 59

center, 7
  of rational homotopy Lie algebra, 287
centralizer, 7
cga, 58
  free, 58
characteristic algebra, 260
characteristic class, 46
Chas–Sullivan loop product, 342
Chern class, 46
classifying space, 38
  cohomology of, 38, 43, 45
  minimal model of, 72
closed
  form, 402, 406
  geodesic, 205
    as critical point, 208
    problem, 207
  manifold, 227
co-closed form, 405, 406
co-differential
  as adjoint of d, 405
  on forms, 405
cofibration, 429
cofiber, 429
cofiber sequence, 430
cohomological shriek map, 319
cohomologically free action, 308
cohomologically symplectic
  cdga, 188
  manifold, 188
cohomology
  de Rham, 402
  equivariant, 276
  Hochschild, 225, 237, 346
  invariant, 125
  of free loop space, 236, 385
  of Lie algebra, 54
  singular, 403
commutative differential graded algebra, 58
commutative graded algebra, 58
  free, 58
complement of an arrangement, 358
  model of, 360
complement of submanifold
  model of, 324
complex
  blow-up, 331
  structure, 196
    generalized, 202
    on tangent space, 150
  manifold, 148
  torus, 149
complex projective space, 149

complexification, 153, 197
configuration space, 351
   Fadell–Neuwirth fibration, 352
   formality of, 355
   model
      of complex projective manifold, 355
   model conjecture, 356
   model of, 356
   of two points, 329
   rational homotopy of, 353
   with labels, 374
conjecture
   Bott, 265, 390
   fibration with torus fiber, 388
   Halperin, 168, 389
   on configuration space model, 356
   toral rank, 271, 283, 388
conjugate point, 229, 232, 237
   multiplicity of, 229
conjugation action, 7
connected sum, 107
   and orientation of fundamental class, 108
   is formal if summands are, 115
   minimal model of, 108
   rational homotopy of, 107
cotangent bundle
   as symplectic manifold, 183
critical value of a homology class, 348
cup length, 88
curvature
   Ricci, 269
   Riemann, 240
   scalar, 269
   and Grassmann manifold, 240
   sectional, 240

de Rham
   cohomology, 402
   forms, 392
   theorem, 403
decomposable elements, 59
dga, 58
diagonal class, 355
diagonal cocycle, 328
   and shriek map, 329
dichotomy theorem
   elliptic case, 85
   hyperbolic case, 85
differential
   decomposition of, 77
   perturbation of, 96
differential forms, 392

differential graded algebra, 58
   morphism, 58
dimension, 84
Dolbeault
   cohomology, 155
      of $S^1 \times S^{2n+1}$, 174
      of $S^{2m+1} \times S^{2n+1}$, 177
   complex, 155
   formality, 169, 180
   model, 169
      of $\mathbb{C}P(n)$, 173
      of Calabi–Eckmann manifolds, 176
double bar construction, 381
double complex, 420
double mapping cylinder, 426

Eilenberg–Mac Lane space, 87
   and non-positive curvature, 241
elliptic, 84
   4-manifolds, 108
energy, 208, 229
entropy, 227
equivalence
   n-, 423
   weak, 423
equivariant cohomology, 47, 51, 276
equivariant minimal model, 123, 221
equivariantly formal, 127, 168
Euler characteristic
   homotopy, 85
   of fixed point set, 128, 292
Euler class, 46, 257
   and flat, 268
exponential, 6
extension of cdga homotopies, 64
exterior derivative, 395, 400

Fadell–Neuwirth fibration, 352
fiber bundle, 26
   along the fibers, 340
   exterior algebra, 398
   pullback of, 27
   symplectic, 190
   tangent, 393
fibration, 79, 430
   c-split, 313
   Hamitonian, 165
   holonomy of, 80
   long exact homotopy sequence of, 79, 81
   nilpotent, 79
   path, 79, 431

relative model of, 81
pure, 189
quasi-nilpotent, 79
relative minimal model of, 81
TNCZ, 164, 313
orientable, 164
fiber, 430
fiber bundle, 431
　Hamiltonian, 314
fiber sequence, 431
filtered
　cdga, 171
　differential graded algebra, 412
　differential module, 412
　model, 96
filtration, 412
　bounded, 412
fixed point, 274
flat
　and vanishing Euler class, 268
　connection, 255
　manifold, 241, 256
　principal bundle, 254
　vector bundle, 255
flow
　associated to a vector field, 399, 408
formal cdga, 92
　criterion for, 102
　obstructions to, 97
formal homology connection, 386
　of $\mathbb{C}P(2)$, 387
formal space, 92
　$F(\mathbb{R}^n, k)$, 354
　and connected sums, 115
　criterion for, 99
　Dolbeault, 169
　equivariant, 126
　Kähler, 166
　Kähler manifold, 365
　nonformal example, 94
　over $\mathbb{Q}$-extensions, 100
　product of, 93
　retract of, 93
　Riemannian symmetric, 144
　strict, 172
　vanishing of all Massey products, 96
　vanishing of triple Massey products, 94
forms
　de Rham, 392
　Sullivan, 67
Four Lemma, 418
Frölicher spectral sequence, 179
　of Kähler manifold, 179
free ccga, 58

free loop space
　cohomology, 385
　model, 210, 369
fundamental 2-form, 157
fundamental vector field, 306

Gelfand–Fuchs cohomology, 375
geodesic
　and Morse theory, 229
　closed, 205
　flow, 227
　　completely integrable, 230
　geometrically distinct, 205
　index of, 229, 232
　invariant, 215
　　as critical point, 216
　on compact Lie group, 241
　totally, 252
geometric lattice, 358
graded Lie algebra, 77
　cochains on, 91
　realization of, 92
Grassmann manifold, 43, 177
group action, 11, 273
　almost free, 271, 274
　and Euler characteristic, 128, 292
　cohomologically free, 308
　effective, 11
　free, 11, 274
　Hamiltonian, 308
　orbit of, 274
　smooth, 274
　symplectic, 306
Grove's question, 242
Gysin sequence, 44, 415

Haefliger model
　for evaluation map, 369
　for mapping space, 369
Halperin conjecture, 168
Hamiltonian
　action, 308
　　criterion for, 310
　　fixed point set of, 310
　　is c-split, 313
　fiber bundle, 314
　　with nilmanifold fiber, 315
hard Lefschetz
　and symplectically harmonic, 185
　and TRC, 285
　for blow-up, 339
　property, 163, 165
　theorem, 163

harmonic form, 25, 158, 407
Heisenberg
  group, 118
  manifold, 117, 162
hermitian
  product, 198
  metric, 156
Hochschild
  cohomology, 237, 346
  complex, 235, 385
  homology, 236
Hodge
  decomposition, 139, 407
  laplacian, 406
  manifold, 160
  star operator, 404
  structure
    mixed, 366
    pure, 364
holomorphic
  form, 155
  map, 148, 149
holonomy, 432
  representation, 80
homogeneous space, 29, 136
  model of, 83
  symplectic, 189
homotopy
  cofiber, 429
  equivalence, 424
  Euler characteristic, 85, 278
  extension, 64
  fiber, 431
  Lie algebra, 106
  lifting, 63
    property, 431
  of cdga morphisms, 61
    left, 62
  pullback, 424, 425
    versus pullback, 432
  pushout, 426, 427
    versus pushout, 430
  real, 69
  type
    real, 143
Hopf
  fibration
    relative model of, 82
  manifold, 159, 161, 174
  surface, 161
Hopf's theorem, 18
Hurewicz image, 306
hyperbolic, 84

indecomposables, 60
  induced map on, 76
independent sequence, 360
index of a geodesic, 229
inner product
  on forms, 404
integrable almost complex manifold, 154
integration
  along a fiber, 160, 378, 379
  of forms, 401
interior multiplication, 398
intersection map, 348
intersection product, 341
invariant cohomology, 13
invariant form, 12
  $\mathcal{L}(X)$, 399
  bi-invariant, 12
  left, 12
invariant geodesic, 215
isometry, 215
  finite order, 217
isotropy subgroup, 274
iterated integral, 377, 378, 385
  product of, 381
Iwasawa decomposition, 3

Jacobi identity, 398, 408

k-invariants, 88
Kähler
  criteria for, 158
  form, 157
  manifold, 157, 166, 172
    formality of, 166, 172, 365
  metric, 157
Killing form, 25, 270
Kodaira–Thurston manifold, 156, 162, 165, 190
  Chern classes of, 340

Lalonde–McDuff question, 314, 391
Laplace
  operator, 158
lattice
  geometric, 358
Lefschetz type, 192, 307
  and TRC, 312
  nilmanifold, 193
left adjoint of cdga tensor product, 371
left homotopy, 62
left invariant form, 12

lemma
  $d\delta$, 186
level-nilpotent, 348
Lie algebra, 3
  2-step, 289
  abelian, 8
  cohomology, 54
  free graded, 78, 92, 386
  free product, 78
  graded, 77
  homomorphism, 4
  homotopy, 106
  nilpotent, 117
  TRC, 289
Lie bracket, 3, 77, 408
Lie derivative, 399
  and symplectic action, 307
  formula, 400
Lie group, 2
  abelian, 8
  action, 11
  classifying space of, 38
  cohomology
    Hopf's theorem, 18
  cohomology of, 16, 43, 45
  connection on, 270
  even dimensional
    as complex manifold, 162
  exponential, 6, 11
  homomorphism, 2
  Killing form of, 270
  left-right translations, 4
  made discrete, 266
  maximal compact subgroup of, 3
  minimal model of, 71
  one-parameter subgroup of, 5
  rank of, 8
  second Betti number, 22, 40
  sectional curvature of, 270
  semisimple, 21, 270
  simple, 21, 26
  third Betti number, 23
Lie subalgebra, 4
Lie subgroup, 3
lifting lemma, 61
lifting of cdga homotopies, 63
localization map, 291
loop product, 342, 343
  dual of, 343
  Hochschild model, 345
  model of, 345
lower grading, 188
Lusternik–Fet theorem, 209
Lusternik–Schnirelmann category, 88

Malcev basis, 121
manifold
  almost complex, 150
  Calabi–Eckmann, 159, 160
  closed, 227
  cohomologically symplectic, 188
  complex, 148
  elliptic four-dimensional, 108
  flat, 241
  Grassmann, 177
  Heisenberg, 117, 162
  Hodge, 160
  Hopf, 159, 161, 174
  integrable almost complex, 154
  Kähler, 157, 166
    strict formality of, 172
  Kodaira–Thurston, 120, 156, 162, 165, 190
  orientable, 397
  P-, 232
  pure, 188
  symplectic, 182
    nonformal, 338
mapping
  cone, 429
    algebraic, 322
  cylinder, 429
  space, 370
    pointed, 373
  theorem, 89
Massey product, 94
  higher order, 95
maximal torus, 8
  and Weyl group, 127
McDuff's blow-up example, 337
minimal cdga, 59
  realization of, 90
minimal model, 59, 64
  of free loop space
    of S^2, 213
  $\mathbb{Q}$, 67
  $\mathbb{Q}$ to $\mathbb{R}$, 68
  $\mathbb{R}$, 67
  canonical, 140, 141
  equivariant, 123, 124, 221
  existence, 64
  of H-spaces, 73
  of a product of manifolds, 73
  of biquotient
    purity of, 138
  of classifying space, 72
  of cofiber, 90

of complex algebraic variety, 367
of complex projective space, 72, 78
of connected sum, 108
of diagonal map, 73
of fiber, 81
of homogeneous space, 131, 133
of Lie group, 71
of map, 66, 67
of nilmanifold, 118, 122
of space, 67
of sphere, 72, 78
of Stiefel manifold, 71
of torus, 72
of wedge of spaces, 73
rational, 67
  of map, 67
real, 67
relative, 66, 67
uniqueness, 64
of Kodaira–Thurston manifold, 162
mixed Hodge structure, 366
model, 74
  r-filtered, 171
  bigraded, 95, 113
  Cartan–Weil, 51
  Cartan–Weil, 309
  Dolbeault, 169
  filtered, 96
  finite, 75
  for blow-up, 336, 337
  for configuration space, 355
  for loop product, 345
  Haefliger, 369
    for mapping space, 369
  of Map($\mathbb{C}P^2, S^6$), 373
  of biquotient, 137
  of blow-up
    example, 334
  of complement of submanifold, 324
  of free loop space, 210, 369
  of homogeneous space, 83, 131, 133
  of map, 74
  of McDuff blow-up example, 337
  of principal bundle, 84
  of space, 74
  Poincaré duality, 109
  pure, 86, 188
  r-bigraded, 170
  semifree, 101
  Sullivan, 64
moment-angle complex, 363
morphism, 58

NDR-pair, 429
Nijenhuis tensor, 155
nilmanifold, 92, 116
  almost complex structure on, 155
  four-dimensional, 192
  Lefschetz type, 193
  minimal model of, 118, 122
  nonformality of, 120, 121
  non-Kähler, 167
  symplectic, 191
    and Brylinski conjecture, 185
  toral rank of, 288
  two-step, 290
nilpotent
  fibration, 79
  fundamental group, 76
  group, 76
  space, 68
    symplectic, 194
non-degenerate basepoint, 429
non-formal space, 94
normalization of the bar construction, 382
normalizer, 7

obstructions to formality, 97
one-parameter subgroup, 5
orbit, 274
orientable, 397
orthogonal group, 9

P-manifold, 232
  rationally elliptic, 234
parallelizable, 5
path fibration, 39, 79, 431
  of S^3, 414
Poincaré duality
  isomorphism, 319
  model, 109
polar decomposition, 3
Pontryagin class, 46
Postnikov tower, 88
  for nilmanifold, 122
primitive element, 18
principal bundle, 32
  flat, 254
  pullback of, 33
  trivial, 33
  universal, 34, 275
problem
  Belegradek–Kapovitch
    biquotients, 390

derivations, 390
curvature
  and cohomology, 390
  elliptic fiber, 389
  exponential growth of Betti numbers, 391
  geometry versus algebra, 390
  Lalonde–McDuff and models, 391
pullback
  of fiber bundle, 27
  of principal bundle, 33
Puppe sequence, 40, 432
pure
  cdga, 419
  fibration, 189
  manifold, 188
  minimal cdga, 86, 188
    splitting of, 87
pure Hodge structure, 364
  and bigradation of model, 365

quasi-isomorphism, 59, 60
quasi-nilpotent fibration, 79
question
  Lalonde–McDuff, 314, 391

r-bigraded cdga, 170
rank, 8, 121
rational
  category, 89
  homotopy
    descent phenomena, 249
    Lie algebra, 77
    of fixed point set, 301
    type, 68
  space, 90
  toral rank, 272
    of nilmanifold, 288
rationalization of space, 68, 91
rationally
  elliptic space, 84
  hyperbolic space, 84
real homotopy type, 69, 143
realization
  of graded Lie algebra, 92
  of minimal cdga, 90
  of model by manifold, 106
relative atomic complex, 359
relative minimal cdga, 66
relative minimal model, 67
  of diagonal map, 73
  of fibration, 81

of Hopf fibration, 82
of path fibration, 81
of sphere bundle, 82
of universal principal bundle, 82
r-bigraded, 170
realization of, 91
relative model
  of Borel fibration, 280
Ricci curvature, 269
Riemann curvature, 240
Riemannian
  connection, 239
    properties of, 239
  metric, 239

scalar curvature, 269
sectional curvature, 240
semifree, 236
  model, 101, 320
  module, 100
shriek map, 319, 329, 342
  cohomological, 319
shuffle, 381
  product, 384
singular cohomology, 403
soul, 252
space
  mapping, 370
  non-nilpotent, 68
  of orbits, 274
  symmetric, 53
  of free paths, 377
special orthogonal group, 9
special unitary group, 10
spectral sequence, 409
  associated to double complex, 421
  collapses, 410
  comparison theorem, 416
  converges, 410
  first quadrant, 410
  in cohomology, 411
  morphism, 409, 411
  odd, 419
  of Hopf fibration, 414
  Serre, 413
  transgression, 411
spinor group, 9
splitting rigid, 259
  and derivations, 263
Stiefel manifold, 42
  cohomology of, 43, 45
  minimal model of, 71
Stokes's theorem, 402

Index

strict formality, 172
subspace arrangement
  formality of, 361
Sullivan
  model, 64
Sullivan cdga, 59, 68
  minimal, 59
  structure of, 60
Sullivan polynomial forms, 67
suspension
  of graded vector space, 57
symmetric space, 53
symplectic
  action, 306
    splitting of, 310
  bundle, 190
  group, 11
  manifold, 182
    almost complex structure of, 187
    nonformal, 338
    nilpotent manifold, 194
    star operator, 184
    vector space, 200
symplectically harmonic form, 184

tangent
  bundle, 393, 396
  map, 393
  space
    complex, 153
    real, 152
theorem
  Allday–Halperin, 279
  Babenko–Taimanov, 338
  Belegradek–Kapovitch, 259
  Blanchard, 165
  Borel localization, 272, 292
  Chen, 388
  Cohen–Jones, 347
  Darboux, 182
  de Rham, 403
  Eschenburg, 243
  existence of equivariant model, 124
  Fang–Rong, 244
  Gromoll–Meyer, 210
  Gromov, 224
  Grove, 216
  Haefliger, 369
  Hodge decomposition, 407
  Hsiang, 276
  Kriz, Totaro, 355
  Lambrechts, 225

  Lusternik–Fet, 209
  mapping, 89
  Miller, 110
    second proof of, 115
  Nomizu, 119, 191
  O'Neill, 242
  of the cube, 428
  on filtered algebras and spectral sequences, 412
  Paternain, 230
  soul, 252
  splitting, 252
  Stasheff, 113
  Stepien, 315
  Stokes's, 402
  Sullivan–Barge, 106
  Tanaka, 222
  Thurston, 190
  Totaro, 250
  Vigué-Poirrier–Sullivan, 213
  Whitehead, 424
  Yuzvinsky, 360
  Zeeman–Moore, 416
  hard Lefschetz, 163
TNCZ, 164, 165, 293, 313
  criterion for, 293
topological blow-up, 332
topological entropy, 227
  and rational ellipticity, 230
toral rank, 271, 277
  of wedge of spheres, 277
  rational, 272, 278
    of homogeneous space, 279
    of Lie group, 279
toric variety, 363
total complex, 421
total degree, 169, 409
totally
  convex, 252
  geodesic, 252
  non-cohomologous to zero, 164, 293
transgressive element, 39
TRC, 283, 388
  algebraic, 284
  and hard Lefschetz, 285
  and Lefschetz type, 312
  for homogeneous spaces, 284
  for Lie algebras, 289
  is toral rank conjecture, 272

unitary group, 10
universal fibration, 45

universal principal bundle, 34, 275
   relative model of, 82
upper central series, 121

vampire, 259
variety
   toric, 363
vector field, 4, 396
   bracket, 4, 398
   flow associated to, 399
   fundamental, 306
   left invariant, 4
vertical sub-bundle, 159

volume form, 396, 397
   bi-invariant, 13

Wang sequence, 415
wedge product
   of forms, 395
Weil algebra, 48
well-pointed, 429
Weyl group, 8–11
Whitehead's theorem, 424

Z/2Z-graded Sullivan algebra, 305

The manufacturer's authorised representative in the EU for product safety is Oxford University Press España S.A. of el Parque Empresarial San Fernando de Henares, Avenida de Castilla, 2 – 28830 Madrid (www.oup.es/en or product.safety@oup.com). OUP España S.A. also acts as importer into Spain of products made by the manufacturer.

 www.ingramcontent.com/pod-product-compliance
Ingram Content Group UK Ltd.
Pitfield, Milton Keynes, MK11 3LW, UK
UKHW022229230426
12048UKWH00016BA/1156